D1611051

Engineering the Complex SOC

Fast, Flexible Design with Configurable Processors

Prentice Hall Modern Semiconductor Design Series

James R. Armstrong and F. Gail Gray
VHDL Design Representation and Synthesis

Jayaram Bhasker
A VHDL Primer, Third Edition

Mark D. Birnbaum
Essential Electronic Design Automation (EDA)

Eric Bogatin
Signal Integrity: Simplified

Douglas Brooks
Signal Integrity Issues and Printed Circuit Board Design

Alfred Crouch
Design-for-Test for Digital IC's and Embedded Core Systems

Tom Granberg
Handbook of Digital Techniques for High-Speed Design

Howard Johnson and Martin Graham
High-Speed Digital Design: A Handbook of Black Magic

Howard Johnson and Martin Graham
High-Speed Signal Propagation: Advanced Black Magic

Farzad Nekoogar and Faranak Nekoogar
From ASICs to SOCs: A Practical Approach

Samir Palnitkar
Design Verification with **e**

Christopher T. Robertson
Printed Circuit Board Designer's Reference: Basics

Chris Rowen and Steve Leibson
Engineering the Complex SOC

Wayne Wolf
FPGA-Based System Design

Wayne Wolf
Modern VLSI Design: System-on-Chip Design, Third Edition

Brian Young
Digital Signal Integrity: Modeling and Simulation with Interconnects and Packages

Engineering the Complex SOC

Fast, Flexible Design with Configurable Processors

Chris Rowen

Edited by Steve Leibson

PRENTICE HALL PTR
UPPER SADDLE RIVER, NJ 07458
WWW.PHPTR.COM

Library of Congress Cataloging-in-Publication Data

A catalog record for this book can be obtained from the Library of Congress.

Editorial/production supervision: *Nicholas Radhuber*
Publisher: *Bernard Goodwin*
Cover design director: *Jerry Votta*
Cover design: *Talar Boorujy*
Manufacturing manager: *Maura Zaldivar*
Editorial assistant: *Michelle Vincenti*
Marketing manager: *Dan DePasquale*

© 2004 Pearson Education, Inc.
PRENTICE HALL PTR
Published by Prentice Hall Professional Technical Reference
Pearson Education, Inc.
Upper Saddle River, New Jersey 07458

Prentice Hall books are widely used by corporations and government agencies for training, marketing, and resale.

Prentice Hall offers excellent discounts on this book when ordered in quantity for bulk purchases or special sales. For more information, please contact:
U.S. Corporate and Government Sales
1-800-382-3419
corpsales@pearsontechgroup.com

For sales outside of the U.S., please contact:
International Sales
1-317-581-3793
international@pearsontechgroup.com

Other product or company names mentioned herein are the trademarks or registered trademarks of their respective owners.

Printed in the United States of America

1st Printing

ISBN 0-13-145537-0

Pearson Education LTD.
Pearson Education Australia PTY, Limited
Pearson Education Singapore, Pte. Ltd.
Pearson Education North Asia Ltd.
Pearson Education Canada, Ltd.
Pearson Educación de Mexico, S.A. de C.V.
Pearson Education — Japan
Pearson Education Malaysia, Pte. Ltd.

To my wife, Anne, and my daughters,
Elizabeth, Catherine, and Leslianne.

CONTENTS

List of Figures

FOREWORD

This is an important and useful book – important because it addresses a phenomenon that affects every industry sooner or later, and useful because it offers a clear, step-by-step methodology by which engineers and executives in the microelectronics industry can create growth and profit from this phenomenon. The companies that seize upon this opportunity will transform the way competition occurs in this industry. I wish that a hands-on guide such as this were available to strategists and design engineers in other industries where this phenomenon is occurring – industries as diverse as operating system software, automobiles, telecommunications equipment, and management education of the sort that we provide at the Harvard Business School. This industry-transforming phenomenon is called a change in the basis of competition – a change in the sorts of improvements in products and services that customers will willingly pay higher prices to get.

There is a natural and predictable process by which this change affects an industry. Chris Rowen, a gifted strategist, engineer and entrepreneur, has worked with me for several years to understand this process. It begins when a company develops a proprietary product that, while not good enough, comes closer to satisfying customers' needs than any of its competitors. The most successful firms do this through a proprietary and optimized architecture, because at this stage the functionality and reliability of such products are superior to those that employ an open, modular architecture. In order to provide proprietary, architecturally interdependent products, the most successful companies must be vertically integrated.

As the company strives to keep ahead of its direct competitors, however, it eventually overshoots the functionality and reliability that customers in less-demanding tiers of the market can utilize. This precipitates a change in the basis of competition in those tiers. Customers will no longer pay premium prices for better, faster and more reliable products, because they can't use those improvements. What is not good enough then becomes speed and convenience. Customers begin demanding new products that are responsively custom-configured to their needs, designed and delivered as rapidly and conveniently as possible. Innovations on these new trajectories of improvement are the improvements that merit attractive prices and drive changes in market share. In order to compete in this way – to be fast, flexible and responsive, the dominant architecture of the products must evolve toward a modular architecture, whose components and sub-systems interface according to industry standards. When this happens, there is no advantage to being integrated. Suppliers of components and sub-systems can begin developing, making and selling their products independently, dealing with partners and customers at arms' length because the key interface standards are completely and clearly specified. This condition begins at the bottom of the market, where functional overshoot occurs first, and then moves up inexorably to affect the higher tiers.

When the architecture of a product or a sub-system is modular, it is conformable. This conformability is very important when it is being incorporated within a next-level product whose functionality and reliability aren't yet good enough to fully satisfy what customers need. Modular conformability enables the customer to customize what it buys, getting every piece of functionality it needs, and none of the functionality it doesn't need.

The microprocessor industry is going through precisely this transition. Microprocessors, and the size of the features from which they are built, historically have not been good enough – and as a result, their architectures have been proprietary and optimized. Now, however, there is strong evidence that for mainstream tiers of the market the basis of competition is changing. Microprocessors are more than fast enough for most computer users. In pursuit of Moore's Law, circuit fabricators have shrunk feature sizes to such a degree that in most tiers of the market, circuit designers are awash in transistors. They cannot use all the transistors that Moore's Law has made available. As a result, especially in embedded, mobile and wireless applications, custom-configured processors are taking over. Their modular configurability helps designers to optimize the performance of the product systems in which the processors are embedded.

With this change, the pace of the microprocessor industry is accelerating. Product design cycles, which in the era of interdependent architectures had to be measured in years, are collapsing to months. In the future they will be counted in weeks. Clean, modular interfaces between the modules that comprise a circuit – libraries of reusable IP – ultimately will enable engineers who are not experts in processor design, to build their own circuits. Ultimately software engineers will be able to design processors that are custom-configured to optimize the performance of their software application.

Clearer interface standards are being defined between circuit designers and fabs, enabling a dis-integrated industry structure to overtake the original integrated one. This structure began at the low end, and now dominates the mainstream of the market as well. The emergence of the modular "designed-to-order" processor is an important milestone in the evolution of the microprocessor industry, but modular processors have broader implications. Just as the emergence of the personal computer allowed a wide range of workers to computerize their tasksfor the first time, the configurable processor will change the lives of a wide range of chip designers and chip users. This new processor-based methodology enables these designers and users to specify and program processors for tasks that are too sensitive to cost or energy-efficiency for traditional processors. It empowers the ordinary software or hardware developer to create new computing engines, once the province of highly-specialized microprocessor architecture and development teams. And these new processor blocks are likely to be used in large numbers – tens and hundreds per chip, with total configurable processors vastly outnumbering traditional microprocessors. The future will therefore be very different from the past.

The design principles and techniques that Chris Rowen describes in this book will be extremely useful to companies that want to capitalize on these changes to create new growth. I thank him for this gift to the semiconductor industry.

Clayton M. Christensen
Robert and Jane Cizik Professor of Business Administration
Harvard Business School
March 2004

FOREWORD

For more than 30 years, Moore's law has been a driving force in the computing and electronics industries, constantly forcing changes and driving innovation as it provides the means to integrate ever-larger systems onto a single chip. Moore's law was the driving force that led to the microprocessor's dominance throughout the world of computing, beginning about 20 years ago. Today, Moore's law is the driver behind the system-on-chip (SOC) paradigm that uses the vastly increased transistor density to integrate ever-larger system components onto a single chip, thereby reducing cost, power consumption, and physical size.

Rowen structures this thorough treatise on the design of complex SOCs around six fundamental problems. These range from market forces, such as tight time to market and limited volume before obsolescence, to the technical challenges of achieving acceptable performance and cost while adhering to an aggressive schedule. These six challenges lead to a focus on specific parts of the SOC design process, from the integration of application-specific logic to the maximization of performance/cost ratios through processor customization.

As Rowen clearly illustrates, the processor and its design is at the core of any complex SOC. After all, a software-mostly solution is likely to reduce implementation time and risk; the difficulty is that such solutions are often unable to achieve acceptable performance or efficiency. For most applications, some combination of a processor (executing application-specific code) and application-specific hardware are necessary. In Chapter 4, the author deals with the critical issue of interfacing custom hardware and embedded processor cores by observing that a mix of custom communication and interconnection hardware together with software to handle decision making and less common tasks often enables the designer to achieve the desired balance of implementation effort, risk, performance, and cost.

Chapters 5 and 6 form the core of this book and build on the years of experience that Tensilica has had in building SOCs based on customized processors. Although the potential advantages of designing a customized processor should be clear—lower cost and better performance—the required CAD tools, customizable building blocks, and software were previously unavailable. Now they are, and Rowen discusses how to undertake the design of both the software and hardware for complex SOCs using state-of-the art tools.

Chapters 7 and 8 address the challenge of obtaining high performance in such systems. For processors, performance comes primarily from the exploitation of parallelism. This is a central topic in both these chapters. Chapter 7 discusses the use of pipelining to achieve higher performance within a single instruction flow in a single processor. Chapter 8 looks to the future, which will increasingly make use of multiple processors, configured and connected according to the needs of the application. The use of multiple processors broadly represents the future in high-performance computer architecture, and not just in embedded applications. I am delighted to see that SOCs are playing a key role in charting this future.

The design of SOCs for new and challenging applications—ranging from telecommunications, to information appliances, to applications we have yet to dream of—is creating new opportunities for computing. This well-written and comprehensive book will help you be a successful participant in these exciting endeavors.

John Hennessy
President
Stanford University
March 2004

Author's Preface

This book is aimed at the architects, designers, and programmers involved with complex SOC design. Managers of companies making significant investments in SOC designs and platforms will also find the essential changes in design process and architecture of platform hardware and software important to understand. These changes may directly or indirectly influence investment strategies, core competencies, and organization structure over time.

The book outlines the major forces changing the SOC design process and introduces the concept of SOC design using extensible processors as a basic design fabric. It teaches the essentials of extensible processor architecture, tools for instruction-set extension, and multiple-processor SOC architecture for embedded systems. It uses examples from Tensilicaís Xtensa architecture and the Tensilica Instruction Extension (TIE) language throughout to give a precise, practical, and up-to-date picture of the real issues and opportunities associated with this new design method. You will find enough information on Xtensa and TIE to understand the methodology, though this book does not attempt to serve as comprehensive product documentation for either.

This book does not offer equal emphasis to all methodologies. Instead, it concentrates on the proposed benefits of this new SOC design methodology, highlighting the opportunities and dealing with issues associated with conversion from a gate-centric to a processor-centric SOC design methodology.

The first part of this book provides a high-level introduction to the many SOC design problems and their solutions. The middle sections give a more detailed look at how extensible processors compare to both traditional processors and hardwired logic. It also discusses how the essential mechanisms of processor extensibility address both the computation and communications needs of advanced SOC architectures. The later sections give a series of detailed examples to reinforce the applicability of the new SOC design method.

This introduction exposes the basic issues in SOC design and the motivation for considering an overhaul of the hardware and software structures and methods used for SOC development.

Chapter 2 provides a current view of SOC hardware structure, software organization, and chip-development flow. This chapter exposes the six basic shortcomings of the current SOC design method and explains why new structures and processes are necessary.

Chapter 3 introduces the new SOC design approach based on use of extensible processors across all control and data-processing functions on the chip. It briefly discusses how this approach addresses the six big problems.

Chapter 4 takes a top-down approach to processor-centric SOC architecture, looking at overall data flow through complex system architectures. The chapter shows how complex functions are decomposed into function blocks which may often be implemented as application-specific processors. Key issues include latency and throughput of blocks, programming models for coordination of parallel functions, hardware interconnect options, and management of complexity across the entire chip design.

Chapters 5 and 6 dig down into the design of the individual tasks. Chapter 5 looks at task design through the eyes of the software developer, especially the process of taking a task originally intended to run on a general-purpose processor and running that task on an application-specific processor. The chapter shows how application-specific processors fit the use model of traditional embedded processor cores while adding simple mechanisms that dramatically improve the performance and efficiency of complex tasks. This chapter includes a simple introduction to the principles of the Xtensa architecture, including Flexible Length Instruction Extensions (FLIX) and fully automated instruction-set generation.

Chapter 6, by contrast, looks at task design through the eyes of the hardware developer, especially at the process of taking a hardware function and translating it into an application-specific processor with comparable performance but thorough programmability. The chapter establishes the basic correspondence between hardware pipelines and processor pipelines and recommends techniques for efficient mapping of traditional hardware functions (including high-bandwidth, low-latency functions) into application-specific processors.

Chapter 7 deals with a series of more advanced SOC-design topics and issues, including techniques for implementing complex state machines, options for task-to-task communication and synchronization, interfaces between processors and remaining hardware blocks, power optimization, and details of the TIE language.

Chapter 8, the final chapter, looks down the road at the long-term future of SOC design, examining basic trends in design methodology and semiconductor technology. It paints a 10 to 15 year outlook for the qualitative and quantitative changes in design, in applications, and in the structure of the electronics industry.

The book uses a number of related terms in the discussion of SOC design. An SOC design methodology is the combination of building blocks, design generators, architectural guidelines, tools, simulation methods, and analysis techniques that together form a consistent environment for development of chip designs and corresponding software. The book generally refers to the recommended method as the advanced SOC design or processor-centric SOC design methodology. Occasionally, we use the phrase MPSOC design methodology for multiple-processor system-on-chip design methodology to emphasize the role of processors, often combined in large numbers, as the basic building blocks for flexible SOCs. The ultimate vision is a role for configurable processors so common, so automatic, and so pervasive that we can rightly call the result a ìsea of processors.î Within a decade, processors could become the new logic gate, with hundreds or thousands of processors per chip and application-specific configuration of processors as routine as logic synthesis is today.

This book touches on a range of hardware, software, and system-design issues, but it cannot hope to cover each of these topics comprehensively. Rather than interrupt the flow with extensive footnotes and technical references, each chapter ends with a section for further reading. These sections highlight significant technical papers in the domain covered by each chapter and list additional books that may augment your understanding of the subject.

This book uses Tensilicaís Xtensa processor architecture and tools to illustrate important ideas. This book does not attempt, however, to fully document Tensilicaís products. Contact Tensilica for more complete details at *http://www.tensilica.com.* Other approaches to automatic processor generation are mentioned in the book, especially in Chapter 3.

ACKNOWLEDGMENTS

This book is the direct result of more than a year's effort. So many people have contributed in so many ways that it is a daunting task to give credit where it is due.

It is clear, however, where to begin. Steve Leibson, my editor and Tensilica colleague, has played a central role in defining and guiding this book. His clear picture of the readers' needs and his relentless dedication to getting both words and concepts right have made this a much better book. The painstaking review of drafts and the lively debates on content have not just added to the book, but have also materially shifted my perspective on long-term needs of chip designers developing new system-on-chip platforms. This book would not exist without Steve's enormous energy and commitment to bring this to life.

Paula Jones has earned special thanks as project manager for the book, working with the publisher, managing internal content reviews, and orchestrating art work and page layout. Her endless optimism helped pull the book through difficult moments when progress seemed painfully slow.

The work of several colleagues deserves particular recognition. The Open Secure Socket Layer case study in Chapter 5 reflects research largely done by Jay McCauley. The ATM segmentation and reassembly example in Chapter 6 is based on work by Jerry Reddington. Himanshu Sanghavi spent long hours with the TIE examples, ensuring not just syntactic but, more importantly, conceptual correctness. Dror Maydan has played a particularly central role in guiding discussion of system and software issues throughout the book and made important material improvements in both the examples and the discussion.

Many other Tensilica colleagues have also spent countless hours reading and reviewing multiple drafts of the book or making important suggestions for content. I warmly thank Beatrice Fu, Eileen Peters Long, Bill Huffman, John Wei, David Goodwin, Eliot Gerstner, Steve Roddy, Gulbin Ezer, Jerry Reddington, David Jacobowitz, Robert Kennedy, Chris Songer, Marc

Gauthier, Jean Fletcher, Morteza Shafiei, George Wall, Larry Pryzwara, Leo Petropoulos, Tomo Tohara, Nenad Nedeljkovic, and Akilesh Parameswar for their kind efforts.

Sincere thanks also go to three particularly important mentors, John Hennessy of Stanford, Richard Newton of Berkeley, and Clayton Christensen of Harvard. These three have played dual roles—encouraging the work and concepts that lie behind this book and offering the important commentary to help put configurable processors and multiple-processor system-on-chip design in a larger business and technical context.

Finally, this book reflects the enormous creative effort invested in the Tensilica processor generator and associated tools for complex SOC design. I thank my Tensilica colleagues for their tireless engineering work and for their support in making this book possible.

Chris Rowen
March 2004

CHAPTER 1

The Case for a New SOC Design Methodology

This book has one central mission: to make the case for a new way of designing complex system-on-chip (SOC) devices and software. This mission reflects both the growing complexity of the end products that might use SOC integration and the limitations of traditional hardware and software components and design methods. This new processor-centric design approach promises both to contain the growth of development cost and to enable system designs with greater adaptability and efficiency.

This book addresses three related needs of readers. First, the book looks at the business challenge of exploiting fundamental changes in basic electronics technology to create more profitable products in rapidly shifting markets. Second, the book explains the diverse range of applications, hardware, and software technologies necessary to advanced SOC design and explores how the complex interaction of applications requirements and underlying hardware/software techniques is evolving. Third, the book looks in depth at the application of configurable processors to SOC design, both at the level of individual tasks and at the level of systems architecture. Together, these three views of SOC design can help managers and engineers make better decisions and create more compelling SOC-based systems.

1.1 The Age of Megagate SOCs

The rapid evolution of silicon technology is bringing on a new crisis in SOC design. To be competitive, new communication, consumer, and computer product designs must exhibit rapid increases in functionality, reliability, and bandwidth and rapid declines in cost and power consumption. These improvements dictate increasing use of high-integration silicon, where many of the data-intensive capabilities are presently realized with register-transfer-level (RTL) hardware-design techniques. At the same time, the design productivity gap—the failure of nanometer semiconductor design productivity to keep up with chip capacity—puts intense pressure on chip designers to develop increasingly complex hardware in decreasing amounts of time. This combination of external forces – more rapid change in complex end-product specifications – and internal forces – greater design risk and complexity triggered by growing silicon density – makes SOC design increasingly daunting.

A few characteristics of typical deep-submicron integrated circuit design illustrate the challenges facing SOC design teams:

- **More gates:** In a generic 130nm standard-cell foundry process, silicon density routinely exceeds 100K usable gates per mm^2. Consequently, a low-cost chip ($50mm^2$ of core area) can carry several million gates of logic today. Simply because it is possible, a system designer somewhere will find ways to exploit this potential in any given market. The designer faces a set of daunting challenges, however.
- **Design effort:** In the past, silicon capacity and design-automation tools limited the practical size of a block of RTL to smaller than 100K gates. Improved synthesis, place-and-route, and verification tools are raising that ceiling. Blocks of 500K gates are now within the capacity of these tools, but existing design description methods are not keeping pace with silicon fabrication capacity, which can now put millions of gates on an SOC.
- **Verification difficulty:** The internal complexity of a typical logic block—hence the number of potential bugs—grows much more rapidly than does its gate count. System complexity increases much more rapidly than the number of constituent blocks. Similarly, verification complexity has grown disproportionately. Many teams that have recently developed real-world designs report that they now spend more than 70% of their development effort on block- or system-level verification.
- **Cost of fixing bugs:** The cost of a design bug is going up. Much is made of the rising mask costs—the cost of a full mask set now exceeds $1M. However, mask costs are just the tip of the iceberg. The larger teams required by complex SOC designs, higher staff costs, bigger NRE (nonrecurring engineering) fees, and lost profitability and market share make show-stopper design bugs intolerable. Design methods that reduce the occurrence of, or permit painless workarounds for, such show-stoppers pay for themselves rapidly.
- **Late hardware/software integration:** All embedded systems now contain significant amounts of software or firmware. Software integration is typically the last step in the system-development process and routinely gets blamed for overall program delays. Late hardware/software validation is widely viewed as a critical risk for new product-development projects.
- **Changing standards and customer requirements:** Standards in communication protocols, multimedia encoding, and security are all undergoing rapid complexity growth. The need to conserve scarce communications spectrum plus the inventiveness of modern protocol designers has resulted in the creation of sophisticated new standards (for example,

IPv6 Internet Protocol packet forwarding, G.729 voice coding, JPEG2000 image compression, MPEG4 video, Rjindael AES encryption). These new protocol standards demand longer implementation descriptions, greater computational throughput, and more painstaking verification than their predecessors. Similarly, electronic systems providers in competitive markets must also evolve their products rapidly, often demanding rapid evolution in the underlying silicon platforms. This combination of rapid change and high complexity puts a premium on design flexibility.

Although general-purpose embedded processors can handle many tasks in theory, they often lack the bandwidth needed to perform complex data-processing tasks such as network packet processing, video processing, and encryption. Chip designers have long turned to hard-wired logic to implement these key functions. As the complexity and bandwidth requirements of electronic systems increase, the total amount of logic rises steadily.

1.1.1 Moore's Law Means Opportunity and Crisis

In 1965, Gordon Moore prophesized that economical integrated circuit density would double roughly every one to two years. The universal acceptance and relentless tracking of this trend set a grueling pace for all chip developers. This trend makes transistors ever cheaper and faster (good) but also invites system buyers to expect constant improvements to functionality, battery life, throughput, and cost (not so good). The moment a new function is technically feasible, the race is on to deliver it. Today, it is perfectly feasible to build SOC devices with more than 100 million transistors, and within a couple of years we'll see billion-transistor chips built for complex applications combining processors, memory, logic, and interface.

The growth in available transistors creates a fundamental role for concurrency in SOC designs. Different tasks, such as audio and video processing and network protocol processing, can operate largely independently of one another. Complex tasks with inherent internal execution parallelism can be decomposed into a tightly-coupled collection of sub-tasks operating concurrently to perform the same work as the original non-concurrent task implementation. This kind of concurrency offers the potential for significant improvements in application latency, data bandwidth, and energy efficiency compared to serial execution of the same collection of tasks with a single computational resource. The exploitation of available concurrency is a recurring theme in this book.

If high silicon integration is a terrific opportunity, then the design task must be recognized as correspondingly terrifying. Three forces work together to make chip design tougher and tougher. First, the astonishing success of semiconductor manufacturers to track Moore's law gives designers twice as many gates to play with roughly every two years. Second, the continuous improvement in process geometry and circuit characteristics motivates chip builders to migrate to new integrated circuit fabrication technologies as they come available. Third, and perhaps most important, the end markets for electronic products—consumer, computing, and communications systems—are in constant churn, demanding a constant stream of new functions and performance to justify new purchases.

As a result, the design "hill" keeps getting steeper. Certainly, improved chip-design tools help—faster RTL simulation, higher capacity logic synthesis, and better block placement and routing all mitigate some of the difficulties. Similarly, the movement toward systematic logic design reuse can reduce the amount of new design that must be done for each chip.

But all these improvements fail to close the design gap. This well-recognized phenomenon is captured in the Semiconductor Research Corporation's simple comparison of the growth in logic complexity and designer productivity in Fig 1-1.

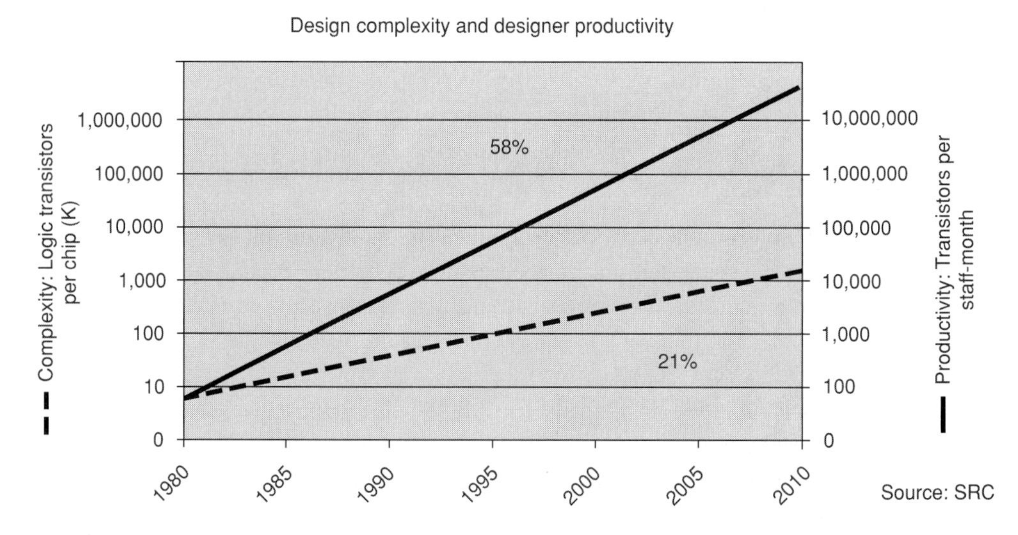

Figure 1-1 Design complexity and designer productivity.

Even as designers wrestle with the growing resource demands of advanced chip design, they face two additional worries:

1. How do design teams ensure that the chip specification really satisfies customer needs?
2. How do design teams ensure that the chip really meets those specifications?

1.1.2 Roadblock 1: Building the Wrong Chip

If the design team fails on the first criterion listed above, the chip may work perfectly but will have inadequate sales to justify the design expense and manufacturing effort. Changes in requirements may be driven by demands of specific key customers or may reflect overall market trends such as the emergence of new data-format standards or new feature expectations across an entire product category. While most SOC designs include some form of embedded control processor, the limited performance of those processors often precludes their use for essential data-processing tasks, so software usually cannot be used to add or change fundamental new features. This is the pitfall of product inflexibility.

1.1.3 Roadblock 2: Building the Chip Wrong

If the design team fails on the second criterion, additional time and resources must go toward changing or fixing the design. This resource diversion delays market entry and causes companies to miss key customer commitments. The failure is most often realized as a program delay. This delay may come in the form of missed integration or verification milestones, or it may come in the form of hardware bugs—explicit logic errors that are not caught in the limited verification coverage of typical hardware simulation. The underlying cause might be a subtle error in a single design element, or it might be a miscommunication of requirements—subtle differences in assumptions between hardware and software teams, between design and verification teams, or between SOC designer and SOC library or foundry supplier. In any case, the design team may often be forced into an urgent cycle of redesign, reverification, and refabrication of the chip. These design "spins" rarely take less than six months, causing significant disruption to product and business plans. This is the pitfall of an inadequate design process.

To improve the design process, simultaneous changes in all three interacting dimensions of the design environment – design elements, design tools and design methodology – must be considered.

- **Design elements** are the basic building blocks, the silicon structures, and the logical elements that form the basic vocabulary of design expression. Historically, these blocks have been basic logic gates (NAND gates, NOR gates, and flip-flops) plus algorithms written in C and assembly code running on RISC microprocessors and digital signal processors.
- **Design tools** are the application programs and techniques that designers use to capture, verify, refine, and translate design descriptions for particular tasks and subsystems. Historically, tools such as RTL compilation and verification, code assemblers and compilers, and standard-cell placement and routing have comprised the essential toolbox for complex chip design.
- **Design methodology** is the design team's strategy for combining the available design elements and tools into a systematic process for developing the target silicon and software. A design methodology specifies what elements and tools are available, describes how the tools are used at each step of the design refinement, and outlines the sequence of design steps. Currently, SOCs are built using four major steps, typically implemented in the following order: hardware-software partitioning; detailed RTL block design and verification; chip integration of RTL blocks, processors and memories; and post-silicon software bring-up.

Changes in any one dimension or piecemeal improvements in RTL design or software development tools cannot solve the two major roadblocks of SOC design: building the wrong chip and building the chip wrong. Instead, it is necessary change the design process itself. The design elements, key tools and the surrounding methodology must all change together. This book is about making a fundamental change in the design process that overcomes these two roadblocks.

1.2 The Fundamental Trends of SOC Design

Several basic trends suggest that the engineering community needs a new approach for SOC design. The first trend is the seemingly inexorable growth in silicon density, which underlies the fundamental economics of building electronic products in the 21st century.

At the center of this trend is the fact that the semiconductor industry seems willing and able to continue to push chip density by sustained innovation through smaller transistor sizes, smaller interconnect geometries, higher transistor speed, significantly lower cost, and lower power dissipation over a long period of time. Technical challenges for scaling abound. Issues of power dissipation, nanometer lithography, signal integrity, and interconnect delay all will require significant innovation, but past experience suggests that these, at worst, will only marginally slow down the pace of scaling. The central question remains: How will we design what Moore's law scaling makes manufacturable?

This silicon scaling trend stimulates the second trend—the drive to take this available density and actually integrate into one piece of silicon the enormous diversity and huge number of functions required by modern electronic products. The increasing integration level creates the possibility of taking all the key functions associated with a network switch, or a digital camera, or a personal information appliance, and putting all of the logic, all of the memory, all of the interfaces (in fact, almost everything electronic in the end product) into one piece of silicon, or something close to it.

The benefits of high-silicon integration levels are clear. Tight integration drives the end product's form factor, making complex systems small enough to put into your pocket, inside your television, or in your car. High integration levels also drive down power dissipation, making more end products battery powered, fan-less, or available for use in a much wider variety of environments. Ever increasing integration levels drive the raw performance—in terms of how quickly a product will accomplish tasks or in terms of the number of different functions that a product can incorporate—ever upward. These attributes are, in fact, likely become even more important product features, ideally enough to make the average consumer rush to their favorite retailer to buy new products to replace the old ones.

1.2.1 A New SOC for Every System is a Bad Idea

The resulting silicon specialization stemming from increasingly higher integration creates an economic challenge for the product developer. If all of the electronics in a system are embodied in roughly one chip, that chip is increasingly likely to be a direct reflection of the end product the designer is trying to define. Such a chip design lacks flexibility. It cannot be used in a wide variety of products.

In the absence of some characteristic that makes that highly integrated chip significantly more flexible and reusable, SOC design moves toward a direct 1:1 correspondence between chip design and system design. Ultimately, if SOC design were to really go down this road, the time to develop a new system and the amount of engineering resources required to build a new system will, unfortunately, become at least as great as the time and costs to build new chips.

In the past, product designers built systems by combining chips onto large printed circuit boards (PCBs). Different systems used different combinations of (mostly) off-the-shelf chips soldered onto system-specific PCBs. The approach worked because a wide variety of silicon components were available and because PCB design and prototyping was easy. System reprogrammability was relatively unimportant because system redesign was relatively cheap and quick.

In the world of nanometer silicon technology, the situation is dramatically different. Demands for smaller physical system size, greater energy efficiency, and lower manufacturing cost all make the large system PCB obsolete. Requirements for high-volume end products can only be satisfied with SOC designs. However, even when appropriate virtual components are available as SOC building blocks, SOC design integration and prototyping are more than two orders of magnitude more expensive than PCB design and prototyping. Moreover, SOC design changes take months, while PCB changes take just days. To reap the benefits of nanometer silicon, SOC design is mandatory. To make SOC design practical, SOCs cannot be built like PCBs. The problem of SOC inflexibility must be addressed.

This potential chip-level inflexibility is really a crisis in reusability of the chip's hardware design. Despite substantial industry attention to the benefits of block-level hardware reuse, the growth in internal complexity of blocks, coupled with the complex interactions among blocks, has limited the systematic and economical reuse of hardware blocks.

Too often customer requirements, implemented standards, and the necessary interfaces to other functions must evolve with each product variation. These boundaries constrain successful block reuse to two categories—simple blocks that implement stable interface functions and inherently flexible functions that can be implemented in processors whose great adaptability is realized via software programmability.

A requirement to build new chips for every system would be an economic disaster for system developers because building chips is hard. We might improve the situation somewhat with better chip-design tools, but in the absence of some significant innovation in chip design methodology, the situation's not getting better very fast.

In fact, in the absence of some major innovation, the efforts required to design a chip will increase more rapidly than the transistor complexity of the chip itself. We're losing ground in systems design because innovation in methodology is lacking. We cannot afford to lose ground on this problem as system and chip design grow closer together.

1.2.2 SOC Design Reform: Lower Design Cost and Greater Design Flexibility

System developers are trying to solve two closely related problems:

- To develop system designs with significantly fewer resources by making it much easier to design the chips in those systems.
- To make SOCs more adaptable so not every new system design requires a new SOC design.

The way to solve these two problems is to make the SOC sufficiently programmable so that one chip design will efficiently serve 10, or 100, or even 1000 different system designs while giving up very few of the benefits of integration. Solving these problems means having chips available off the shelf to satisfy the requirements of the next system design and amortize the costs of chip development over a large number of system designs.

These trends constitute the force behind the drive for a fundamental shift in integrated circuit design. That fundamental shift will ideally provide both a big improvement in the effort needed to design SOCs (not just in the silicon but also the required software) and it will increase the intrinsic flexibility of SOC designs so that the design effort can be shared across many system designs.

Economic success in the electronics industry hinges on the ability to make future SOCs more flexible and more highly optimized at the same time. The core dilemma for the SOC industry and for all the users of SOC devices is this simultaneous management of flexibility and optimality.

SOC developers are trying to minimize chip-design costs and trying to get closer to the promised benefits of high-level silicon integration at the same time. Consequently, they need to take full advantage of what high-density silicon offers and they need to overcome or mitigate the issues created by the sheer complexity of those SOC designs and the high costs and risks associated with the long SOC development cycle.

The underlying scalability of silicon suggests that two basic concepts are important to successfully reform the SOC design approach: concurrency and programmability. Concurrency (or parallelism) allows SOC designers to directly exploit the remarkable transistor and interconnect density of leading-edge silicon fabrication. Programmability allows SOC designers to substantially mitigate the costs and risks of complex SOC designs by accelerating the initial development effort and by easing the effort to accommodate subsequent revisions of system requirements.

1.2.3 Concurrency

One of the key benefits of high silicon density is the ability to design highly parallel systems on one chip. The ever-shrinking size of the transistor produces both faster and smaller transistors. The impact of faster transistors on performance is obvious. However, much of the historical performance increase in electronic systems has come not from faster transistors but from exploitation of the large numbers of transistors available. System architects have been amazingly and consistently clever in devising ways of using more transistors to do more work by executing several operations concurrently, independently of transistor speed.

For example, consider a high-bandwidth network router. The router system may handle hundreds of network connections, each of which streams millions of Internet packets per second. Every packet on every incoming connection must be analyzed and routed to an outgoing connection with minimal latency. The capacity of high-end routers has grown dramatically in the past decade into many terabits per second of aggregate bandwidth. This growth is far steeper than the growth in basic transistor speed. Clever architects have developed scalable routing fabric architectures to build large routers from a large number of individual building blocks. These routing blocks are implemented as routing chips, each of which contains an array of routing engines or network processors. Each processor handles only one packet at a time, but the complete router may contain thousands of network processors, all analyzing and forwarding packets concurrently.

From the designer's perspective, this system concurrency is often implicit in the nature of the system's definition. The various subsystems can run simultaneously, allowing greater seamlessness in the user experience. Examples of highly parallel digital machines now abound even in embedded system designs. If you look at some multimedia system products, you will likely find an audio subsystem and a video subsystem, and you may find an encryption subsystem, a networking interface, and a user interface. All of these subsystems operate in parallel.

Functional concurrency is now essential to the user's experience. We naturally expect that a cell phone will be able to maintain a radio connection, encode and decode voice conversations, and run a user interface that lets us find phone numbers. As real-time multimedia becomes ubiquitous, we may find that simultaneous video encoding and decoding, encryption, and gaming will become mandatory in handheld devices.

In the past, when transistors were expensive, many systems created the illusion of parallelism by sharing the capability of one fast processor, programmed to perform parts of different tasks in narrow slices of time. In many cases, sequential time slicing of a common computing resource may be less efficient that direct parallel implementation because parallel implementation eliminates the task-switching overhead of sequential sharing.

Parallel implementation effectively exploits the plummeting transistor costs delivered by high-density silicon. The availability of cheap transistors does not relieve pressure for greater design efficiency, however. Effective execution concurrency results when all of a design's subsystems and function units are performing useful work simultaneously. Parallel execution techniques apply to both system architecture and subsystem implementation. These system and subsystem organizations include heterogeneous processor groups, homogeneous processor arrays, function-unit pipelining, processor architectures with multiple execution pipelines, wide data paths performing the same operation on multiple operands, and task engines implementing multiple dependent operations using application-specific instruction sets. Each of these design approaches serves not just to use more of the available transistors, but more importantly, to ensure that each transistor is engaged in useful work much of the time. Exploitation of concurrency is an essential element of any solution that addresses the challenges of SOC design.

1.2.4 Programmability

Rising system complexity also makes programmability essential to SOCs, and the more efficient programming becomes, the more pervasive it will be. The market already offers a wide range of system programmability options, including field-programmable gate arrays (FPGAs), standard microprocessors, and reconfigurable logic.

Programmability's benefits come at two levels. First, programmability increases the likelihood that a pre-existing design can meet the performance, efficiency, and functional requirements of the system. If there is a fit, no new SOC development is required—an existing platform will serve. Second, programmability means that even when a new chip must be designed, more of the total functions are implemented in a programmable fashion, reducing the design risk and effort. The success of both embedded FPGA and processor technology is traceable to these factors.

The programming models for different platforms differ widely. Traditional processors (including DSPs) can execute applications of unbounded complexity, though as complexity grows, performance typically suffers. Processors typically use sophisticated pipelining and circuit-design techniques to achieve high clock frequency, but achieve only modest parallelism—one (or a few) operations per clock cycle. FPGAs, by contrast, have finite capacity—once the problem grows beyond some level of complexity, the problem simply will not fit in an FPGA at all. On the other hand, FPGAs can implement algorithms with very high levels of intrinsic parallelism, sometimes performing the equivalent of hundreds of operations per cycle. FPGAs typically show more modest clock rates than processors and tend to have larger die sizes and higher chip costs than processors used in the same applications.

1.2.5 Programmability Versus Efficiency

All these flavors of programmability allow the underlying silicon design to be somewhat generic while permitting configuration or personalization for a specific situation at the time that the system is booted or during system operation. Traditional programmability leaves a tremendous gap in efficiency and/or performance relative to a hardwired design for the same function. This efficiency and performance gap—the "programmability overhead"—can be defined as the increased area for implementation of a function using programmable methods compared to a hardwired implementation with the same performance. Alternatively, this overhead can be seen as the increase in execution time in a programmable solution compared to a hardwired implementation of the same silicon area. As a rule of thumb, the overhead for FPGA or generic processor programmability is more than a factor of 10 and can reach a factor of 100. For security applications such as DES and AES encryption, for example, hardwired logic solutions are typically about 100 times faster than the same tasks implemented with a general-purpose RISC processor. An FPGA implementation of these encryption functions may run at 3-4 times lower clock frequency than hardwired logic and may require 10-20 times more silicon area.

These inefficiencies stem from the excessive generality of the universal digital substrates: FPGAs and general-purpose processors. The designers of these substrates are working to construct platforms to covers all possible scenarios. Unfortunately, the creation of truly general-purpose substrates requires a superabundance of basic facilities from which to compose specific computational functions and connection paths to move data among computation functions. Silicon efficiency is constrained by the limited reuse or "time-multiplexing" of the transistors implementing an application's essential functions.

In fact, if you look at either an FPGA or a general-purpose processor performing an "add" computation, you will find a group of logic gates comprising an adder surrounded by a vast number of multiplexers and wires to deliver the right data to the right adder at the right moment. The circuit overhead associated with storing and moving the data and selecting the correct sequence of primitive functions leads to much higher circuit delays and a much larger number of required transistors and wires than a design where the combination or sequence of operations to be performed is known at the time the chip is built. General-purpose processors rely on time-multiplexing of a small and basic set of function units for basic arithmetic, logical operations, and memory references. Most of the processor logic serves as hardware to route different operands to the small set of shared-hardware function units. Communication among functions is implicit in the reuse of processor registers and memory locations by different operations.

FPGA logic, by contrast, minimizes the implicit sharing of hardware among different functions. Instead, each function is statically mapped to a particular region of the FPGA's silicon, so that each transistor typically performs a single function repeatedly. Communication among functions is explicit in the static configuration of interconnect among functions. Both general-purpose processors and general-purpose FPGA technologies have overhead, but an exploration of software programmability highlights the hidden overhead of field hardware programmability.

Modern software programmability's power really stems from two complementary characteristics. One of these is abstraction. Software programs allow developers to deal with computation in a form that is more concise, more readily understood at a glance, and more easily enhanced independently of implementation details. Abstraction yields insight into overall solution structure by hiding the implementation details. Modest-sized software teams routinely develop, reuse, and enhance applications with hundreds of thousands of lines of source code, including extensive reuse of operating systems, application libraries, and middleware software components. In addition, sophisticated application-analysis tools have evolved to help teams debug and maintain these complex applications.

By comparison, similar-sized hardware teams consider logic functions with tens of thousands of lines of Verilog or VHDL code to be quite large and complex. Blocks are modified only with the greatest care. Coding abstraction is limited—simple registers, memories, and primitive arithmetic functions may constitute the most complex generic reusable hardware functions in a block.

The second characteristic is software's ease of modification. System functionality changes when you first boot the system, and it changes dynamically when you switch tasks. In a software-driven environment, if a task requires a complete change to a subsystem's functionality, the system can load a new software-based personality for that subsystem from memory in a few microseconds in response to changing system demands.

This is a key point: the economic benefits of software flexibility appear both in the development cycle (what happens between product conception and system prototype power-on) and during the expected operational life of a design (what happens between the first prototype power-on and the moment when the last variant of the product is shipped to the last customer).

Continuous adaptability to new product requirements plays a central role in improving product profitability. If the system can be reprogrammed quickly and cheaply, the developers have less risk of failure in meeting design specifications and have greater opportunity to quickly adapt the product to new customer needs. Field-upgrading software has become routine with PC systems, and the ability to upgrade software in the field is starting to find its way into embedded products. For example, products such as Cisco network switches get regular software upgrades. The greater the flexibility (at a given cost), the greater the number of customers and the higher the SOC volume.

In contrast, hardwired design choices must be made very early in the system-design cycle. If, at any point in the development cycle—during design, during prototyping, during field trials, during upgrades in the field, or during second-generation product development—the system developers decide to change key computation or communication decisions, then it's back to square one for the system design. Hardwiring key design decisions also narrows the potential range of customers and systems into which the SOC might fit and limits the potential volume shipments.

Making systems more programmable has benefits and liabilities. The benefits are seen both in development agility and efficiency and in post-silicon adaptability. Designers don't need to decide exactly how the computational elements relate to each other until later in the design cycle, when the cost of change is low. Whether programmability comes through loading a net list into an FPGA or software into processors, the designer doesn't have to decide on a final configuration until system boot.

If programmability is realized through software running on processors, developers may defer many design decisions until system bring-up, and some decisions can be deferred until the eve of product shipment. The liabilities of programmability have historically been cost, power-efficiency, and performance. Conventional processor and FPGA programmability carries an overhead of thousands of transistors and thousands of microns of wire between the computational functions. This overhead translates into large die size and low clock frequency for FPGAs and long execution time for processors, compared to hardwired logic implementing the same function. Fixing the implementation hardware in silicon can dramatically improve unit cost and system performance but dramatically raises design risk and design cost. The design team faces trying choices. Which functions should be implemented in hardware? Which in software? Which functions are most likely to change? How will communication among blocks evolve?

This tradeoff between efficiency and performance on one hand, and programmability and flexibility on the other hand, is a recurring theme is this book. Fig 1-2 gives a conceptual view of this tradeoff.

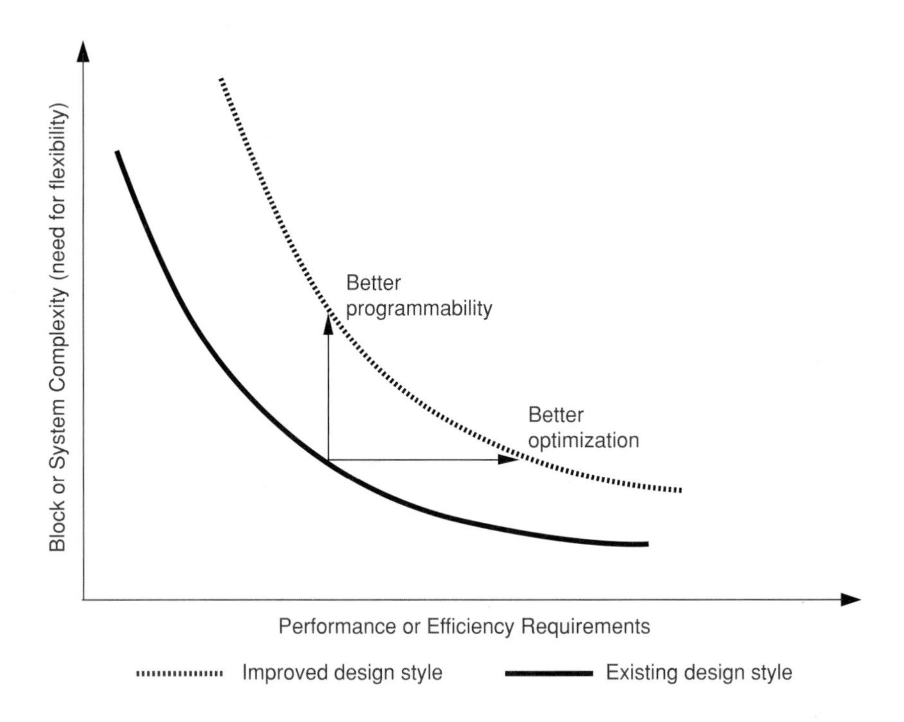

Figure 1-2 The essential tradeoff of design.

The vertical axis indicates the intrinsic complexity of a block or of the whole system. The horizontal axis indicates the performance or efficiency requirement of a block or the whole system. One dilemma of SOC design recurs: solutions that have the flexibility to support complex designs sacrifice performance, efficiency, and throughput; solutions with high efficiency or throughput often sacrifice the flexibility necessary to handle complex applications.

The curves represent overall design styles or methodologies. Within a methodology, a variety of solutions are possible for each block or for the whole system, but the design team must trade off complexity and efficiency. An improved design style or methodology may still exhibit tradeoffs in solutions, but with an overall improvement in the "flexibility-efficiency product." In moving to an improved design methodology, the team may focus on improving programmability to get better flexibility at a given level of performance or may focus on improving performance to get better efficiency and throughput at a given level of programmability.

In a sense, the key to efficient SOC design is the management of uncertainty. If the design team can optimize all dimensions of the product for which requirements are stable and leave flexible all dimensions of the product that are unstable, they will have a cheaper, more durable, and more efficient product than their competitors.

1.2.6 The Key to SOC Design Success: Domain-Specific Flexibility

The goal of SOC development is to strike an optimal balance between getting just enough flexibility in the SOC to meet changing demands within a problem, still realizing the efficiency and optimality associated with targeting an end application. Therefore, what's needed is an SOC design methodology that permits a high degree of system and subsystem parallelism, an appropriate degree of programmability, and rapid design.

It is not necessary for SOC-based system developers to use a completely universal piece of silicon—most SOCs ship in enough volume to justify some specialization. For example, a designer of digital cameras doesn't need to use the same chip that's used in a high-end optical network switch. One camera chip, however, can support a range of related consumer imaging products, as shown in Fig 1-3.

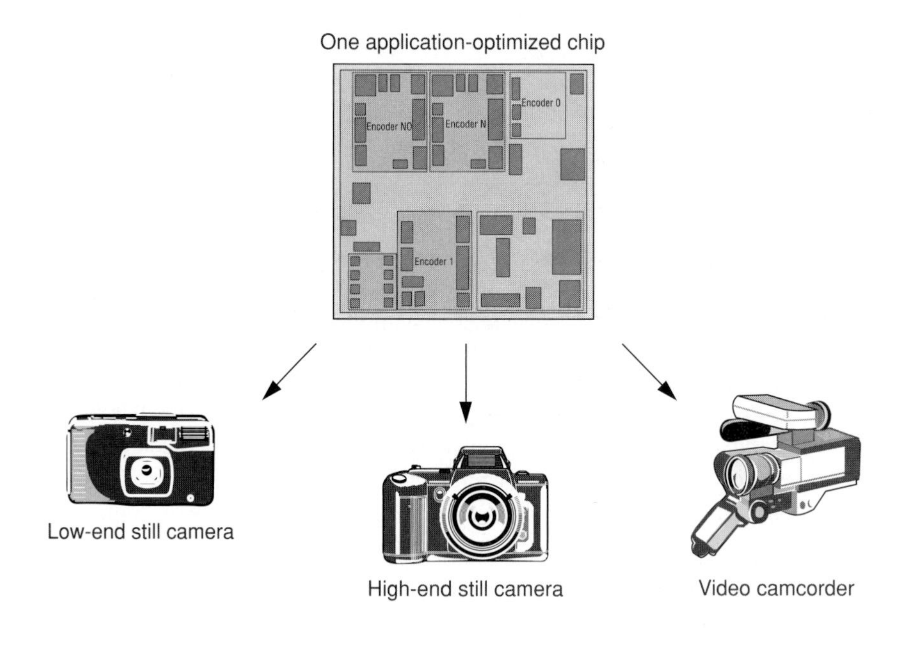

Figure 1-3 One camera SOC into many camera systems.

The difference in benefit derived from a chip shared by 10 similar designs versus one shared by 1,000 designs is relatively modest. If each camera design's volume is 200,000 units, and the shared SOC design costs $10 million, then the SOC design contributes $5 to final camera cost (~5%). Sharing the SOC design across 1000 different designs could save $5 in amortized design cost, but would almost certainly require such generality in the SOC that SOC production costs would increase by far more than $5. SOC designs need not be completely universal—high-volume products can easily afford to have a chip-level design platform that is appropriate to their application domain yet flexible within it.

If designers have sufficient flexibility within an SOC to adapt to any tasks they are likely to encounter during that design's lifetime, then they essentially have all the relevant benefits of universal flexibility without much of the overhead of universal generality. If the platform design is done correctly, the cost for this application-specific flexibility is much lower than the flexibility derived from a truly universal device such as an FPGA and a high-performance, general-purpose processor.

In addition, a good design methodology should enable as broad a population of hardware and software engineers as possible to design and program the SOCs. The larger the talent pool, the faster the development and the lower the project cost. The key characteristics for such an SOC design methodology are

1. support for concurrent processing,
2. appropriate application efficiency, and
3. ease of development by application experts (developers who are not necessarily specialists in SOC design).

1.3 What's Wrong with Today's Approach to SOC Design?

Complex digital chips combining processors, memory, and application-specific logic have been designed for more than a decade. With each improvement in silicon density, increasingly complex designs are attempted. It is natural to hope that simple scaling of that approach will allow the semiconductor-design community to continue to deliver compelling electronic solutions.

Unfortunately, just using bigger processors and more blocks of application-specific logic does not scale. The computational demands of emerging communications and consumer applications, the complexity of product requirements, and the design cost pressures are outgrowing the capacity of simple semiconductor scaling and incremental design productivity improvement. Major changes are needed in the role of processors, the design of application-specific hardware, and the integration of all the pieces into complete SOC hardware and software.

For example, supporting higher data rates and more channel capacity in wireless communications requires better and better information encoding. Better encoding allows communications over noisy communication channels to come closer to the information theory limits outlined by Shannon in the 1940s. (Claude Shannon proved an upper bound on the maximum possible data communication rate that can be achieved in the presence of noise in the communication channel. "Shannon's law" is a foundation of modern information theory.) Better encoding requires exponentially increasing computational effort. Ravi Subramanian used his substantial experience with both EDA tool development and processor architecture to compare the growth in the computational demands of succeeding generations of mobile wireless communications to the growth of transistor-level performance, as implied by the traditional Moore's law. Fig 1-4 shows how the estimate of wireless communications computation, as embodied in 1G, 2G and 3G wireless standards, outgrows silicon speed by almost 25% per year.

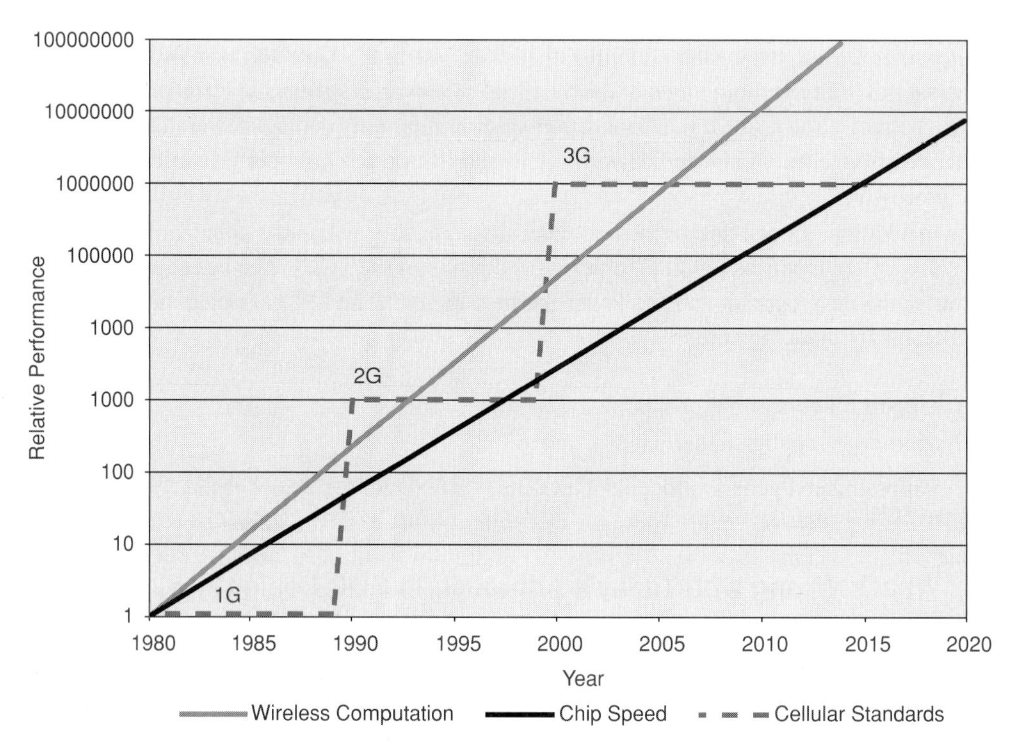

Figure 1-4 Wireless computation outstrips transistor performance scaling.

Worse yet, this model assumes no slowdown in the growth of silicon density and performance. As we shall see in Chapter 8, the International Roadmap for Semiconductors does not forecast an end to Moore's law, but it does suggest a gradual reduction in the rate of progress. So, Moore's-law scaling for processors or logic circuits won't solve our core problem. We must look elsewhere to close the gap between the demands of electronics applications and the capabilities of raw semiconductor technology.

1.3.1 What's Wrong with Traditional Processors?

Traditional RISC processor cores have many uses, and it's certainly possible to build big SOCs using large numbers of traditional processor cores. But the needs of SOC design are fundamentally diverging from the design intent of the traditional processor. This divergence stimulates the bold claim that traditional processors are not the answer for complex SOC design.

This iconoclastic position is based on one central observation: general-purpose processors are simply not fast or efficient enough to do much of the hard work embedded in SOCs. Data throughput requirements are climbing, and power budgets are shrinking faster than simple Moore's-law processor scaling can support. For many data-intensive applications, the only solution for high performance or high energy efficiency is hardwired logic design, but designing complex functions in hardwired logic is fraught with risks.

General-purpose processor architectures are not fast enough because they implement only generic operations—typically primitive arithmetic, logical, shift, and comparison operations on 8-, 16-, and 32-bit integers—and because they perform little more than one basic operation per cycle. This generality and modest performance level is perfectly adequate for some applications. User-interface applications, where the application only needs to keep up with the user's fingers, may perform admirably with current RISC processor performance (a couple of hundred million instructions per second in 2003). The latest RISC processors can also handle protocol stacks, low-resolution image manipulation, and low-bandwidth signal processing. When the application work fully fits within the performance range of a generic processor, the system design process benefits. Software-based systems design is widely understood and highly tolerant of late design changes and evolving customer requirements.

Unfortunately, generic processors lack the performance to do the most demanding signal, image, protocol, security, and other data-intensive processing tasks. Today's digital cameras, cell phones, DSL modems, high-performance routers, and digital televisions all use special-purpose processors or hardwired logic circuits for their most demanding image-, signal-, and packet-processing functions. Moreover, the data-processing demands in many application domains are growing significantly more rapidly than basic processor performance.

General-purpose processor performance roughly tracks with the performance increase of transistors as they scale in size. Larger transistor budgets help by allowing more aggressive parallel execution of instructions, but this benefit is largely offset by increases in relative interconnect delay on large designs, by increases in power dissipation, and by the slower improvement in off-chip memory-access performance. A leading-edge pipelined RISC processor in the late 1980s could be implemented with a few hundred thousand logic transistors and could sustain execution of roughly 0.5 and 0.75 instructions per cycle. A state-of the-art RISC processor today requires more than 10 million logic transistors and may still not sustain execution of as much as two instructions per cycle.

Traditional processors can't keep up with the demands of data-intensive applications because they are designed to be generic—the architect can have no particular application domain in mind. Extreme processor-design techniques—speculative, out-of-order execution of instructions, aggressive high-power circuit designs, and use of high-current transistors—can raise the absolute processor performance level, but at a large price in silicon area and power dissipation. As a result, general-purpose processors fall farther and farther behind the performance and efficiency needs of data-intensive embedded applications.

By contrast, the designers of embedded subsystems often know the type and degree of computing power required in their application quite explicitly. They think in terms of the real applications' essential computation and communication. The drive for efficient throughput and the inadequacy of generic processors appear to leave the SOC designer with only one choice for data-intensive applications: hardwired design. Hardwired designs can exploit the potential parallelism of the underlying application far more directly than a standard microprocessor. If the subsystem is operating on a video stream, or a sequence of voice samples, or a block of encrypted data, for example, designers can incorporate into the hardware the essential media decoding, voice filtering, or stream-decryption features that are intrinsic to that data type and function. SOC designers can't afford the energy and delay of long instruction sequences.

For example, think what it means to process an Internet Protocol (IP) packet header. An IP packet header contains a number of different fields and each of those fields has a special meaning. Operations to be performed on that packet include extracting fields (for example, the time-to-live field), updating fields, and using the destination field for routing-table look-up and level-of-service handling. To map these packet-processing functions to a conventional microprocessor, the designer's application code must transform the packet header into a collection of 8-bit, 16-bit, and 32-bit integer quantities and then perform a series of 8-bit, 16-bit, and 32-bit integer operations on these quantities to emulate the actual work to be done. Using a general-purpose microprocessor to execute this long sequence of individual integer operations is inherently inefficient.

The efficiency and performance gap between integer code running on a generic RISC processor and an application-specific sequence of operations tuned for the task can be dramatic. Across a wide range of embedded data-processing applications—signal, protocol, security, and multimedia—a factor-of-five or factor-of-10 difference in potential throughput is routine, and factor-of-100 differences are not unheard of. This performance gap provides the primary motivation for hardwired logic design.

The liabilities of hardwired design, however, are growing with the sheer complexity of applications. Application-specific processors, whose instruction sets and data paths directly implement to the application's natural operations, provide a potent alternative. They merge the programmability and tools of state-of-the-art processors with the efficiency and throughput characteristics of dedicated logic designs.

1.3.2 What's Wrong with Traditional SOC Methodology?

Conventional SOC design methodology consists of three phases:

1. Partition the design's functions or tasks into hardware and software.
2. Develop the components:

- Program pre-existing, fixed-function, legacy processors for lower-performance functions.
- Develop RTL for functions where performance, power, or area efficiency are important.

3. Integrate the diverse set of hardware and software pieces into system models, physical implementation, and final software. If cost, power, performance, and flexibility goals are not met, go back to step 1.

All would be well if one pass through this flow guaranteed success. All phases interact, so small changes in RTL, for example, ripple through to revised partitioning, revised software, and revised system models. Worse yet, RTL design and verification of complex blocks is slow and labor-intensive, even for highly skilled teams.

For example, to implement a function in hardware (the traditional ASIC or SOC implementation methodology), designers go through the detailed development of that RTL hardware using an HDL (hardware description language) such as VHDL or Verilog. They compile and verify the RTL on an HDL simulator, which runs much slower than real time. A large block of RTL code might run at 100 cycles/sec on an HDL simulator. That's roughly a million times slower than real time.

Consequently, it takes a long time to completely test even simple RTL blocks, and it's often impossible to run verification tests that require a large number of cycles. Some development teams prototype individual SOC functions with FPGAs, but that involves overhead: finding or making the appropriate FPGA board, defining the right interface, and visualizing what's happening inside the circuit. Even with FPGA emulation, hardware-verification tests don't run at full speed. FPGA emulations typically run 10 to 100 times slower than real hardware. Designing and verifying hardware is hard work.

When design teams try to integrate all of the pieces of an SOC design, they face an even tougher problem because they must somehow combine software running on a processor with an RTL hardware simulation. There are some cosimulation tools available, but RTL simulation is still the long pole in the verification tent for the reasons mentioned above. The combined simulation speed of the processor simulator and RTL simulation is dominated by the RTL so, at best, the system cosimulation still runs at only hundreds of cycles per second.

Moreover, the level of visibility into what's really going on in a block of RTL is governed by what's available in hardware simulation/visualization tools. Usually, such visibility is limited to waveform viewing, making it difficult to get the application-level picture of what's happening in the overall simulation.

RTL logic errors are so damaging to product delivery schedules that hardware teams must devote the bulk of their development resources to verification. The question, "Can we verify it?" can even become a major consideration in setting SOC hardware specifications. Even with this obsession with avoiding bugs, almost half of all ASICs still show RTL functional bugs that require silicon revision according to Aart de Geus, CEO of Synopsys, in his keynote address to the ESNUG Conference in September 2003. As the designs become more complex, the hardware-verification risk increases, and the importance of tight hardware/software cooperation increases. This combination—long, complex hardware development, high risk of failure, and growing emphasis on tight team coordination—is a recipe for stress in the SOC design effort.

Consequently, hardware/software codesign (partitioning and generating the detailed design of the embedded software running on a processor and the other SOC hardware) and co-simulation (getting the hardware and software simulations to both run in a virtual-prototyping environment including modeling all of the interactions between them) are painful, slow, and expensive.

It is also worth highlighting here that the process of optimizing a design is not simply a process of adding whatever hardware is needed to efficiently execute an application. Hardware optimization is also a process of omitting unnecessary functions. The common, desirable practice of hardware or software design reuse leads to an accumulation of features. Functions are originally designed (or they evolve over time) to include all the features that any user might require. In software development, this superset principle is generally benign because unused software routines do not consume execution cycles or even memory space in caches. In hardware design, however, unused functions are expensive and they consume silicon area, power, and design-generation time. Once a feature has been incorporated into a hardware block, the potential benefit of saving silicon resources by later removing the function must be weighed against the risk and cost of respecifying and reverifying the block without the unused features.

For example, processor instruction sets usually grow over time. New processor variants maintain upward compatibility from previous processor generations and add new programmer state and instructions to handle new application domains or to improve performance in well-established application classes. Processor vendors, compiler developers, and processor users together spend millions of dollars upgrading compilers, operating systems, simulators, and other software-development tools to accommodate the additions made to just a single processor generation. As features are added, however, utilization of existing features inevitably declines. For example, high-performance standalone RISC processor integrated circuits typically include floating-point units, reflecting the origins of RISC processors in engineering workstations. When a high-performance processor is used in a network router (now a common usage) the floating-point unit is completely idle even though it may consume more silicon area than the integer execution unit. Slow evolution of software tools makes removal of these hardware features difficult even when old features are effectively superseded by new features. New feature additions may increase the absolute processor performance, but the performance efficiency – performance per unit of silicon area, per unit of power, or per dollar of design effort – may degrade. As we develop the rationale for configurable processors in this book, the opportunity to omit unused features plays a substantial role–sometimes as significant as the opportunity to add new application-specific features–in driving the efficiency of processor-based design toward that of hardwired designs.

An important caveat on processor-based design methodology should be mentioned here. No methodology is going to eliminate all hardwired logic. It makes no sense to use a 25,000-gate processor to perform a function when 1000 gates of logic will suffice. When the problem is simple or stable, design at the level of RTL, or even transistor net lists, makes perfect sense. The challenge comes when the task is neither stable nor simple. Then the attraction of RTL-based design rests on performance and efficiency in the implementation of complex functions. This is where RTL design and verification methods tend to falter.

1.4 Preview: An Improved Design Methodology for SOC Design

A fundamentally new way to speed development of megagate SOCs is emerging. First, processors replace hardwired logic to accelerate hardware design and bring full chip-level programmability. Second, those processors are extended, often automatically, to run functions very efficiently with high throughput, low power dissipation, and modest silicon area. Blocks based on extended processors often have characteristics that rival those of the rigid RTL blocks they replace. Third, these processors become the basic building blocks for complete SOCs, where the rapid development, flexible interfacing, and easy programming and debugging of the processors accelerate the overall design process. Finally, and perhaps most importantly, the resulting SOC-based products are highly efficient and highly adaptable to changing requirements. This improved SOC design flow allows full exploitation of the intrinsic technological potential of deep-submicron semiconductors (concurrency, pipelining, fast transistors, application-specific operations) and the benefits of modern software development methodology.

1.4.1 The SOC Design Flow

This book introduces and teaches a new SOC design approach, the cornerstone of which is flexibility. Flexibility helps produce a chip that is right the first time (for lower development cost) and remains right over time (for greater manufacturing volume and revenue). A sketch of the new SOC design flow is shown in Fig 1-5.

The flow starts from the high-level requirements, especially the external input and output requirements for the new SOC platform and the set of tasks that the system performs on the data flowing through the system. The computation within tasks and the communication among tasks and interfaces are optimized using application-specific processors and quick function-, performance-, and cost-analysis tools. The flow makes an accurate system model available early in the design schedule, so detailed very-large-scale-integration (VLSI) and software implementations can proceed in parallel. Early and accurate modeling of both hardware and software reduces development time and minimizes expensive surprises late in the development and bring-up of the entire system.

Using this design approach means that designers can move through the design process with fewer dead ends and false starts and without the need to back up and start over. SOC designers can make a much fuller and more detailed exploration of the design possibilities early in the design cycle. Using this approach, they can better understand the design's hardware costs,

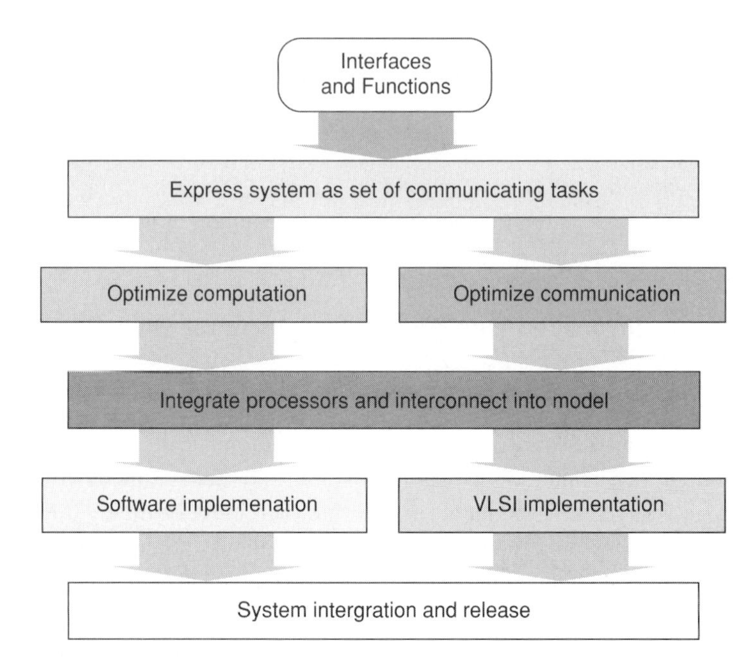

Figure 1-5 MPSOC design flow overview.

application performance, interface, programming model, and all the other important characteristics of an SOC's design.

Taking this approach to designing SOCs means that the cone of possible, efficient uses of the silicon platform will be as large as possible with the fewest compromises in the cost and the power efficiency of that platform. The more a design team uses the application-specific processor as the basic SOC building block—as opposed to hardwired logic written as RTL—the more the SOC will be able to exploit the flexibility inherent in a software-centric design approach.

1.4.2 Configurable Processor as Building Block

The basic building block of this methodology is a new type of microprocessor: the configurable, extensible microprocessor core. These processors are created by a generator that transforms high-level application-domain requirements (in the form of instruction-set descriptions or even examples of the application code) into an efficient hardware design and software tools. The "sea of processors" approach to SOC design allows engineers without microprocessor design experience to specify, evaluate, configure, program, interconnect, and compose those basic building blocks into combinations of processors, often large numbers of processors, that together create the essential digital electronics for SOC devices.

To develop a processor configuration using one of these configurable microprocessor cores, the chip designer or application expert comes to the processor-generator interface, shown in Fig 1-6, and selects or describes the application source code, instruction-set options, memory hierarchy, closely coupled peripherals, and interfaces required by the application. It takes about one hour to fully generate the hardware design—in the form of standard RTL languages, EDA tool scripts and test benches, and the software environment (C and C++ compilers, debuggers, simulators, real-time operating system (RTOS) code, and other support software). The generation process provides immediate availability of a fab-portable hardware design and software development environment. This quick delivery of the hardware and software infrastructure permits rapid tuning and testing of the software applications on that processor design. The completeness of software largely eliminates the issues of software porting and enables a rapid design-iteration methodology. The tools can even be configured to automatically explore a wide range of possible processor configurations from a common application base to reveal a more optimal hardware solution as measured by target application requirements.

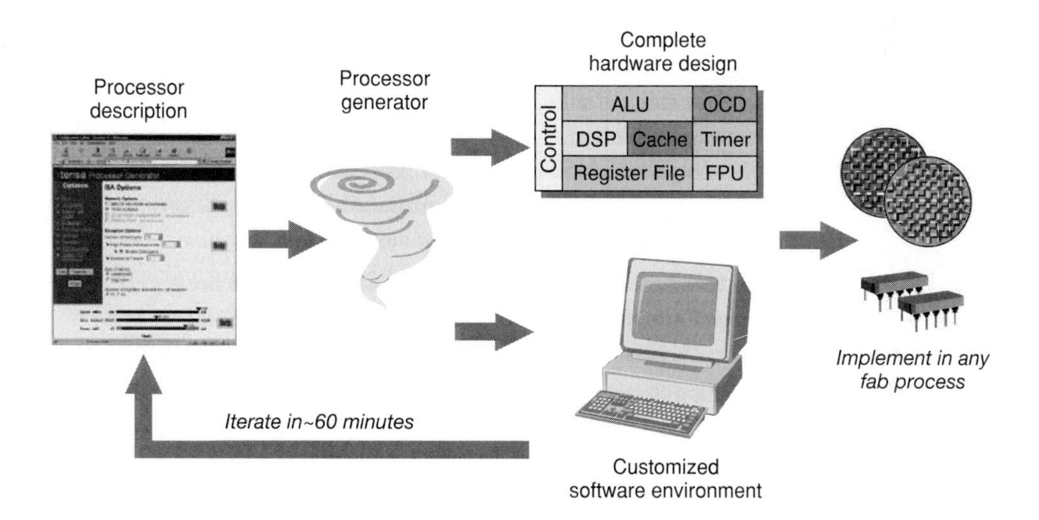

Figure 1-6 Basic processor generation flow.

The application-specific processor performs all of the same tasks that a microcontroller or a high-end RISC processor can perform: run applications developed in high-level languages; implement a wide variety of real-time features; and support complex protocols stacks, libraries, and application layers. Application-specific processors perform generic integer tasks very efficiently, even as measured by traditional microprocessor power, speed, area, and code-size criteria. But because these application-specific processors can incorporate the data paths, instructions, and register storage for the idiosyncratic data types and computation required by an embedded application, they can also support virtually all of the functions that chip designers have historically implemented as hardwired logic.

1.4.3 A Trivial Example

The creation and use of application-specific processors is a major topic of this book, but a short example may make the basic concept more concrete. Consider the following fragment of C code for reversing the order of bytes within a sequence of 32-bit words:

```
int i,a[],b[];
for (i=0; i<4096; i++) {
  a[i] = (((b[i] & 0x000000ff) << 24) | ((b[i] & 0x0000ff00) << 8) |
          ((b[i] & 0x00ff0000) >> 8)  | ((b[i] & 0xff000000) >> 24));}
```

This code sequence requires about a dozen basic integer operations per loop iteration. It could be a major bottleneck in an embedded application—this sort of byte swapping is often required in networking connect systems or protocol processing tasks with different order packing of bytes into words. On an unmodified Xtensa processor, for example, this code segment requires about 50,000 cycles, including the time needed to fetch the 4096 words of data (`b[i]`) from a remote memory and store the results (`a[i]`).

An instruction to reverse the bytes in a 32-bit word might not be useful across all applications, but it would be extremely useful in this application. The designer might define a new instruction using, for example, the following definition written in the Tensilica Instruction Extension (TIE) format (an instruction-set description language that is explored in greater depth throughout the book):

```
operation swap {in AR x, out AR y}{} {y ={x[7:0],x[15:8],x[23:16],x[31:24]};}
```

This one-line description drives all the processor hardware and software modification and extension, so the swap operator becomes a native part of a new application-specific processor. The C code can directly use this new operator as a basic function, so the previous code snippet may be rewritten as follows:

```
int i,a[],b[];
for (i=0; i<4096; i++) a[i] = swap(b[i]);
```

For this code and improved processor, the algorithm needs only about four operations per iteration, a factor-of-three improvement in performance with additional benefits in energy efficiency, binary code size, and even C code readability. This particular example is explored further in Chapter 5, but the basic impact should be clear. Tailoring the processor to the computational needs of the application increases processor performance and efficiency and enables processors to fit into high-performance, high-efficiency roles where general-purpose processors fall short.

1.4.4 Results of Application-Specific Processor Configuration

Automatically generated application-specific processors have attractive performance and efficiency characteristics. Fig 1-7 compares the typical power dissipation and application performance for Tensilica's Xtensa V application-specific processors and a leading RISC processor, the ARM1026EJ-S. The chart includes results from three application suites of the popular EEMBC benchmarks (*http://www.eembc.org*): Telecom, Consumer, and Networking. For the Xtensa processors, results are shown in two forms for each benchmark suite—one without use of application-specific extensions ("out-of-box") and one optimized to take advantage of application-specific extensions (one set of extensions for each suite). In all cases, the measure of performance is application throughput per cycle, relative to performance on a reference processor. The reference is a 32-bit MIPS processor for Telecom and Networking (reference processor has performance of 0.01 Marks per MHz) and a 32-bit ST20 processor for Consumer (reference processor has performance of 0.02 Marks per MHz). This chart compares performance on a per-MHz basis, but the Tensilica and ARM processors achieve quite similar clock frequencies in comparable silicon technology. All power dissipation figures are normalized to typical 130nm foundry technology at 1.0V operating voltage. The relative power advantage of the configurable processors stems from two factors: power management via fine-grained clock gating in the hardware implementation and overall area efficiency via omission of unneeded features. The ARM1026EJ-S processor and Xtensa V processor base configuration implement broadly similar core instruction sets and pipelines. However, every implementation of the ARM1026EJ-S includes Java instruction-set extensions, operating system support, and sophisticated bus interfaces, even though these features are not used by these applications.

Three results stand out. First, even without application-specific extensions, the configured processors are highly efficient—slightly faster than the traditional RISC processor but with significantly lower power dissipation. Automatically generated processors are built with just the features required by the application, so the base instruction-set architecture is not burdened with extraneous features unused by the target application. Eliminating extra baggage reduces processor size and power.

Second, the extension of the application-specific processor adds only modestly to the total power dissipation. Sophisticated power management, including fine-grained clock-gating in application-specific extension hardware, minimizes the increase in dynamic power dissipation with extended hardware.

Third, the application-specific processor configurations are much faster—from 7.5 times faster on networking to 39 times faster on consumer tasks in these EEMBC benchmark suites. The combination of lower power and higher performance means much higher energy efficiency. Overall, the application-optimized configurations average about 50 times the energy efficiency of the RISC processor across these tasks. These size, power, and performance advantages make application-specific processors a compelling building block for a new MPSOC design methodology.

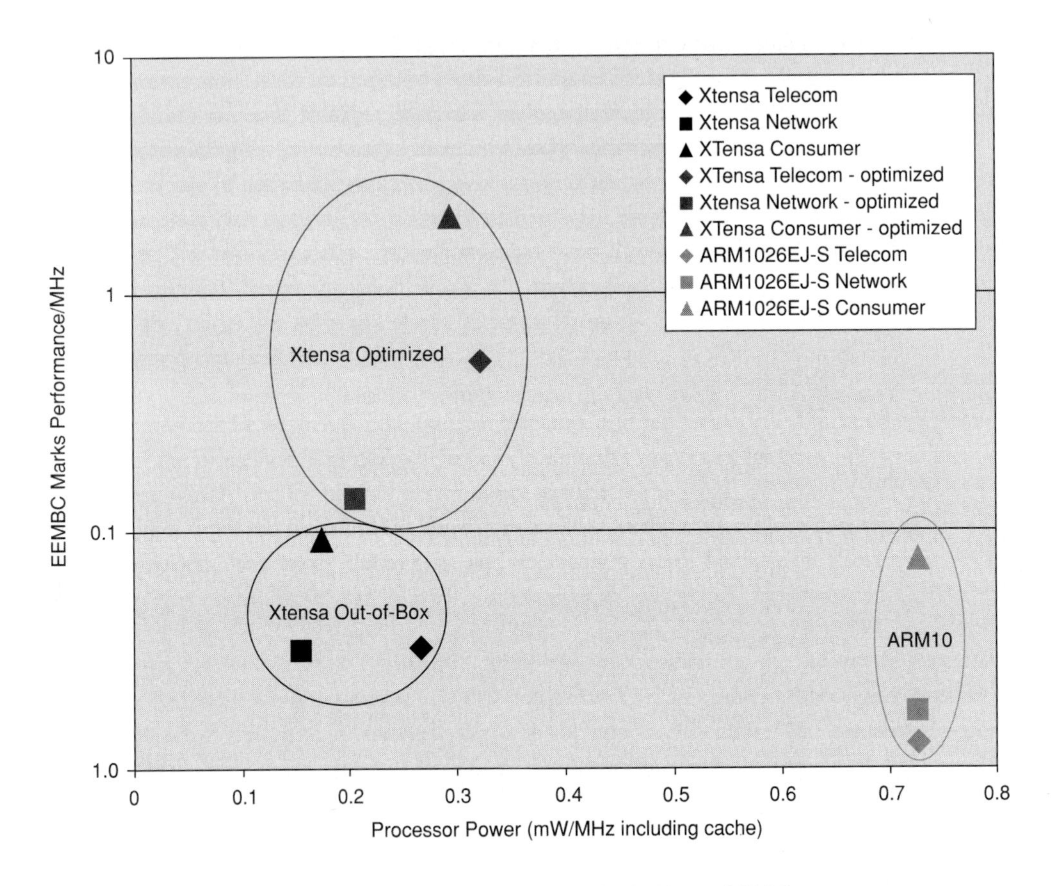

Figure 1-7 Performance and power for configurable and RISC processors.

These results don't just show that configurable processors are a potentially attractive alternative to RISC—they suggest that these new processors may bring full programmability to a range of tasks where performance and power efficiency are at a premium.

1.4.5 The Processor as SOC Building Block

This book develops a core proposition: The use of tailored, application-specific processors as building blocks is a far more effective way to create SOC designs. Processors that work well with SOC designs have the following characteristics:

a. Especially small, fast, or efficient at executing the class of functions required by a particular subsystem
b. Easily adapted to changes (via software programming)
c. Easily combined to solve larger problems.

These types of processors are nearly perfect raw material for SOC design. This book describes a system, architecture, and development methodology that comes much closer to meeting that ideal than conventional SOC design methods in broad use today.

The direct effect of generating an application-specific processor instead of traditional processors or hardwired logic is a superior combination of performance, efficiency, and reprogrammability. The indirect, but equally important, effect is simplification of the overall SOC design process, including both lower design costs and wider use of the chip design.

For the packet-processing example mentioned previously, an application-specific processor implements native operations that directly perform typical, intrinsic operations on IPV4 packet headers. For example, an application-specific processor tailored for packet processing could update the time-to-live field in one operation. It might include an instruction to perform a table lookup on a destination address. It could read a packet-header field and dispatch the packet to any of the appropriate routines in a single cycle.

Because these application-specific functions on application-specific data types correspond more directly to the intrinsic hardware-supported data path functions in the SOC, a processor can execute these tasks more quickly, in a more parallel fashion, using less power. Configuration and programming of processors makes SOC development easier than if the designers needed to decompose the problem into general-purpose integer or bit operations.

So, tailoring the processor to fit the application's intrinsic data types and intrinsic operations on those data types increases system efficiency, boosts system performance, and improves the flexibility of the design. All three of these factors shorten the SOC development cycle.

1.4.6 Solving the System Design Problem

Beyond this basic unit of computation (the application-specific processor), a set of more global design issues loom: How will all the hardware and software pieces come together to solve the end-system requirements? Building a network router is not just a question of how to forward a series of IPV4 packet headers and building a handheld, high-resolution digital television is not just a question of how to decode an MPEG4 video stream. These complex SOC designs must deal with the system-level issues in addition to individual subsystems (such as video decode, audio encode, network-packet forwarding, and DES encryption).

An effective system-design methodology must allow designers to more easily, quickly, and cheaply pull the various subsystems together into the whole design and to reliably verify that the assembly is correct. After all, the general SOC design problem is to get each of the subsystems to do the right thing and get all of the subsystems to work together effectively.

SOC developers need a design methodology that allows them to easily create each of the various SOC subsystems—both subsystems that have historically been implemented in software running on general-purpose processors (typically because these applications didn't require much performance so that even conventional processors with no application-specific enhancements were fast enough) or subsystems that were implemented as blocks of hardwired logic (because these blocks were so small and simple or their function was changing so slowly that they could justifiably be implemented in hardware).

At the conceptual level, the entire system can be treated as a constellation of concurrent, interacting subsystems or tasks. Each implements a set of interfaces through which it communicates with other subsystems and shares common resources (memory, shared data structures, network ports). At a minimum, any modern electronic system contains at least one software-based task—hence at least one block of processor hardware plus data and instruction memories—and at least one input and one output interface device. In practice, a system can be viewed as dozens or even hundreds of interacting tasks or blocks. Software tasks communicate with other software tasks through software abstractions such as application programming interfaces (often abstracted into messages or synchronized access to shared memory). Hardware blocks communicate with other hardware blocks over wires (often abstracted into buses). Software blocks typically communicate with hardware blocks through memory-mapped control registers (often abstracted into device drivers). A hypothetical system is shown in Fig 1-8, with each subsystem mapped either to a software or a hardware block. Moreover, the blocks are mapped onto both a scale of relative complexity and relative computational throughput demands.

Complex but undemanding tasks or infrequently executed tasks naturally gravitate toward software implementation. Simple but high-throughput functions, especially heavily used functions at the heart of the system application, naturally gravitate toward hardware implementation. The tough design decisions revolve around two questions:

1. What is the right implementation (hardware vs. software) for each block?
2. What is the best implementation for the interfaces between a block and those with which it communicates?

In the figure, few issues arise over software blocks A, B, C, D and the communications among them. Standard software methods for task-to-task communication are probably adequate. The performance demands are modest, so a traditional processor core may serve admirably. Similarly, hardware blocks G, J, and K are simple, so hardware design and verification may not be difficult and changes are unlikely. Communication among simple blocks is probably also simple. Blocks H and I, and especially E and F, present bigger challenges to traditional methodology. Here the combination of complexity and performance may both increase the effort required inside the block and complicate the interfaces among them.

The hardware/software interfaces in Figure 1-8 (C:E, D:F, D:H, and D:G) also present challenges. Matching the programming model of the interface, as seen by software, and the wire implementation in hardware is intrinsically complex and error-prone. Two representations, one written in C, for example, describing sequential operations on data-structures, and one written in

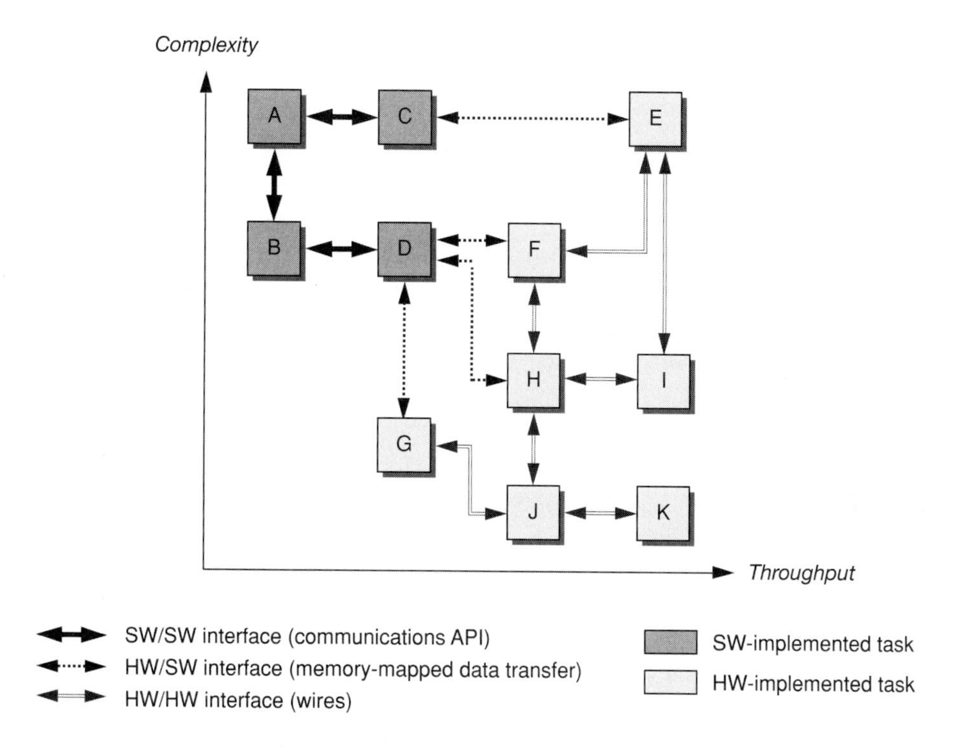

Figure 1-8 Conceptual system partitioning.

Verilog, for example, describing parallel operations on signals, must be synchronized. A small industry—hardware/software coverification—has emerged just to address this deep-seated incompatibility.

As system requirements evolve over the course of a project and from product generation to product generation, both complexity and throughput requirements inevitably grow. More blocks are added to the system and many blocks move up and to the right.

The introduction of configurable, extensible processors changes the SOC design equation. Essentially, these tailorable, application-specific processors significantly increase the potential subsystem-design space that can be covered by processors. Even very small processors can now deliver very high performance. By tailoring the processor for the intended application, and by leaving out hardware not needed by the application, processor efficiency improves dramatically. The performance per gate, performance per square millimeter of silicon, performance per watt, and performance per clock of these processors can often rival the performance of hardwired logic blocks that they replace.

Efficient application-specific processors open up a world in which all but a handful of the SOC subsystems and functions can be implemented in software. In this scenario, several different functions can often share a single processor, effectively time-slicing it. In other cases, different tasks require dedicated application-specific processors. The distribution of processors in an SOC becomes just a function of system partitioning and most SOCs employ many processors to implement the majority of the SOC's subsystems.

The leverage of the complex SOC design methodology on the partitioning problem is particularly important to understand. Previously, when a system designer looked at SOC design partitioning, it was important to settle on a partition between hardware and software early in the project. Once the partition was established, the task of backtracking (of saying, "Gosh, I was wrong about the cost, performance, or function") became difficult. Designers sometimes discovered that planned hardware subsystems were too complex and had to be implemented in software to take advantage of software's better ability to manage complexity.

Conversely, tasks slated to be implemented in software sometimes required more performance than the general-purpose processor could provide, so designers had to figure out how to move the function into hardware. Each change between hardware and software implementation necessitated a change in all the interfaces to that function, so every hardware or software block that interacted with the modified function needed redesign and reverification. Often, many iterations were required to meet the system's performance and functionality goals.

These difficult partitioning choices are a central and critical task associated with the current method of SOC design. The tools and design methods available to help designers make these partitioning choices and changes have been quite limited. Migrating a task between hardware and software has been very painful, especially because the hardware and software task representations are so different (high-level languages versus hardware-description languages). Further, it's more painful to verify the proper interaction between the SOC's hardware and software prior to building the chip. It's more painful still to find out that something's wrong with the design after the chip has been built.

Fig 1-9 revisits the system partitioning example. Blocks E, F, H, and I are implementable as application-specific processors. This means that intertask communications are implementable in software and can evolve easily and inexpensively, even after the chip is built. Not all hardware blocks are eliminated, of course, but the number of hardware/software interfaces, especially complex interfaces, is reduced. In addition, as we shall see in Chapter 6, configurable processors can include optimized application-specific interfaces, allowing key interfaces such as H:J to be simply and directly implemented as a native part of the processor's interface definition. The low-throughput tasks, A, B, C, and D, also map efficiently onto configurable processors, so all software can run on a single family of processors with a common set of tools, models, and development methods.

So, the true leverage of the application-specific processor really arises from the way it enables the designer to do more of the total work in a software-friendly form and to move more easily between the hardware and software worlds. When a much wider variety of subsystems all fit within the capabilities of a processor, the effort to move a software task running on a generic processor to an application-specific processor is very low, because the functional specification remains primarily the software, generally written in a high-level language such as C or C++.

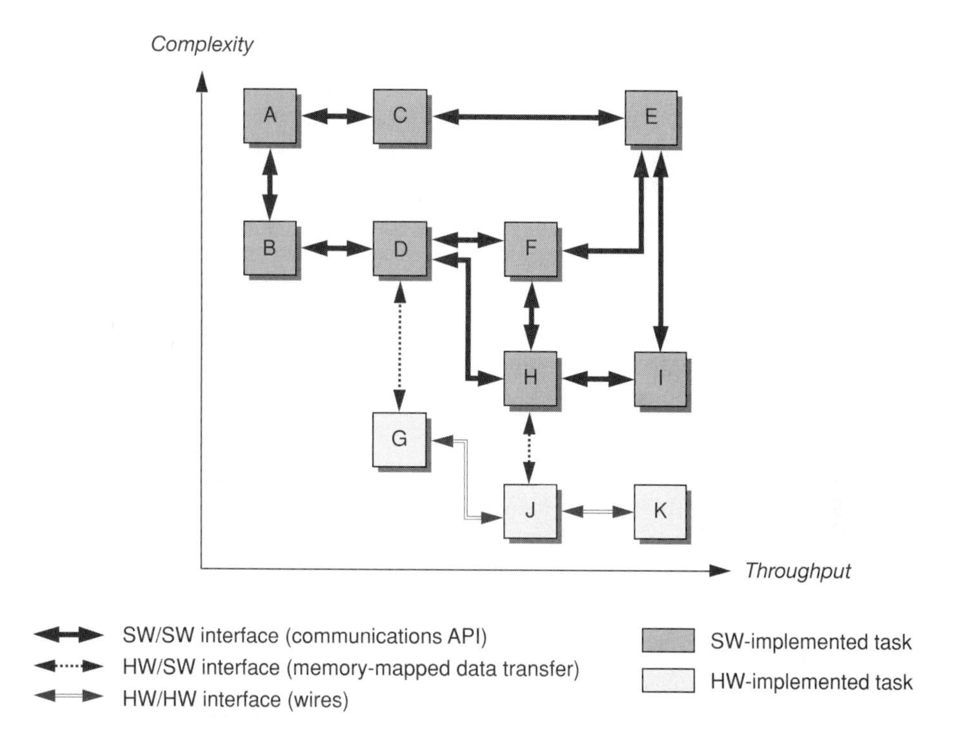

Figure 1-9 Conceptual system partitioning with application-specific processors.

As SOC designers seek even more subsystem performance, they need make only minor changes to the definition of the affected application-specific processors (adding facilities to improve execution speed and efficiency) and minor changes to the program running on that processor to take advantage of the processor's new enhancements. Thus, the effort needed to move a function onto or off of a particular processor, to split a function across processors, or to combine functions, is much lower than the Herculean effort required to move a task from a software onto one processor to a hardware representation, an effort that requires fundamentally rethinking the design and completely rewriting that function using, for example, Verilog instead of C.

The advanced SOC design methodology also affects simulation and validation of the individual system functions and combinations of these functions. The world of electronic design already offers many facilities for modeling a piece of embedded software running on a processor. The program can either run on a hardware prototype of the processor or on an instruction-set simulator (ISS) for that processor. Software simulation has gotten so efficient now (around 1 million simulation cycles/second versus hundreds of cycles per second for gate-level hardware simulation) that in many cases ISS speed is perfectly adequate for the prototyping of significant pieces of embedded software. Moreover, new modeling tools enable rapid description of tightly communicating groups of processors, memories, and other blocks. These tools, described in

Chapter 5, make design of complex multiple processor systems fast and simple. Better modeling takes the pressure off hardware prototyping—the software may never run on real hardware until the SOC prototype is powered on.

1.4.7 Implications of Improved SOC Methodology

This design methodology builds a closer tie between the two basic phases of design: partitioning and integration. Coordinating the two phases has important implications for the larger process of designing electronic systems. A number of long-standing challenges associated with SOC design start to have more obvious solutions. Some of the problems the complex SOC design methodology addresses include the following:

1. **Design Reuse:** Intellectual Property (IP) reuse (or more accurately "design artifact reuse") has been a catch phrase in the SOC and EDA industry for the last decade because people recognized that designs with more than 1 million gates cannot be built economically from scratch. Designers hoped to routinely reuse previously designed and verified blocks. These blocks would be created as a useful function in one chip-design generation and simply be imported as a subsystem in the next-generation product.

   Designers often discovered, unfortunately, that most RTL blocks are so highly specialized that it is often easier to design from scratch than it is to somehow accommodate the limitations of the pre-existing block. Consequently, RTL logic blocks turned out to be much less reusable than expected because they were much more idiosyncratic than most people had hoped.

   If, however, the basic functional block becomes a program running on an application-specific processor, that software block is much easier to reuse because software programmability carries with it the required and inherent adaptability to tailor the block for reuse. As software-block interfaces need to change or as incremental new functions need to be incorporated, the inclusion of those features, their verification, and the integration of this block into the whole system becomes more straightforward. A software-based functional block paired with a microprocessor (the ultimate reusable IP block) form a much more reusable kind of block than an inflexible, nonprogrammable RTL block. This complex SOC design methodology extends the microprocessor's applicability and usability to a much wider range of SOC functions.
2. **Efficient use of memory blocks:** Efficient memory use is another problem that has appeared in many SOC designs. SOC designs consist of more than just logic—they also contain memory. SOC memory content is increasing steadily and dominates silicon area for some SOC designs. Many SOC designs now have dozens of isolated RAM blocks scattered throughout. Access to each of these RAM blocks is peculiar to the logic circuits that surround the RAM. The sharing of these memories is very difficult and the testing and verification of systems containing a large number of RAMs is often unstructured and difficult. The processor-based SOC design approach makes most of the memories in the sys-

tem processor-visible, processor-controlled, processor-managed, processor-tested, and processor-initialized. In addition, this SOC design approach creates greater opportunities for flexible sharing and reuse of these on-chip memories, increasing total usable memory bandwidth.

The processor-based SOC design approach can also reduce the amount of memory needed on the SOC because the memories need not be narrowly dedicated to a single function within the SOC, especially if the associated processor performs several tasks. Further, the testing and verification of those on-chip memories is now performed within the context of controlling and operating that memory with a processor, which makes on-chip memory testing easier and less costly. Consequently, memory-centric or memory-intensive SOC design is actually synergistic with processor-based SOC- and software-based design.

3. **System modeling:** No SOC design should be considered successful until systems using the design ship to customers. Early system availability depends on early completion of system hardware and software. Historically, fast, accurate system modeling was constrained by the speed and availability of adequate chip models. Verifying proper system behavior demands both accurate system behavioral models and efficient, flexible, high-speed implementations of individual chip models. More complete SOC modeling, including multiple-processor modeling, enables full system modeling, including effective evaluation of interacting, concurrent software running on the intended hardware platform. The substantial speed advantage of processor-based system simulation over RTL-based system simulation enables easier instantiation of individual chip models into a system model and the execution of very large numbers of complex system exercises.

1.4.8 The Transition to Processor-based SOC Design

The transition from conventional SOC design methodology to a new SOC design methodology offers two fundamental benefits. First, it brings flexibility and speed of design to traditionally hardwired functions. The performance and energy efficiency of configurable processors far exceed that of conventional processors and rivals capability of hardwired logic functions, but with simpler, faster initial design and thorough post-silicon programmability. For many design teams, automatic generation of processors is displacing RTL design for their more complex, data-intensive function blocks. This transition has greatest impact in SOC applications where system complexity is on a collision course with bandwidth or bandwidth efficiency. Second, the more pervasive use of configurable processors as the basic building block of choice simplifies system-level design. Across all the functions implemented with application-specific processor configurations, hardware and software teams use one set of hardware interfaces, software tools, simulation models, and debug methods. This unification reduces design time, misunderstandings between hardware and software developers, and risk of failure.

As a by-product of broad adoption of configurable processors as a basic SOC building block, configurable processors are also taking on the roles once played by software on traditional processors. This transition is driven by two factors:

a. Application-specific processors already offer the features and tools expected to accompany traditional processors. These processors have the complex caches, the sophisticated interrupt and exception handling, the memory management, the real-time operating systems, and the software-development environments that accommodate development of very large and complex applications—the traditional sweet spot for embedded processor cores.
b. Unifying the design around a single family of processors has significant economic and technical benefits. Engineers need to be familiar with just one set of tools, and all the various subsystems can be integrated into a single seamless framework of programming tools and hardware interfaces.

Only the existence of legacy assembly code and precompiled binaries—a meaningful barrier in some cases—may impede broad substitution of application-specific processors for generic processors in non-performance-critical processor roles.

The prospect of a single family of processors, each automatically configured to fit the task at hand and used in large numbers per chip, forms the basis for our vision of the sea of processors. This book teaches an SOC design method using configurable processors to fit that vision.

1.5 Further Reading

A broad assortment of good books has been written on SOC design in recent years. Here are just a few that touch on the central topic:

- Gordon Moore's 1965 article on integrated-circuit evolution has proven remarkably insightful, not only for the scaling law that is attached to his name, but also for his forecasts on the applications of integrated circuits: Gordon E. Moore, "Cramming more components onto integrated circuits," *Electronics*, April 1965
- Michael Keating and Pierre Bricaud, *Reuse Methodology Manual*, Kluwer Academic Publishers, 1998, is a basic introduction to current SOC hardware design methodology, with particular emphasis on guidelines for hardware block reuse. While reuse of hardware blocks plays a role in reducing design cost and risk, it cannot be considered a comprehensive solution to full hardware/software system development.
- Henry Chang, Larry Cooke, Merrill Hunt, Grant Martin, Andrew McNelly, and Lee Todd, *Surviving the SOC Revolution* Kluwer Academic Publishers, 1998, takes a broader look at the SOC design flow, especially the issues of chip integration.
- A new book on embedded processors, Friedrich Mayer-Lindenberg, *Dedicated Digital Processors*, John Wiley & Sons, 2004, promises a good introduction to processor—especially digital signal processor—architecture, programming, and interfacing.

- A very basic and accessible introduction to semiconductor technology and the semiconductor industry can be found in Jim Turley's book: *The Essential Guide to Semiconductors*, Prentice Hall, 2002.
- Subramanian's comments on computational requirements in communication have been widely circulated: Ravi Subramanian "Shannon vs. Moore: Digital Signal Processing in the Broadband Age" In *Proceedings of the 1999 IEEE Communication Theory Workshop*, 1999.
- The widely cited "productivity gap" analysis was done by the Semiconductor Research Corporation on behalf of the Semiconductor Industry Association's National Technology Roadmap for Semiconductors (NTRS) in 1997. The NTRS evolved into the International Technology Roadmap for Semiconductors (ITRS) after 1997.

CHAPTER 2

SOC Design Today

To understand where SOC design must go, we first look at the current state of design needs and methodology. Different design teams have different goals and use somewhat different design approaches, but a number of common themes and serious problems stand out. This chapter looks at the basic issues and exposes a set of key unsolved problems.

2.1 Hardware System Structure

Traditionally, embedded systems have made a simple distinction between subsystems primarily used for processing application data and subsystems used for housekeeping functions. A simple view of this structure is shown in Fig 2-1. This partitioning into control and data flows works well when the data processing is simple (but very demanding in bandwidth or efficiency) and when the control functions are complex (but not performance demanding). In that case, the control functions can be implemented as software tasks running on a general-purpose embedded processor. The data processing can be implemented as hardwired logic for speed.

The control and data-processing subsystems communicate through memory, perhaps in the form of commands passed from the control processor to the data-processing logic, and status or results passed from the data-processing logic to the control processor. When the data-processing tasks are simple, they may be entirely implemented with a processor.

For moderate performance embedded applications, 8- and 16-bit microcontrollers and 32-bit RISC processors deliver adequate performance and efficiency. Similarly, digital signal processors (DSPs) broaden the range of tasks that can be implemented with software running on a standard processor. However, the potential parallelism, bandwidth, and power efficiency of hardwired logic makes optimized custom logic the data-processing method of choice.

The likelihood of design changes is a key issue when making a partition between control

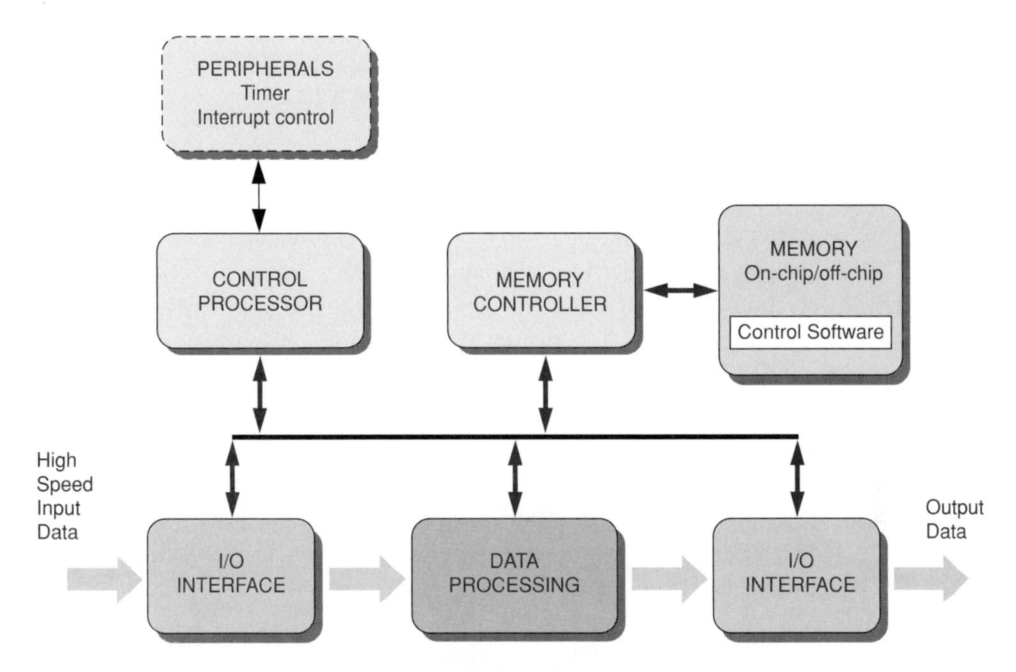

Figure 2-1 Simple system structure.

and data-processing functions. These changes may be driven by new specifications, evolving market requirements, or the need to fix bugs to correct a mismatch between specification and implementation. Changes in the control subsystem associated, for example, with new user interface functions, can be easily accommodated by changes in the software task loaded when the system powers up. Changes in the hardware-based data-processing function are harder to achieve, because the physical structure of the system must be changed. If the data-processing functions are implemented in an integrated circuit, the chip design must go back through much of the design and prototyping flow.

The system architect's task includes the basic partitioning of all system functions into two buckets: complex functions with a high probability of change are mapped to software; simple stable functions with high computation requirements are mapped to hardware. The tools and techniques of current SOC design are largely built around the assumption that these two buckets are adequate, that complex functions are relatively undemanding computationally, and that high-throughput functions are relatively simple.

2.1.1 How Is RTL Used Today?

In the past 10 years, the wide availability of logic synthesis and ASIC design tools has made RTL design the standard for hardware developers. Reasonably efficient compared to custom transistor-level circuit design, RTL-based design effectively exploits the intrinsic concurrency of many data-intensive problems. RTL design methods can often achieve tens or hundreds of times the performance of a general-purpose processor. This performance advantage arises from two essential characteristics of dedicated logic:

1. Custom-designed logic can precisely implement the sequence of operations required by the desired function. Operands flow from the output of one primitive operation (e.g., add, multiply, compare) directly to the inputs of the next. Some operations, such as bit-field selection or bit reordering, are trivial because they can be implemented with simple wires. Moreover, operations need only be as wide as required by the algorithm. This type of design is efficient because there is no waste. The hardware is only as large as needed by the algorithm. Arbitrary expansion to the next power-of-two in bit width, as required in the design of most fixed-instruction-set processors, is unnecessary
2. Natural concurrency is expressed directly. If one operation is not dependent on another, the two can be executed in separate logic blocks at the same time. Moreover, designers can often restructure algorithms to more clearly expose intrinsic parallelism. For applications with very high intrinsic parallelism, performance is limited only by the throughput of the longest path through the logic and by the hardware budget.

On the other hand, the power of RTL-based design heavily depends on the ability of the hardware developer to comprehend and implement the entire functional specification. For a function entirely implemented in hardware, all setup, error handling, and little-used cases must be entirely implemented in logic gates. The numerous interactions within the function block and with other function blocks can overwhelm the hardware designer with complexity.

Even if the designer off-loads functions that do not have critical performance requirements to a processor, the hardware interface to the processor and the corresponding software driver must still be designed and verified. This design partitioning may not closely match the algorithm's intrinsic partitioning. Such artificial interfaces are frequent sources of bugs.

For all or most functions, the hardware designer implements the function using some combination of finite state machines, data-path logic, and local memory blocks, as shown in Fig 2-2. It is useful to look at the characteristics of these subcomponents of a hardware design to understand the real partitioning issues.

In some cases a single finite state machine controls an entire data path. In other case, the control is implemented as a set of interacting state machines (either state machines interacting as peers or a master state machine controlling slave state machines). In any of these cases, the aggregate state machine complexity creates intrinsic design fragility. Getting the design right in the first place, and updating it as design requirements evolve, becomes increasingly difficult.

2.1.2 Control, Data Path, and Memory

In most RTL designs, the data path consumes the vast majority of the gates in the logic block. A typical data path may be as narrow as 8, 16, or 32 bits, or it may be hundreds of bits wide. The

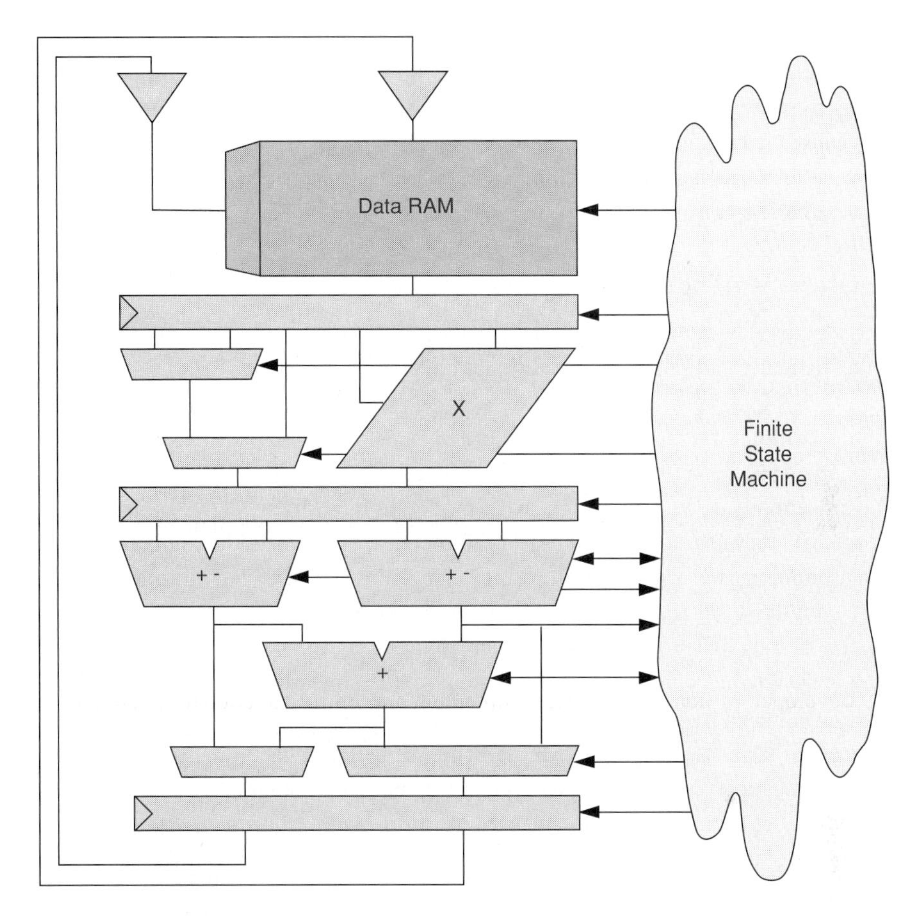

Figure 2-2 Hardwired RTL function: data path + finite state machine.

data path typically contains many data registers, representing intermediate computational states, and often has significant blocks of RAM or interfaces to RAM blocks that are shared with other RTL blocks. These basic data-path structures reflect the nature of the application data and are largely independent of the finer details of the specific algorithm operating on that data.

By contrast, the RTL logic block's finite state machine contains nothing but control details. All the nuances of the sequencing of data through the data path, all the handling of exception and error conditions, and all the handshakes with other blocks are captured in the RTL block's state machine. This state machine may consume only a few percent of the block's gate count, but it embodies most of the design and verification risk due to its complexity.

If a late design change is made in an RTL block, the change is more likely to affect the state machine than the structure of the data path. This situation heightens the design risk. Any design method that reduces the risk of state-machine design also reduces the overall design risk for an SOC that contains a significant number of RTL-based blocks.

Moreover, the design risk of a hardwired block is not limited to interactions between the finite state machine and the corresponding data path. The complexity of the interface between blocks tends to grow with complexity of the blocks themselves. Design changes in one block commonly cause changes in the interface, which then trigger mandatory changes in other blocks. The discovery of a single state-machine bug late in the design cycle can propagate changes throughout much of a complex chip.

2.1.3 Hardware Trends

As semiconductor device density increases, five trends in hardware design are emerging to increase the difficulty and risk of this traditional development model:

1. **Higher complexity:** Competitive pressure in end-product markets and more plentiful availability of cheaper transistors conspire together to push the average chip complexity up exponentially. Even with a forecasted moderation in Moore's law scaling, the industry expects silicon integrated circuit density to increase by almost 30% annually. This increase in transistor capacity will result in larger logic blocks and more blocks per SOC. Moreover, the design and verification effort increases more than linearly with the number and size of the blocks. Without significant changes in SOC design methodology, the cost of logic design on a chip will also increase more than linearly with transistor or gate count.
2. **Greater concern over power dissipation:** Aggressive scaling of transistor technology has enabled faster, denser circuits, but power dissipation is becoming a critical issue. High power dissipation pushes the limits of heat dissipation in integrated circuit and system packaging, degrades battery life, and compromises overall system energy efficiency. Even in line-power applications such as server farms, the difficulty of pulling heat out of the electronics may limit the total capacity of the facility. Active power—dissipation due to circuit switching—is already a major concern in SOCs but, with 90nm technology, static or standby power also poses major problems.
3. **New deep-submicron effects:** Transistors and wires scale differently with line width. Even with improvements in the number of routing layers and interconnect materials, interconnect wiring will consume a growing fraction of the clock period. In addition, a growing fraction of the total capacitive wire load will couple to adjacent wires. This increased coupling increases the risk of degraded signal integrity and data-dependent propagation delays. Local variations in fabrication processing cause increasingly important variations in the electrical characteristics of transistors and wires. These issues necessitate more accurate 3D circuit analysis, more pervasive statistical modeling and more sophisticated delay-estimation tools.
4. **Heavier simulation load:** Changes in the basic function of electronic systems force changes in the design process. More systems are continuously connected to networks, especially wireless networks. These networks typically have complex access protocols. Validating correct system operation when it's attached to a network may require large

numbers of long test sequences. This expanded testing requirement may force the number of simulation cycles to increase by many orders of magnitude to achieve adequate confidence in test coverage and design correctness.

5. **More fabrication choices:** The structure of the electronics industry is changing, with more focus on manufacturing outsourcing by semiconductor vendors and more silicon foundries appearing around the world. This trend promises to reduce raw silicon cost, but puts new priority on design portability. The SOC designer can make fewer assumptions about the underlying silicon performance and must design for easy migration across foundry suppliers and scalability across process generations.

Together these trends will force chip designers to reexamine their basic methodologies and design styles.

2.2 Software Structure

While a discussion of SOC design often focuses on hardware architecture and VLSI design, software structure and flow often has just as much impact on schedule, cost, and performance. Fig 2-3 sketches a typical software environment, including the software components on the actual target system and the development tools running on a host development system. The picture includes analog circuit blocks, which implement important physical interfaces used for network communication and user input and output, even though these blocks may not be directly visible to software. The picture also includes a number of hardware function blocks. These may play an important role in overall system behavior (including erroneous behavior due to hardware bugs), but they are not directly visible to software or controllable by the programmer.

In many embedded systems, the path of data flow through the hardware and software determines system efficiency and performance. Fig 2-3 shows a typical data flow through the software layers, highlighted by lines with arrows. Data bits flow into the system through physical input interfaces into hardware input device controllers. Software devices drivers typically copy the data into operating system data structures, and then to the memory space of an application. Communicating applications exchange the data directly through shared global memory or pass the data as messages via the operating system. On output, the application typically passes the data through the operating system to a device driver, which copies it to a hardware output controller, which then moves the data through an analog circuit interface to the outside world. Generally the more layers of software that touch the data, the higher the latency and lower the data bandwidth.

The exact software structure of complex systems is highly variable and application-dependent, but typical target-system software components include:

- Low-level device reset and exception-handling code.
- Standard operating system services for resource management, task initialization, scheduling, and communication.

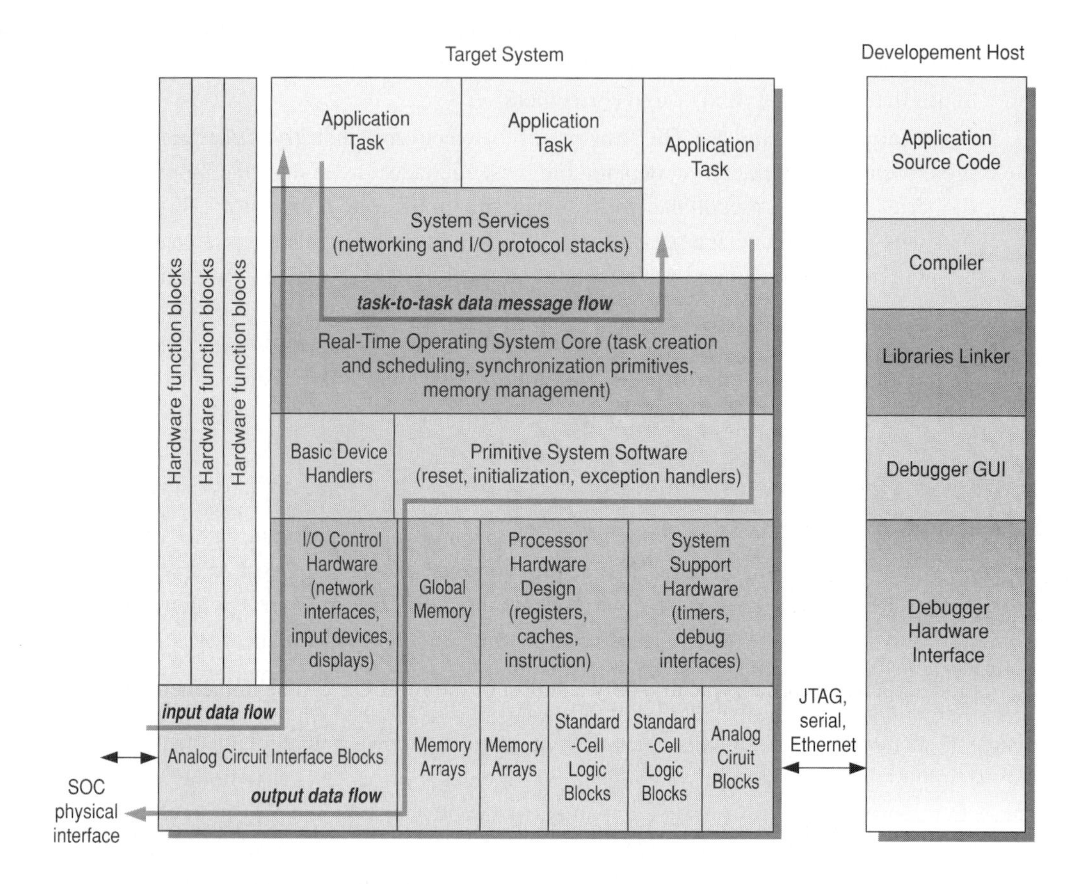

Figure 2-3 Typical software runtime and development structure.

- Networking and other protocol stacks.
- A number of application tasks that implement the major externally visible system functions.

All of these software components generally couple to the underlying hardware architecture in two ways. First, all the software is generally written or compiled specifically for the instruction-set architecture of the processors used in the hardware platform. Second, many software components may explicitly depend on the particular I/O interfaces, memory map, system peripherals, and processor-control functions of the hardware. This is especially true for lower level software such as exception handlers and real-time operating systems. In fact, the abstraction or hiding of these important implementation details from the application tasks is a key role of these layers.

In some systems, certain software layers are more truly hardware-independent. Intermediate, architecture-neutral software formats such as Java bytecodes or scripting languages run on top of processor-specific interpreters to provide additional platform-independent functions.

In most cases, the binary machine code of these software components will ultimately be loaded into flash memory on the SOC or elsewhere in the system, or made available over a network. The relatively low cost of changing boot-time software stored in flash memory or on a network makes software the natural vehicle for any function that is complex or likely to change rapidly, and thus more susceptible to design errors. Such functions are best implemented in software as long as the software-based design running on the available processor meets performance and efficiency requirements.

Given the complexity of real-time system software, the development environment for software creation, debugging, performance tuning, and system verification becomes an important engineering consideration. The development system includes the tools needed to build and test individual software components and to run and debug the combination of components on target-system simulation models. The nature of individual SOC models varies widely, but some of the common forms include:

- Physical breadboards built with discrete off-the-shelf integrated circuits (processors, memories, and FPGAs) corresponding to the intended subsystems of the target SOC.
- FPGA prototypes implementing the SOC's logic and memories, albeit in a slower and more expensive form. Typically, only a subset of the full SOC can be implemented in one FPGA due to large differences in capacity between FPGA and ASIC implementations. Hardware emulators, which use arrays of FPGAs or special logic-emulation chips, serve a similar role for SOC designs with higher gate counts.
- Simulated SOC hardware built from a mix of RTL logic models and higher level models for processors (instruction-set simulators) and memory subsystems. Hardware simulation accelerators have similar characteristics.
- Analog interface functions are not typically modeled as separate design elements. Instead, analog functions are modeled as part of the on-chip I/O controller logic, as part of the external environment that produces or consumes simulated incoming and outgoing data streams, or ignored altogether for this level of modeling.
- Fast, purpose-built simulators that model subsystems at a high abstraction level.

Without good modeling, many of the most critical flaws in specification, architecture, and hardware or software implementation will not be discovered until final product integration. The cost in time and money of fixing these errors can be so great as to put the entire product-development project at risk.

The potential benefits of good hardware modeling and early software testing are widely accepted, but the actual adoption of robust methods still lags behind the recognition of the problem. These problems are magnified by the rapid increases in both hardware and software complexity. Increasing hardware complexity makes models more expensive to develop and expands the number of hardware/software interfaces to consider. Increasing software complexity drives the need for sophisticated layering of operating systems and system services, and mandates more sophisticated debugging tools that can provide more insight into interactions within and among software subsystems.

Ironically, the growth in system complexity tends to both increase the number of execution cycles necessary to cover all the end system's interesting operating modes and to decrease the performance of the platform on which the models are tested.

2.2.1 Software Trends

SOC complexity is growing, and software generally handles complexity more gracefully than hardware. Nevertheless, two software trends that have long been on "just on the horizon" bear watching:

1. **Virtual prototyping:** Higher integration and tighter coupling between hardware and software subsystems are forcing a fundamental change in prototyping techniques. While some isolated software components can be developed on standalone development boards, SOC prototypes are increasingly implemented as simulations. Simulated prototypes are slower than hardware prototypes, but they are much more convenient and reliable. As workstation and PC processors get faster, so do prototype simulators. In addition, prototype simulators can model more of the system and more easily allow gradual refinement of the SOC during design. The tools for debugging simulated prototypes, for interfacing them to real-time peripherals, and for analyzing simulated-prototype performance are consequently undergoing rapid improvement.
2. **Multiple processors:** Even without the advent of application-specific processors, the drive for high integration is putting multiple processors, often of diverse architectures, together on a single chip. Managing software development on a platform with, perhaps, both a RISC core and a DSP core, presents new problems for SOC designers. Developers of software for multiprocessor SOCs must generally work with multiple compilers, multiple debugging interfaces, incompatible simulators, and tricky software migration across architectures.

The trend toward greater software content is inexorable, but the broadening role and complexity of software systems stretches the limits of current methodology.

2.3 Current SOC Design Flow

The typical SOC design flow, shown in Fig 2-4, reflects a historical separation between hardware and software development. Many of the key architectural decisions are made quite early in the design process, long before any detailed performance or implementation feedback is available. The long delay from key algorithm selection and system-performance measurement on hardware sharply increases the risk of surprises and cost of fixing issues. Moreover, the likelihood of many iterations and delays between hardware partitioning and VLSI timing closure also leads either to substantial overdesign of blocks (higher cost and power) or to unexpected performance and yield problems.

Three dimensions of this flow are worth noting:

- **Architecture design:** The process of refining high-level product requirements into detailed technical requirements for all hardware and software. One key phase of architecture design is the partitioning of the design into a collection of hardware blocks and software tasks, including the specification of interfaces between components. As the development proceeds, the architecture must almost always evolve to overcome limitations in the initial design. The evolving architecture must also accommodate changing end-product requirements and exploit new system-design and architectural insights to improve performance, cost, reliability, and functionality.
- **Software design:** The process of implementing, testing, tuning, and integrating functions implemented as programs running on processors. Often the software team's early participation in the design is focused on development and validation of only the most important algorithms or new software components. Limitations in model capacity and authenticity typically delay all hardware/software system-integration until hardware prototypes are available.
- **Hardware design:** The hardware design process really consists of two interacting flows—the design, verification, and integration of the various hardware blocks and the integration of all hardware components into one final, physical VLSI design. The pace of growth in both system complexity and silicon capacity demands a substantial increase in the reuse of existing hardware blocks, even though many hardware blocks are too rigid and specialized to allow wide reuse.

The necessity of getting hardware blocks and VLSI implementation absolutely correct makes the design and integration a long, iterative process. Designers want an SOC that is as small, fast, and efficient as possible. Thus, hardware engineers are motivated to include every necessary feature, but no more than are absolutely necessary. Designers must evolve block designs to allow the VLSI implementation to meet cycle-time targets; to fit gate-count and area goals; and to satisfy placement, routing, signal-integrity, power-distribution, and total power-dissipation requirements. Preliminary estimates from block diagrams, floor plans, and initial RTL code are typically crude and optimistic, so block design and block interfaces must change in response to simulation results from actual block layouts. Changes in the interface, size, timing, and power characteristics of one block will propagate to other blocks, triggering even more design changes.

The inevitable growth in complexity of system functions and silicon has particularly dire consequences on the hardware design flow. Increased block complexity means that the SOC designer faces a greater challenge in understanding and implementing each function. Block-level tools such as block-level simulation, logic synthesis, placement, and routing tools run slower. Typically, the speed degradation is worse than linear as a function of block size.

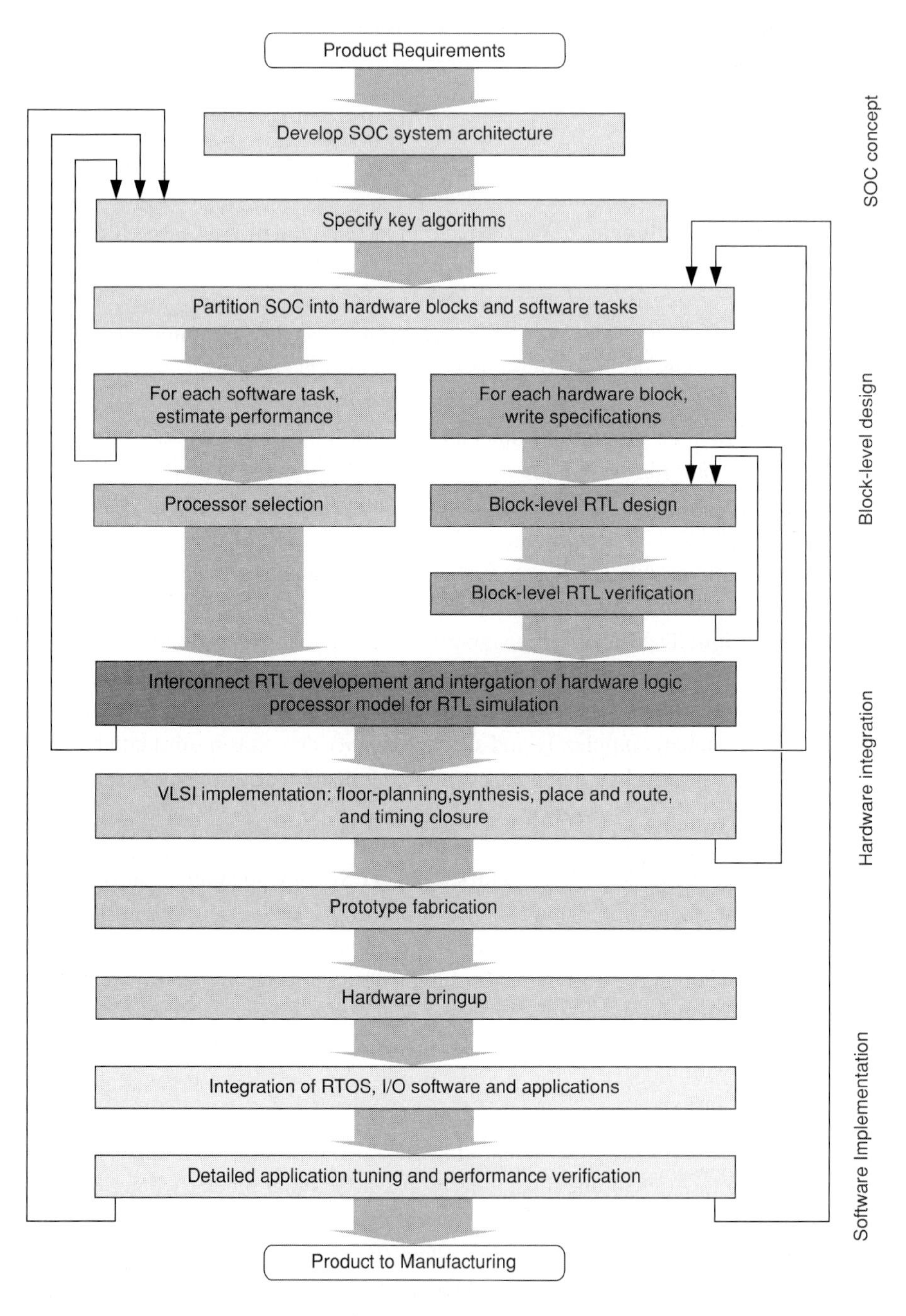

Figure 2-4 Today's typical SOC design flow.

Growing chip complexity also mandates more blocks, with more interfaces to document and test. When different designers are responsible for different blocks, the increased communication requirements impede problem resolution and increase the risk of misunderstandings among members of the design team.

Increasing complexity also has a cascade effect on VLSI design characteristics. More complex blocks require more gates and wires; higher gate and wire counts increase average wire length; longer wire means that more of the total clock cycle is spent in wire-delay; longer wire-delay means that placement and routing optimization increasingly influence overall circuit size, performance, and power dissipation.

The dominance of physical design issues for megagate SOCs mandates a new set of placement-aware design methods and a new generation of more complex and expensive design-automation tools because more effort is required to meet clock-frequency goals. Longer wires on the chip also increase the likelihood of greater capacitive coupling between wires, which produces more crosstalk. New signal-integrity tools and design checks push out VLSI design schedules and increase design budgets still further.

All these factors—growing block complexity, slower simulation, new interactions between logic design and physical layout—create a vicious cycle of development delay. Each complication in the design phase stretches out the start date for VLSI implementation. Each complication in the VLSI flow delays feedback on possible enhancements or bug fixes in the logic. Increasing overall complexity multiplies the number of interactions required between logical and physical design before a chip design can be confidently released for prototyping.

These basic design trends are well understood by SOC design teams and their technology suppliers. A number of useful tools and techniques have emerged to smooth this flow, though without changing the basic process and its liabilities. Important incremental tool enhancements include the following:

- **Hardware/software cosimulation:** Simulation languages such as SystemC and tools such as Coware's ConvergenSC and Mentor's Seamless provide an environment for running fast high-level simulation models (especially for standard processor cores and memories) in conjunction with more detailed models of logic implemented in RTL or C. These tools typically allow fast simulation of code running on the processor in support of software development. They also provide basic verification of the hardware and software interface between the processor and other system logic.
- **Floorplanning and physical synthesis:** The growth in hardware design size and the growing role of physical effects, especially wire delay, crosstalk and other deep-submicron silicon characteristics, mandates a new generation of physical design tools. Synopsys' Physical Compiler and Astro place-and-route system, Cadence's SOC Encounter flow, and Magma's Blast all represent important incremental product improvements, especially for 90nm designs.

- **Integrated software development environments:** As the software content of embedded systems grows, software development productivity becomes an important factor in overall system-development cost and timeliness. Integrated code development, project management, source-level debug, and execution visualization help software teams develop, tune, and maintain complex real-time software. Wind River's Tornado environment, for example, smoothly integrates software development with operating-system-aware debug. Multicore debug interfaces, based on JTAG (Joint Test Action Group) hardware interfaces, such as ARM's MultiICE, permit software debug of chips with more than one embedded processor core.

Despite the value of these improved tools, they rely heavily on the same basic building blocks: general-purpose processor cores and hardwired logic blocks developed as RTL.

2.4 The Impact of Semiconductor Economics

The business context directly influences countless technical decisions in SOC design. The electronics industry is characterized by fierce global competition and rapidly changing market requirements across a diverse set of applications and product types. In addition, the semiconductor segment of the electronics business is highly cyclical due to the heavy investment in fabrication equipment and the resulting waves of overcapacity and product shortage. Between 1998 and 2003, the semiconductor industry endured the most dramatic boom and bust of its history. The material results and painful memories of that recent experience form the context for current SOC business thinking.

The decision to design an SOC is an investment decision. The design team hopes that the future profit from the sale of chips, or the sale of systems that require the chip, will well exceed the cost of designing the chip. The cost of SOC design is dominated by the costs of deploying the right engineering manpower with the right design tools. The good news for design teams is that the global market for semiconductors is currently expected to grow by an average of 10% per year over the next few years, reflecting substantial new opportunities for volume chip shipments. The bad news is that the costs of design are also fated to grow sharply as well.

The research firm International Business Strategies (IBS) has looked at engineering effort for hardware and software for major SOC design projects. IBS recognizes that the growing capacity of silicon enables much more complex hardware platforms and encourages development of much richer software for these new platforms. In the absence of a significant change in design approach, this combination drives the total cost up by more than a factor of two with each major technology generation, as shown in Fig 2-5.

If this trend continues, the profit per design—volume times profit per chip—must grow at the same rate to maintain the viability of design. Clearly, not all conceivable designs can meet this standard of return on investment. Teams are choosing not to design new SOCs for functions that do not meet this economic standard. Such functions are either implemented with off-the-shelf devices or are not implemented at all.

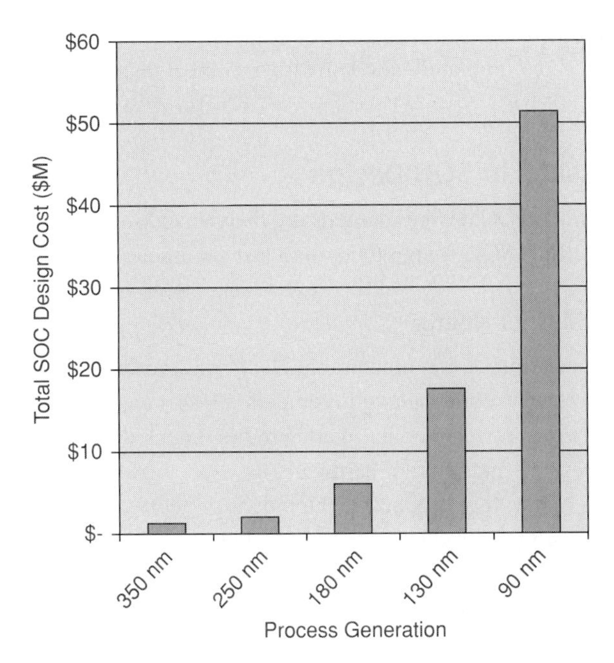

Figure 2-5 Total SOC design cost growth.

Already, the availability of off-the-shelf, programmable chips has affected the number of low-complexity, application-specific chip designs. New designs for low-end ASICs have fallen sharply in the past five years. FPGAs and microprocessors can sometimes be good substitutes for ASIC designs when logic complexity is not high (a few hundred thousand logic gates) and volume is modest (such that chip cost is swamped by design engineering cost).

Programmable general-purpose chips are typically larger and more expensive per unit than chips focused on a narrower application segment, but building a system by programming an off-the-shelf device has lower development cost than developing a chip design from scratch. For example, suppose a $10 ASIC is be able to perform the same function as a $200 FPGA, but requires $10 million to design and prototype, compared to just $2 million for the FPGA-based design. In high volume, this unit cost advantage is compelling—design cost plus chip cost for two million units is $30M for the ASIC compared to $602M for the FPGA. In low volume, however, the situation is reversed. For 20,000 units, the total ASIC cost is $10.2 million, but the total FPGA cost is just $6 million.

These economic pressures are vivid for a system design team facing a "make/buy" decision. If the team takes on an SOC design, they may invest tens of millions of dollars in engineering time, tools, fees, and prototyping costs to get a silicon and software platform ideally suited to market needs. The target chip's cost and specification may be compelling, but the team faces the risk that sales will be too low to adequately recoup expenses. On the other hand, if the team abandons the new SOC design and instead employs off-the-shelf components, high component costs, inadequate performance, and high power dissipation may cause the resulting system to be merely undifferentiated or completely uncompetitive. A great deal is at stake for the SOC design team today.

2.5 Six Major Issues in SOC Design

The challenges of current SOC design methods are diverse and interlocking, but it is possible to distill the hard problems of SOC design today into just six essential issues.

2.5.1 Changing Market Needs

Translating market needs into a manufacturable electronic product takes a long time. Existing customers and new prospects contribute requirements. Market experts offer data on trends. Strategic visionaries throw out novel product ideas. Product teams distill these inputs into product requirements. Engineering teams design hardware, develop software, build prototypes, and verify the solution. As a result, few product-development projects take less than a year. The time between concept and available product might stretch to two or three years, or more for complex, high-volume products.

Three years is a very long time in the life of an electronic product's market these days. Assumptions about features, pricing, performance, and form factor can all change significantly during that time. Entire product segments can virtually die out, and new categories rise in their places. Even modest changes in the supported data formats and communication interfaces can obsolete a product before it's ready for sale. If the market does change, how can product developers recover?

Almost all electronic system products contain some sort of control processor, so certain features are changeable via software. Any features implemented in hardware, especially those features cast in silicon inside an SOC, present a bigger design risk and may cause a long delay in bringing the reworked product to market. Ironically, many painstaking hardware optimizations made to improve performance and power efficiency, intended to make the product more attractive in the market, ultimately increase the design risk and delay product introduction if those optimized features must change. Product definition is fragile.

2.5.2 Inadequate Product Volume and Longevity

The real cost of building a system includes both the direct manufacturing costs—parts plus manufacturing labor and services—and the indirect costs of defining, designing, and marketing the product. To achieve sustained profitability, a product developer must choose a design and implementation strategy that balances the design and manufacturing costs appropriately for the projected manufacturing volume, manufacturing processes, and design capabilities of the enterprise.

High-integration, nanometer CMOS technology offers spectacular performance, power, and density opportunities for digital electronics and can achieve remarkably low unit manufacturing costs. The up-front design costs for large SOCs, on the other hand, can also be spectacular. Much has been made of the rising cost of photolithographic masks for SOC design—often exceeding $1 million for 90nm technology—and the capital costs of semiconductor fabs—exceeding $2 billion for a high-capacity facility. These numbers tell only part of the story.

The emergence of silicon foundries and standard design representation formats has allowed fab costs to be spread across a large number of different designs from different design teams sharing common design rules. The high mask costs are associated with a single design are real, but represent just the tip of the design-cost iceberg.

The sheer logic complexity of leading-edge chips and the rising demands of chip-level verification and physical design require design efforts that routinely consume many tens of engineer-years and often top 100 engineer-years. When combined with the cost of design tools, prototyping charges, and productization, design costs for a single SOC development project typically exceed $10 million. High silicon-design cost seems to go hand-in-hand with spectacular silicon functionality.

If an SOC is used in just one system, then the full SOC design cost must be amortized across one system's manufacturing volume. Some products have the cost margin and volume to comfortably accommodate $10 million of additional design costs, but many do not.

SOC development may become economically untenable unless the design costs can be reduced or the manufacturing volume can be increased. Increasing the manufacturing volume is especially difficult for an inflexible chip, because the SOC may not be sufficiently adaptable to serve the needs of multiple products in a product line or multiple customers in an industry. Sometimes the cost of optimizing of an SOC design creates serious economic problems because of inadequate volume.

2.5.3 Inflexibility in the Semiconductor Supply Chain

The wild success of electronic products over the past quarter century has been driven, in part, by the uniformity and ubiquitousness of digital MOS semiconductor technology. Commonality of processing equipment, engineering methods, and design tools has driven substantial commonality in design representation, circuit techniques, and physical-design rules. This design standardization gives the silicon designer commodity fabrication pricing and increases the security of supply by enabling multiple sources for any chip.

The trend toward design commoditization flies in the face of the economic interests of semiconductor suppliers. Suppliers seek to differentiate products and lock in customers by offering unique capabilities such as higher performance, extra design support, and proprietary building blocks. Their customers must delicately balance the technical advantages of these proprietary offerings against the loss of business leverage that comes from adopting a sole-source technology.

Historically, microprocessors have been recognized as the building block with the most potent lock-in value. This characteristic stems both from the difficulty and effort required to create state-of-the-art processors and software tools and from the rising cost of switching architectures as the customer develops a growing software library wedded to that processor architecture.

Multiple-source processors do not fully address this inflexibility. Common techniques, such as using a synthesizable logic implementation, make the processor design independent of the specific semiconductor supplier. However, when compared to hardened processor cores, synthesizable processor designs typically run at lower operating frequencies with greater power dissipation and silicon area. Moreover, availability of processor implementations is generally restricted to a small number of lowest-common-denominator architectural configurations.

Some system designers respond to the lock-in risk of sole-source processors from semiconductor suppliers by developing their own processors. While this approach returns control to the SOC designer, the difficulty and distraction of from-scratch processor development creates a raft of other problems such as the need for development tools and development-tool maintenance, documentation, and application support. Moving more functions into synthesizable RTL blocks can also be seen as a move to reduce lock-in risk—though at the cost of increased risk of design fragility.

Ultimately, the SOC designer wants seemingly incompatible powers—the freedom to use commodity silicon fabrication for minimum cost and maximum supply flexibility and the leverage to create highly differentiated performance, silicon efficiency, functional richness, and flexible reprogrammability in optimized system designs.

2.5.4 Inadequate Performance, Efficiency, and Cost

The pace of technology improvement creates an opportunity for electronic product companies, but rampant competition transforms opportunity into necessity—build faster, cheaper, lower-power products… or die. The current standard partitioning between processors and hardwired logic blocks impedes success by forcing significant compromises in SOC throughput, efficiency, and cost. Any task too complex to implement exclusively in hardware must run on a control processor.

Generally, these processors rarely exceed a few hundred MIPS of performance in leading-edge 130nm or 90nm process technology. The resulting performance of these processors may be adequate for user interfaces and simple system-control tasks, but essential features for network, security, signal, and multimedia processing often require much higher throughput, both individually and in aggregate, for the whole set of product features.

Processors embedded in an SOC may have better power dissipation specifications on a per-MHz basis than standalone processor chips, but often they must run at high clock frequencies to satisfy system-performance requirements. High clock frequency affects power dissipation in two ways. First, the active power in any given CMOS VLSI circuit is proportional to the operating clock frequency. Second, designers must often adopt aggressive circuit-design methods and a higher operating voltage to increase the maximum clock rate. Aggressive design methods that boost clock rate also increase power dissipation. Consequently, a performance increase typically requires a more-than-proportional increase in power dissipation. When operating-frequency and operating-voltage effects are combined, the power often scales rapidly—roughly with the cube of operating frequency.

When product requirements demand very high performance, or even just headroom in available performance, the design also requires more silicon. High-end processors capable of simultaneous speculative execution of many possible instructions increase area out of proportion to the increase in useful throughput. Higher clock rates often require larger transistors and longer wires. The push for performance often drives a transition to more advanced process technology, increasing manufacturing costs. Together, the demand for higher processor performance typically means larger die size, higher silicon costs, and higher power dissipation.

2.5.5 Risk, Cost, and Delay in Design and Verification

SOC design is full of uncertainties. The complex interactions of subsystems, the variety and complexity of design representations, and the long layout and fabrication process all combine to increase the risk and the cost of errors. Typical SOC design teams consist of architects, software developers, hardware designers, VLSI engineers, design-tool experts and other technical support engineers. Verification alone often consumes 70% or more of the entire effort.

Typical SOC designs consume tens of people working for two to three years from design concept to manufacturing release. Total development costs, including design tools, prototyping charges, and outside services routinely top $10 million today, with estimates that complex SOC designs using 90 nanometer technology may cost $30 million. The most ambitious SOC efforts may reach even higher.

SOC hardware-design errors are particularly dangerous. The discovery of a hardware bug in prototype silicon may require six months to fix—three months to identify the proper modification and finalize the design change plus three months to respin the silicon. The risk of error grows disproportionately with the complexity of the design because the number of interactions between design elements increases far faster than the number of elements.

Moreover, as system functionality increases, the number of test cases needed to verify correct behavior explodes. As a result, SOC design is caught in a bind. The size of the blocks increases, the number of blocks increases, the number of test cases per block increases, simulation time for each test case increases, and the cost of overlooking a bug increases. The consequences of bug-related delays and costs for the hardware-design and verification teams are daunting. Finally, product developers increasingly understand that the impact of bug fixes and silicon respins is more than just program delay and direct engineering costs. Late market entry also reduces the time over which the product generates revenue and reduces profitability due to increased competition.

The two critical questions are

- How can developers reduce the risk of “killer bugs”?
- How can the entire design and prototyping strategy reduce the time to incorporate fixes?

2.5.6 Inadequate Coordination Between Hardware and Software Teams

The software content of embedded systems is increasing rapidly because many system functions are too complex for direct hardware implementation. As the total software content increases, the breadth and importance of interactions between the hardware and software teams becomes critical.

Historically, systems could be rigidly partitioned between relatively simple control functions and implemented with standard microprocessor chips and relatively simple hardware functions. They could also be implemented with data-path ASICs or programmable, off-the-shelf ICs. This traditional design approach has allowed embedded software programmers either to work independently of the hardware development or to start software development only after the hardware team delivered prototype systems.

Increased system complexity, tighter schedules, narrower end-market windows, and growing use of SOCs all put pressure on this loosely coupled model. Too often, the hardware architecture and software architecture efforts are separated in either space or time. Hardware and software departments often occupy different buildings; sometimes, different sites; occasionally, different continents. Even with good electronic communications, project coordination can be the weak link. Complex, ad hoc hardware/software interfaces often confuse design team members, and confusion leads to expensive rework. Often, design issues caused by ad hoc interfaces is only resolved during final system integration, when it may be too late for effective or economical redesign.

Misunderstanding of performance issues can create processing bottlenecks that can only be corrected by substantial hardware redesign. Hardware architectures designed without sufficient software consideration often lack adequate visibility and controllability. When hardware features are insufficiently controllable by software, workarounds become difficult or impossible.

Even when hardware and software teams work together in close physical proximity, dependence on hardware prototypes for early software development forces a pipelined development model in which the software team is still finishing the last project while key hardware architecture decisions are made on a new project.

Pipelined development can create a vicious cycle of delay, where the late discovery of inadequate interfaces and hardware shortcomings forces lengthy and elaborate software delays, ensuring that the software team has even less opportunity to apply their experience to the next project's architecture. Without some means to realign and improve the hardware/software interfaces, project schedules and product efficiency will continue to suffer.

2.5.7 Solving the Six Problems

These six issues represent a critical hurdle for more universal adoption of SOC designs. If adequate solutions to these problems are not found, SOC design costs will climb along with SOC transistor densities until only a handful of large SOCs can be built each year. With only a few basic platforms to choose from, end-product designs will become less differentiated, compelling, and efficient. If, on the other hand, new design methods can make these chips easier to develop and more easily reused, then new electronic products will continue to proliferate.

2.6 Further Reading

- Source data for the SOC design costs discussed in Section 2.4 can be found in Dr. Handel Jones, IBS, Inc.. *Economics of time-to-market in chip design, IBM Engineering & Technology Services*," June 2003; and *Analysis of the Relationship Between EDA Expenditures and Competitive Positioning of IC Vendors, A Custom Study for EDA Consortium* by International Business Strategies, Inc., 2002.
- The ARM family of conventional RISC cores serves as the most common control processor for SOCs today. A good reference text is: Steve Furber's *ARM System Architecture*, Addison-Wesley, 1996.
- The Semiconductor Industry Association tracks historical trends in semiconductor markets and makes annual forecasts looking out three to four years. See *http://www.semi-chips.org*.
- A number of the basic ideas of SOC design were developed in Felice Balarin, Editor. *Hardware-Software Co-Design of Embedded Systems: The POLIS Approach.* Kluwer 1997.
- A very good introduction to the current SOC processor based on fairly rigid hardware-software partitioning is found in this book, especially in Chapter 7 ("Hardware/Software Co-Design" by Dirk Jansen): Peter Marwedel. *Embedded System Design.* Kluwer Academic Publishers, Dordrecht, 2003.
- Pragmatic examples in applying current SOC design methodology can also be found in Grant Martin and Henry Chang, Editors. *Winning the SOC Revolution: Experiences in Real Design.* Kluwer Academic Publishers, Boston, 2003.

Vendor Web sites give good overviews of recent improvements in hardware and software design tools:

- Synopsys: *http://www.synopsys.com*
- Cadence: *http://www.cadence.com*
- Magma: *http://www.magma-da.com*
- ARM: *http://www.arm.com*
- Wind River: *http://www.windriver.com*
- Coware: *http://www.coware.com*
- Mentor: *http://www.mentor.com*

CHAPTER 3

A New Look at SOC Design

This book focuses on a particular SOC design technology and methodology, here called the *advanced* or *processor-centric SOC design method*. The essential enabler for this technology is automatic processor generation—the rapid and easy creation of new microprocessor architectures, complete with efficient hardware designs and comprehensive software tools. The high speed of the generation process and the great flexibility of the generated architectures underpin a fundamental shift of the role of processors in system architecture. Automatic generation allows processors to be more closely tailored to the computation and communication demands of an application. This tailoring, in turn, expands the roles of processors in system architecture. The processors may serve both as a faster or more efficient alternatives to processors used historically in programmable roles and as a programmable alternative to other forms of digital logic, such as complex state machines. The continued rapid pace of transistor scaling allows a rich and diverse set of functions to be implemented together on SOCs. When these SOCs are designed with configurable processors, the resulting electronic systems have better performance, better efficiency, and better adaptability.

This chapter introduces the basics of multiple-processor system design and automatic processor generation, shows the impact of processor tailoring on tasks historically implemented as software on general-purpose processors, and sketches the influence of tailored processors on tasks traditionally implemented as nonprogrammable hardware blocks. The use of tailored processors is the foundation of the advanced SOC method first, the mapping of a system's functions into a set of communicating tasks; second, the combined optimization of each task and the processor architecture on which it runs; and third, the integration of the processors and task software into a properly operating system model and corresponding integrated-circuit implementation.

3.1 The Basics of Processor-Centric SOC Architecture

Deep-submicron silicon opens up the possibility of highly parallel systems architectures—architectures that improve overall throughput, latency, and efficiency by executing many tasks and operations in parallel. The target applications for volume SOC designs—communications, consumer, and computation systems—often show high degrees of intrinsic parallelism, including parallelism at the task level (major subsystems that can run in parallel with others) and at the data and instruction level (individual operations within one task that can run in parallel with one another). The processor-centric SOC design methodology seeks to exploit parallelism at the task level. Use of optimized, automatically generated processors seeks to exploit the parallelism inherent at the data and instruction level.

3.1.1 Processor Generation

Chapters 5, 6, and 7 focus on the single most important component of the new SOC design flow—the design of the individual processors that comprise the heart of the SOC design. A processor design has three important dimensions:

1. **Architecture:** Defines resources available to the programmer, including the instruction set, registers, memories, operators, interrupt- and exception-handling methods, and system-control functions. The architecture can be considered a contract between the hardware design and the software environment. In SOC design, the "goodness" of a processor architecture is not measured by abstract notions of performance and generality. Instead, architecture quality can be quantified by measurements of efficiency and performance of applications that are expected to run on the processor.
2. **Hardware implementation:** The logic and circuit design that implements the fundamental process of fetching and executing instructions. The hardware implementation includes the microprocessor pipeline; data paths and widths; and the arrangement of memories, buses, and other interfaces to circuits outside the processor. The mapping of this logic into transistor circuits and integrated circuit structures plays a central role in determining the cost, performance, and power dissipation of the final hardware.
3. **Software environment:** The collection of software tools used to develop high-level-language programs and to translate those programs into a correct sequence of instructions, plus runtime software components to support application execution on the architecture. Tailoring the processor (or set of processors) to the intended application is central to the SOC development methodology, so the software environment must include cycle-accurate instruction set simulators, code performance profilers and interprocessor communications modeling tools.

Sometimes it is also useful to speak of a *microarchitecture*, meaning key hardware implementation choices such as the arrangement of the execution pipeline, the size and width of cache memories, and other major determinants of processor performance.

Architecture, but not microarchitecture, is directly visible to the programmer. Microarchitecture, however, may have an important indirect impact in that it will affect application performance and hardware cost. Both architectural and microarchitectural changes may be important levers for optimizing the processor to meet cost and performance goals.

The essence of automatic processor generation is the automated translation of a high-level, abstract processor description into a complete hardware implementation and tailored software environment. The processor description may start from a base architecture—a simple standard set of architectural features useful across all applications—and add feature enhancements needed by the target set of applications. Starting from a simple base-processor architecture offers two important benefits:

1. Programs can be written for the base architecture and run unmodified on any configuration of the processor. This characteristic is potentially important for basic software utilities that are widely needed but not performance-critical.
2. Having a small but functional minimum processor configuration simplifies the generation of compilers, simulators, debuggers, and operating systems because the most basic operations—load, store, add, branch—are known to exist in all configurations.

The processor description constitutes sufficient information for the automatic generation of a quality hardware implementation and tailored compilers, assemblers, debuggers, and simulators and to adapt real-time operating systems to the new architecture. The automatic generation of the hardware design and software tools also enables rapid assessment of the specified architecture's suitability for the target set of applications by enabling quick prototyping and measurement of real implementation characteristics.

Rapid measurement of hardware size, power, and speed—and rapid performance profiling of the application set on the target architecture—yield the essential insight into design suitability with regard to cost, power, and speed. These traits of the advanced SOC design method allow the designer both to rapidly iterate through many potential processor architectures and to start the integration of this processor subsystem into the SOC. The basic structure of the processor optimization flow is shown in Fig 3-1.

The two major inputs into the processor-generation flow are the application source code and sample input data for that application. The generator's output is the processor design and configured software tools tuned for that processor configuration. The basic process consists of generating an initial processor with its software environment, then compiling the application source code using that software environment to create a machine-code binary program, and then simulating the application with that binary program and sample application input data. This process yields a performance profile and corresponding output data.

The performance profile indicates the number of cycles required to execute the application: one of the key measures of processor effectiveness. This design process also generates the processor's RTL implementation and scripts for gate-level design tools, so the netlist and physical placement and routing of gate-level cells can be produced. Clock frequency, silicon chip

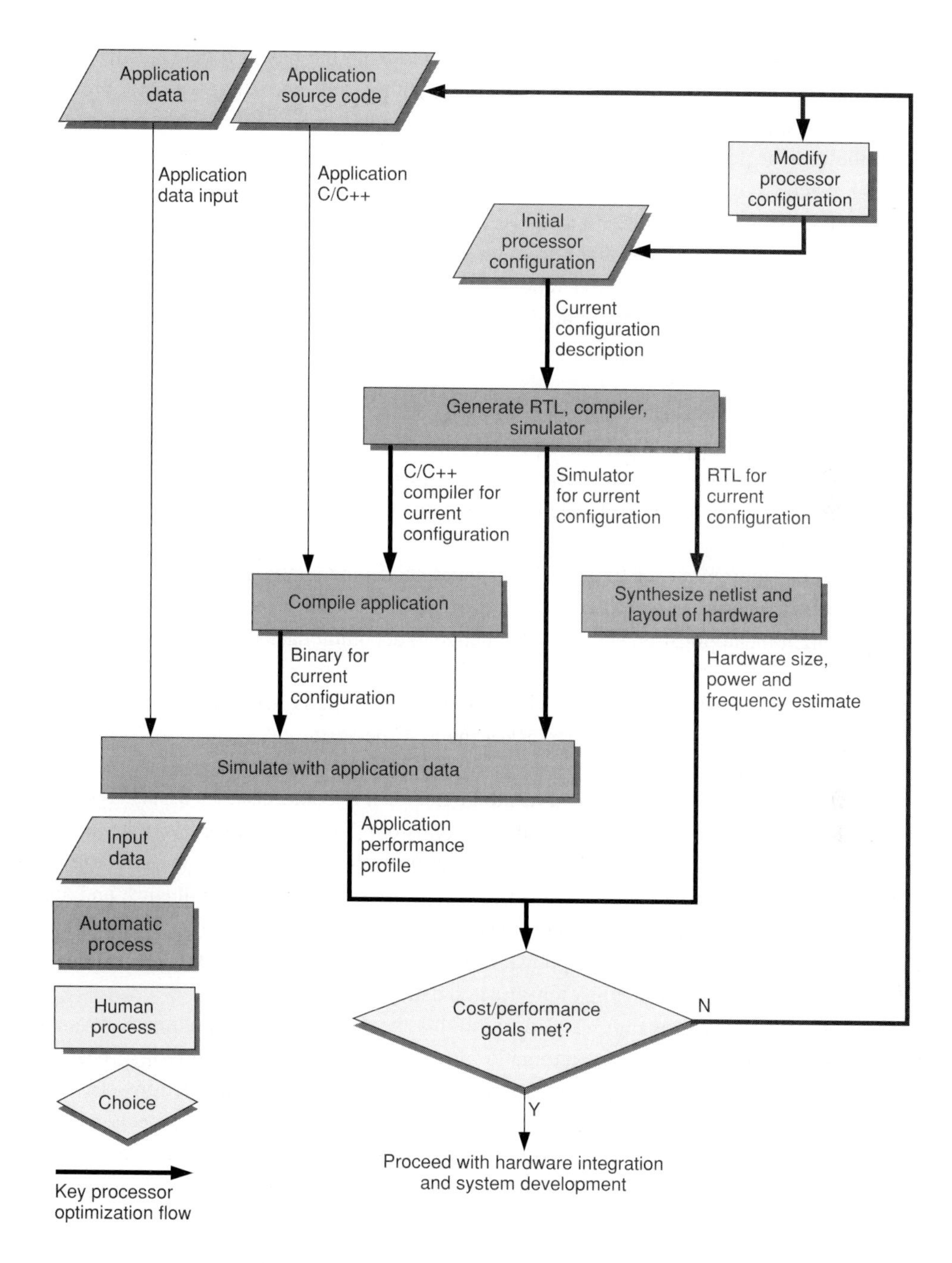

Figure 3-1 Basic processor generation flow.

area, and power dissipation can be accurately estimated from the physical design. These hardware implementation characteristics represent the other key measures of the processor configuration's appropriateness.

If the cost or power dissipation is too high, the configuration can be downsized. If the number of execution cycles is too high, adding new instructions, increasing memory capacity or bus width, or making other changes to the processor's definition can enhance the configuration's efficiency to meet performance goals. This enhanced processor configuration serves as the seed for generating updated compilers, the RTL description, and all their by-products so that the design team can revise its estimates of application performance and implementation cost.

This processor design flow introduces a number of key questions:

- What is specified in a processor configuration?
- What are the measures of processor performance for this specific application?
- What are the cost goals for this processor implementation?
- What kind of cost and performance improvement is feasible for different classes of applications?
- What is the appropriate set of application source code and application data that will drive the processor optimization process?
- How might the application source code change to accommodate processor configuration changes?
- How does automation of processor generation reduce design time and design errors?
- How does the processor fit into the rest of the SOC design?

These key questions are addressed in the course of this book. Chapter 4 uses Tensilica's processor generator and the Tensilica Instruction Extension (TIE) language to illustrate how processors can be configured and extended. These tools form the backbone of a methodology that includes concrete metrics of gate count, die area, power dissipation, clock frequency, and application cycle count as primary measures of performance and cost. Various processor configurations dictate different changes in application source code—ranging from no change to significant exploitation of new application-directed datatypes and operators.

The process pictured in Fig 3-1 also highlights the opportunity for automatic processor optimization. Chapter 4 discusses automatic processor-optimization technology, based on advanced compilers, that identifies application hot spots and candidate instruction sets, evaluates the impact on both application performance and processor hardware cost, selects a Pareto-optimal solution, and generates both an optimized hardware design and software tools for the chosen processor configuration. The role of these compiler-based methods is to automate the path shown in bold in Fig 3-1, allowing thousands of candidate architectures to be evaluated in minutes.

Before launching into a detailed discussion of the mechanisms of the new SOC methodology, a short discussion of the impact of processor configuration on application performance will show the core benefits of processor optimization. The hardware-centric and software-centric views of SOC design require parallel discussion, because these two traditions often have wildly different perspectives on development flow, measures of goodness, and mechanisms for change. In fact, hardware and software engineers often see distinctly different motivations for migrating specific functions into an application-specific processor block.

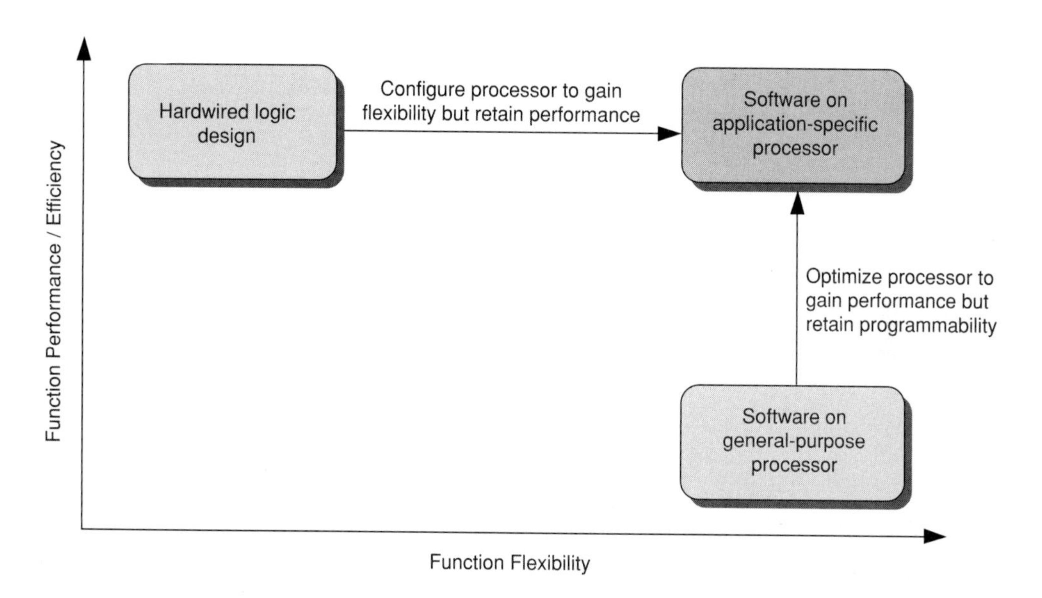

Figure 3-2 Migration from hardwired logic and general-purpose processors.

SOC designers face a major dilemma as they architect and implement a system. Many subsystems can be implemented, at least in theory, as either a block of hardwired logic or as software running on a general-purpose processor. For many data-intensive functions, the hardwired logic implementation can be more than 100 times smaller or faster than the software implementation because the logic contains only the data-path elements required, and these elements can often be easily arranged to run in parallel for high throughput.

By contrast, processors show their superior flexibility when function changes are required. Modifying and retesting a software change may be more than 100 times faster and cheaper than modifying the circuits on a large integrated circuit. This contrast is shown in Fig 3-2. The implementation of the function on an application-specific processor retains all of the programming flexibility of a general-purpose processor but with functional performance or efficiency that rivals that of hardwired logic. Fig 3-2 also suggests the benefits of using application-specific processors seen by hardware designers and by software developers.

Hardware designers move from hardwired logic to application-specific processors to reduce development time and ensure continuous and easy upgradeability as the application evolves. Customizing and extending predefined base-processor architectures instead of designing complex logic blocks reduces the initial SOC development effort, because specification of processor configurations and software is easier than specification of logic design at the gate or RTL level. Processor-based design of logic functions reduces risk in two ways. The initial design is easier to verify, because the specification is more concise and can be simulated in the system context more directly and at far higher speeds. Second, the resulting design is more tolerant of specification changes because far more of the design, especially the control flow of the design, is implemented in software and can be changed at any time, before or after fabricating the SOC. Simplified design verification is a major incentive for adoption of configurable processors over traditional RTL design.

Software developers move to application-specific processors from general-purpose processors to reach performance and efficiency goals without giving up the well-known benefits of rapid software evolution provided by a programmable processor. Software-based SOC design gets the software team involved earlier and gives the team a greater role in the overall project. In many cases, the tools for processor generation give the software developer sufficient insight into both application performance and hardware implementation that they can play a central role even in jobs formerly reserved for specialized hardware designers. This migration from hardwired logic and from software running on general-purpose processors is governed by different situations and offer different, but important, benefits to their users.

3.2 Accelerating Processors for Traditional Software Tasks

As the complexity of embedded systems rises, software naturally takes on a more important role in the system. As the product design evolves and requires more performance or more functions running in software, every software component must run faster. Not every line of code, of course, is equally important. Some tasks, and some task sections, dominate the software performance profile. If those sections and those tasks can be effectively optimized, the overall system performance improves.

3.2.1 The Evolution of Generic Processors

Most microprocessor architectures were created with general-purpose applications in mind. These architectures are often good targets for a broad range of generic integer applications written in C or C++, but they are typically capable of only modest data-manipulation rates. Such one-size-fits-all architectures may be appropriate for high-level control tasks in SOC applications, but they are significantly less efficient for more data-intensive tasks such as signal, media, network, and security applications.

The origins and evolution of microprocessors further constrain their use in traditional SOC design. Most popular embedded microprocessors, especially the 32-bit architectures, descend directly from 1980s desktop computer architectures such as ARM (originally the Acorn RISC Machine, a British desktop), MIPS, 68000/ColdFire, PowerPC, and x86. Designed to serve general-purpose applications, these processors support only the most generic data types such as 8-, 16-, and 32-bit integers. Likewise, they support only the most common operations such as integer load, store, add, shift, compare, and bitwise logical functions.

Their general-purpose nature makes these processors well suited to the diverse mix of applications run on computer systems: their architectures perform equally well when running databases, spreadsheets, PC games, and desktop publishing. However, all these general-purpose processors suffer from a common bottleneck. Their need for complete generality requires them to execute an arbitrary sequence of primitive instructions on an unknown range of data types. Put another way, general-purpose processors are not optimized to deal with the specific data types of any given embedded task. Inefficiencies result.

Of course, general-purpose processors can emulate complex operations on application-specific datatypes using relatively long sequences of primitive integer operations. For example Chapter 5, Section 5.1, shows how the basic pixel blend operation, arguably a single operation for an imaging-oriented architecture, requires 33 RISC operations to execute. Many embedded applications can be expressed and implemented most naturally in something other than 32-bit integer operations. Security processing, signal processing, video processing, and network protocols all have unique computational requirements with, at best, a loose fit to basic integer operations. This "semantic gap" has long inspired efforts in application-directed processor architecture, but it has rarely been economical to commercialize such processors because of the high costs and specialized skills involved.

Compared to general-purpose computer systems, embedded systems comprise a more diverse group and individually show more specialization. A digital camera must perform a variety of complex image-processing tasks but it never executes SQL database queries. A network switch must handle complex communications protocols at optical interconnect speeds but it doesn't manipulate 3D graphics.

The specialized nature of each individual embedded application creates two issues for general-purpose processors in data-intensive embedded applications. First, the critical data-manipulation functions of many embedded applications and a processor's basic integer instruction set and register file are a poor match. Because of this mismatch, these critical embedded functions require many computation cycles when run on general-purpose processors.

Second, more focused embedded products cannot take full advantage of a general-purpose processor's broad capabilities. Expensive silicon resources built into the processor go to waste because the specific embedded task that's assigned to the processor doesn't need them. Unused features that might be tolerable within the cost and power budgets of a desktop computer are a painful extravagance in low-cost, battery-powered consumer products.

Many embedded systems interact closely with the real world or communicate complex data at high rates. A hypothetical general-purpose microprocessor running at tremendous speed could perform these data-intensive tasks. This is the basic assumption behind the use of multi-GHz processors in today's PCs: throw a fast enough processor at a problem (no matter the cost in dollars or power dissipation) and you can solve any problem. For many embedded tasks, however, no such processor exists as a practical alternative because the fastest available processors typically cost orders of magnitude too much and dissipate orders of magnitude too much power to meet embedded-system design goals. Instead, embedded-system hardware designers have traditionally turned to hardwired circuits to perform these data-intensive functions.

3.2.2 Explaining Configurability and Extensibility

Changing the processor's instruction set, memories, and interfaces can make a significant difference in its efficiency and performance, particularly for the data-intensive applications that represent the "heavy lifting" of many embedded systems. These features might be too narrowly used to justify inclusion in a general-purpose instruction set, hand-designed processor hardware, and handcrafted software tools. The general-purpose processor represents a compromise where features that provide modest benefits to all customers supersede features that provide dramatic benefits to a few. This design compromise is necessary because the historic costs and difficulty of manual processor design mandate that only a few different designs can be built. Automatic processor generation reduces the cost and development time so that inclusion of application-specific features and deletion of unused features suddenly becomes attractive.

We use the term *configurable processor* to denote a processor whose features can be pruned or augmented by parametric selection. Configurable processors may be implemented in many different hardware forms, ranging from ASICs with hardware implementation times of many weeks to FPGAs with implementation times measured in minutes. An important superset of configurable processors is *extensible processors*—processors whose functions, especially the instruction set, can be extended by the application developer to include features never considered by the original processor designer.

For both configurable and extensible processors, the usefulness of the configurability and extensibility is strongly tied to the automatic availability of both hardware implementation and software environment supporting all aspects of the configurations or extensions. Automated software support for extended features is especially important, however. Configuration or extension of the hardware without complementary enhancement of the compiler, assembler, simulator, debugger, real-time operating systems, and other software support tools would leave the promises of performance and flexibility unfulfilled because the new processor could not be programmed.

3.2.3 Processor Extensibility

Extensibility's goal is to allow features to be added or adapted in any form that optimizes the cost, power, and application performance of the processor. In practice, the configurable and extensible features can be broken into four categories, as shown in Fig 3-3.

A block diagram for a configurable processor is shown in Fig 3-4, again using Tensilica's Xtensa processor as an example. The figure identifies baseline instruction-set architecture features, scalable register files, memories and interfaces, optional and configurable processor peripherals, selectable DSP coprocessors, and facilities to integrate user-defined instruction-set extensions. Almost every feature of the processor is configurable, extensible, or optional (including the size of almost every data path, the number and type of execution units, the number and type of peripherals, the number and size of load/store units and ports into memory, the size of instructions, and the number of operations encoded in each instruction).

Instruction Set	• Extensions to ALU functions using general registers (e.g., population count instruction) • Coprocessors supporting application-specific data types (e.g. network packets, pixel blocks), including new registers and register files • Wide instruction formats with multiple independent operation slots per instruction • High-performance arithmetic and DSP (e.g., compound DSP instructions, vector/SIMD, floating point), often with wide execution units and registers • Selection among function unit implementations (e.g., small iterative multiplier vs. pipelined array multiplier)
Memory System	• Instruction-cache size, associativity, and line size • Data-cache size, associativity, line size, and write policy • Memory protection and translation (by segment, by page) • Instruction and data RAM/ROM size and address range • Mapping of special-purpose memories (queues, multiported memories) into the address space of the processor • Slave access to local memories by external processors and DMA engines
Interface	• External bus interface width, protocol, and address decoding • Direct connection of system control registers to internal registers and data ports • Arbitrary-width wire interfaces mapped into instructions • Queue interfaces among processors or between processors and external logic functions • State-visibility trace ports and JTAG-based debug ports
Processor Peripherals	• Timers • Interrupt controller: number, priority, type, fast switching registers • Exception vectors addresses • Remote debug and breakpoint controls

Figure 3-3 Processor configuration and extension types.

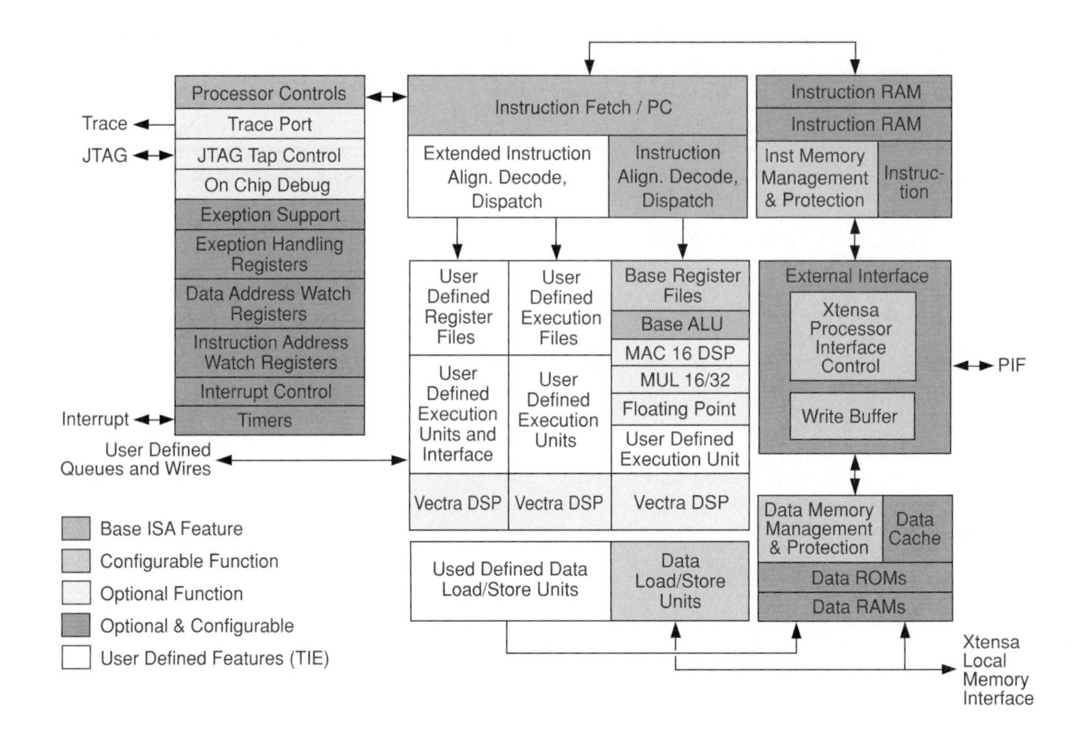

Figure 3-4 Block diagram of configurable Xtensa processor.

3.2.4 Designer-Defined Instruction Sets

Processor extensibility serves as a particularly potent form of configurability because it handles a wide range of applications and is easily usable by designers with a wide range of skills. Processor extensibility allows a system designer or application expert to directly exploit proprietary *insight* about the application's functional and performance needs directly in the instruction-set and register extensions.

This shift of insight, however, also makes special demands on the processor-generator environment. The means of expressing extensions must be flexible—to accommodate a wide range of possible instruction-set extensions—and the instruction-set description tools must be intuitive and bullet-proof—to ensure that someone who is not a processor designer can easily create a new instruction set with full software support while preventing the creation of instructions with subtle bugs that could prevent the basic processor from functioning correctly.

A simple example written in the TIE language illustrates the potential simplicity of this approach. Fig 3-5 shows the complete TIE description of an instruction that takes two sets of four 8-bit values packed into two 32-bit entries (`a` and `b`) in the AR register file, multiplies together the corresponding 8-bit values, and sums adjacent 16-bit multiplication results into a pair of 16-bit accumulated values. The resulting value (`c`) is written back into a third 32-bit AR register file entry:

```
operation mac4x8 {in AR a, in AR b, out AR c} {} {
assign c = {a[31:24]*b[31:24] + a[23:16]*b[23:16], a[15:8]*b[15:8] +
a[7:0]*b[7:0]};}
```

Figure 3-5 Simple 4-way MAC TIE example.

From this instruction description, the following actions take place automatically:

- New data-path elements, including four 8 x 8 multipliers and two 16-bit adders are added to the processor hardware description.
- New decode logic is added to the processor to decode the new `mac4x8` instruction using a previously unallocated operation encoding.
- The integrated development environment, including instruction-set simulator, debugger, profiler, assembler, and compiler are extended to support this new `mac4x8` instruction.
- Plug-in extensions for third-party tools, including debuggers and RTOS are generated so these tools also include support for the new instruction.

Instruction-set-description languages such as TIE represent an important step beyond traditional hardware description languages (HDL) such as Verilog and VHDL. Like HDLs, the semantics of instruction-set-description languages are expressed as a combination of logical and arithmetic operations on bits and bit-fields transforming input values into output values.

The TIE language, for example, uses pure combinational Verilog syntax for describing instructions. Unlike HDLs, key software information about instruction names, operands, and pipelining must be succinctly expressed. This form of expression raises the level of abstraction for design by direct incorporation of new hardware functions into the processor structure and directly exposes the new instruction(s) in the high-level programming environment available to the software team.

3.2.5 Memory Systems and Configurability

Even today, on-chip memories consume more silicon area than any other single type of circuit. Increases in data bandwidth, data-encoding complexity, and data resolution are all likely to increase the fraction of chip area consumed by memory. As a result, efficient use of memories becomes even more important over time.

The data-intensive role for configurable processors further emphasizes the criticality of memory systems and interfaces. Six key characteristics of processor memory systems are addressed by memory-system configurability:

Local bandwidth: The performance of many embedded applications (especially in signal, network, and media processing) are limited by the basic data rate of transfers between processor memories and processor-execution units. By extending the fundamental width for application-specific data words, the processor can transfer data at higher speeds—typically 8 or 16 bytes per cycle—between processor registers and execution units and local data RAMs and caches. Moreover, combining data operations with load/store operations in a single instruction often doubles the effective local processing bandwidth.

Interface bandwidth: For many data-intensive applications, high local-memory bandwidth is necessary but not sufficient. The interface between the processor and the rest of the system must also carry high-bandwidth data streams. Increased interface bandwidth is particularly important as the processor's computational bandwidth increases. The extensibility of the interface for wide data words, fast block transfers, high-speed DMA operations, and application-specific direct data interfaces all contribute to useful interface bandwidth that can easily exceed 10x that of 32-bit, general-purpose RISC processors.

Latency: Many applications have tight interaction between the fetch and use of data. Data-access latency, not just the overall data bandwidth, may be decisive in overall application throughput. Single-cycle access to local memories, and even to surrounding logic, enables an application to generate an address, load new data, and act on the data in a single instruction over just two or three pipeline stages.

Memory scalability: Pipelined processors depend on the ability to tightly couple instruction and data memories into the microarchitecture. Different applications have vastly different memory requirements, yet the configurable processor design must scale across a wide spectrum of memory sizes and types. The ability to scale from local pipelined memories as small as 1 Kbyte to up into hundreds of Kbytes, and to make arbitrary combinations of RAM, ROM, caches, and specialized memories (queues, logic registers), expands the processor's usefulness and guarantees high SOC clock rates even with complex memories.

Multiple operations: Some applications are less sensitive to the number of bytes transferred than to the rate of unique memory operation completions (the number of loads and stores made from and to multiple memory and non-memory locations). Extensible processors can implement multiple load/store units, multiple memory ports, and additional interfaces to specialized memories and registers. These extended interface ports allow two, three, or more independent memory and data-movement operations per cycle.

Support for concurrency: Small, configurable processors are the natural building block for SOCs, but this approach implies task concurrency across a number of processors. To cooperate on a single application, the processors must communicate data in some form. Often, the most logical and efficient communications mechanism is shared memory. For correctness, however, the SOC must implement some hardware-based mechanism to ensure that multiple processors sharing data have the same view of memory at critical points in their computations. While hardware support for memory synchronization can be implemented outside the processor, architectural support simplifies and unifies SOC development.

3.2.6 The Origins of Configurable Processors

The concept of building new processors quickly to fit specific needs is almost as old as the microprocessor itself. Architects have long recognized that high development costs, particularly for design verification and software infrastructure, forces compromises in processor design. A processor had to be "a jack of all trades, master of none."

Early work in building flexible processor simulators, especially ISP in 1970, in retargetable compilers, especially the GNU C compiler in the late 1980s, and in synthesizable processor in the 1990s, stimulated hope that the cost of developing new processor variants might be driven down dramatically. Research in application-specific instruction processors (ASIPs), especially in Europe in code generation at IMEC, in processor specification at University of Dortmund, in micro-code engines ("transport-triggered architectures") at the Technical University of Delft, and fast simulation at University of Aachen all confirmed the possibility that a fully automated system for generating processors could be developed. The VLIW processor work at Hewlett-Packard has also produced forms of processor configurability, found in both the PICO (Program-In-Chip-Out) and LX projects of HP Labs. However, none of these projects has demonstrated a complete system for microprocessor hardware design and software environment generation (compiler, simulator, debugger) from a common processor description. (Note: The last section of this chapter suggests further reading on this early research work.)

While all of this work contributes to the principles of fully automated processor generation, none constitutes a full or automated processor-generation system. Automatic processor generators can be characterized as meeting all of the following criteria:

1. Both hardware description and software tools are generated from a single processor description or configuration.
2. Automatically generated software tools include, at a minimum, a C compiler, assembler, debug, simulator, and a runtime environment or bindings for a standard real-time operating system.
3. The automatically generated processors are complete in the sense that all generated processors are guaranteed to support arbitrary applications programs written, for example, in C. Put another way, all of the automatically generated processors are capable of running programs. They are not merely sequencers.
4. The processor configuration or description is significantly more abstract than an HDL description for an equivalent processor, so hardware and software engineers without deep knowledge of processor architecture can nevertheless understand and develop new processors.
5. The processor-generation process is automated: software programs alone generate detailed processor implementations from the processor description or configuration, without human intervention, in minutes or hours.

Tensilica's processor generator is not the only commercial product in this domain, though arguably it's the most complete. ARC International's ARCTangent processors and ARChitect tool provide a form of processor extension based on RTL modification without direct hooks to software tools. MIPS Technologies' "User Defined Instructions," supported by the company's Pro series of processors, also provides some means for structured additions to the processor hardware design.

3.3 Example: Tensilica Xtensa Processors for EEMBC Benchmarks

To assess the impact of configurable processing, we must look at processor performance over a variety of applications. Public performance information on large embedded applications for a wide range of processors is scarce. We therefore turn to the best available benchmark data to prove that configurable processors deliver significant performance and efficiency benefits compared to general-purpose processors over a wide range of embedded applications. No single benchmark can accurately capture the full diversity and wide range of embedded applications. Consequently, *EDN* magazine sponsored the creation of a comprehensive embedded benchmark suite that would be more informative than the simple synthetic Dhrystone benchmark once used to rate embedded processor performance. Since 1997, the resulting organization—dubbed EEMBC for EDN Embedded Microprocessor Benchmark Consortium (and pronounced "embassy")—has attracted participation by more than 40 leading processor and software companies. Working together, these companies developed both a set of benchmarks and a fair process for running, measuring, certifying, and publishing test results.

These benchmarks cover a wide range of embedded tasks, but the bulk of the certified results are available for three suites: consumer, telecommunications, and networking. EEMBC rules allow reporting of two types of results: "out-of-the-box" and "full fury." The out-of-the-box scores are based on compiling and running unmodified C code on the processor as realized in silicon or using a cycle-accurate processor simulator. Full-fury scores allow modifications to the C code and hand-coding in assembly language. The full-fury optimizations also allow the use of application-specific configurations to demonstrate the abilities of processor-extension technology.

Each EEMBC benchmark suite consists of a number of different programs written in C. Many of the component programs are run with several data sets or parameter values. The performance for each benchmark within the suite is compared to a reference processor, and the performance ratios for all of the components are averaged using a geometric mean—the accepted method of averaging performance ratios. For each of the benchmark suites in the following sections, we show both the benchmark performance per MHz of clock and the full performance at the rated clock speed of the processor. The clock speeds are taken directly from the corresponding vendor specs for the processor.

In all cases, the data in this section is taken directly from the certified results on the EEMBC Web site at *http://www.eembc.org*, as of April 2003. In each case, we compare results for the Xtensa V processor to the fastest processor from each relevant vendor or architecture for which complete results are available. The bulk of the data represents licensable processor-core architectures, though in a few cases, additional architectures are included in the comparison. The NEC VR4122 implements a 32-bit MIPS instruction set and uses a pipeline and interface roughly similar to that of the Xtensa processor. The MIPS 20Kc implements a 64-bit MIPS instruction set and incorporates a pipeline capable of issuing multiple instructions per clock. The ARM architecture is represented in the EEMBC benchmarks by the ARM1026EJ-S. This processor core implements the 32-bit ARMv5TEJ instruction set architecture and uses a dual 64-bit bus interface.

To demonstrate the impact of extensible processing on each suite, we include the results for a baseline configuration of the Xtensa processor, with no added TIE instructions, running simple compiled C code, and for an optimized configuration, including TIE instructions and optimized code. A different optimization of the Xtensa processor is used for each benchmark category, just as an SOC designer would develop individual configurations for each set of related application tasks on an SOC. We include the best results available for the other processors. For most, only baseline results are available.

3.3.1 EEMBC Consumer Benchmarks

Video processing lays at the heart of many consumer electronics products—in digital video cameras, in digital televisions, and in games. Common video tasks include color-space conversion, 2D filters, and image compression. The EEMBC Consumer benchmarks include a representative sample of all these applications. Fig 3-6 and Fig 3-7 show the performance on EEMBC's Consumer benchmark suite for the target set of processors plus the Philips Trimedia TM1300, a very-long-instruction-word (VLIW) processor that is optimized for multimedia applications.

Optimized EEMBC scores can include both hand-tuning of code and tuning of the processor configuration for this class of consumer tasks. For Xtensa processors, roughly 200,000 gates of added consumer instruction-set features are implemented in TIE. Fig 3-6 shows the Consumer benchmark scores (ConsumerMarks) per MHz, where the STMicroelectronics ST20 at 50MHz sets the performance reference of 1.0. These results show that the configured Xtensa processor provides about three times the cycle efficiency of a good media processor (the Philips Trimedia TM1300), about 33 times the cycle efficiency of a good 64-bit RISC processor (the MIPS 20Kc), and 50 times the performance of a basic 32-bit RISC processor (the NEC VR4122). Fig 3-7 shows the benchmark performance including the impact of clock frequency. These results confirm that even substantial processor optimizations do not significantly reduce clock frequency, so the high instruction/clock efficiency achieved through processor extension translates directly into high absolute performance.

3.3.2 Telecommunications

Telecommunications applications present a different set of challenges. Here the data is often represented as 16-bit fixed-point values, as densely compacted bit-streams or as redundantly encoded channel data. Over the past 10 years, standard digital signal processors (DSPs) have

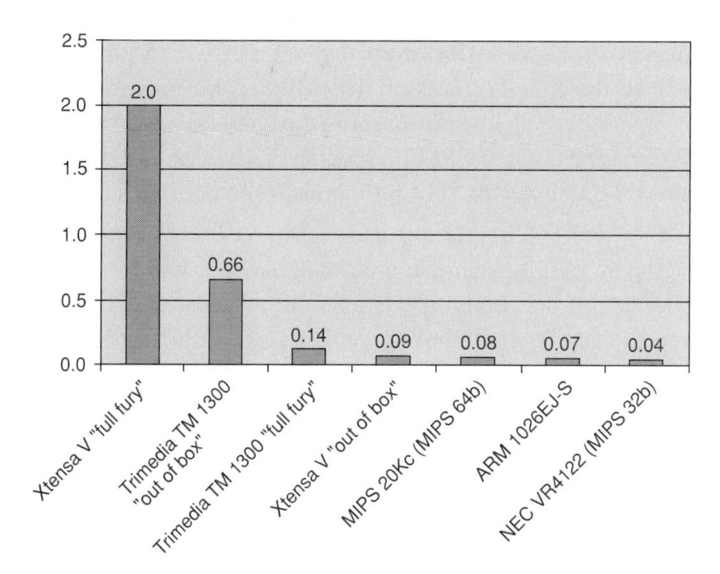

Figure 3-6 EEMBC ConsumerMarks—per MHz.

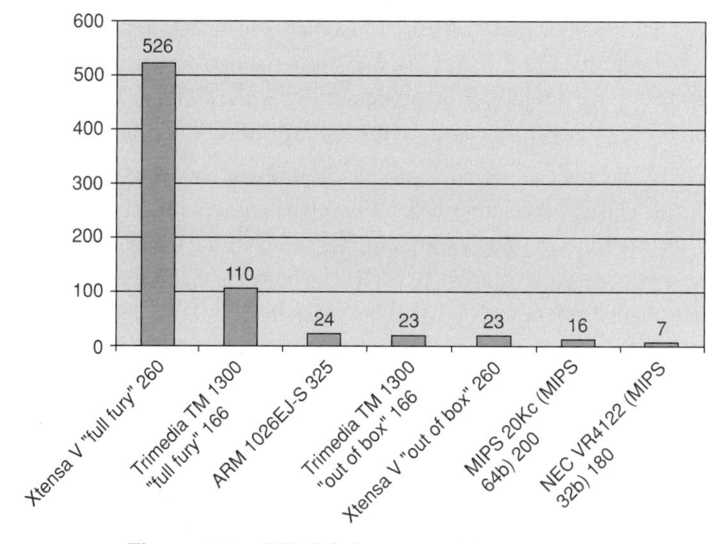

Figure 3-7 EEMBC ConsumerMarks—performance.

evolved to address many filtering, error correction, and transform algorithms. The EEMBC Telecom benchmark includes many of these common tasks.

Fig 3-8 shows the efficiency of the Xtensa processor (including the Vectra V1620-8 vector-coprocessing option) and various architectures, on a performance per MHz basis, where the

IDT 32334 (a 32-bit MIPS architecture) at 100MHz sets the performance reference score of 1.0. The other architectures shown include a high-end DSP (the Texas Instruments TMS320C6203) plus the ARM and MIPS processors. For the "full fury" Telecom scores (TeleMarks). The figure includes results for optimized configurations of the Xtensa V core and hand-tuned code for the TI 'C6203. The Xtensa processor has been specifically optimized for two of the benchmarks in the telecom suite: convolutional coding and bit allocation. Fig 3-9 shows the dramatic impact of application-directed instruction sets and code tuning for each of the optimized DSP architectures relative to conventional 32- and 64-bit processors. The configured processor is 30 times faster than all of the conventional RISC cores.

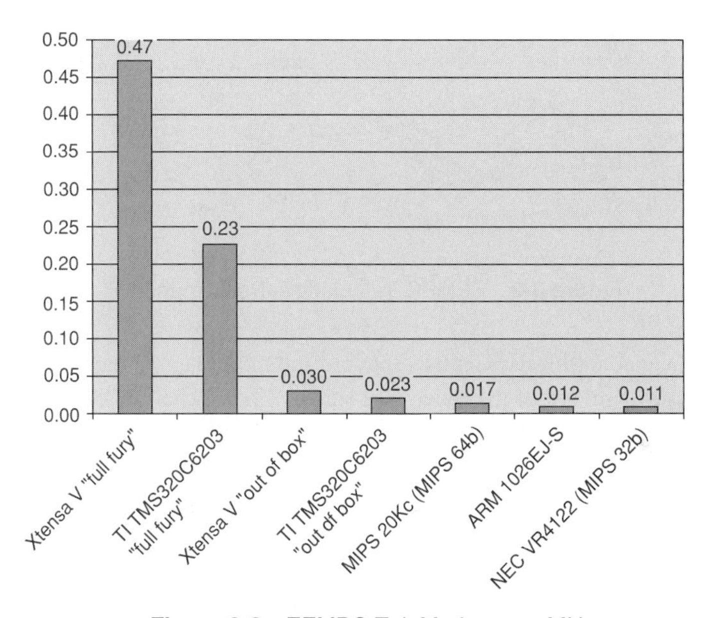

Figure 3-8 EEMBC TeleMarks—per MHz.

3.3.3 EEMBC Networking Benchmarks

Networking applications have significantly different characteristics than consumer and telecommunications applications. They typically involve less arithmetic computation, generally contain less low-level data parallelism, and frequently require rapid control-flow decisions. The EEMBC Networking benchmark suite contains representative code for routing and packet analysis. Fig 3-10 compares EEMBC Network performance (NetMarks) per MHz for Xtensa V processors with and without extensions and several other architectures, where the IDT 32334 (a 32-bit MIPS architecture) at 100MHz sets the performance reference score of 1.0. The optimizations for the Xtensa processor are small, adding less than 14,000 additional gates (less than 0.2mm^2 in area) to the processor. The extended Xtensa processor achieves about seven times the

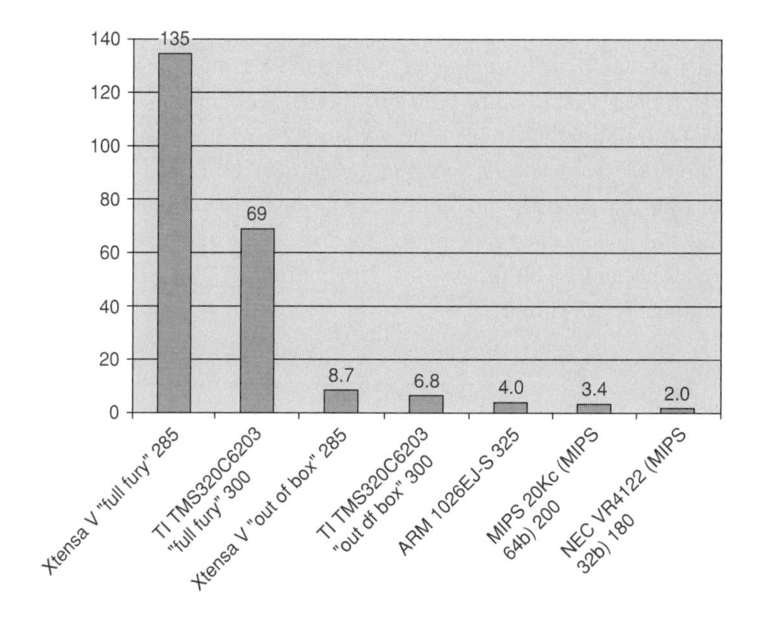

Figure 3-9 EEMBC TeleMarks—Performance

cycle efficiency of a good 64-bit RISC processor core and more than 12 times the efficiency of a 32-bit RISC processor core. All of the processors shown run at generally comparable clock frequencies, but performance varies widely, giving the overall NetMark performance shown in Fig 3-11.

The EEMBC benchmark results show that extensible processors achieve significant improvements in throughput across a wide range of embedded applications, relative to good 32- and 64-bit RISC, DSP, and media processor cores.

Interestingly, when EEMBC certified and released the optimized Xtensa processor benchmark results, some observers complained, "That's not fair—the Xtensa processors have been optimized for the benchmarks." That's exactly the point of configurable processors— in embedded applications, much of the workload is known before the chip is built, so optimizing the processors to the target software is an appropriate strategy. The EEMBC benchmark codes are simple but reasonable stand-ins for types of applications at the heart of real embedded systems. The process of optimizing the processors for the EEMBC benchmarks—all the Xtensa data presented here—was the result one graduate student's summer project and should closely match the real-world process of optimizing processors for SOC applications.

3.3.4 The Processor as RTL Alternative

Extensible processors can play another fundamental SOC role in addition to serving as a faster processor running software tasks. In many cases, the instruction set can incorporate complex logic, bit manipulation, and arithmetic functions that look strikingly like the data paths of hard-

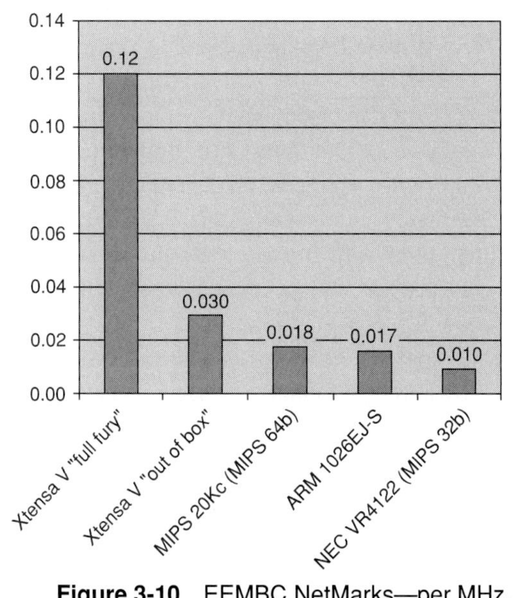

Figure 3-10 EEMBC NetMarks—per MHz.

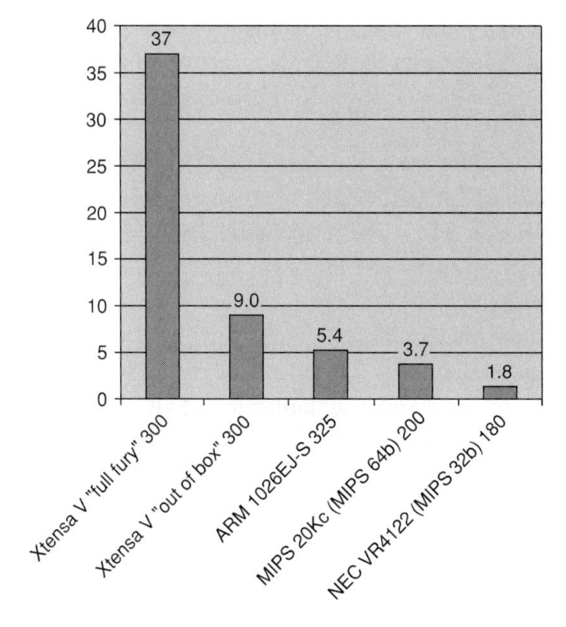

Figure 3-11 EEMBC NetMarks—Performance.

wired logic blocks. Extensible processors are easily configured to have similar throughput, similar interfaces, and similar silicon area and power dissipation characteristics when compared to RTL logic blocks.

The big differences between a dedicated function block implemented as logic and an application-specific processor may boil down to just two. Both turn out to be advantages for the extensible processor. First, the processor structure and generation process make the function easier to specify, model, and debug than the equivalent logic block developed just using RTL. Second, the processor is easily reprogrammed for new tasks using standard programming tools, limited only by the basic functions captured in its application-specific instruction set and configuration.

The opportunity to replace complex RTL blocks with processors depends on some important differences between this class of configurable processor and traditional general-purpose processors. The differentiating attributes of processors used as RTL alternatives include the following:

- Support for wide, deep, application-specific data paths as extended instructions.
- Support for very simple system-programming models without requiring interrupts and exception handing.
- Support for novel memory systems, not just caches, including wide memories, multiported data RAMs, instruction ROM, direct mapping of queues and special registers.
- Support for direct connection of external signals as inputs to instructions and direct connection of instruction results to external output signals.

With these capabilities, the extensible, application-specific processor can efficiently substitute for blocks of RTL. Like RTL-based design, configurable processor technology enables the design of high-speed logic blocks tailored to the assigned task. The key difference is that RTL state machines are realized in hardware and logic blocks based on extensible processors realize their state machines with firmware.

Hardwired RTL design has many attractive characteristics—small area, low power, and high throughput. However, the liabilities of RTL (difficult design, slow verification, and poor scalability for complex problems) are starting to dominate as SOCs get bigger. A design methodology that retains most of the efficiency benefits of RTL but with reduced design time and risk has a natural appeal. Application-specific processors as a replacement for complex RTL functions fit this need.

An application-specific processor can implement data path operations that closely match those of RTL functions. The equivalent of RTL data paths are implemented using the integer pipeline of the base-processor, plus additional execution units, registers, and other functions added by the chip architect for a specific application. A high-level instruction-set description is much more concise than an RTL description because it omits all sequential logic, including state-machine descriptions, pipeline registers, and initialization sequences.

The new processor instruction definitions are available to the firmware programmer via the same compiler and assembler that serve the processor's base instructions and register set. All sequencing of operations within the processor's data paths is controlled by firmware through the processor's existing instruction-fetch, decode, and execution mechanisms. Firmware for these application-specific processors can be written either in assembly code or in a high-level language such as C or C++.

Extended processors used as RTL-block replacements routinely use the same high-throughput techniques as traditional data-path-intensive RTL blocks: deep pipelines, parallel execution units, problem-specific state registers, and wide data paths to local and global memories. These extended processors can sustain the same high computation throughput and support the same low-level data interfaces as typical RTL designs.

However, control of the extended-processor data paths is very different from the control functions implemented in RTL-based designs. Cycle-by-cycle control of the processor's data paths is not fixed in hardwired state transitions. Instead, the sequence of operations is explicit in the firmware executed by the processor. Compare Fig 3-12 with Fig 2-2 in the previous chapter. In processor-based design, control-flow decisions are made explicitly in branches, and memory references are explicit in load and store operations, computation sequences are explicit sequences of general-purpose and application-specific processor instructions.

Implementing a complex logic function using a processor provides important new structure for both the design process and the final design. The essential, defining characteristics of the function are embodied in the processor's extended instruction set.

The transitory details become the program that runs on the processor. This separation aids rapid discovery of a natural and efficient compromise between the seemingly incompatible goals of narrow tuning to the exact needs of the application and sufficient flexibility to accommodate changing needs.

Using processor generation as an alternative to RTL design also brings to bear a powerful set of software-related tools and methodologies. The essential function description is a behavioral C or C++ program rather than a structural logic description in a Verilog or VHDL file. Processor simulation is orders of magnitude faster compared to RTL simulation and waveform viewers. The full spectrum of software-development tools, from RTOSes and standard interconnect buses to software project management and source-level debuggers, all map directly from the traditional control-processor world to this new domain of processor-as-logic-replacement.

Thus, easy and complete generation of new processors and tools creates two distinct benefits—higher performance and efficiency for processors in traditional processor roles and easier design and reprogrammability for processors in RTL-alternative roles. Additional benefits come from using the same class of processors and tools for both processor and hardwired logic roles. For many complex system problems, all major logic functions can be implemented with processors. Using processors to implement logic blocks means that all simulation can be done with fast processor simulators; all functions can be described at a high level (as C/C++ programs and processor configurations); debugging can be performed with C/C++ source-level; all memories are managed, and potentially shared by software; all functions can be continuously updated

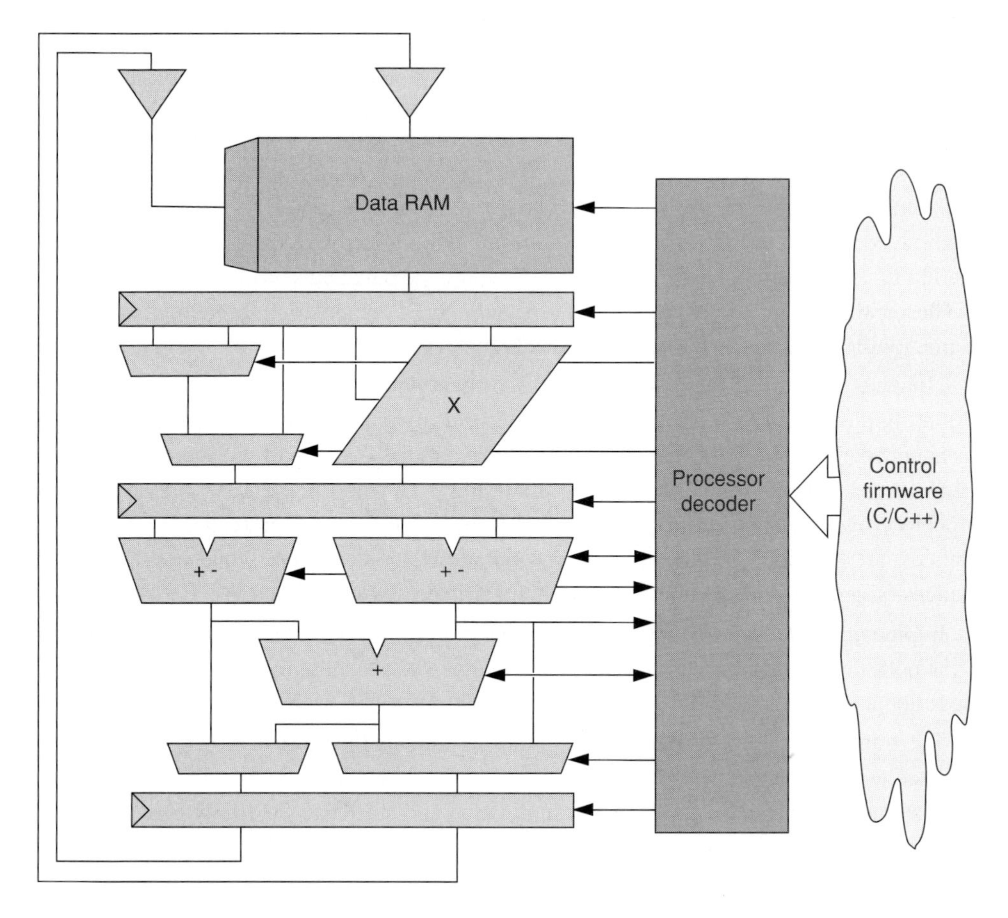

Figure 3-12 Configurable processor as RTL alternative.

just by dynamically changing the software; and the entire design is commonly portable across silicon fabrication processes and easily reusable in parts or in whole. This unification makes the design of subsystems, and the system as a whole, quicker and less costly.

3.4 System Design with Multiple Processors

Semiconductor device scaling creates tremendous opportunity for high density and high parallelism for SOC devices. The effectiveness of extensible processors both in traditional processor roles and as replacements for hardwired logic enables a design style in which large numbers of processors operate in parallel to efficiently implement the rich functions of the system. The mere availability of such small fast processors does not solve the system-design problem, however. We must consider how multiple processors can work together on complex problems.

3.4.1 Available Concurrency

The steady proliferation of digital electronics in computing, communication, and consumer applications is strongly tied to the steady progress of semiconductor device scaling. Electronic product performance benefits from improving the speed of individual transistors. The bigger benefit, however, derives from integration of large numbers of transistors together in each integrated circuit. Progress in digital electronics, then, depends on the ability of the chip or system designer to find ways to use many transistors, working in parallel, to more efficiently implement the system's functions. The designer can exploit concurrency at many different levels, but all of these levels can be reduced or generalized to three:

- **Bit-level parallelism:** Almost all digital systems operate on data (text characters, pixels, voice samples, network packet headers) with more than one bit of precision. Many transistors—many gates—operate in parallel to perform the basic operations on this data—add, shift, bitwise and, load, store, compare, and so forth. The density of today's silicon technology is already so high that we rarely need to think about the bit-level parallelism except when the natural computation is unusually wide. Both existing processors and logic design methods are well-equipped to deal with bit widths of 32 or 64 bits. However, some encryption and networking cases arise where the natural bit width of basic operations is much wider.
- **Operation-level parallelism:** Any complex function consists of a series of operations performed on groups of data values. Some of those operations are intrinsically dependent on one another and must be performed serially. In almost any function, however, some dependencies are somewhat looser, and it is theoretically possible to perform loosely dependent operations in parallel. The system designer may still choose to perform some nondependent operations serially, either to share the hardware or to simplify the computation. Microprocessors have traditionally served as a simple way to serially execute almost any digital function. Processor configuration and extension principally serve as a simple, structured way to increase operation-level parallelism.

  That operation-level concurrency can often be exploited by applying the same operator to a number of different operands at the same time for single-instruction multiple data (SIMD) parallelism. Additional operation-level concurrency can be exploited by executing several independent operations at the same time, using a very-long-instruction word (VLIW) processor architecture or superscalar processor implementation. Pipelined implementation of execution units also increases realized concurrency – combinations of dependent operations can form pipelined execution units, while independent sequences of operations flow down the pipeline in parallel with one another.
- **Task-level parallelism:** Most systems perform more than one essential function. These functions or subsystems may communicate intermittently with others, but they are sufficiently independent that their functions can be described and implemented separately from other functions in the system. The collection of communicating functions in a

system—user interface, video processing, audio processing, wireless channel processing, and encryption functions in a cell phone, for example—are sufficiently independent that the system user wants the effect or illusion of simultaneous parallel operation by all. Semiconductor device scaling enables the combining of all these functions into a single device, so task-level parallelism is increasingly important to the chip architect. Task-level parallelism is most easily exploited when it is already explicit in the functions of the system—for example, the audio and video subsystems obviously run in parallel. Task-level parallelism can also be extracted from applications originally developed as a single sequential task. No universal extraction method exists, but appropriate tools and building blocks help in discovery of hidden task-level parallelism.

3.4.2 Parallelism and Power

Parallelism also provides one key to power efficiency in SOC devices. Consider a CMOS circuit in which most of the power dissipation is active power. In this case, the power dissipation for the circuit is roughly

$$P \propto CV^2 f$$

where C is the switched capacitance of the circuit, V is the supply voltage, and f is the effective switching frequency of the nodes in the circuit. We can take advantage of the fact that the maximum operating frequency of the circuit is roughly proportional to voltage (a good assumption for typical digital CMOS circuits). If the function implemented in the circuit can be scaled through parallel logic implementation by some factor $s(>1)$, so that the circuit has more transistors (for more parallelism) by a factor of s, but can achieve the same throughput at frequency reduced by s we can reduce the operating voltage by roughly s as well, so we see a significant reduction in power:

$$P_{new} \propto (sC)\left(\frac{V}{s}\right)^2\left(\frac{f}{s}\right) = \frac{P_{original}}{s^2}$$

In practice, s cannot be increased arbitrarily because of the operation of CMOS circuits degrades as operating voltage approaches transistor-threshold voltage. Put another way, there's a practical upper limit to the size of s. On the other hand, the scaling of switched capacitance (sC) with performance may be unrealistically conservative. As shown in the EEMBC benchmarks cited earlier, small additions to the processor hardware, which slightly increase switched capacitance, often lead to large improvements in throughput. This simplified analysis underscores the important trend toward increased parallelism in circuits.

3.4.3 A Pragmatic View of Multiple Processor Design Methodology

In the best of all possible worlds, applications developers would simply write algorithms in a high-level language. Software tools would identify huge degrees of latent parallelism and generate code running across hundreds or thousands of small processors. Hardware design would be trivial too, because a single universal processor design would be replicated as much as necessary to balance cost and performance goals.

Unfortunately, this view is pure fantasy for most applications. Latent parallelism varies widely across different embedded systems, and even when parallelism exists, no fully automated methods are available to extract it. Moreover, a significant portion of the available parallelism in SOC applications comes not from a single algorithm, but from the collection of largely independent algorithms running together on one platform. Developers start from a set of tasks for the system and exploit the parallelism by applying a spectrum of techniques, including four basic actions:

1. Allocate (mostly) independent tasks to different processors, with communications among tasks expressed via shared memory and messages.
2. Speed up each individual task by optimizing the processor on which it runs. Typically, this process involves processor extension to create an instruction set and a program that performs more operations per cycle (more fine-grained parallelism).
3. For particularly performance-critical tasks, decompose the task into a set of parallel tasks running on a set of communicating processors. The new suite of processors may all be identical and operate on different data subsets or they may be configured differently, each optimized for a different phase of the original algorithm.
4. Combine multiple tasks on one processor by time-slicing. This approach degrades parallelism but may improve SOC cost and efficiency if a processor has available computation cycles.

These methods interact with one another, so iterative refinement is probably essential, particularly as the design evolves. As a result, quick exploration of tradeoffs through trial system design, experimental processor configuration, and fast system simulation are especially important. Chapter 4 discusses the refinement method in greater depth.

3.4.4 Forms of Partitioning

When we partition a system's functions into multiple interacting function blocks, we find several possible forms or structures. A quick overview of these basic types illustrates a few key partitioning issues.

- **Heterogeneous tasks:** As described above, many systems contain distinct, loosely coupled subsystems (e.g., video, audio, networking). These subsystems share only modest amounts of common data or control information and can be implemented largely independently of each other. The chief system-level design concern will be supplying adequate resources for the sum of all the requirements of the individual subsystems. Likely needs include memory capacity, bus bandwidth, and access from a system-control processor. In many cases, the required capacity can come from a common pool of on-chip resources to simplify design and encourage flexible sharing, especially if the resource needs are uncertain. Fig 3-13 shows one plausible topology for such a system. This system design assumes that networking, video, and audio processing tasks are implemented in separate processors, sharing common memory, bus, and I/O resources.

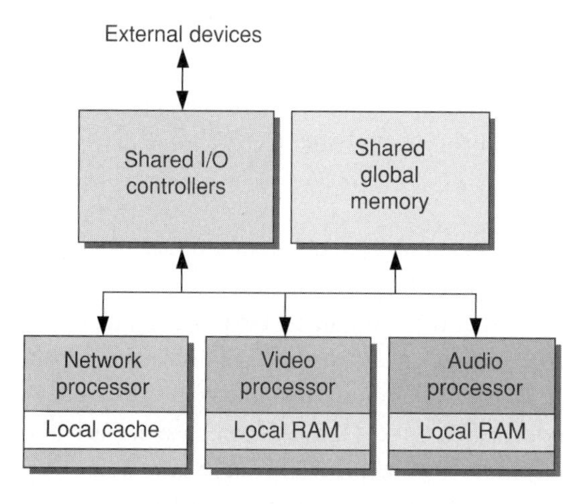

Figure 3-13 Simple heterogeneous system partitioning.

- **Parallel tasks:** Some embedded tasks are naturally and explicitly parallel. Communications infrastructure equipment, for example, often supports large numbers of wired communications ports, voice channels, or wireless frequency-band controllers. These tasks are easily decomposed into a number of identical subsystems, perhaps with some setup and management from a controller, as shown in Fig 3-14. These parallel processors may share common resources as long as the shared resource doesn't become a bottleneck. Even when the parallelism is not obvious, many system applications still lend themselves to parallel implementation. For example, an image-processing system may operate on a dependent series of frames, but the operations on one part of a frame may be largely independent of operations on another part of that same frame. Creating a two-dimensional array of subimage processors may achieve high parallelism without substantial algorithm redesign.
- **Pipelined tasks:** Some embedded system functions may require a long series of dependent operations that precludes use of a parallel array of processors as discussed above. Nevertheless, the algorithms can be naturally organized into phases, such that one phase of

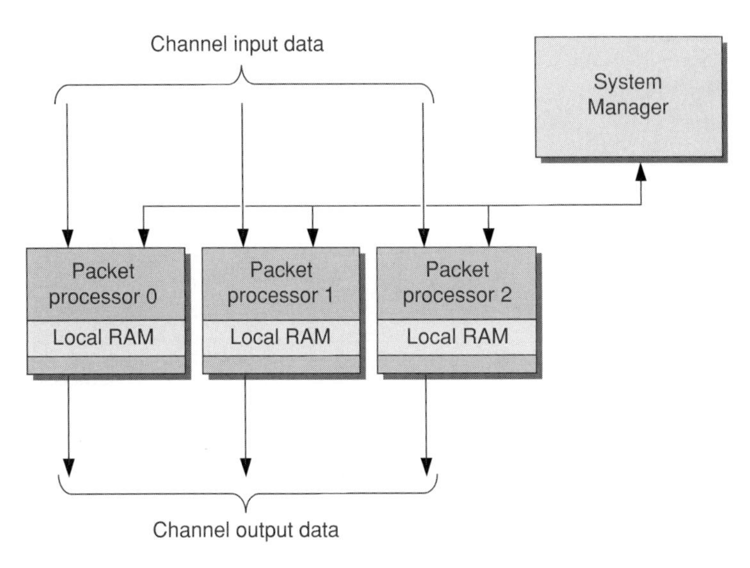

Figure 3-14 Parallel Task System Partitioning

the algorithm can be performed on one block of data while a subsequent phase is performed on an earlier block. (This arrangement is called a systolic-processing array.) Data blocks move down the pipeline of processors, and each processor in the pipeline is dedicated and optimized for a particular subsequence of the entire function.

Fig 3-15 shows an example of a systolic array for decoding compressed video. The Huffman decode processor pulls an encoded video stream out of memory, expands it, and passes the data through a dedicated queue to the inverse discrete cosine transform (iDCT) processor, which performs a complex sequence of tasks on the image block and passes that block through a second queue to a motion-compensation processor, which combines the data with previous image data to produce the final decoded video stream.

- **Hybrids:** The above three system-partitioning cases are unrealistically simple. Real systems often require a rich mixture of these partitioning styles. For example, one of the independent subsystems of Fig 3-13 might be best implemented in a group of parallel processors. Or perhaps the system manager in Fig 3-14 might be implemented by another processor as supervisor. The nature of the application and the ingenuity of the system architect drive the parallel system structure.

Complex systems may generate correspondingly complex topologies. The combination of optimizing the architecture of individual processors and optimizing system organization, including use of parallel processors, provides a new rich palette of possibilities for the system designer.

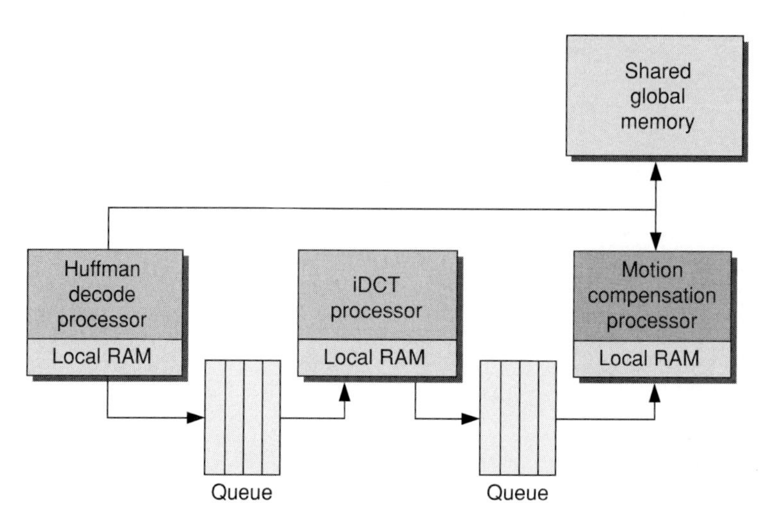

Figure 3-15 Pipelined task system partitioning.

3.4.5 Processor Interface and Interconnect

Full exploitation of the potential for parallel processors depends strongly on the characteristics of the processor's hardware interface and programming support for communication. Different applications make different demands on communication between blocks. Four questions capture the most essential system-communications performance issues:

1. **Required bandwidth:** To sustain the required throughput of a function block, what sustained input data and output data bandwidths are necessary?
2. **Sensitivity to latency:** What response latency (average and worst-case) is required for a functional block's requests on other memory or logic functions?
3. **Data granularity:** What is the typical size of a request—a large data block or a single word?
4. **Blocking or nonblocking communications:** Can the computation be organized so that the function block can make a request and then proceed with other work without waiting for the response to the request?

For extensible processors to be effective in a broad range of roles, replacing both traditional processors and traditional hardwired logic blocks, they must provide a basic vocabulary of communications interfaces that suit applications across these four dimensions. Fig 3-16 highlights the three basic forms of interface natural to processors tuned for SOC applications:

1. Memory-mapped, word-sized interface—typically implemented as a local-memory-like connection.
2. Memory-mapped, block-sized connection—typically implemented as a bus connection.
3. Instruction-mapped, arbitrary-sized connection—typically implemented as a direct point-to-point connection. Instruction-mapped connections can range from a single bit to thousands of bits.

Note that traditional processor cores, which evolved from standalone processors, typically provide only the block-oriented, general-bus interface. Low bandwidth, long latency, or inappropriate interface semantics would either degrade performance or complicate development for parallel applications. Each of these interfaces has particular strengths, as outlined in Fig 3-17.

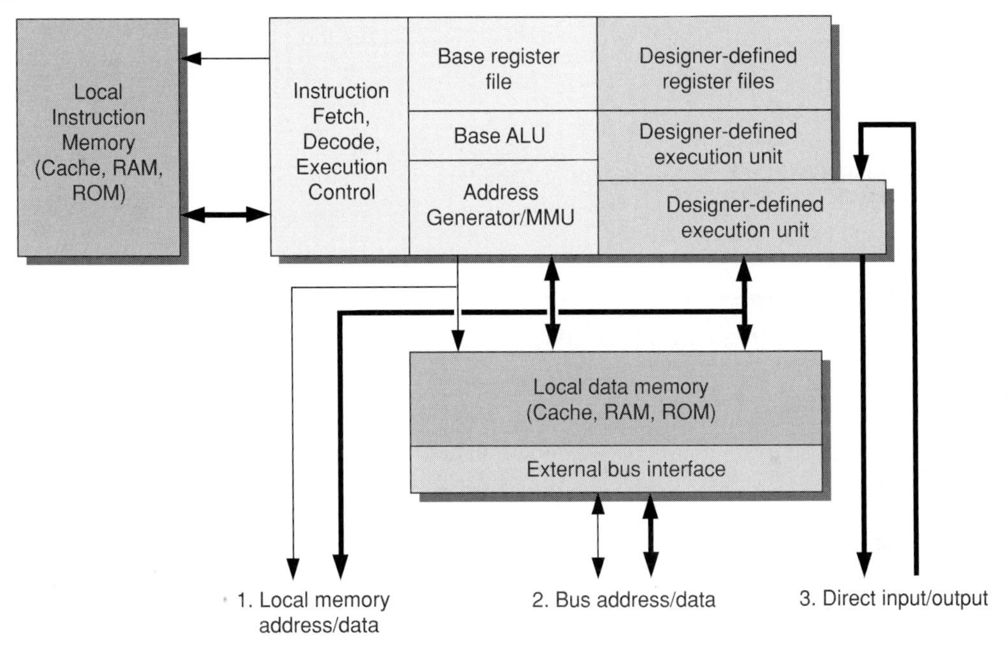

Figure 3-16 Basic extensible processor interfaces.

Interface Type	Interface Characteristics	Interface Uses
Local memory interface	Memory mapped, word oriented, fixed latency	• Word access to shared local memory • Word access to memory-mapped mailbox register or data queue
Bus interface	Memory mapped, block-data oriented, variable latency	• Block access to shared remote memory area • Bus access to another processor's local memory
Direct port interface	Instruction mapped, word oriented, dedicated connection	• Control and status input and output • Data operand transfers

Figure 3-17 Interface characteristics and uses.

3.4.5.1 Bus Interface

General bus interfaces are quite familiar to designers of microprocessor-based systems. These interfaces support a wide range of memory-mapped read and write operations, including both word transfers and block transfers for cache-line fills and dirty-cache-line write-backs. A processor typically uses this interface to access system wide shared resources such as main memory, I/O device-control registers, and other hardware functions. Because there is often contention for these shared resources—either directly at the resource or for any shared bus that connects these resources to the processor—the interface must support variable or indeterminate latency. Often, this interface will support split transactions (separation of data request and data response), bus errors (signaling to the processor that the request cannot be correctly satisfied), and transactions in support of cache coherency (hardware support for maintaining the illusion to software that each memory address holds exactly one value).

Normally, the memory and registers addressed over the bus are separate from the processors that access them. With a more sophisticated bidirectional multimaster bus interface, even the local resources of one processor become visible to other processors on the bus. A processor's private local data RAM, for example, can be made public and available as a source and destination of both I/O-device-to-processor traffic (direct memory access, or DMA) and processor-to-processor data traffic.

3.4.5.2 Local-Memory Interface

Memory hierarchies play an important role in high-performance embedded processors. The big performance gap between large, shared, off-chip memories and small, dedicated, on-chip memories makes it natural for processor pipelines to require that the first level of the instruction and data storage is closely coupled to the processor and has high bandwidth and a latency of roughly one processor clock cycle. These local memories may be structured as caches (data movement from off chip is automatic), or as local memory spaces (data movement from off chip is directed explicitly by software).

Even though these memories are accessed through optimized private interfaces, these interfaces can be extended to support other high-bandwidth, low-latency, memory-mapped functions. A local interface can support the same range of data-word sizes as the private memory (8, 16, 32, possibly 64 and 128 bits, or wider). This local-memory interface assumes the same quick access time (typically one cycle) as the private memory, but often includes some stall mechanism to allow arbitrated access to a shared resource.

This high speed makes a local interface suitable for functions where high-bandwidth and low-latency data access is required. In some cases, random-access shared memories are connected to this local-memory interface. In other cases, the local-memory interface will be used to connect to a queue, the head or tail of which is mapped into a particular memory address. In some cases, memory-mapped system registers representing the control or data interface of some external function are connected to the local-memory interface.

3.4.5.3 Direct Connection

The unique characteristics of extensible processors and on-chip interconnect create the opportunity for third, fundamental form of interface—direct port connections. This interface exploits two phenomena. First, the intrinsic on-chip bandwidth of deep submicron semiconductor devices is surprisingly huge. For example, in a six-layer-metal process, we might expect that two layers will be available in the X dimension and two layers in the Y dimension for global wiring. In 90nm technology, we can expect 500MHz processor clock frequencies and roughly four metal lines per micron per metal layer.

Therefore, the theoretical bandwidth at the cross-section of a 10mm x 10mm die is roughly:

$$BW_{\text{cross-section}} = \left(500 \bullet 10^{6}\,\text{Hz}\right) \times \left(4\ \text{lines}/\mu\text{m}\right) \times \left(2\ \text{layers}\right) \times \left(10\text{mm}\right) \times \left(10^{3}\,\mu\text{m/mm}\right)$$

$$= 4 \bullet 10^{13}\ \text{bits/sec}$$

$$= 40\ \text{terabits/sec (Tbps)}$$

This high theoretical bandwidth highlights the potential for parallel on-chip processing, even for applications with very high bandwidth demands between processing elements. Note that 40 Tbps is much greater than the bus bandwidth or internal operand bandwidth of any conventional, general-purpose 32-bit or 64-bit processor core. The fact that conventional processors use only a small fraction of this available bandwidth motivates the search for alternatives that can better exploit this potential.

Second, extensible processors support wide, high-throughput computational data paths that vastly increase the potential performance per processor. These processors can be configured with pipelined data paths that are *hundreds* or, in extreme cases, *thousands* of bits wide. Consequently, each data path can potentially produce or consume several operand values per cycle, so the total data appetite for a single extensible processor can scale to literally terabits per second. If only a fraction of this bandwidth is consumed by processor-to-processor communications, only a few dozen processors are required to use the available 40 Tbps of cross-section bandwidth. It's therefore feasible to exploit the unique bandwidth characteristics of advanced semiconductor technology using multiple processors with high-bandwidth interfaces.

To best exploit the available bandwidth, direct processor-to-processor connections are not mapped into the address space of the processor. Instead, processors send or receive these data values as a byproduct of executing particular instructions. Such communications are *instruction-mapped* rather than memory-mapped because the processor-to-processor communications are a consequence of executing an instruction that is not an explicit load or store instruction. Conventional processors use inputs from processor registers and memory and produce outputs in processor registers and memory. Direct processor-to-processor connections allow a third category of source or destination, potentially used in combination with register- and memory-based sources and destinations. Four important potential advantages characterize direct connections versus memory-mapped connections:

- The operational use of the direct-connection interface is implicit in the instruction so, additional instruction bits are not required to explicitly specify the target address of the transfer.
- Neither an address calculation nor a memory load/store port is required, saving precious address and memory bandwidth.
- Transfer sizes are not limited to the normal power-of-two memory word widths, allowing both much larger and application-specific transfers.
- Because transfers are implicit in instruction execution, an unlimited number of parallel transfers can be associated with a single instruction.

Two forms of direct-connect interface—operand queues and control signals—are discussed in detail in Chapter 5, Section 5.6. This expanded vocabulary of basic interface structures offers two fundamental capabilities to the system architect:

1. Richer, wider, and faster interfaces among small processors, which allow higher bandwidth and lower latency among groups of processors on a chip.
2. The combination of both flexible memory-mapped and flexible instruction-mapped interfaces makes a broader range of communications functions directly available to the software.

Together these interface structures create a basic, adaptable computing fabric that serves as a foundation for small, fast, programmable SOC designs.

3.4.6 Communications between Tasks

To effectively use multiple processors, software developers must write programs that communicate appropriately. Several basic modes of communication programming have emerged to handle a variety of circumstances. In some cases, the choice of programming model depends on the data-flow structure among the tasks that comprise the total application. In other cases, the choice depends on the underlying hardware used to implement the function. There are three basic styles of communication between a task producing data on one processor and a task consuming that data on another processor: shared memory, device drivers, and message passing.

3.5 New Essentials of SOC Design Methodology

We have now covered all the basic pieces of the advanced SOC design methodology:

- How extensible processors accelerate traditional software-based tasks.
- How extensible processors flexibly implement hardware-based tasks.
- How to connect multiple processors together to create very sophisticated SOCs.

The emergence of multiple-processor SOCs using configurable and extensible processors as basic building blocks significantly affects three dimensions of the SOC design flow:

1. **Partitioning:** In the traditional SOC-design methodology, architects use two types of functional building blocks: processors with modest performance running complex algorithms in software and hardware with modest complexity performing tasks at high speed. The new design vocabulary of extensible processors gives designers more flexible choices, allowing highly complex functions to be efficiently implemented for the first time. Moreover, the generation of new processor variants is so fast that architects can change their partitioning easily in response to new discoveries about overall system design efficiencies and new market requirements. Greater partitioning choice and greater repartitioning ease significantly expand SOC design horizons.
2. **Modeling:** When the core of the SOC design consists of both RTL blocks and processors, much of the modeling must be done at the lowest-common-denominator level of RTL simulation. In the absence of special hardware accelerators, software-based RTL simulation of

a large subsystem may only achieve hundreds of cycles per second. A complex cluster of processor and RTL logic might run at simulation speeds of only a few tens of cycles per second. By contrast, a fully cycle-accurate extensible-processor instruction-set simulator runs at hundreds of thousands of cycles per second.

A simulation of a complex processor cluster of would routinely sustain tens of thousands of cycles per second on such an instruction-set simulator. This three-order-of-magnitude performance gap means that SOC design teams can run entire software applications to verify their system designs instead of being restricted to running some simple verification tests on a slow HDL simulator. In addition, processor-centric design also makes it far easier to convert early, abstract system-level models written in C or C++ into the final SOC system implementations. Architects can directly evolve early system-level behavioral models into final implementation of task software and configured processors.

3. **Convergence to implementation:** In the traditional SOC design flow, growth in design complexity can severely degrade progress toward a final verified implementation. A small change in requirements can cause major redesign of RTL in one block. Changes to one block often propagate to adjacent blocks and throughout the design. Each block change requires reverification and potential reassessment of the cost, performance, and power consumption of the entire set of blocks. Consequently, much iteration is sometimes required to restabilize the system design after a supposedly "simple" change.

   When extensible processors are used as a basic building block, changes are more easily tried and verified via software. The interface methods are more structured and more rapidly modeled, reducing long redesign cycles. Even if significant repartitioning is required, design elements in the form of software algorithms and processor configurations can be migrated transparently from one processor block to the next. In addition, all the building blocks, whether used as control processors or RTL-replacement engines, are generated in the same portable form from the same tool set. This approach brings uniformity to the VLSI design processor and accelerates floor planning, synthesis, cell placement and routing, signal integrity, design-rule checking, and tape-out. Similarly, with all software built for a single family of compatible processors under a common development environment, development, testing, and packaging of the embedded software suite is unified and simplified.

3.5.1 SOC Design Flow

Compare the advanced SOC design flow in Fig 3-18 to the more traditional SOC flow sketched in Fig 3-14. The fundamental inputs to the flow are SOC interface and functional requirements. The essential input/output interface specification includes physical layer electrical specifications and higher-level logical specifications, including protocol conformance. The essential computational requirements include both specification of the algorithms that relate inputs to outputs and the temporal constraints on algorithmic behavior, including maximum latency, minimum bandwidth, and other "quality of service" characteristics.

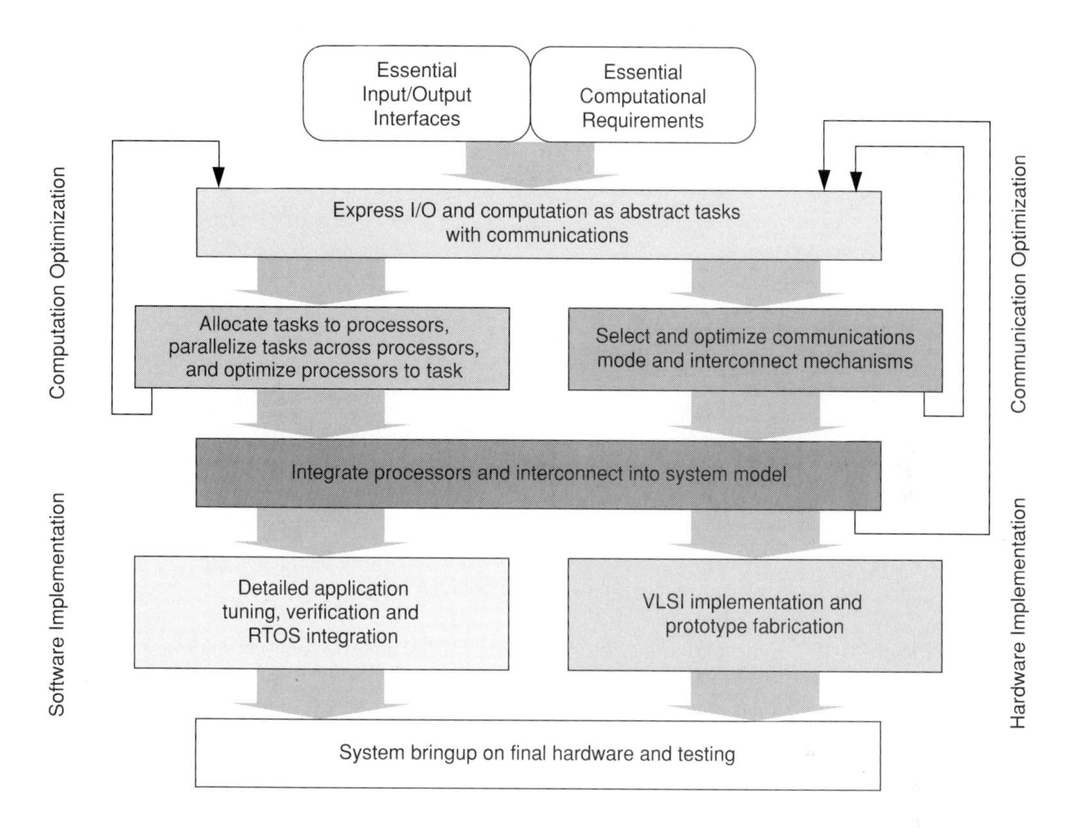

Figure 3-18 Advanced SOC design process.

Within the overall flow are four major subflows: computation optimization, communication optimization, software implementation, and hardware implementation. Almost all feedback loops are found within these blocks (the details are discussed in Chapter 4). Optimization of computation involves finding the right number and type of process (and where necessary, other blocks) to implement the complex nest of computational requirements of the system. Communication optimization focuses on providing the necessary communication software programming model and hardware interconnects among processors and to memories and input/output interfaces, to enable low-cost, low-latency communications, especially for the most critical dataflows in the system. This flow relies on two important principles. First, early system-level simulation is a key tool for creating detailed designer insight into function and performance issues for a complex system architecture. Second, rapid incremental refinement from initial implementation guesses into final application, processor configuration, memory, interconnect, and input/output implementations lets the design team learn from cost and performance surprises and take advantage of new insights quickly. Chapter 4 outlines the issues and describes the recommended system design flow in detail.

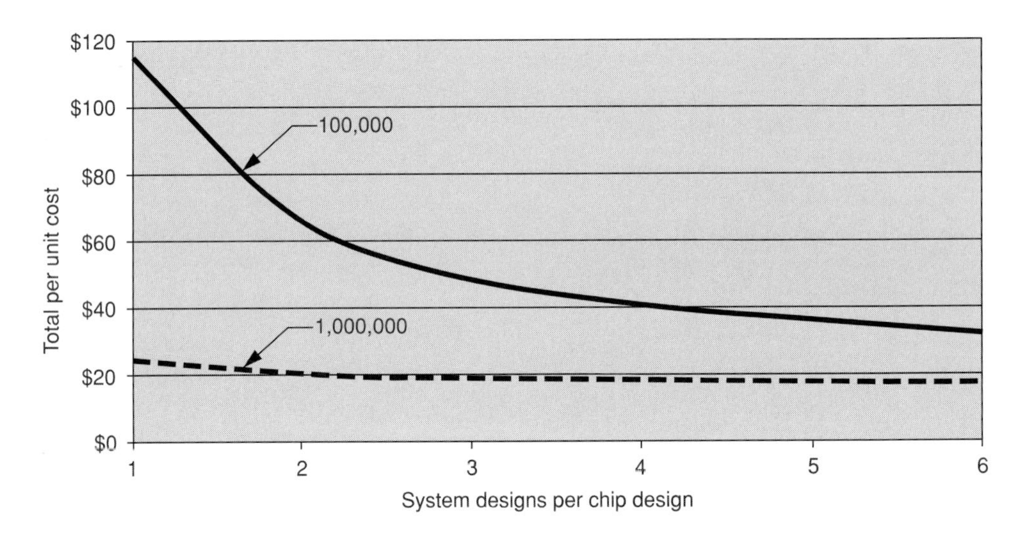

Figure 3-19 Example amortized chip costs (100K and 1M system volumes).

Crucially, this design flow does not make new demands on total design resources or mandate new fundamental engineering skills. Experience with a large number of design teams suggests that both traditional RTL designers and embedded software developers can be comfortable and effective in the design of new processor configurations and application-specific instruction sets using processor-generation tools. Familiarity with C and Verilog are both useful foundations for effective and efficient use of the processor-centric SOC design method. The flow closely depends on the expanded role of configurable processors. Availability of these more flexible SOC building blocks mean fewer design bugs, fewer hardware design iterations, wider exploration of block-partitioning options, earlier block software testing and validation, and in all likelihood, earlier product availability.

3.5.2 The Essential Phases of the New Flow

Every complex SOC design project is different, but the new SOC design flow shown in Fig 3-18 highlights five basic process steps. The SOC design process is intrinsically interdependent, so design teams may perform some of these steps in parallel and may iterate back to an early step after making key observations in the course of later design steps. Nevertheless, this design flow leverages the inherent flexibility of the extensible processor building block and the inherent development productivity of a software-centric design method:

1. **Develop a functional model of system essentials**—Starting from a reference software implementation of key tasks or from an algorithmic description of mission-critical functions, develop a C or C++ program that captures the most performance-demanding or most complex set of system activities. In practice, the architect typically understands that many of the system's functions are quite generic and make few unique demands on processor features, I/O capability, interconnect bandwidth, or memory resources. Only those features expected to drive key design decisions need be modeled at this stage.
2. **Partition the system into multiple processors**—Identify obvious task-level parallelism in the underlying system structure. Decompose tasks with multiple phases into separate pipelined tasks with data communications between tasks. Subdivide each task into parallel computations, where each processor works on a subset of the datastreams in the system. Compile and run the core task or tasks on the baseline processor to establish a reference performance level and to identify major computational bottlenecks. Refine the system model using multiple processors running various software tasks with communications among the processors.
3. **Optimize each processor for its respective task**—For each processor, profile the task performance and identify hotspots within each task. Configure each processor's memories, system peripherals, and interfaces and describe or select instruction-set extensions. Update area and power estimates. Regenerate the software tools and simulators using the new processor extensions. Update applications to utilize more efficient data types and application-specific functions. Refine the processor configurations until the system design meets the per-task cost, power, and performance goals.
4. **Verify system-level cost, performance, and function**—Reintegrate each of the processor models and optimized tasks into the system-level model. Confirm that major partitioning and architecture choices are appropriate. Simulate the system to measure key performance characteristics. Check adequacy of shared resources (memory capacity, interconnect bandwidth, total silicon area, and power budget). Deploy software-development tools and the system model to final application developers.
5. **Finalize the implementation**—Use the final RTL for each processor, the generated memories, and other logic to floor plan, synthesize, place, and route each block. Stitch together global clocks, debug and test scan-chain hardware and the power infrastructure. Run full-chip logic, circuit, and physical design checks. Release to fabrication. Meanwhile, prepare the final application and system software on the simulated or emulated system model. Perform initial system power-up testing and release the design to final system testing and production.

3.6 Addressing the Six Problems

Experienced SOC designers implement many variations of the above design flow but the development history of many advanced multiple-processor SOCs supports this basic flow. Configurability expands the role of processors in SOC design and enables a new SOC design flow. The advanced SOC design method advocated here appears to deliver on the promise of significantly easing the six key problems of existing SOC design methods, as introduced in Chapter 2:

1. Changing market needs
2. Inadequate product volume and longevity to support a single-use SOC design
3. Inflexibility in the semiconductor supply chain
4. Inadequate SOC performance, efficiency, and cost
5. Risk, cost, delay, and unpredictibility in design and verification
6. Inadequate coordination between hardware and software design teams

3.6.1 Make the SOC More Programmable

Using this new SOC design method, processors implement a wide range of functions: both traditional software-based functions and traditional hardware-based functions. Extensible processors used as basic building blocks bring greater programmability to all these functions. In traditionally hardwired function blocks, control from high-level-language applications running from RAM displaces finite-state-machine controllers implemented with logic gates. When replacing traditional, processor-based functions, extensibility increases performance headroom and allows incremental features to be incorporated without hitting a performance ceiling.

The processor-centric SOC design approach increases the adaptability of the design at every stage of its lifecycle: during early definition, during detailed implementation, at product launch, and for subsequent generations of end products. Increased SOC adaptability gets the product to market earlier and more profitably and increases the likelihood that the design remains the right product as market requirements evolve.

3.6.2 Build an Optimized Platform to Aggregate Volume

Inadequate volume is the bugbear of SOC economics. SOC integration can easily drive down electronics manufacturing costs compared to board-level integration, but design and prototyping costs can be significant, especially for designs with less than a million units of volume. The intrinsic flexibility of designs using programmable processors as their basic fabric increases the number of systems that can be supported by one chip design.

Fig 3-19 shows a simple model of the total chip cost, including amortized development costs, and the impact of increasing the number of systems design supported by one SOC design. The model assumes $10,000,000 total development cost, $15 chip manufacturing cost, and shows results for 100,000-unit and 1,000,000-unit production volumes. This model also assumes an incremental chip cost (5%) for the small overhead of programmability required to allow the SOC to support more than one system design. As Fig 3-19 shows, programmability is useful for all SOCs and is economically essential for lower volume SOCs.

3.6.3 Use Portable IP Foundations for Supply Leverage

The globalization of markets and the emergence of silicon foundries increase the number of silicon supply options available to chip designers. SOC designers working at systems companies have many ASIC suppliers from which to choose. They also have the option of designing the complete SOC themselves using a customer-owned-tooling (COT) VLSI implementation and a silicon foundry. Semiconductor companies are generally moving away from operating their own fabs and are moving to fabless models, to joint venture fabs, or to a mixture of operating models. Flexibility in fabrication choices offers pricing leverage and ensures better fab availability to the chip designer. Making the chip design portable increases the ease of moving between fabs and increases the longevity of the design over process generations.

Portable SOC designs are less silicon-efficient than hand-optimized, fab-specific designs, but automated tools (including both logic compilers and processor generators) generate these designs much more cheaply and quickly. A faster, easier design process lets the designer tune the chip more thoroughly and allows a much wider range of applications to be powered by optimized chips. Application-directed architectures, even in portable design form, are generally much more efficient than more generic designs that rely solely on circuit optimization. Use of extensible processors as basic building blocks complements the leverage associated with process portability. Each programmable block is generally reusable at the design level and aids the generality and reusability of the design as a whole.

3.6.4 Optimize Processors for Performance and Efficiency

The impact of using application-specific processors on data-intensive tasks is usually significant and sometimes dramatic. Five-fold to ten-fold improvement over standard RISC architecture performance is routine. Designers can commonly speed up applications with especially high bit-level and operand-level parallelism by 50 or 100 times using the processor-centric SOC design approach. Application-specific processor architectures sharply reduce the required number of processing cycles required for a given level of application throughput.

In most cases, the expanded configuration has somewhat increased power dissipation on a per-cycle basis but the actual, overall power consumption for any given application drops with cycle count. In fact, a lower operating frequency allows the necessary performance to be reached at lower operating voltage too. As described earlier, this frequency reduction leads to better than linear improvement in real power efficiency. This kind of performance headroom and power efficiency, combined with easy programmability, translates directly into improved end-product attributes, including greater functional capabilities, longer battery life, and lower product cost.

3.6.5 Replace Hard-wired Design with Tuned Processors

No modern digital design is bug-free—large systems are simply too complex to avoid all flaws. The designer's real goal is to reduce the number and severity of bugs and increase the speed and flexibility of fixes. The more a system's functions can be put in soft form—dynamically loadable during operations or upgradeable at system power-on—the more likely it is that bugs can be fixed without a silicon respin. Just as FPGAs offer fast, cheap upgrades and patches to basic logic functions, the RAM-based code in processor-centric SOCs enables rapid fixes for these more complex, high-volume electronics platforms. In addition, processor-based design offers faster and more complete modeling of the system functions, reducing the number of potential bugs that need fixing. The net result is a higher rate of first-time-functional chip designs and greater longevity for those designs in the marketplace.

3.6.6 Unify Hardware and Software with Processor-Centric SOC Methodology

The hardware and software views of a complex SOC commonly diverge. Three aspects of the new SOC approach mitigate this tendency. First, the development tools allow rapid creation of an early, inexact but functional model of key system behavior. This reference model removes much ambiguity from the early specification and allows gradual refinement of both the software-based functions and the underlying multiple-processor hardware platform.

Second, the description of a system as a set of C/C++ tasks, a set of processor configurations, and the interconnections among the processors constitutes a more abstract, portable, and comprehensible description than the more traditional ad hoc collection of programs, Verilog–coded logic functions, and other block-boxes. Hardware and software developers and system architects can understand and share this system representation and manipulate the representation using a common set of tools that offer both hardware- and software-oriented views.

Third, the integration of subsystems—the Achilles heel of many complex projects—is naturally organized around a processor-based perspective. Integration does not require welding together two innately dissimilar forms of matter (code and gates) but rather involves composition of a set of components of a closely related nature (configured processors with associated task software). The result is smoother integration with fewer communications troubles and fewer surprises late in the design cycle.

3.6.7 Complex SOC and the Six Problems

This new SOC methodology is not a panacea, but it is an effective tool. A more detailed discussion of processor-centric SOC design methods appears in Chapters 4, 5, 6, and 7. Chapter 8 looks to the long-term implications of this new system-design style, projecting deep changes in both the evolution of SOC design and the structure of the industry that creates new chip designs.

3.7 Further Reading

- A quick introduction to Tensilica's Xtensa architecture and basic mechanisms for extensibility can be found in the following: Ricardo Gonzalez, "Configurable and Extensible Processors Change System Design." In *Hot Chips 11*, 1999. The Tensilica Web site (*http://www.tensilica.com*) also offers a wealth of information on the technology.

The following articles are particularly relevant to the history of processor description and tool automation:

- G. Bell and A Newell. "The PMS and ISP descriptive systems for computer structures." In *Proceedings of the Spring Joint Computer Conference*. AFIPS Press, 1970.
- The GNU C Compiler emerged in the 1980s as the most widely visible effort to create an architecture-neutral compiler technology infrastructure. Its goal was simplified, and it was not designed for fully automated support of new processor architectures. See *http://gcc.gnu.org*.
- J. Van Praet, G. Goosens, D. Lanner, and H. De Man, "Instruction set definition and instruction selection for ASIP." In *High Level Synthesis Symp.* pp. 1116, 1994.
- G. Zimmermann, "The Mimola design system—A computer-aided digital processor design method." In *Proceedings of the 16th Design Automation Conference*, 1979.
- H. Corporaal and R. Lamberts, "TTA processor synthesis." In First Annual Conference of ASCI, May 1995.
- S. Pees, A. Hoffmann, V. Zivojnovic, and Heinrich Meyr. "LISA—Machine description language for cycle-accurate models of programmable DSP architectures." In *Proceedings of the Design Automation Conference*, 1999.
- Marnix Arnold and Henk Corporaal. "Designing domain specific processors." In *Proceedings of the 9th International Workshop on Hardware/Software Codesign*, pp. 61–66, Copenhagen, April 2001.
- John R. Hauser and John Wawrzynek. "Garp: A MIPS processor with a reconfigurable coprocessor." In *IEEE Symposium on FPGAs for Custom Computing Machines*, pp. 12–21, 1997.
- Alex Ye, Nagaraj Shenoy, and Prithviraj Banerjee. "A C compiler for a processor with a reconfigurable functional unit." In *Proceedings of the 8th ACM International Symposium on Field-Programmable Gate Arrays*, pp 95–100, February 2000.

These two articles concentrate on instruction-set description for architectural evaluation rather than for actual hardware generation:

- George Hadjiyiannis, Silvina Hanono, and Srinivas Devadas. "ISDL: An instruction set description language for retargetability." In *Proceedings of the Design Automation Conference*, 1997.
- Stefan Pees, Andreas Hoffmann, Vojin Zivojnovic, and Heinrich Meyr. "LISA machine description language for cycle-accurate models of programmable DSP architectures." In *Proceedings of the Design Automation Conference*, 1999.

Some degree of configurability has become popular in commercial processor and DSP cores. The following vendor Web sites describe configurable features:

- ARC International's ARCTangent processors and ARChitect tools are described on the company's Web site: *http://www.arccores.com*.

- MIPS Technologies' features for user-modification of processor RTL are described on the company's Web site: *http://www.mips.com.*
- CEVA's CEVA-X DSP architecture offers the option of user-defined extensions. See *http://www.ceva-dsp.com/*

HP Labs researchers have written a number of papers on processor generation, including the following:

- Shail Aditya, B. Ramakrishna Rau, and Vinod Kathail. "Automatic architectural synthesis of VLIW and EPIC processors." In *International Symposium on System Synthesis*, pp 107–113, 1999.
- J. A. Fisher, P. Faraboschi, and G. Desoli. "Custom-fit processors: letting applications define architectures." in *29th Annual IEEE/ACM Symposium on Microarchitecture (MICRO-29)*, pp 324–335. 1996.

The exploration of application-specific processors has also included notions of field-configurable processor hardware:

- Rahul Razdan and Michael D. Smith. "A high-performance microarchitecture with hardware-programmable functional units." In *Proceedings of the 27th Annual International Symposium on Microarchitecture*, 1994.
- Michael J. Wirthlin and Brad L. Hutchings. "DISC: The dynamic instruction set computer." In *Proc. SPIE*, 1995.

A more detailed analysis of the impact of parallelism on power efficiency can be found in A. P. Chandrakasan, S. Sheng, and R. W. Brodersen, "Low-power CMOS digital design." In *IEEE Journal of Solid-State Circuits*, 27(4) April 1992, pp 473–484.

Application benchmarking stimulates rich debate.

- Historically, the most widely cited (and most commonly misused) benchmark used on embedded processors is Reinhold Weicker's Dhrystone benchmark. R. P. Weicker, Dhrystone: A synthetic systems programming benchmark. Communications of the ACM, 27:1013–1030, 1984.
- The EEMBC benchmark suite has become the most widely used modern benchmark for embedded processors. The benchmarks span consumer (mostly image processing), telecom (mostly signal processing), networking (mostly packet processing), automotive, office automation (text and image rendering), Java and 8-/16-bit microcontroller applications. See http://www.eembc.org for a more comprehensive discussion and results.

CHAPTER 4

System-Level Design of Complex SOCs

This chapter drives to the heart of the SOC design challenge: How can you use multiple processors to implement complex system functions? To build this type of concurrent system, you must work both at the architectural level—identifying the best partitioning and communication among subsystems—and at the implementation level—creating the right interfaces, task software, and processor definitions to realize an efficient parallel system.

The concurrent architecture challenge is twofold:

- How to best take advantage of overt concurrency. For example, where is concurrence best applied in the independent processing of a media stream containing both audio and video components?
- How to identify and extract concurrence within tasks that are less overtly parallel. For example, how should the design spread the video-stream processing across a set of processors that each work on a sub-image?

Similarly, the implementation challenge has two components:

- How do the tasks, implemented as software running on the subsystems, communicate with one another reliably and with the required efficiency?
- How should global interconnect, shared memories, and sub-system interface hardware be designed to enable appropriate high-bandwidth, low-latency, low-cost communication?

The issues of a communications programming model, interconnect architecture, and overall system integration recur throughput this chapter. Chapters 5 and 6, in turn, drive down to the next level of detail: how to use extensible processors to implement these subsystems from a software developer's or a hardware designer's perspective.

This chapter combines the hardware and software view and explicitly builds connections between a hardware architect's key concerns—chip cost, power dissipation, clock frequency, design hierarchy, and logical verification—and the software architect's key concerns—programming model, development tools, application throughput, real-time system response, code and data footprint, and upgradeability. While the use of multiple task-specific processors as basic building blocks eases the conflict between these perspectives, tradeoffs still persist. Understanding and reducing these tradeoffs is an ongoing theme in this chapter.

Why is concurrency so central to SOCs for embedded applications? Almost any logical set of tasks could be performed serially on a single processor, often ignoring the high degree of latent concurrency in the set of tasks. A more parallel design, however, exploits concurrent task execution to both reduce application latency (by overlapping operations) and to improve energy efficiency (by permitting smaller, lower-power circuits to achieve the necessary throughput). Whether the potential latency reduction and efficiency improvement are achieved, however, depends on choices made in the design of intertask communications. The architect must address a series of questions in the course of design:

- Where is concurrency significant and obvious, where is it significant but not obvious, and where are the tasks inherently serial?
- Where concurrency exists, how sensitive is the overall system latency and throughput to the latency and throughput of the communications among the tasks that make up the overall system?
- How stable and well-understood is the application? Does substantial uncertainty remain in the individual tasks, in the pattern of communication among tasks, or both?
- What team design and skill legacies should the architect work to accommodate? Are there major hardware blocks, programming models, predefined applications, or interconnect structures that can (or must) be reused?
- How much design headroom is required with the evolving design specification?

This book systematically advocates a particular SOC design approach—centered on using programmable processors (and associated software) as the basic, logical building blocks to improve productivity and flexibility and maximize design success. Even so, complex SOCs are not built solely from processors. I/O peripherals, functions implemented on other chips, blocks of on-chip RTL-based logic and, most especially, memories also play important roles. Therefore, the SOC design problem is not just one of how to design, interconnect, program, and verify processors—the real issue is how to best combine all types building blocks into an efficient and complete whole.

Processors play lead roles in this collection of different building-block types for two reasons. First, the processor-based SOC design method moves much of the system's computational functions into processors instead of hardwired logic. The result is an increase in the number of processors on the chip, especially for data-intensive roles. Second, software running on the processors naturally controls all the other functions on the chip: software device drivers manage the peripherals; software routines initialize and control hardware logic functions; processors use the memories for instruction storage, for I/O data buffering, and for intermediate computation results. With processors in such central roles, the issues of how processors interact with other building blocks are crucial to the overall design flow.

This chapter explains the new SOC design methodology. The first two sections discuss the principles of advanced SOC system architecture—concurrency and system/task organization and the major architectural decisions for processor-centric SOC design. The next two sections discuss communications: the software view of communications—message passing, shared memory, applications programming interfaces (APIs), and hardware interconnect mechanisms—buses, crossbars, queues, and other implementation structures. The fifth section presents performance-driven design methods and modeling, including examples. The sixth section pulls it all the together as a recommended SOC design flow. The seventh section discusses the role of other building blocks in SOC design: memory, I/O, and hardwired logic. The final section traces the evolution and implications of processor-centric SOC architecture.

4.1 Complex SOC System Architecture Opportunities

Two distinct challenges drive the movement toward more parallel architectures for SOC designs.

First, the electronic systems implemented by SOCs are becoming more complex. A greater diversity of functions—networking, video processing, audio, wireless communications, and security—must be combined to deliver competitive functionality and compelling user experience. These attributes generally make quite different demands on the underlying computation and communication structure, both in the fundamental types of computation—for example, signal processing, packet processing, or image manipulation—and in the degree of concurrency.

Many systems explicitly support a number of parallel channels. Communications-infrastructure products—cellular base stations, network switches, and voice-over-IP concentrators—are often built around large numbers of independent channels. In these cases, concurrency (or the illusion of concurrency) is intrinsic to the product concept. In fact, the traditional role of the systems designer was often to find ways to efficiently time-multiplex the large number of intrinsically parallel tasks onto a serial-processing architecture. Potential concurrency is typically easy to identify in these systems, though the implementation still offers challenges. The growing capacity of semiconductor devices reduces the pressure to share hardware among intrinsically parallel functions and creates new opportunities to use concurrency to deliver more throughput across multiple tasks.

Second, parallel implementation of a single task may be important to achieving the latency, throughput, or efficiency demanded by customers. In this case, concurrency is a means to an end rather than an end in itself. Fortunately, most algorithms can be implemented in a variety of ways, including versions that expose significant degrees of concurrency, even for individual tasks.

The process of uncovering concurrency revolves around identifying dependencies in the algorithm—stages where the results from one part of the computation are required as inputs to another part. Complex algorithms often show both *fine-grained* dependencies—where a primitive computation or low-level control decision depends on a closely preceding computation or control decision—and *coarse-grained* dependencies—where an entire phase or section of the algorithm depends on results of an earlier phase or section. Sometimes the dependency is a simple implementation artifact—it is an accident of coding rather than a necessity of function.

For example, an algorithm might process independent blocks of data in arrival order, even though the overall application makes no explicit demands on ordering and no data is communicated from one block to the next. These incidental dependencies are often easy to eliminate.

Sometimes, however, the dependencies are more deeply entrenched. Then the process of parallelization involves understanding the dependencies in detail and finding places in the algorithm where computations or phases can be separated with a minimum of intrinsic data or control dependencies between them. The variety of possible partitionings and the subtlety of control and data dependencies make task parallelization complex.

4.1.1 The Basic Process of Parallel Design

Regardless of the source of concurrency—explicit and obvious in the system definition or painstakingly extracted from an originally sequential algorithm—the form of concurrency and the requirements of the product drive key system architecture decisions. These decisions can be grouped into three overriding design tasks:

1. **Partitioning:** How is the application best divided into a set of communicating tasks?
2. **Block design:** What form of underlying hardware block or software algorithm on a processor should be used to implement each task?
3. **Communication design:** What communication channel is required between blocks?

These three design tasks interact strongly. The form and granularity of partitioning depend on the relative performance of different kinds of hardware blocks (general-purpose processors, application-specific processors, and hardwired logic blocks) and the latency, bandwidth, and cost of communications links between the blocks. Moreover, these three design tasks may be applied recursively to decompose a partitioned task into yet smaller communicating sub-tasks.

The design of one block depends on the tasks mapped onto it, the design of surrounding blocks, and the demands of communications links. The design of communications channels depends on the capabilities of the individual processors; worst-case latency and bandwidth requirements; the communication model of the system application; and other cost, power, and scalability goals.

Moreover, any uncertainty in the set of tasks deeply influences the design. The chip architect must take this uncertainty into account in all design decisions, especially if the overall system application is subject to change or if individual tasks are evolving rapidly. Robustness in the face of change may come at a price in terms of hardware cost, system efficiency, and design time, but may improve the overall project and product success. A little bit of extra computational headroom or communication capacity can save a design project when the spec changes late in the game or when design bugs are discovered and must be fixed quickly to meet shipment deadlines.

4.1.2 The SOC as a Network of Interacting Components

The SOC plays a dual role in the hierarchy of system-design abstractions. On one hand, the entire SOC—along with its dedicated off-chip support (memories, analog communication interfaces, and other support chips) generally serves as one node in a larger system of communicating elements. This role as a node is obvious for an SOC that forms the heart of a network router, line-card, or home gateway box. The system architecture and software functionality are closely tied to the function of the larger distributed system. The role of SOC as node is sometimes equally valid for a digital camera or a PDA that connects to a PC or the Internet only intermittently. In all these cases, the SOC implements a range of communication-based functions and must be designed and verified in the context of a complex web of standards and product requirements for cost and bandwidth.

Fig 4-1 shows an SOC with its support infrastructure of human interface devices and storage—*the immediate system*—as part of a larger network environment, including Internet infrastructure and a Web server with remote application data—*the extended system*.

On the other hand, the SOC itself can also be viewed as a network of interacting nodes, where each node performs some computation, storage, or external interface task. The definition of some of those nodes is explicit in the role of the SOC in the larger system context. For example, the Ethernet, IEEE 802.11 WiFi, and DSL modem functions may be quite explicit in the definition of a home-gateway SOC. In other cases, the definition of some nodes is a by-product of the particular architectural partitioning and implementation chosen by a design team.

For example, one candidate design may require off-chip DRAM and an associated on-chip DRAM controller, while another candidate design uses only on-chip RAM. At the most abstract level, the SOC really consists of nothing but a set of communicating tasks and a set of external interfaces to other entities, including other nodes in the network or mandatory external devices. At this stage, each task is explicit in the definition of the system. Fig 4-2 shows an abstract model of an SOC for a networked disk drive.

The SOC design process starts from this abstract system model and evaluates, combines, and implements all of the tasks and interfaces. Tasks can be implemented in many forms, and the implementation choices reflect both the internal requirements of the particular task and the demands for communication with neighboring tasks and interfaces. In practice, the implementation choice for a task may also reflect the starting point for design—available pre-

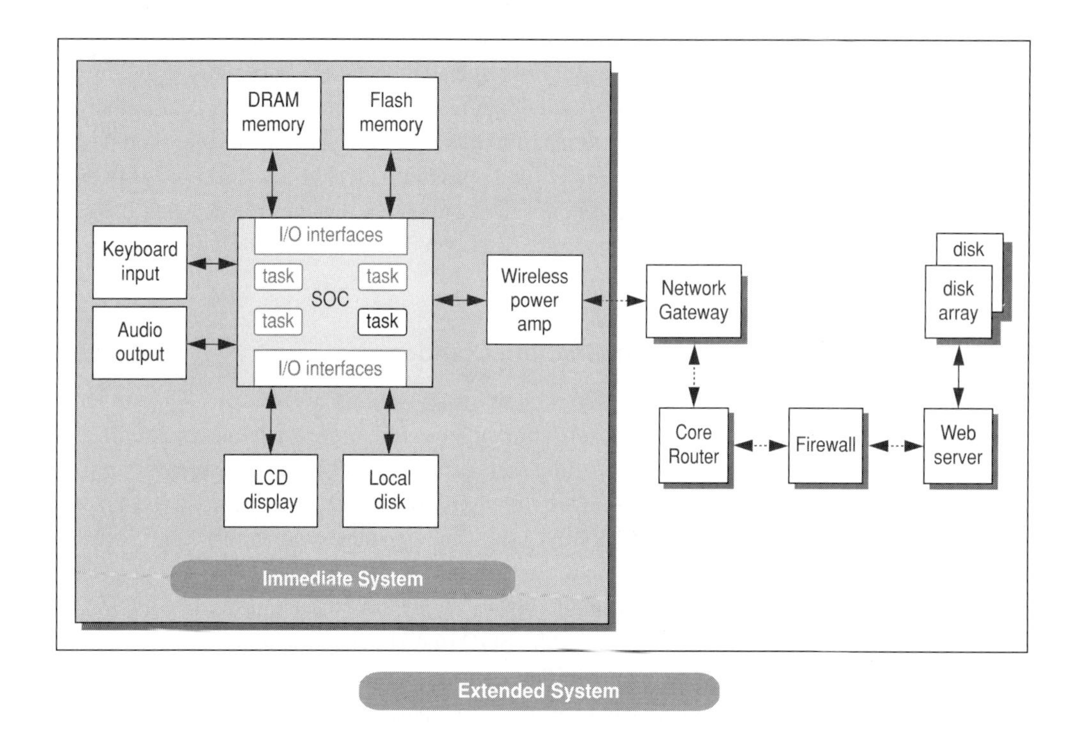

Figure 4-1 SOC as component within a large system.

existing hardware or software for that task or a similar task—and the skills of the design team, including biases in preferred implementation style. If the benefits of the new methodology (and its associated building blocks and tools) are sufficiently clear, teams can readily overcome this design inertia.

The SOC architecture process is really a process of refining the design, implementing the tasks, and developing the communications among tasks. Every proposed design solution represents one possible compromise between two opposing actions:

- **Partition:** The designer splits one task into two subtasks, implemented in separate blocks with communication between them. Separation generally increases the concurrency (potentially reducing task latency and increasing throughput) but also increases resource demands (more logic area, more wiring, more memory for data buffering and instructions, and greater congestion on shared buses).
- **Combination:** The designer combines two tasks into a larger task within the same block, turning node-to-node communication into internal communication. Combination generally reduces concurrency (potentially increasing latency and reducing throughput) but

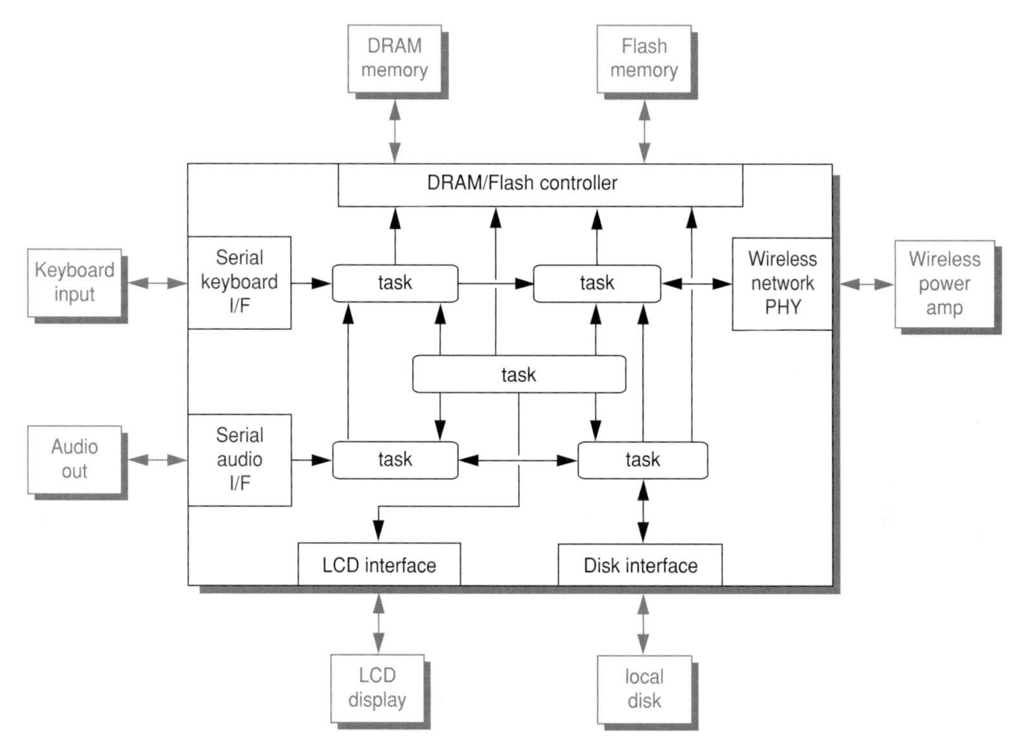

Figure 4-2 SOC as a network of communicating nodes.

may reduce resource demands (less area, less memory, and less contention for shared resources).

In a complex SOC, the discovery of the best balance of forces is a complex process. The problem, including the criteria for solution "goodness," is certainly too complex to allow one truly optimal solution. Nevertheless, systematic design methods do enable good decision-making among conflicting goals of cost, performance, design time, and flexibility.

4.1.3 Impact of Silicon Scaling on System Partitioning

As the density of silicon chips increases, the range of design choices within the SOC also expands. Three distinct trends stand out:

- **Growing transistor density:** With progress in semiconductor technology comes higher integration, where the number of tasks per SOC and the allowable transistor budget per task continue to grow exponentially. This trend may broaden the ambitiousness of the end-product goals—the number of tasks in the initial product definition—but it certainly opens the door to more parallel implementations.

- **Widening gap between on-chip and off-chip connections:** The speed of deep-submicron wires within integrated circuits scale up over time more rapidly than the speed of connections between chips (primarily through reduced capacitance due to smaller physical dimensions). Local connections within a region of an SOC are expected to scale roughly with the transistor speed. Cross-chip, global connections do not scale as closely, but they still scale better than chip-to-chip connections. (See "The Future of Wires" by Ho, Mai, and Horowitz.) This increasing performance gap increases the benefit of replacing chip-to-chip connections with intrachip connections whenever possible. Reduced capacitance sharply reduces the energy requirements of intrachip communications compared to inter-chip communications. (The growing gap between local and global wiring energy will also put more emphasis on modular SOC design, where the module-to-module connections are managed and tuned more carefully than intramodule communications.) Moreover, forming connections during integrated circuit fabrication is generally far more reliable than forming connections during chip packaging, PCB soldering, or system assembly.
- **Shift in applications:** The electronic systems market understands and exploits the seemingly inexorable increase in electronic functionality. These competitive demands drive requirements for better "always-on" network connectivity, battery life, interactive responsiveness, physical portability, and user-interface sophistication.

Together, these three trends appear to make higher integration and the consequent on-chip concurrency inevitable. Even if a given digital electronics product category is not yet appropriate to SOC integration, it is usually just a matter of time before it is. The primary potential benefits of a more parallel design style—higher system throughput, better energy efficiency, and greater interactivity—all ride the silicon integration trend.

4.1.4 Why Multiple Processors

Concurrency may be good in principle, but what makes multiple-processor SOC design a good embodiment of the principle? What aspects of the SOC design problem might favor the implementation of tasks using processors over other forms of task implementation? Two reasons stand out:

1. **Flexibility at the individual task level:** Complex systems perform a variety of complex tasks. Each task is subject to change due to shifts in market requirements and late discovery of implementation bugs. Software programmability serves to insulate the design team from the delays and costs of these changes. It also eases initial development efforts by permitting late design changes to be made even after the chip is fabricated.
2. **Flexibility in concurrency and communication among tasks:** Complex systems have complex interactions among tasks. Even if the definition of the individual tasks is comparatively stable, the mix and interaction of the tasks may shift. As the workload changes—either from one product variant to the next or from one moment of execution to

the next—the efficient balancing of task loads and routing of data may shift also. This variability favors dynamic allocation of tasks and dynamic buffering and forwarding of application data through the system. Processors serve as effective foundations for dynamic task allocation and flexible data movement.

In principle, these forms of flexibility apply to all kinds of processors including fixed-instruction-set architectures such as general-purpose RISC processors and DSPs. In practice, the limitation of standard instruction sets to generic 32-bit integer operations creates a performance ceiling for standard architectures, restricting their applicability to tasks that need the flexibility benefit of processors but make only modest demands for application performance or efficiency. For many tasks, application-specific processors raise that performance or efficiency ceiling by an order of magnitude or more, vastly expanding the range of tasks for which processor-based flexibility is available. The full exploitation of the application-specific processor for an individual task is the focus of Chapters 5 and 6. But first we explore ways to exploit concurrency through use of multiple processors.

4.1.5 Types of Concurrency and System Architecture

Concurrency comes in many forms and at many levels within the suite of tasks that make up an SOC. We can categorize forms of concurrency by looking at two workload characteristics: the *granularity* and the *uniformity* of the concurrency.

Granularity refers to the scale at which we recognize that operations are independent and can occur in parallel. Concurrency may appear at the *fine-grained* level of individual data operations—where basic operations on the data (loads, adds, compares) are independent of one another and can be performed at the same time. Concurrency may also appear at a *coarse-grained* level where entire procedures are independent of one another and can execute at the same time without respect to ordering of operations among them. Note that these two terms—fine-grained and course-grained—are not absolutes. They suggest a continuum, in which concurrency may be found among the most basic operations within a single iteration of an algorithm, among iterations of an algorithm, among major phases of computation of an algorithm, or among a set of different algorithms making up the SOC workload.

Uniformity refers to the similarity of sets of independent operations. When the operations are *homogeneous*, the operations within the sets are basically the same, but the operands—the data—are different among the sets. When the operations are *heterogeneous*, the operations within the sets are quite different from one other.

These two characteristics define the two axes of Fig 4-3, which categorizes combinations of granularity and uniformity in task concurrency and the types of processing that best exploit that concurrency. The figure gives four examples of parallel approaches with processors, though there are many variations and combinations possible:

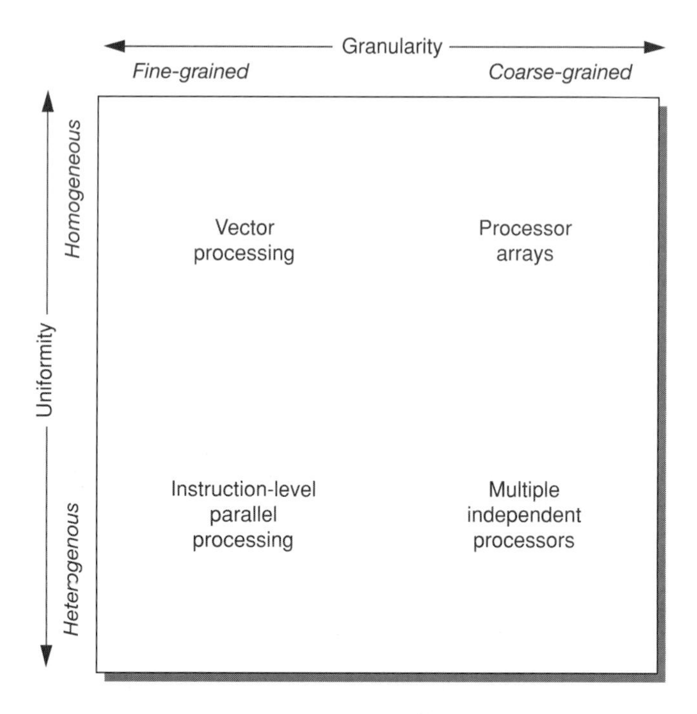

Figure 4-3 Characteristics for parallelism in processors.

1. **Instruction-level parallelism:** Almost all tasks contain intrinsic concurrency among operations, even at the lowest level. For example, the inner loop of an algorithm may compute operand addresses, move operands between memory and registers, perform operations on data in registers, calculate loop conditions, and perform branches to complete the loop. The calculation of addresses, loading of operands, and operations on one operand may be dependent, but in many cases, the operand-address calculations and operand loads for different operands are independent and can execute in parallel. In addition, the calculation of loop conditions and branch targets are often independent of the other operations in the loop. Different tasks offer different degrees of instruction-level parallelism.

   Various processors offer a wide variety of ways to exploit this common phenomenon. *Superscalar processors* implement fairly simple instructions, but attempt to execute more than one of these simple instructions per cycle by detecting the independence of operations dynamically in hardware. *Long-instruction-word processors* use sophisticated compilers to find independent operations and combine them into wide instructions. *Application-specific processors* take advantage of the particular set of independent operations in the target applications and combine them into application-specific instructions (often resulting in narrower, simpler instructions than those used by long-instruction-word processors). Techniques for exploiting instruction-level parallelism are discussed in depth in Chapters 5, 6, and 7.

In addition, processors with deep execution pipelines also implement instruction-level parallelism. Deep pipelines that execute a long sequence of operations—often operations that depend on one another—allow significant overlap among the operations of different instructions. Where the operations of each pipe-stage are different, pipelining neatly fits the category of fine-grained, heterogeneous concurrency.

2. **Vector:** With vector or single instruction, multiple-data (SIMD) methods, the same operation or sequence of operations is performed on a linear stream of operands. The absence of dependencies between the results of one sequence and the inputs to another sequence allow these operations to be performed in parallel. Many signal- and image-processing tasks fall naturally into this category of simple operations on long data streams.

   Vector processing can be implemented either with wide execution units using shallow pipelining, where the all of the operations of the instruction are executed at the same time, or with narrower execution units using deep pipelining, where a single execution unit operates in series on the independent operands of the vector, or even with a combination—wide and deep execution units. In all cases, significant concurrency is possible among these operations or between the operations on the vector and other operations in the processor. Chapter 5 explains vector methods in more detail, especially for application-specific processing.

3. **Processor arrays:** In many applications, the same complex algorithm is performed on many different data streams. Significant data dependencies exist within each stream, but the streams are largely independent of one another. In these cases, the designer can replicate the processor and assign different data streams to each processor to achieve a high degree of concurrency, even when the processing of each stream resists concurrency at the level of individual operations. This pattern occurs commonly in communications-infrastructure applications in which a large number of independent communication channels are concentrated or routed through one chip. SOCs with large arrays of very similar processors—from fewer than ten to hundreds—fit these architectural requirements quite neatly.

4. **Multiple processors:** When the set of tasks in the SOC application is less uniform, a group of processors (each assigned to a different type of task) often serves well. Sometimes, the set of tasks are highly independent. Sometimes there is significant sharing of data among the tasks, where each processor is responsible for one phase of the overall computation creating a high-level task pipeline. In either case, if the dependencies are slight or if the data can be buffered to insulate one task from the other, substantial parallelism can be exposed. Often the determining factor is the presence of tight data dependencies or feedback among the phases of the algorithm. For example, a wireless digital-television SOC might have a data flow for wireless-access signal processing, network protocol operations, and video and audio decoding as shown in Fig 4-4. While all four tasks

operate on the same data stream, each task depends only on one upstream task and not on downstream computations. Moreover the types of operations are quite different so different processor configurations might be appropriate for each task.

The coarse-grained processing models are sometimes called multiple-instruction multiple-data (MIMD) but this fails to make the important distinction between homogenous and heterogeneous processor designs.

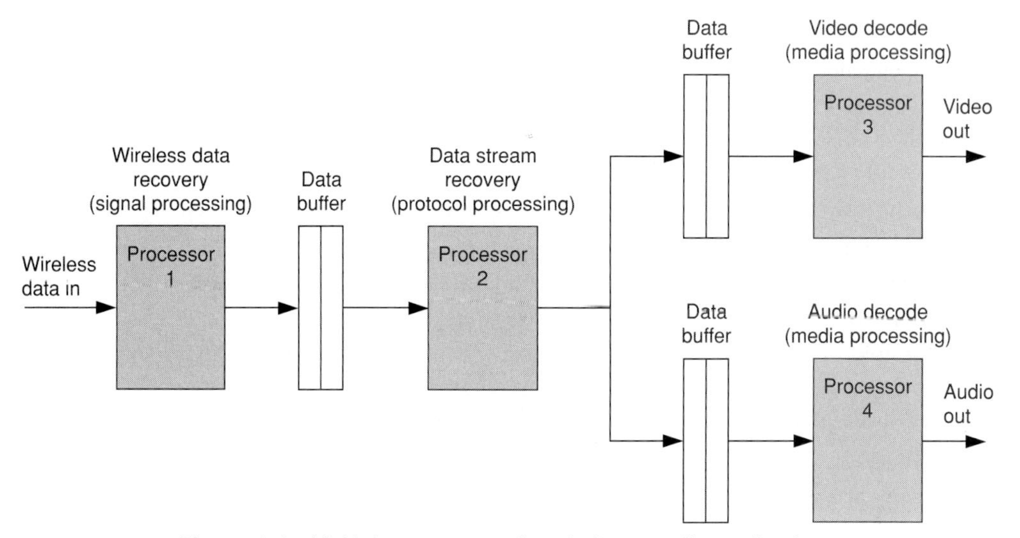

Figure 4-4 Multiple processors for wireless media application.

Sometimes the coupling among the tasks is even less direct than in this example, creating yet more freedom to use multiple processors to achieve high parallelism. These four categories represent only common forms of parallelism. SOC designers are discovering an enormous range of variants to serve the particular requirements of their task mix. Two principal missions of this chapter are to demonstrate how application-specific processor flexibility opens up a wide range of new parallel system organizations and to show how designers can use the specific leverage of new processor interfaces, new processor software communications primitives, and new application-specific instruction sets to implement these parallel-processing systems.

4.1.6 Latency, Bandwidth and Communications Structure

The design of an efficient SOC system organization rests firmly on the foundation of efficient communications among the component tasks (and the processors on which the tasks execute). In cases where the concurrency is simple and obvious, the design of the communications structure is easy. System performance is limited solely by the performance of the individual tasks (though some interdependencies may still exist due to contention for shared resources such as memory). In other cases, the communication among tasks is more complex or intense. In these cases, the latency, bandwidth, and structure of communications links among tasks become central design concerns. Consider two core issues: the impact of latency and bandwidth on task performance and the impact of aggregate communication structure of a chip on the latency and bandwidth of individual connections. The sensitivity of system behavior to internal latency depends on the presence of feedback in the functions of the system. When the system implements no important feedback paths, the system is latency-insensitive or can be reogranized to make it latency-insensitive.

From the standpoint of an individual task, communications looks like a series of input transfers (getting data from some source) and output transfers (sending data to some destination). Note that we sometimes speak of transfers directly between tasks; we sometimes speak of transfers between a task and memory. A memory is not a task, but may serve as the intermediary or buffer area for task inputs and outputs. An input transfer looks like a data request to another task (a read-address transfer to memory), followed by a data response (a read-data transfer from memory). An output transfer looks like a data transmission to another task (write-address and write-data transfers to memory), followed by a data acknowledgment (a write-complete acknowledgment from memory).

Using memory as an intermediary between a source task and a destination task has benefits and liabilities. Using memory decouples the execution of the two tasks by allowing the producer to get far ahead of the destination task if rates of data production and data consumption are highly variable. This approach often increases overall system throughput. On the other hand, using memory as an intermediary between tasks may also increase communication latency. The destination task doesn't just wait for the source task to create the data, but also for the transfer from source to memory and from memory to destination. When there is heavy contention for the memory or for the paths between processors and memories, this latency can become significant. Moreover, when the system implements a memory hierarchy, with local memories or caches closely tied to each processor and one or more levels of shared memories or caches, the memory latency includes both the transfer time between processor and local memory, and the transfer time between local memory and global, shared memory.

The basic timelines for input and output transfers are shown in Fig 4-5. These timelines suggest that the input-transfer response latency is often an important consideration, but the output-transfer acknowledgment latency is only important if the task must wait for acknowledgment before proceeding to the next computation stage.

Optimizing the processor has an important effect on I/O bandwidth. Configuring the processor to reduce number of execution cycles—for example, by adding new instructions that compress a number of operations into one cycle—increases the potential demands on the communication structure. Most application optimizations do not affect the number of input operands required or output operands produced. System requirements establish the number of required operands. As a processor is tuned to reduce the number of execution cycles, the interval between input or output requests is reduced, and the I/O bandwidth must consequently increase

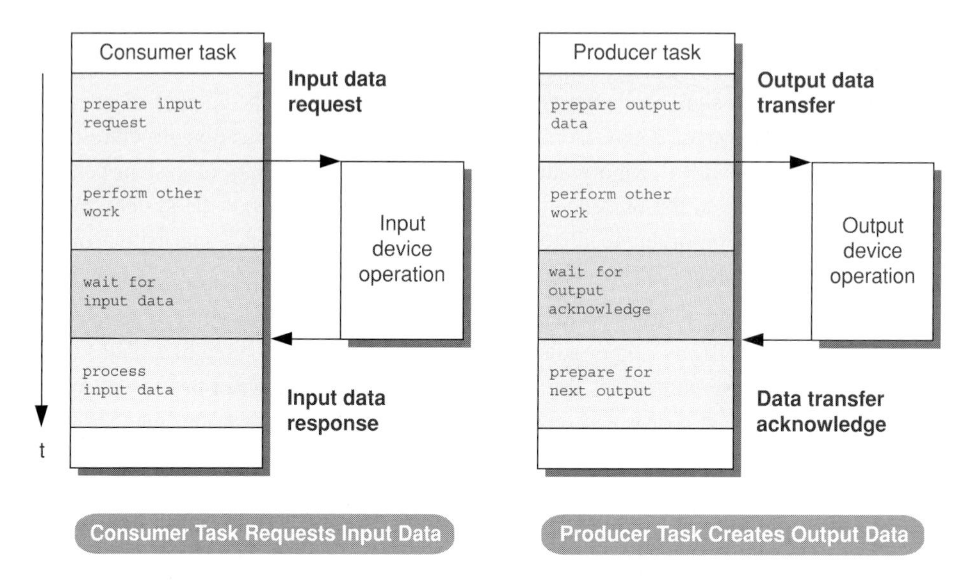

Figure 4-5 Basic I/O request and response flow.

to keep up with the faster processor. In the extreme case, all the computation is hidden within the latency of I/O operations. At this point, no further processor optimizations make sense. As a result, optimized processors often need optimized communications to maintain balance.

4.1.6.1 How Latency and Bandwidth Impact Task Performance

Latency is the delay from the point in time when data is ideally needed to the time when data is actually available for use. If data is flowing directly from another task, the latency usually reflects the interval between some form of request ("I'm done with previous data, please send the next data set") and the receipt of the necessary data. If the data is buffered in memory, the transfer latency reflects the interval between the memory request ("read this address or set of addresses") and the arrival of the necessary data in the processor's registers (or other hardware). The total latency between the source task and the destination task then consists of transfer latency from source to memory and from memory to destination, and often includes some overhead for the handshake between source and destination tasks that indicates that data is available in memory.

For tasks with feedback or limited buffering, shorter latency is better because it avoids stalling the task. While the processor cannot proceed without needed input data, many methods are used to shift the data request earlier and to perform other useful work, independently of the requested data, while waiting for the response. Latency-reduction methods focus on reducing the delay between the production of necessary data and its availability at the point of use. Reducing latency typically involves optimizing the communication mechanism, including the hardware interconnection between the source and destination. Latency-hiding methods focus on decoupling the overall performance of a set of tasks from the behavior of the communication method. Hiding latency typically involves optimizing the tasks and the communication protocol, including the buffering of communication data, so that the system performance becomes less sensitive to the actual data-transfer characteristics. Latency-hiding and latency-tolerance methods include a broad range of mechanisms:

- Prefetching data from memory to hide memory latency.
- Building processors with dynamic, out-of-order instruction execution to defer execution of just those operations dependent on incoming data.
- Multithreaded execution that switches processing to other tasks during long processor-to-processor or memory-to-processor communication delays.
- Software that overlaps the execution of different loop iterations.
- Deep buffering of communications so the source task can get far ahead of the destination task, on the average.
- Use of a separate agent, such as a DMA (direct memory access) engine to manage transfer of data between source and destination processors.

Even when the request is advanced in time and the point of use is deferred, the processor may still run out of work before the requested data is available. Eventually, the task stalls until data finally arrives.

Latency unpredictability complicates assessing the impact of latency on task performance. If the algorithm assumes the best-case communication latency, stalls are likely whenever actual latency is worse. Opportunities to hide the extra latency are lost. If the algorithm assumes the worst-case latency, data often arrives well before it can be used and sits idle. That's a lost opportunity for improved performance.

While latency describes the characteristics of individual transfers, bandwidth describes the aggregate behavior of a set of transfers. A task's algorithm—a transformation of incoming data into outgoing data—implicitly bounds the quantitative relationship between the data input and output rates and the task's execution rate. The I/O transfer rate bounds the task throughput. The throughput slows proportionally if either the input or the output data-transfer requirement is not met.

Data bandwidth is a rate—measured in operands per unit of time—but what unit of time is relevant? In practice, both I/O transfer rate and the task throughput vary over time due to the nature of the I/O device, due to contention for a shared resource between the I/O system and some other task, or due to the algorithm. Temporary discrepancies between the I/O rate and the task's internal execution rate typically matter only to the extent that data buffering is inadequate between input and data consumption or between data production and output. One exception is systems where the worst-case delay for computation and communication latency must be controlled. For example, the real-time system controlling a disk drive may have strict limits on the interval between the signal indicating the arrival of the disk head at the start of a disk sector and the computation of data to be written into that sector.

Bandwidth variability has much the same effect as latency variability for individual transfers. If the input rate falls below the task's internal consumption needs, the task starves and must eventually slow to the reduced input rate. If the output rate falls below the task's production rate, the output is congested and the task must eventually slow to the reduced output rate. Deeper input and output buffering insulates the task from longer periods of inadequate input and output bandwidth. Deeper buffering, on the other hand, often increases hardware cost, either in the form of larger data memories or in deeper hardware queues.

4.1.6.2 How Communication Structure Affects Latency and Bandwidth

Even when the tasks, communications mechanisms, and processors are designed to hide some latency, systems remain sensitive to latency to some degree. If low latency, high bandwidth, and minimal variability are important goals, how do we design the communication structure? What affects latency and bandwidth? You can approximate the latency of a simple stream of data transfers using three components:

1. The **intrinsic latency** of the device (the memory, the registers of an I/O device controller, or another processor). This latency is a function only of the internal device operations and is insensitive to any other system activity. For example, an off-chip static RAM has a fundamental access time.
2. The **transfer latency** of the connection to and from the device. In the absence of any other system activity, all systems have some data-propagation delay between the I/O device and the block demanding the transfer. Sometimes this transfer latency is as simple as the propagation delay of signals down wires. However, complex systems often require some overhead cycles to register the data and to move data between buses and interfaces.
3. The **contention latency** caused by interaction with other uses of a shared resource. The contention might be found in the request path (in getting a transfer request from the requesting task to the I/O device), in the response path (in getting a transfer request back from the I/O device), or within the device itself (if it is a shared resource). Contention latency can be a big issue because the effect can be quite large, and it may appear intermittently during traffic peaks.

In a simple queuing model in which requests for a shared resource have a Poisson or geometric distribution (the probability of a new request in each time interval is uniform), the average delay is easily estimated. Let's set the rate of satisfying requests at 1 and call the average arrival rate of requests α, where $0 < \alpha < 1$. (If $\alpha > 1$, the service rate—the number of waiting requests—will grow without bound.) Queuing theory for single-server systems (in queuing theory "M/M/1" for <arrival rate>/<service rate>/<# of servers>) says that and average waiting time (and the average queue depth) is then

$$\frac{\alpha}{1-\alpha}$$

As shown in Fig 4-6, the total service time—the queue delay plus the service time (1)—grows dramatically as α approaches 1.0 (as the requests come almost as fast as they can be serviced). In other words, if the system architect attempts to fully utilize a shared resource, the latency may increase significantly. Once utilization exceeds 50% of the available bus bandwidth, service time starts to grow significantly. At 90% utilization, latency has grown by a factor of 10.

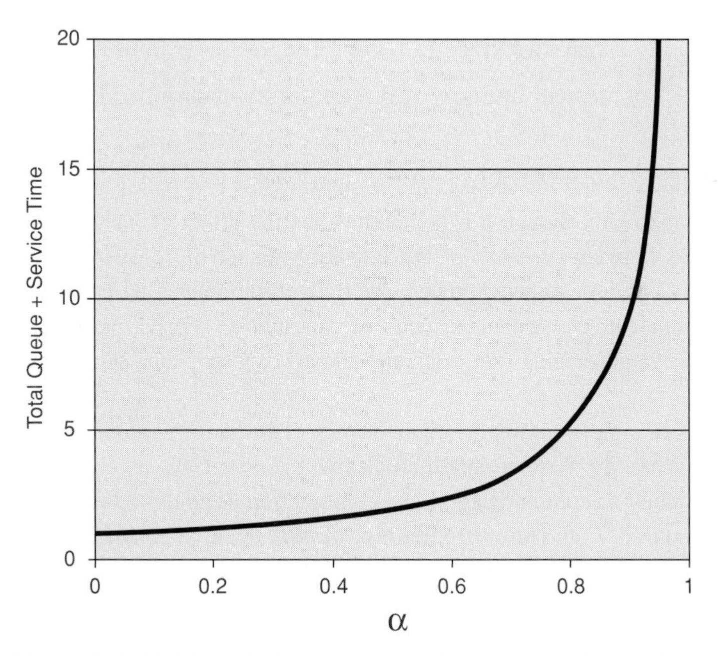

Figure 4-6 Total contention + service latency for queuing model.

For the simple case, the total input or output latency is sum of the parts: the transfer latency of the request, the intrinsic latency of the device, the transfer latency of the response, plus any contention latency associated with request, device, or response. For many output transfers, especially where there is buffering between the processor and the output device, the acknowledgment is implicit or combined with the request handshake, so the output latency is invisible to the processor except when there is unexpected contention for the transfer path or device. Bandwidth is also bounded by these same effects. If the I/O operation is unbuffered—that is, if no request or response is pipelined—then the bandwidth is set directly by the overall latency:

$$\frac{\text{transfer size}}{(\text{request latency} + \text{device latency} + \text{response latency})}$$

plus any added latency due to contention in the request, device, or response. If the transfer is buffered at the boundary between the request path and the device, and between the device and the response path, the bandwidth on each section of the transfer is higher and the bandwidth is set by the minimum of these three bandwidths:

$$MIN\left(\frac{\text{transfer size}}{\text{request latency}}, \frac{\text{transfer size}}{\text{device latency}}, \frac{\text{transfer size}}{\text{response latency}}\right)$$

Both the devices and the request and response paths can be further pipelined to increase the minimum bandwidth, though this sometimes has the effect of increasing the total latency. However, buffers between sections of the transfer path have a cost in terms of silicon area, power dissipation, or both area and power. The tradeoffs in managing latency and bandwidth in complex communication structures are often subtle. Early, accurate simulation of communications requirements and actual delays can prevent big surprises in system development.

The issues of simple communications latency also show up in more complex interactions between tasks. When the flow of data through two or more tasks involves feedback or cycles, then the total latency of communication and computation around the loop becomes significant. As shown in Figure 4-7, an output from a task (Task 1) can be considered a "request" and an input to that task can be considered a "response," if that input depends, perhaps through a series of dependent tasks, on the original output.

In this case, the effective throughput of Task 1 depends on the communication latency to Task 2, the computation latency within Task 2, the communication latency to Task 3, the computation latency within Task 3, and the communication latency back to Task 1.

4.1.7 Reliability and Scalability in SOC Communications Architecture

If the future evolution of product requirements were completely known at the start of an SOC design project, the design of the optimal communications structure would be a fairly simple, mechanical process. Unfortunately, the inevitable changes in market requirements (as well as

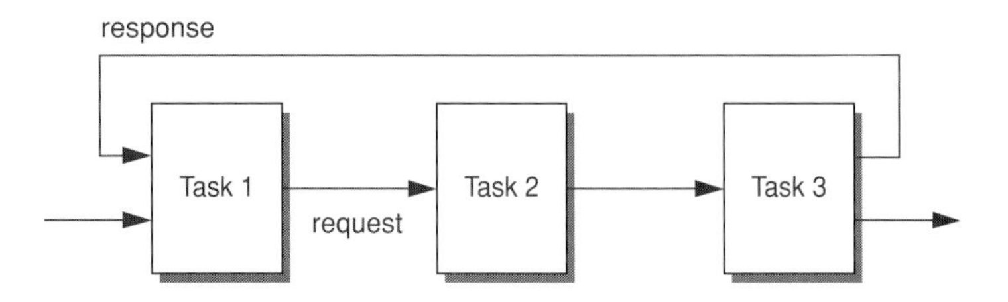

Figure 4-7 Dependency loop among tasks.

inevitable ambiguities in the original specification) guarantee that the pattern and volume of communications among tasks will change, just as the details of the tasks themselves evolve. Building a communication structure that performs acceptably as the overall application evolves is just as important as building processors that perform acceptably as the underlying algorithms evolve. Three potential shortcomings must be avoided:

1. The communications structure just barely meets bandwidth and latency requirements. If the load on some resource increases because a task, or worse, a set of task requests (from one task or a combination of tasks), grows beyond expectation even slightly, the latency rises and the throughput of all the affected tasks falls.
2. The communication structure has so much headroom that silicon area, power, and design time are wasted. For example, some SOC applications have just a few tasks or sections of code for which latency is critical, but the communications structure is designed for strictly bounded latency everywhere. Such a design might involve dedicating unique wires or bus slots to every communication path or providing such high service rates on shared resources that only a few percent of the available capacity is used in practice ($\alpha << 0.5$). Headroom is important for product flexibility, but too much headroom, or headroom in the wrong places can undermine the initial product's viability.
3. The communication structure is so inflexible that either important communication paths are missing or sharing of resources such as memories is impossible. These design flaws would hamper the applicability of the chip to new tasks or would mandate indirect communications. For example, if a new requirement needs Task A to communicate with Task B (a communication path not foreseen in the original design), then a common intermediary Task C must be found so that Task A communicates with Task C and Task C communicates with Task B.

Avoiding these pitfalls makes a more robust design. This robustness shows up both as product reliability—rare combinations of resource requests don't cause excessive queuing delays and missed deadlines—and as product scalability. Even as the workload increases or changes, the design has sufficient flexibility and headroom to accommodate the changes.

4.1.8 Communications Programming Flexibility

Design based on software and application-specific processors enable a wide range of communications choices. This range of possible implementations for communication connections drives important characteristics of the resulting chip and system design, including silicon area, communication latency and bandwidth, and the degree of programming flexibility. The benefits of low silicon area and latency and high bandwidth are clear. The role of programming flexibility is worth explaining.

Programming—rather than hardwiring—of communications brings two distinct benefits. First, programming gives *dynamic choice* of communication sources, destinations, transfer sizes, priorities, and timing. The program establishes a logical connection between two tasks only if and when needed for the particular stage in the task and particular data inputs. This is especially important if the number of possible communication connections across for all possible applications for the SOC is much larger than the number of communications connections used at any one time in any one system. For example, the program may allocate memory buffers as needed to hold input and data (and may pass an address pointer from one task to the next). This dynamic buffer allocation can be very efficient from the standpoint of memory allocation because the total memory required is set by the maximum number of buffers active at any one point in the program, not by the total number of buffers needed across the entire program or across all possible programs in all systems.

On the other hand, dynamic allocation of communications incurs some overhead and risks. Finding available buffers and calculating communications routes, for example, require extra processor execution cycles, especially if the application task must make a system call to the RTOS for service. In addition, dynamic communication may make determination of worst-case latency and bandwidth quite difficult. Finally, if the worst-case analysis of activity is wrong, the system may run out of memory under some rare unanticipated circumstance and fail. The cumulative uncertainty from hardware contention, task execution, and RTOS service response may create unacceptable maximum delays for hard-real-time communication needs. As a gross generalization, the greater the dynamic flexibility of communications programming, the greater the performance uncertainty.

Second, programming offers *abstraction* in communications programming. If the programmer can develop a task without locking down the detailed communications implementation, the task's program will be easier to test, more portable, and more easily adapted to new requirements. This abstraction—the API—typically follows the structure shown in Fig 4-8 where the data producer and data consumer tasks are written in terms a general-purpose output and input facility, in this example, `send()` and `receive()`.

The details of the mechanism than connects the producer API to the consumer API are hidden from both tasks. Those details include two major components:

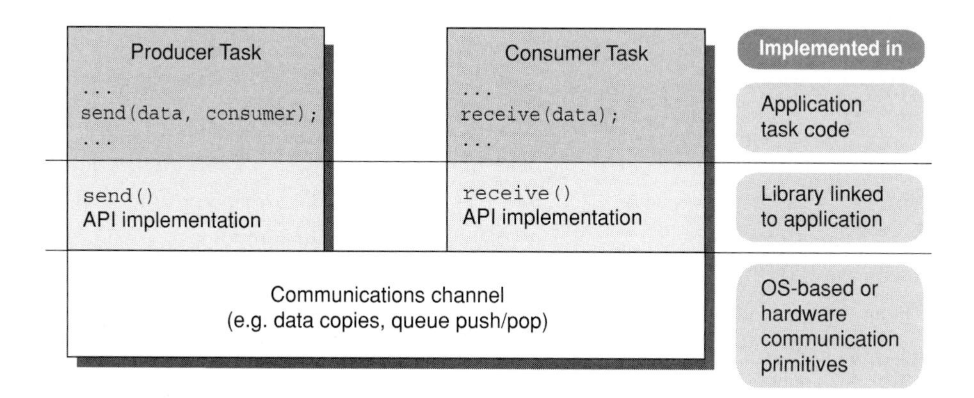

Figure 4-8 Structure of producer—consumer communications abstraction.

1. *The choice of underlying communications:* Possibilities include direct transfers over a dedicated set of wires ("ports"), transfer through a queue, buffering through on-chip or off-chip memory shared between producer and consumer, or double copying of data—from producer to RTOS memory space and from RTOS memory space to the consumer.
2. *The implementation of the API within the task code:* This mechanism involves mapping the data structures used in the API to the chosen communication mechanism. This mapping is often implemented as library functions linked to the application code after the communication channel implementation is chosen.

Abstraction in communication is particularly valuable if uncertainties about communications patterns and task implementations are high, but the sensitivity to communication latency is low. For example, if the developer cannot decide whether two communicating tasks will run continuously in parallel on different processors or will time-share a single processor, then communications abstraction is essential. If hundreds of different tasks must communicate but no one communication channel needs more than a few thousand bytes per second, the benefit of program flexibility probably outweighs the liability of program overhead.

Abstraction goes hand in hand with runtime software robustness. If tasks can directly access the memory of other tasks, they can accidentally or maliciously corrupt that memory and bring down the system. Often, developers decide to trust the tasks (and test them to assure no accidental memory corruption). In other cases, a task cannot be trusted, and incorrect memory references will disrupt other tasks. In this case, tasks can be allowed only to communicate with an RTOS kernel or central communication manager.

The kernel or communications manager can enforce rules for communications (transfer lengths, address ranges for references to shared memory regions and other rules). The hardware may even implement memory protection to guarantee that each task can access only those sections or pages of memory allowed by the RTOS. The RTOS also implements the communications API via a set of system calls. For example, a data-producing task may write its data into a private memory area, then pass the outgoing address to the RTOS. The RTOS can copy this data into a buffer, whose address is later passed to the data-consuming task. The producing task never gets direct access to any memory shared with another task. Memory is only shared with the RTOS kernel. Similarly, the consuming task has access only to its own private memory and to a data buffer allocated and controlled by the RTOS.

The RTOS kernel, of course, must be trusted. The individual tasks should still be tested for correctness, but they interact only with the kernel, not directly with each other. This simplification sharply reduces the number of combinations to test. The greater the number communicating tasks, the greater the potential improvement in robustness that comes from memory-space protection.

The abstraction of a communication API helps the software engineer manage the complexity of system development. The communications libraries that implement these APIs are complex and subject to subtle errors, so using thoroughly tested and documented standards aids system correctness. Implementing task-to-task communications in the application in terms of a communications API both insulates the application from the details of the communication mechanism and relieves the developer from the effort (and risk) of low-level communications code development.

4.1.9 Early vs. Late-Binding of Interaction Mechanisms

Application-specific processors extend the upper bound of communications bandwidth achievable in a single processor (more than 10GB per second at 300MHz) through the use of multiple wide I/O buses compared to a practical limit of less than 1GB per second for conventional 32-bit RISC cores running at the same clock speed. Previously, tasks needing gigabyte-per-second I/O required a nonprogrammable hardware implementation regardless of complexity. The high-bandwidth capability of application-specific processors extends the range of communication implementation choices with processor-centric SOC design.

For each instance of task-to-task communication, the developer must choose the implementation, but the developer has a wide range of options on when to choose the detailed implementation. The table in Fig 4-9 outlines the three basic choices on the time of binding the task to the communications mechanism—binding in hardware, binding at system boot, or dynamic binding. It sets very approximate bounds for the latency for minimal (single-operand) messages, the rate of sending small messages, and the data bandwidth for large messages (large enough for data transfer to dominate bandwidth).

Binding Time	Typical Communications Method	Small Message Latency	Small Message Rate	Large Message Bandwidth	Key Benefit
Chip Design	Direct, dedicated, wide port connection between processors	<10 cycles	<10 cycles per message	>10 GB/s (Example: Tensilica processor direct-connect ports support up to 1024-bit width)	High bandwidth and low latency enable greater parallelism
System Build or Boot	Communication by direct task access to device registers or shared memory with statically assigned buffer addresses	10-100 cycles	<10 cycles per message	< 1GB/s (Example: task-to-task 128-bit wide memory copies)	Good combination of efficiency and post-silicon flexibility
Dynamic Invocation	Communication by indirect task access to dynamically allocated shared memory	>100 cycles	>100 cycles per message	<250MB/s (Example: task-to-RTOS and RTOS-to-task 32-bit memory copies)	Maximum flexibility, abstraction and robustness

Figure 4-9 Early, middle or late binding of communications.

Early binding of communications gives the highest potential bandwidth and lowest latency because optimized communications channels can be built entirely into the application-specific processor's instruction set and interface. These optimized channels, however, typically demand dedicated interconnect between processors and rigid binding of the sending and receiving software tasks to those processors. If the SOC is expected to span a very wide range of communications rates and patterns, these dedicated, optimized paths may go underutilized. The application-specific processor ports described in Section 4.4 provide a very high-bandwidth, low-latency communication primitive. While these point-to-point links are most often used for dedicated processor-to-processor communications, the links (especially the hardware queues) can also be used to create general communications meshes with dynamic data routing built into either the task software that uses the links or the physical implementation of the hardware queues themselves.

Binding communication at the time of system software compilation or system boot allows for more flexible reconfiguration of communications on a system-by-system basis. The system software can initialize static addresses to point to shared memory buffers and optimize the implementation of the communication API in each task to use those buffers directly. The implementation uses software synchronization to ensure that no consumer task will use the data until the producer task is finished creating the data. Performing the synchronization and transferring data to and from a shared memory generally has longer latency and lower bandwidth than direct processor-to-processor transfers.

One useful compromise between hardware build-time binding and system boot-time binding is the use of queues that are mapped into shared memory space. Queues eliminate the need for software synchronization because the producing processor automtically stalls when the queue is full and the consuming processor stalls when the queue is empty. If more than one processor can read or write the queues, or if more than one memory-mapped queue is implemented, the mapping of queues into memory lends important flexibility to the system. Choosing the appropriate queues and assigning the corresponding queue head and tail addresses configures the pattern of high-speed data flows through the system. Memory-mapped communication is less efficient than hardwired communications channels, but more efficient than completely dynamic runtime binding where communications are mediated by a real-time operating system (RTOS).

Late binding is the most general and flexible approach to communications configuration. Each task uses an abstract API to send and receive data streams. The abstraction includes synchronization, buffer allocation, and error handling. The API insulates one task from another. The software controls the choices of source and destination of data transfers, message size and type, and error handling. A RTOS mediates the communication, at least in setting up communications between tasks and sometimes for each transfer or message. The RTOS kernel is often responsible for all data transfers between the task data spaces. This mediation has a price however. The system call to the RTOS incurs significant overhead. The context switch between the task and the RTOS kernel may require hundreds of cycles at a minimum.

4.2 Major Decisions in Processor-Centric SOC Organization

The system architect faces a number of important decisions in creating the best SOC structure. Good choices early in the design process reduce silicon cost and power, increase system performance, and improve development and verification efficiency. This section explains the major decisions in architecting an SOC and guides the designer to a systematic approach to design structure. The flow encourages wide use of processors as the default for implementing tasks and focuses on how to balance cost, performance, and flexibility within an SOC design framework. The foundations for the design flow are these:

1. Work top-down from the system's essential I/O interfaces and computation requirements.
2. Use processors pervasively to implement tasks:

    a. When tasks have specific computational patterns, optimize the processor to fit the tasks.

   b. When a task exceeds the capacity of an optimized processor, parallelize the task across processors.
   c. When a group of tasks fits together within a processor's capacity limit, map the tasks together onto one processor to minimize hardware cost, power, and communications overhead.

3. Measure the communications traffic patterns and optimize the software and hardware interconnects around those patterns.
4. Start with early, rough simulation of communication tasks and refine the system into detailed implementations of processors, software, and other blocks, all running in increasingly accurate simulations.

4.2.1 The Starting Point: Essential Interfaces and Computation

The first system-design step is identification of the chip's essential I/O interfaces and computation. The target product's marketing requirements usually establish the mandatory physical interfaces and necessary functions. These mandatory elements form the starting point for all other decisions— implementation decisions about these functions and decisions about inclusion of other supporting functions.

Not all interfaces and computations are equally fundamental to the design, of course. For example, in an integrated disk-drive controller, the external interface to the read-head and the servo motor are essential to the function, but external interface to buffer memory is not. Buffer memory could be implemented either on-chip or off-chip depending on more detailed analysis of cost and bandwidth tradeoffs. Similarly, a Secure Socket Layer security chip effectively requires implementation of the RSA (Rivest-Shamir-Adelman) algorithm for public/private key encryption, but may or may not include other TCP/IP protocol-processing functions.

In the early stage of the design, the product requirements often continue to change. Target customers want more features, engineering wants simplification, product managers want lower costs. As needs are better understood, the feature list tends to grow so more of the SOC functional requirements come into focus. At the same time, initial analysis of each proposed implementation gives early insight into the repercussions of the feature list and development plan. The development manager is probably asking questions such as

- Does initial analysis suggest that the chip meets performance requirements?
- What is the die size?
- What is the power dissipation?
- How long is the development and prototyping schedule?
- What engineering resources must be allocated?
- How predictable is the rest of the development schedule?
- How well have the chip I/O and functional specifications been defined by the customer?
- How will customers use (and program) the chip?

Inevitably, not all feature requests (cost, power, and development budget constraints) can be accommodated with confidence, so priorities must be established. A clear set of requirements and decision-making priorities are both important to the successful refinement and implementation of the SOC design.

This process of identifying mandatory computation and interfaces, plus setting priorities for other functions, should produce two outputs: a set of target tasks and an initial chip interface. This explicit problem statement allows the next two major sets of decisions to proceed: the allocation of tasks to processors (or other computational blocks) and the organization of communication among the tasks and I/O interfaces.

Early System Modeling

If the tasks can be represented in executable form—as algorithms expressed in a programming language such as C—the architect should consider early system modeling. At this stage, tasks have not been allocated to processors, and communications among tasks is still expressed abstractly, either through a message-passing or a shared-memory programming paradigm (discussed in Section 4.3.1: Software Communication Modes). The tasks are all implemented as processes or threads running under the operating system of the host development platform—a PC or UNIX workstation. The input and output interfaces are also modeled as tasks, perhaps generating simulated input flows from a file and capturing output to another file.

The functions of a complex real-time operating system may be hard to emulate, but basic behavior of the set of tasks is readily modeled. This multitask model serves to verify the functionality of the set of tasks, even though real-time behavior is lost. In addition, the system model can measure the size, type, and number of messages or data operands transferred between tasks. This information can be represented as a traffic profile, such as the hypothetical results shown in Fig 4-10, where each data source represents a row and each data destination represents a column. The value in each cell shows the number of transfers (or, alternatively, the number of bytes of data or other information relevant to system design). Two simulated inputs are included in the set of source tasks. One simulated output is included in the table. Note that no intra-task communication (e.g., Task 1 to Task 1) is logged in the table. Also note that most of the data flows from inputs through computational tasks to outputs, but some feedback loops exists (e.g., Task 2 sends data back to Task 1).

The architect can use this information to build a picture of the principal intertask connections in the system, permitting a sketch of overall traffic flow, such as in Fig 4-11, where the heavy lines represent high-volume connections and fine lines represent incidental connections.

Early System Modeling

	Destinations				
	Task 1	Task 2	Task 3	Task 4	Output
Input A	17,700,000	0	0	0	0
Input B	4,700,000	0	0	0	0
Task 1	-	21,400,000	45,000	45,000	0
Task 2	5,600,000	-	45,800,000	100	0
Task 3	0	0	-	87,900,000	1,000
Task 4	0	3,500	37,400,000	-	13,400,000
Input A	17,700,000	0	0	0	0

Figure 4-10 Traffic profile for abstract system model

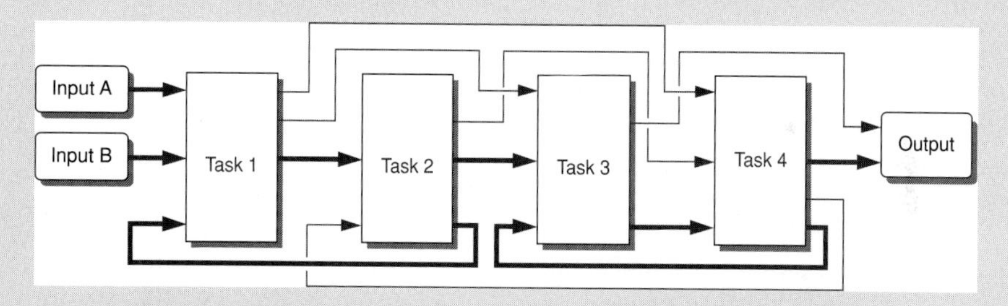

Figure 4-11 Traffic flow graph for abstract system model.

In theory, the same abstract system simulation model can serve as the basis for sizing the computational demands of each task. This information is less exact, however, because the processor architecture (and the code) of the host development platform are unlikely to look like architecture (and code) for the target SOC. Moreover, use of application-specific architecture will improve the performance of some tasks dramatically, skewing the real system results compared to the performance on the host development platform. Nevertheless, early, high-level modeling can yield important insights into both computational and communications hot spots.

4.2.2 Parallelizing a Task

When a single task shows particularly high computational demands, the designer must either use a faster general-purpose processor or pursue a more parallel hardware implementation. Fast general-purpose processors often require very high clock rates—for example, greater than 1 GHz. High clock-rate processors are likely to be both unacceptably power-hungry and difficult to design and integrate into an SOC without specialized skills. For embedded applications, an approach using parallel hardware resources is more likely to fit the embedded SOC's needs.

Historically, hardwired logic using RTL-based design was the only viable choice for parallel computation, but this design approach typically limited the complexity of the algorithms that could be implemented. Extensible processors open up a simple but highly efficient means to exploit concurrency, especially for fine-grained (instruction-level) parallelism. Techniques for accelerating task performance using application-specific processor extensions are the primary subject of Chapters 5, 6, and 7. When the task's available concurrency is more coarse-grained, however, parallelizing the task into a set of communicating subtasks can complement or substitute for fine-grained parallelization.

Task parallelization can take many forms. The goal is to recode the task into a set of tasks that runs on multiple processors so that the resulting task group either reduces the latency or improves the computation throughput compared to the single-processor task. Parallelization is constrained by computational dependencies within the task. If each operation depends on the immediately preceding one, no concurrency is possible. Even limited dependences may require significant communication of intermediate results between the two tasks. The overhead for this communication can offset the benefit of parallelization.

Splitting a complex task across several processors does require some thought on communication, but may ultimately simplify the system implementation. Jack Ganssle, an experienced observer of real-world embedded system development, argues that the benefits of simplifying the constituent tasks far outweigh the overhead of performing the partitioning into tasks.

Many tasks have sufficient latent concurrency, unconstrained by dependences, to make the effort worthwhile. Moreover, extensible processors offer low-overhead communications mechanisms such as direct queues and efficient shared local memories. These low-overhead communications alternatives expand the set of tasks for which parallelization is worthwhile.

Two quick examples illustrate some of the available parallelization techniques:

Parallelize task phases: Some tasks are organized into computational phases, with complex dependencies or *recurrences* within each phase, but relatively little data passing from one phase to the next. Often, these phases can be mapped to different processors, with phase-to-phase data passing between processors. In this code fragment, the two phases are split across two tasks so that one complex task consisting of two complex operations (`complex_op1` and complex_op2): A monolithic task with two inner loops

```
int A[],B[],C[];
Task() {
 int i, j;
 for (i=0; i<NBlocks,i++) {
  for (j=0;j<M;j++) B[i][j] = complex_op1(A[i][j]);
  for (j=M-1;j>=0;j--) C[i][j] = complex_op2(B[i][j]);
 }
}
```

becomes two somewhat simpler tasks. The results of the first phase are sent as messages to the task implementing the second phase, which receives each result to use as an input:

```
int i,A[],C[];
Task1() {
int i, j;
 for (i=0; i<NBlocks,i++) {
  for (j=0;j<M;j++) {send(complex_op1(A[i][j]);}
}
}
Task2() {
int i,j,B[N];
 for (i=0; i<NBlocks,i++) {
  for (j=0; j<M, j++) B[j] = receive();
  for (j=M-1;j>=0;j--) {C[i][j] = complex_op2(B[j])
}
}
```

The second task cannot start its processing until the first has completed its first outer-loop iteration (for *i*), but from that point, the two tasks operate in parallel, creating a high-level pipeline of processors. This transformation is efficient if the overhead for sending and receiving data is less than the cost of the function complex_op2().

Parallelize loop iterations: Some tasks perform complex computations within each loop iteration, but the data in each iteration is independent of other iterations. In other cases, different loop iterations can be mapped to different processors. For example, this simple initial C code fragment:

```
int A[],B[],C[];
Task() {
 int i
 for (i=0; i<N, i++) C[i] = complex_operation(A[i],B[i]);
}
```

can be split into two tasks, each performing half the iterations, as follows:

```
int A[],B[],C[];
Task1() {
 int i;
 for (i=0; i<N/2, i++) C[i] = complex_operation(A[i],B[i]);
 }
}

Task2() {
 int I;
 for (i=N/2; i<N, i++) C[i] = complex_operation(A[i],B[i]);
}
```

The two tasks can start as soon as the input values (`A[]` and `B[]`) are available. This approach works particularly well if the input and output data is available in a common memory or if the amount of computation per iteration swamps the amount of input and output data per iteration. If the iterations are truly independent, then high degrees of concurrency may be possible—even to the point of assigning each loop iteration to a different processor. This example is trivial, but many algorithms, particularly ones that with large data sets, often have some subtasks or phases with high data parallelism that allow such task-level decomposition.

The designer's insight into the nature of data dependencies is crucial to recoding tasks to exploit parallel processors. In some cases, the best algorithm for a single processor can be replaced with a different algorithm that is less efficient on a single processor, but more easily parallelized. A few simple parallelization methods can be automated, but human designers continue to play a central role in task parallelization.

4.2.3 Assigning Tasks to Processors

The mapping of tasks to an SOC implementation raises some complex issues. Depending on the task and the goals of the design, the implementation of some tasks is purely in hardware—data paths, state machines and associated memory. Other tasks are implemented as software running on processors, either generic or application-specific. While this book's new SOC design methodology emphasizes configurable processors for a wide range of tasks, some designs implement a combination of hardwired logic, software using generic instruction sets, and software using application-specific instruction sets. The implementation choice depends on the design's performance, cost, and flexibility goals.

The first step is clear: all significant tasks must be identified and specified. But what is a "significant" task? The simplest definition is this: a task is significant if it requires specific implementation hardware. Any task implemented directly in hardwired logic should be identified quite early in the design process. Similarly, the designer should highlight tasks with heavy or unique computational requirements. These tasks require either a specialized processor configuration (with application-specific configuration of instruction set, memories, or interfaces) or the dedication of a substantial fraction of a processor's computational capacity to these tasks. Other, lesser tasks can implemented in software running on the existing processors. While the aggregate demands of the minor tasks may ultimately affect the number and types of processors on the chip, these tasks can often fit into the unused cycles of the processors used for major tasks.

4.2.3.1 Two Guidelines for Mapping Tasks to Processors

Ideally, the complete set of tasks can be fully implemented using processors (application-specific or otherwise). Dedicated hardware is required solely in device interfaces, such serial ports, network physical-layer interfaces, external memory and bus controllers, and digital/analog converters. The designer's next step is to determine the number of processors and assign tasks to those processors. Two considerations drive the mapping process:

1. The set of processors must have sufficient computational capacity to handle all of the tasks. If the computational load seems too high for the processors, the developer has five choices:

    - Add processors to share the load.
    - Use faster processors, often by enhancing the processors with application-specific extensions.
    - Migrate tasks into hardwired logic (and face the liabilities of inflexibility).
    - Cut back on the computation requirements by changing the spec.
    - Find new ways to tune the algorithm or the code implementation to improve performance.

2. Tasks with similar requirements should be allocated to the same processor as long as the processor has the computational capacity. These tasks then leverage the common instruction set, memory, and interface configuration. The tasks will time-share the available computational cycles of that processor, either through dynamic allocation by a real-time operating system or by manual, static task management by the programmer.
   Three important task-mapping issues stand out:

    - The configured performance capacity of a processor depends heavily on the nature of the assigned tasks.
    - Assigning tasks with similar application data types and operation requirements to the same processor configuration saves costly silicon.
    - The pattern of communication between tasks influences how tasks must be assigned.

4.2.3.2 Different Processor Types Have Different Capacity

The process of determining the right number of processors cannot be separated from the process of determining the right processor type and configuration. If the computational demands of all tasks were precisely known and all processors were identical, the sizing of computation capacity would be simple. Traditionally, a real-time computation task has been characterized with a "MIPS requirement": how many millions of execution cycles per second are required. If all tasks were precisely characterized at the outset and if all processor MIPS were created equal, the MIPS requirements could be summed to find the total system load. The designer would simply include enough processors in the SOC to ensure that total computational capacity exceeded total computational demands.

Fig 4-12 shows a set of tasks with the initial estimate of the MIPS requirement for a hypothetical SOC platform that includes 3G wireless connectivity, simultaneous Motion-JPEG compression and decompression (1.2-Mpixel frames at 15 frames/second), audio encoding and decoding, network protocol stack, user interface, and a couple of layered applications. The initial performance estimates are very approximate.

Task	Millions of execution cycles/sec
3G wireless channel coding (UTMS)	3,000
MJPEG compression	5,000
MJPEG decompression	3,500
Audio encode	100
Audio decode	50
User Interface	10
Application 1	50
Application 2	50
RTOS	10
TOTAL	**11,770**

Figure 4-12 Baseline task performance requirements.

If all processors provided the same generic performance level, the SOC would need to provide more than 11 billion instructions/second of computing capacity (or shift the demanding functions into hardwired logic). However, different processors actually have quite different computational characteristics. SOC designers face this diversity of processor options even without the additional flexibility of automatic, application-specific processor generation.

The computational capacity of a 32-bit RISC control processor is not interchangeable with the computational capacity of a 16-bit DSP. A control task is likely to need substantially more cycles if it's running on a simple DSP than if it's running on a RISC processor. A numerical task is likely to need substantially more cycles running on a RISC CPU than if it's running on a DSP. RISC and DSP MIPS are not equivalent. That's why different processors are assigned different roles, even on today's SOCs. For example, all cell phones in use today combine a RISC processor and a DSP on one chip. The RISC processor handles the user interface and housekeeping chores. The DSP handles voice coding.

The advantage of using different processors for different task sets is the ability to better match the processors' performance capabilities to the tasks' needs. The disadvantage of mixing RISC processors and DSPs is the added burden of working with more than one set of software-development tools. The application-specific processor widens the set of available processor choices (because it can be configured to efficiently process any given task set), makes processor selection quick and systematic (because one base configurable processor architecture can efficiently serve many task sets), and reduces the number of required software-development tool suites to one.

Application-specific processors sharply reduce the number of cycles required to execute a task, but the added capability introduces a new wrinkle in processor sizing and balancing. A task's MIPS requirements on a generic processor may be a poor guide to the MIPS requirement after instruction-set optimization. As shown in Chapter 3, using an application-specific instruction set may reduce the number of cycles required to execute a task by 10 to 50 times. The overall allocation of tasks to processors should take this load reduction into account during architectural planning. Even if the final acceleration is not exactly known, an initial estimate provides a better starting point for task allocation than baseline RISC performance. Fig 4-13 revisits the simple allocation of tasks with a hypothetical acceleration of each task to reduce MIPS requirements.

Task	Millions of execution cycles/sec (original)	Example application speedup	Millions of execution cycles/sec (configured)
3G wireless channel (UTMS)	3,000	5.6	535
MJPEG compression	5,000	9.4	531
MJPEG decompression	3,500	9.2	291
Audio encode	100	3.0	33
Audio decode	50	1.9	26
User Interface	10	1.0	10
Application 1	50	1.5	33
Application 2	50	1.6	31
RTOS	10	1.0	10
TOTAL	**11,770**		**1,090**

Figure 4-13 Task requirements after processor configuration.

From this analysis, the designer, using an estimate of 300M execution cycles per second per processor, might conclude that two processors are required for the 3G wireless channel, two for the Motion JPEG compression, one for the MJPEG decompression, and one for everything else, including audio, for a total of six processors.

Early in the development process, estimating the accelerated performance of a task sometimes requires approximation. If the degree of acceleration is underestimated, the system will simply have more performance headroom once full acceleration is achieved. Chapter 5 discusses the process of accelerating applications in detail. Architects can develop reasonably conservative acceleration approximations by analyzing and accelerating just the most critical loops in the task (rather than looking at all the tasks that contribute significantly to performance) or by using compiler-based automatic processor configuration, as discussed in Section 5.8.

4.2.3.3 Assign Tasks with Similar Requirements to the Same Processor Configuration

Assigning tasks with similar computational requirements to the same processor has obvious advantages. If a set of tasks benefits from similar instruction set characteristics, those instruction features need only be implemented in the processor that runs that set of tasks. This clustering of tasks improves performance and reduces silicon cost. The same clustering is found in systems that combine a RISC controller and a DSP—the set of tasks is partitioned into those that are most "DSP-like" and those that are most "RISC-like." Simple two-way partitioning, however, tends to force expensive generality onto those processors. The DSP is expected to be optimal for every signal- and media-processing task, and the RISC controller is expected to be optimal for everything else. Consequently, both processor architecture types tend to become bloated with very rich instruction sets, which are not fully exploited by most designs.

The application-specific processor eliminates the hard choice between two generic architectures. Different application-specific processors (based on the same configurable foundation architecture) can be specified for each cluster of tasks using both the similarity of technical requirements and the load-balancing of the tasks across the processors as processor-selection and task-allocation criteria. Of course, the tasks may not be so easily bucketed into neat categories. Compromises arise both in the closeness of fit between task and processor architecture and in the degree of utilization of each of the processors. However, application-specific processors enable the features required by different tasks to be freely and safely combined on a single processor, so the differences in task requirements can be accommodated.

Using application-specific processors built from a common foundation offers a further degree of flexibility. While some tasks run best with highly optimized processor configurations, other tasks are less specialized and run well with more basic instruction-set features. These more generic tasks can be allocated with fewer constraints—any processor in the system has the necessary capabilities. Such tasks can fill unused processor cycles on any available processor because they can be compiled for the base architecture and allocated at system-boot time or at run time to any application-specific processor that shares that base configurable architecture. This task allocation flexibility is one key benefit of building all the processors in a system from one base instruction-set architecture.

4.2.3.4 Assign Tasks to Manage Latency

Allocating tasks based on maximizing processor utilization is good for throughput and cost but may not satisfy latency goals. In particular, if the system requires especially low latency between input and output, special care is needed. Often, optimization of one processor's instruction set can reduce the computation cycles sufficiently to meet the latency goal. Sometimes, however, the task must be distributed across multiple processors, each operating in parallel on data subsets or operating in pipelined fashion on computation subsets. In these cases, the tasks are assigned to different processors even when the computational load and the similarity of processor configuration requirements might otherwise suggest putting them together on a single processor. The processors overlap the computation and reduce the total delay from input to output, using the method of parallelizing phases of the task, discussed in Section 4.2.2. The principle is shown by the execution timeline in Fig 4-14 and Fig 4-15. When the two processing tasks (A and B) execute on the same processor, the end-to-end latency is greater than when the two tasks execute on different processors. In both of these cases, input and output data transfers are partially overlapped with computation, but Task A and Task B can only overlap if they are mapped to different processors. Depending on the nature of the tasks, A or B might be further partitioned to achieve additional overlap.

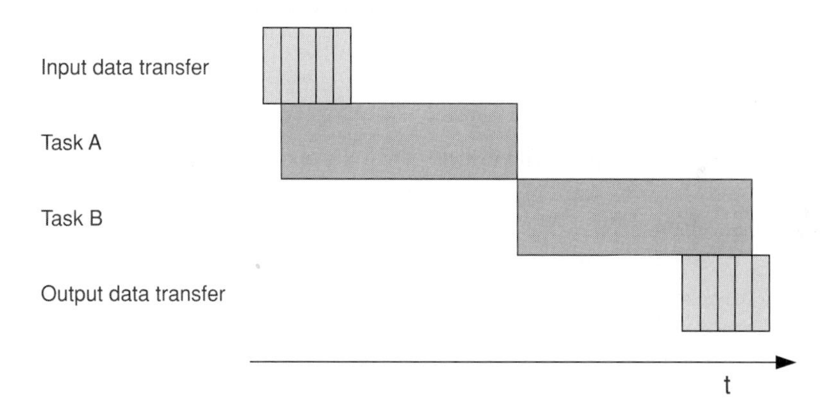

Figure 4-14 Latency of sequential task execution.

4.2.4 Choosing the Right Communications Structure

Once the rough number and type of processors is known and tasks are tentatively assigned to the processors, basic communication structure design starts. The communication structure depends entirely on the pattern of communication among the tasks and between tasks and I/O interfaces. The designer's goal is discovery of the least expensive communication structure that satisfies the bandwidth and latency requirements of the tasks, including changes in the task load that may occur as the SOC's use evolves over time or across a variety of target systems.

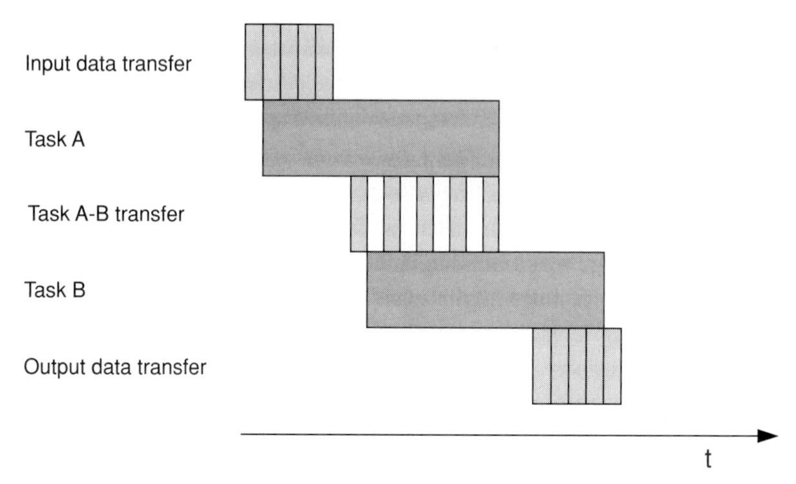

Figure 4-15 Latency of overlapped task execution.

The communications architecture is built from a simple set of basic *software communication modes*. These communication mechanisms are complemented by a set of *interconnect hardware mechanisms*. The next section discusses the communication modes and interconnection mechanisms in greater detail, but architects gravitate toward a few basic styles of overall communication structure. The range of styles reflects the range of system challenges faced by the design team:

Challenge: Lowest cost, good flexibility. When the biggest challenge in the design is low cost and high generality of the chip architecture, cheap and flexible communication is mandatory. A shared-bus architecture, in which all resources are connected to one bus, may be most appropriate. A shared bus provides the simplest possible mechanism for arbitrating access to all resources. All resources are available on the bus and the bus master (a processor or other logic) controlling the bus at any point in time also controls the memories and other devices attached to that bus. The basic structure of this system style is shown in Fig 4-16.

Buses have two significant advantages: they tend to have low hardware complexity and bus-design issues are familiar to most designers who have worked at the board level. The high cost of printed circuit board and backplane interconnect mandate heavy use of shared buses, but those cost pressures are much lower in SOCs because of silicon's remarkable wiring density.

The glaring liability of the shared bus is long and unpredictable latency, particularly when a number of bus masters contend for access to different shared resources connected to the bus. The bus can be made wider to reduce transfer latency, but achieving high bandwidth may have a high hardware cost.

Challenge: Good throughput, good flexibility. When the biggest challenge is total communications throughput with flexibility, the preferred structure is a general-purpose parallel communication network. This structure provides more than one physical communication path

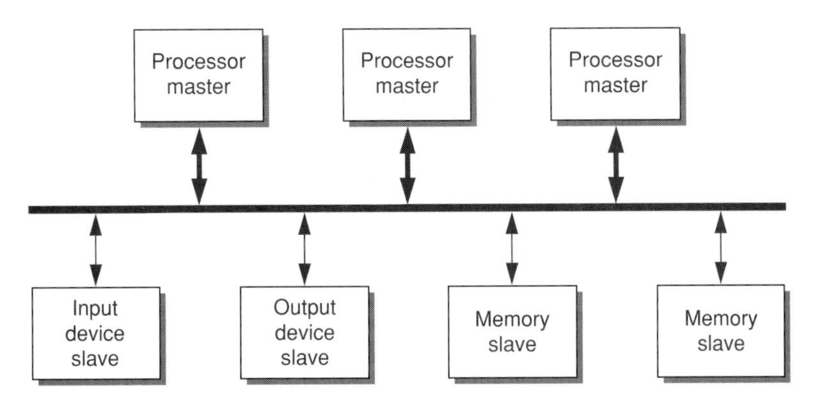

Figure 4-16 Shared bus communications style.

between tasks (or between tasks and I/O) and allows different sets of tasks to communicate simultaneously. Semiconductor scaling provides continuous improvement in silicon density, so substantial concurrency among on-chip communication channels is increasingly the norm.

The theoretical capabilities of different parallel routing networks have been exhaustively studied, but an exhaustive discussion of these network topologies is beyond the scope of this book. Two simple, common examples—a full crossbar connection and a two-level hierarchy of buses—are shown in Fig 4-17 and Fig 4-18 respectively. Both of these network topologies allow tasks to reach any other task or resource on the chip under software control, and both allow multiple data transfers to be active simultaneously. The crossbar implements a dedicated path between each master and slave. Arbitration may still be required at each slave if several masters request access to the same slave simultaneously.

A hierarchy of buses provides some of the same benefits as a crossbar. Different processors can access their own local slaves simultaneously, but access to global resources still requires access to a common bus, so not all transfers are possible in parallel.

On-chip networks or mesh connections of processors have been an important topic for academic research and are likely to become more important in commercial SOC designs. A simple example of mesh topology, with nine processors, each with local memory and a connection to a routing node, is shown in Fig 4-19. The routers allow data communication, in the form of messages or shared-memory transfers, to be sent from one processor to another by dynamically chosen paths at a cost proportional to the number of router-to-router hops through the mesh. This topology is quite scalable and is particular effective if most processor-to-processor communication is local. However, the mesh must support unpredictable and global communication as well. Note that the router mesh is also connected to global resources such as I/O interfaces and memory.

The full range of potential network topologies offers the architect a rich set of tradeoffs in implementation cost, best- and worst-case latency, and degree of concurrency. The wide range of possible architectures exposes both the advantages and the liabilities of this style:

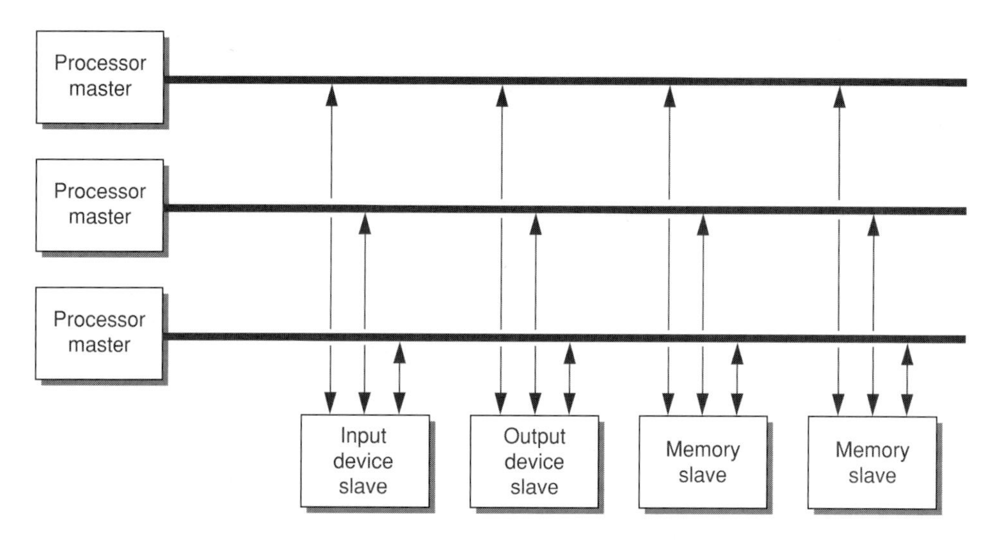

Figure 4-17 General-purpose parallel communications style: cross-bar.

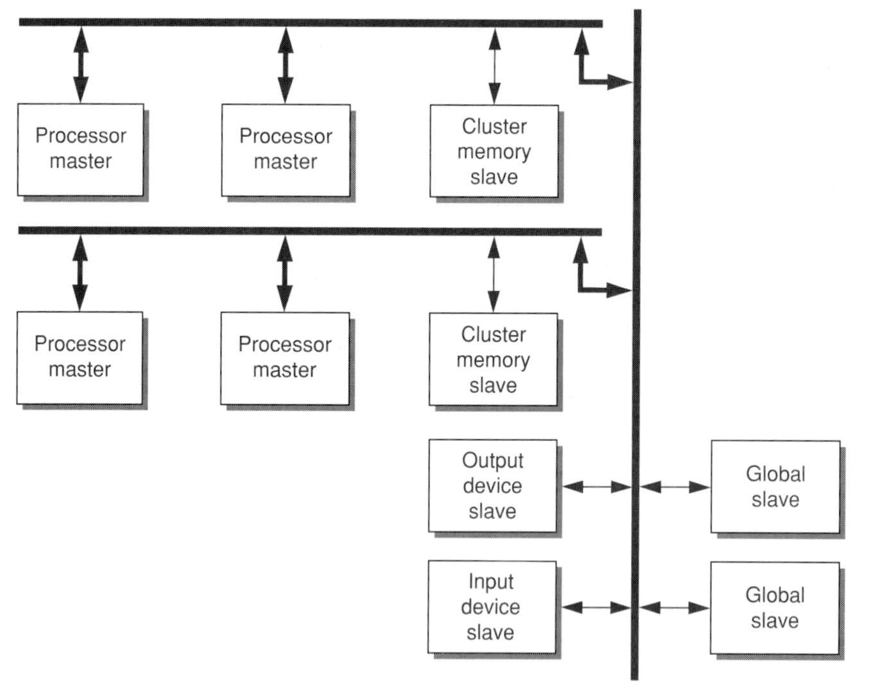

Figure 4-18 General-purpose parallel communications style: two-level bus hierarchy.

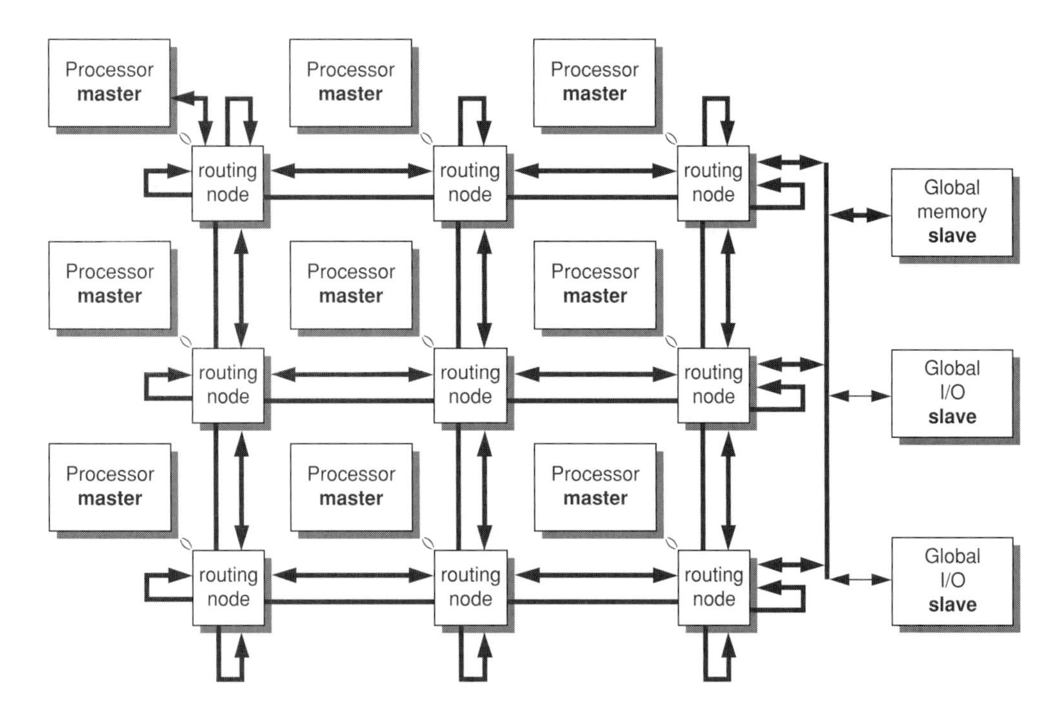

Figure 4-19 General-purpose parallel communication style: on-chip mesh network.

implementation cost, generality, and performance tend to be strongly interrelated, so the architect must remain aware of how much performance and flexibility are necessary, or the resulting cost may be higher than originally anticipated.

Challenge: Best latency and throughput. When the communication pattern is well known at design time and likely to be stable, the architect can optimize the communications structure around that particular pattern.

In these cases, the structure is less general-purpose but more efficient, because its basic components are dedicated connections among the tasks that must communicate the most, as shown in Fig 4-20. In some cases, the connections are bidirectional: data can flow in either direction between tasks. In other cases, data flows in only one direction and the hardware is optimized for unidirectional flow. If other tasks require additional, lower-bandwidth communication, the system's communication needs can be satisfied either through an auxiliary low-bandwidth shared interconnect (such as a bus) or by programming each processor to forward data between tasks that are not directly connected.

These three communication styles represent conceptual starting points, rather than specific solutions for SOC design. The architect must look closely at the real communication requirements of the set of tasks to find a suitable communication structure. The architect may

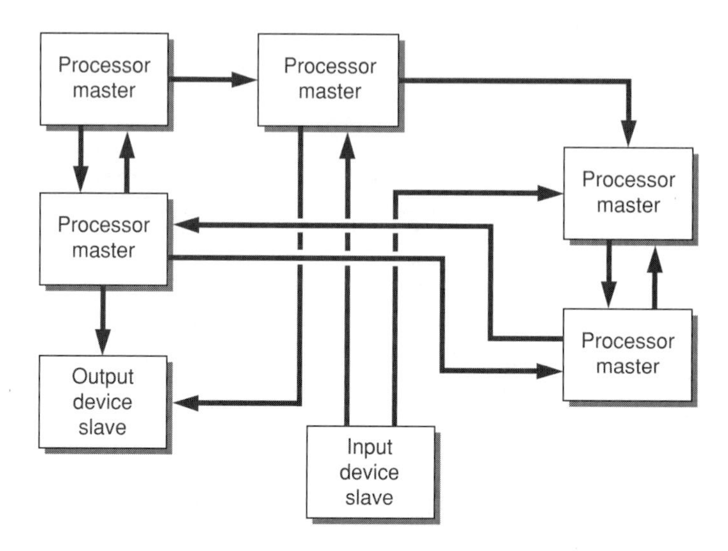

Figure 4-20 Application-specific parallel communications style: optimized.

choose to closely follow one of these styles, or may build a hybrid that uses a combination of styles to accommodate the diversity of processors and tasks in different parts of the system.

These basic communication structures are just a starting point. When the set of tasks is easy to simulate and the communications structure is easy to change, a quantitative, predictable, application-driven design methodology becomes possible.

4.3 Communication Design = Software Mode + Hardware Interconnect

Intertask communications are built on two foundations: the choice of the programming paradigm or software communication mode and the choice of the corresponding hardware or interconnect mechanism that actually ensures that data is transported from one task to the next. Each software mode can span a wide range of performance and hardware cost levels, and each software mode can be implemented with a variety of different hardware mechanisms. This section describes the major modes and mechanisms and highlights the benefits and limitations of each one.

4.3.1 Software Communication Modes

Most programming models for communication fall into two broad categories: *explicit communication*, in which data transfers between tasks are handled by some form of `send` and `receive` message, and *implicit communication*, in which data transfers between tasks occur through read and write accesses to shared memory.

Both modes are quite general—almost any task can be coded with each method—and each has implementation variants that may affect cost, performance and efficiency in expressing the desired task functions. A third mode—the *device driver interface*—combines the characteristics of message passing and shared memory. This third mode serves particularly well when the design team that wants to upgrade traditionally hardwired functions into software tasks running on a processor.

Effective exploitation of SOC technology closely depends on building fast, programmable, parallel-processing designs with good task-to-task communications. A more flexible range of hardware interfaces and a rich vocabulary of communications programming paradigms minimize cost, increase data bandwidth, and control software complexity. Designing with these communications methods creates electronic systems with higher throughput, lower power, and greater adaptability to changing needs.

4.3.2 Message Passing

Message passing makes all communication between tasks overt. All data is private to a task except when operands are sent by one task and received by another task. The basic semantics of message passing are simple:

`send(data_out, to_task)`—send the message **data_out** to the task **to_task**.

`receive(data_in, from_task)`—receive the message, **data_in** from the task **from_task**.

The send/receive model implies a queue between the sender and receiver. Messages are received in the order sent. Messages cannot be sent if the output queue is full and, of course, messages cannot be received if the input queue is empty.

Send and receive operations often come in two flavors: blocking and nonblocking. A blocking send suspends task execution at the point of the send if the system is unable to hold the data pending the receive operation. Similarly, a blocking receive suspends task execution until data arrives if it is not immediately available. Nonblocking send and receive operations typically return a Boolean value that indicates whether the operation was successful. The task can then either wait or go on to some other operation and retry the send or receive operation later.

Send and receive operations may either specify the data directly—as a single operand value—or point to a data structure that may contain a large amount of data. In either case, the data is called a message. The interpretation of the message is up to the software designer. Send/receive operations may specify a channel or message queue, rather than a task. This allows a wider range of message passing (many different message connections between two tasks can be active simultaneously), and the messages are not confused. Similarly, message channels can be shared so that a number of different tasks can produce data and put it into a common queue where the data is received by one or more consuming tasks. Shared queues prove useful especially if the producer or consumer task is parallelized across multiple processors.

Message passing also makes the points of task-to-task communication more obvious because the communication is easily distinguished in the source code from other memory references. This distinction often makes task code easier to understand and less prone to subtle bugs. Message passing is generally easier to code than shared memory when the tasks are largely independent, but is often harder to code efficiently when the tasks are very tightly coupled.

The message-passing approach also abstracts the communication programming from the hardware implementation. Message passing can be implemented by a variety of hardware mechanisms, including both direct hardware queues and emulation of queues in global shared memory. Hardware queues give the lowest latency and processor overhead, especially for small, fixed-length messages such as single operands. Global memory implementation gives the lowest cost for deep buffering and large message sizes. Section 4.4 compares these alternatives in detail.

4.3.2.1 Message Passing API Example: Wind River VxWorks

Wind River's VxWorks operating system offers a wealth of choices in its APIs, including single-processor and multiprocessor programming models, shared-memory and message-passing libraries, task management and scheduling support, file systems, and comprehensive middle-ware support for networking, graphics, and other systems programming.

For message passing, the essential operations are sending and receiving of messages through dynamically allocated queues. The principal functions are the following:

Function	Purpose	Arguments
`msgQSend()`	Send a message to a message queue	**`msgQID:`** identifier of message queue on which to send **`buffer:`** pointer to message to send **`nBytes:`** length of message **`timeout:`** time to wait if queue is full (special cases: `WAIT_FOREVER` and `NO_WAIT`) **`priority:`** `URGENT` or `NORMAL`
`msgQReceive()`	Receive a message from a message queue	**`msgQID:`** identifier of message queue from which to receive **`buffer:`** pointer to buffer to receive message **`nBytes:`** length of buffer **`timeout:`** time to wait if queue is empty (special cases: `WAIT_FOREVER` and `NO_WAIT`)

Function	Purpose	Arguments
`msgQCreate()`	Allocate and initialize a message queue	**`maxMsgs:`** max messages that can be queued **`maxMsgLength:`** max bytes in a message **`options:`** messages queued in FIFO or priority order
`msgQDelete()`	Terminate and free a message queue	**`msgQID:`** identifier of message queue to delete

A variety of other API calls allow queries about queue status and implementation of shared queues. This facility is fully dynamic. Queues of arbitrary capacity can be created and destroyed at any time. This flexibility carries a price in communication overhead (high latency, low bandwidth, potential for unbounded memory usage).

Many of the optimizations for improving communications throughput are equivalent to restricting or fixing the arguments of the library calls. For example, in a hypothetical multiple-processor implementation of this library, fixing the queue ID of the send and receive channels, the message length, and the maximum queue capacity create a communication model that could be directly implemented as a hardware queue between two processors, where the message length is the hardware queue width and the maximum message capacity is the hardware queue depth.

4.3.3 Shared Memory

The shared-memory communication mode enables implicit communication between tasks. All memory or a significant region of data memory is shared and visible to two or more tasks. The processors arbitrate for access to that memory block so that only one task reads from or writes to the data buffer at a time. When one task writes a data value to an address, subsequent reads of that address by other tasks get the new value. Fig 4-21 shows the basic interaction between two tasks communicating through shared memory.

While the data transfer itself is implicit in the common address, successful use of shared memory requires explicit access synchronization. A destination task must know when the sourcing task has written valid data. Otherwise, the destination task may read old data. Similarly, the destination task must somehow signal to the source task that it has finished reading the old data and that new data can be written to shared memory. Fig 4-22 shows the necessary handshake between two tasks communicating through shared memory.

Languages for embedded software typically include features that ease shared memory programming. In C, for example, the key variables in shared memory are tagged as volatile to force the compiler to generate object code that orders the variable references as indicated by the source code and ensures that the reads and writes actually reference shared memory. Explicitly

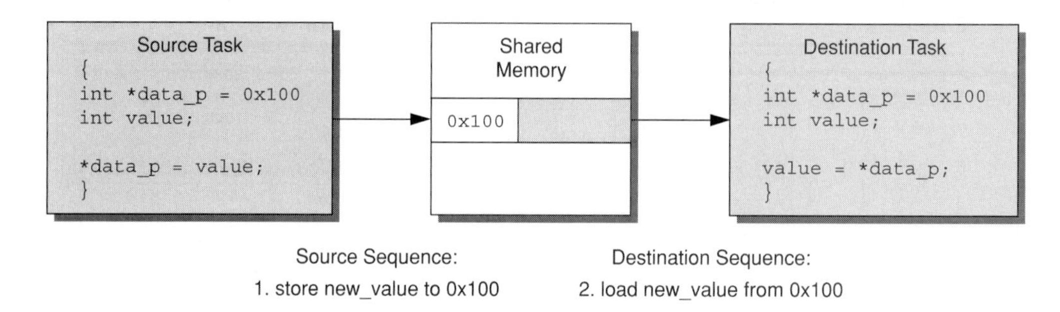

Figure 4-21 Idealized shared-memory communications mode for simple data transfer.

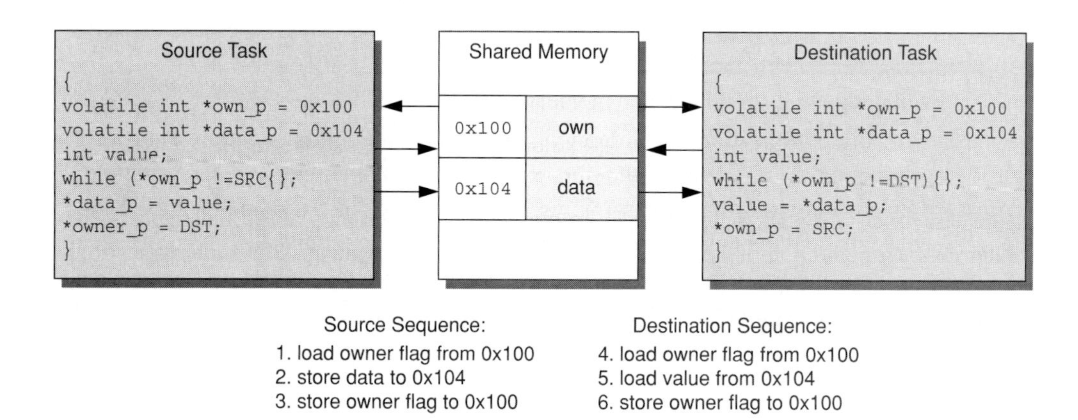

Figure 4-22 Shared memory communications mode with ownership flag.

ordering and flushing the values to shared memory (for Tensilica's xcc compiler, with "#pragma flush") or making a call to a communication library through a standard API may be a preferred method for enforcing these shared-memory access rules.

Coordination of one source and one destination through one location in shared memory is simple. A single memory address can hold a value that indicates whether the source task ("SRC") or the destination task ("DST") should have access to the shared memory. So long as the memory operations, especially between references to the flag and references to the data, are actually performed in the memory in the same order as implied by the task code, this code is robust.

Coordination becomes more complex when more than one source task or more than one destination task share memory. In these cases, the synchronization mechanism must deal with the possibility of simultaneous access to the flag that controls data access. Consider a situation in which two destinations task are competing for access to data just written by a source. The flag—a sample semaphore that controls data access—must have at least three states: owned by the source task ("SRC"), available to a destination task ("RDY"), and owned by one destination task ("DST"). The destination tasks might try to take ownership of the shared memory by executing a simple test:

```
{
...
while (*flagp != RDY){}; /* check flag */
*flagp = DST; /* set flag */
...
/* use data */
}
```

If two destination tasks (Task A and Task B) are executing this same code sequence, there are four possible orderings of the reads (*flagp != RDY) and writes (*flagp = DST), shown in Fig 4-23. (Assume that all processors see exactly the same ordering of reads and writes. Even further complications arise if memory-event ordering is perceived differently in different processors.)

Case	1	2	3	4

Memory reference order	`A: check flag`
	`A: set flag`
	`B: check flag`
		`A: check flag`
		`A: set flag`
		`B: check flag`
			`A: check flag`
			`A: set flag`
			`B: check flag`
				`A: check flag`
				`A: set flag`
				`B: check flag`

Result	Task A gets access to the data and Task B waits	Both Task A and Task B see the data as available and erroneously use the same data	Task B gets access to the data and Task A waits	Both Task A and Task B see the data as available and erroneously use the same data

Figure 4-23 Unpredictable outcome for simultaneous shared-memory access by two tasks.

The possibility of unpredictable results drives the need for locks when multiple processors share memory. A number of different mechanisms have been used, but all provide a means to ensure that one task can check and set a flag without worrying that another task is checking at the same time. The most efficient locks use some hardware support to defer other tasks' accesses to a memory location, to a memory, or to all addresses on a bus until the check-and-set operation is complete. Some common locking mechanisms are described in more detail in Chapter 7.

4.3.3.1 Shared Memory API Example: VxWorks

WindRiver's VxWorks also supports a shared memory-programming model. The essential functions, shown in Fig 4-24, include mechanisms for allocating memory, designating it as shared, and controlling simultaneous access to shared data with semaphores.

VxWorks' API also implements a range of semaphore variants (including semaphores optimized for mutual exclusion and counting semaphores) and includes calls to delete semaphores, free shared memory, and to examine and manipulate the tasks waiting for semaphores to be available.

Function	Purpose	Arguments
`smMemMalloc()`	Allocate a block of memory from the shared-memory system partition, returning a pointer to the allocated block	`nBytes:` number of bytes to allocate
`semTake()`	Take (acquire) a semaphore	`semID:` semaphore ID to take `timeout:` time to wait if semaphore is already taken (special cases: `WAIT_FOREVER` and `NO_WAIT`)
`semGive()`	Give (release) a semaphore	`semID:` semaphore ID to give
`semBCreate()`	Create and initialize a binary semaphore, returning ID	`options:` queuing style for blocked (waiting) tasks: `PRIORITY` or `FIFO`) `initialState:` initial semaphore state (`SEM_FULL` or `SEM_EMPTY`)

Figure 4-24 VxWorks shared-memory API functions.

4.3.4 Device Driver

One routine form of concurrency used in computer systems is parallel execution of the CPU and the I/O controllers. This approach to task distribution has led to a common master—slave programming mode in which a control task writes and reads a small set of control registers in the slave device. The slave, traditionally implemented as hardwired data path and finite-state machine, takes command information from those control registers and sends or receives data through a simple interface register. The hardware-device-plus-software-device-driver model was most commonly used with complex I/O interfaces, such as networks or storage devices such as disk drives. The device-driver software itself often serves as part of the real-time operating system to consolidate requests from a variety of tasks, limiting and coordinating access to privileged system resources such as network and storage devices.

If a slave device's responsiveness is slow or irregular, interrupts can enhance basic communication between master and slave. Interrupts allow the master to continue with other work while waiting for a response from the slave. The interrupt from the slave forces the master to execute the interrupt service routine, to complete a data transfer from the slave, or to start the next operation between slave and master. Without an interrupt mechanism, the master must poll the slave intermittently, which consumes processor cycles that might be used for other processing.

The device-driver mode combines elements of message passing and shared-memory access. The interrupt from slave to master implements a preemptive message. Other data and control transfers from master to slave are mapped into registers as shared-memory operations. Those few registers are dedicated to the slave's operations, and the slave has no access to other shared memory. Usually, only one master attempts to control the slave, so synchronization is trivial, requiring only a simple "start" command with corresponding "finished" status returned by the slave.

The principles of the device driver can be applied to almost any pair of communicating tasks, especially where the interface between tasks looks like series of requests and responses. Fig 4-25 shows the device-driver interface with four slave registers: master command register (written by master, read by slave), slave response register (written by slave, read by master), to-slave data register (written by master, read by slave), from-slave data register (written by slave, read by master).

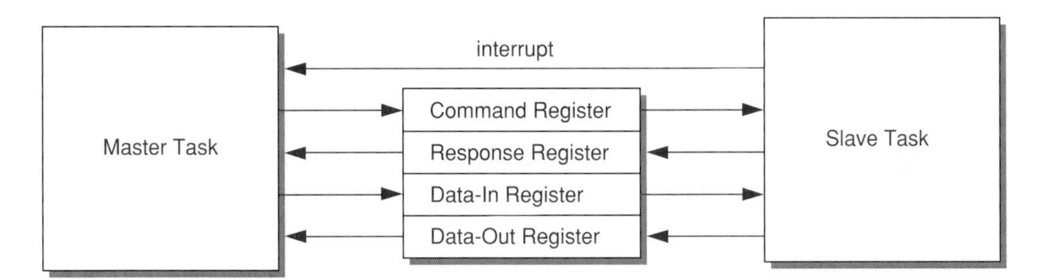

Figure 4-25 Device-driver master/slave interface handshake.

4.4 Hardware Interconnect Mechanisms

The choice of hardware-interconnection mechanism affects communication performance and silicon cost. The message-passing-software communications mode has a natural correspondence to data queues, but message passing can be implemented using other types of hardware, such as bus-based hardware with global memory. Similarly, the shared-memory software-communications mode has a natural correspondence to bus-based hardware, but shared-memory protocols can be physically implemented even when no globally accessible physical memory exists. This implementation flexibility allows chip designers to implement a spectrum of different task-to-task connections in ways that optimize performance, power, and cost together.

A short description of the common hardware mechanisms—buses, direct connections, and data queues—is appropriate here. Except where explicitly noted in this discussion, we assume a one-to-one correspondence between tasks and processors. In fact, multiple tasks can be mapped onto one time-sliced processor and tasks can be implemented by other nonprogrammable hardware blocks.

Note that in many cases, the task-to-task connection is not made directly, but between the task and a memory. If that memory can be reached by more than one task, then communication between the tasks becomes possible. Memory sharing may be hidden from each task's software developer by a software layer, so the presence of shared memory in hardware is equivalent to a shared-memory communication mode.

4.4.1 Buses

A bus is a shared-access hardware mechanism allowing one or more processors to communicate with slave memories and I/O interfaces. In the simplest case, each slave is accessible only from one bus, so the processor that owns the bus also owns the slaves. Different processors must arbitrate for the bus, but this is the sole arbitration mechanism. Processors and slaves may have a range of bus-transfer requirements based on hardware limitations (e.g., an 8-bit UART slave device may not allow any 16- or 32-bit transfers) or traffic patterns (the processor may maximize performance with cache-line-sized block transfers—16 bytes or more).

Moreover, some transfers may be quite latency-sensitive (the task doesn't need much data, but it needs that data immediately), and others may be more bandwidth-sensitive (the task must get some average sustained bandwidth, but the latency of any one transfer is inconsequential).

4.4.1.1 Bus Design Tradeoffs

Bus designs may use a range of strategies to satisfy conflicting goals among the processors, memories, and other devices they connect. Three classes of design decision stand out:

Bus Width and Clock Rate: The bus width and clock rate determine the peak transfer rate over the bus. These factors impact cost, power, and technology requirements.

Arbitration: The arbitration mechanism affects tradeoffs between total bus utilization and the latency seen by any one bus master. Round-robin arbitration gives all masters equal access to the bus, but even the most important requests may face long contention delay. *Round-robin* arbitration is fair in that all masters have an equal chance to get bandwidth, and it is efficient in that bus cycles are utilized if any master needs them. *Strict-priority* arbitration gives the most critical bus master preferential treatment all the time so that it sees minimum contention latency. *Reserved-bandwidth* arbitration gives a bus master a minimum guaranteed bandwidth over a time interval, but the master can also compete for additional bandwidth on a round-robin basis. The choice of arbitration mechanism is driven by the system bandwidth and latency requirements but may be constrained by a pre-defined bus protocol.

Transfer Types: Simple buses may implement just a few transfer types such as 8-, 16-, and 32-bit reads and writes. More complex buses may implement any of a number of more advanced transfer types:

- *Fixed-block transfers:* Power-of-two sized blocks, often used for cache-line refills and write-backs.
- *Variable-block transfers:* Arbitrary length transfers, often used to move data in streams with application-dependent block sizes.
- *Split transactions:* The decomposition of a bus request, usually a read, into two transfers: one to convey an address from the master to the slave and a second to return a response data block from the slave to the master. The bus is relinquished to other masters during the interval between the request and response. Split transactions are particularly important for maintaining high bus bandwidth with long memory-device latencies and multiple bus masters.
- *Atomic transactions:* When two or more masters are competing for access to a shared resource, some locking mechanism is required to support arbitration mechanisms. Sometimes this mechanism is implemented as a bus lock, in which certain read operations retain bus mastership after the read data is returned, so that the processor can perform a write without risk that another processor may read the same location. Bus locking is not efficient, however, in a system with many processors, many separate memories, and frequent locking operations.

4.4.1.2 Bus Implementation with Configurable Processors

Configurable processors offer significant flexibility in supporting arbitrated access to shared devices and memory. The full spectrum of processor interface options is discussed in Chapter 6, but the basic topologies for shared-memory buses are enumerated here:

1. **Remote global memory accessed over a general processor bus:**

   The processor implements a general-purpose interface that allows a wide variety of bus transactions. If the processor determines that the corresponding data is not local during a read (based on the address or due to a cache miss), the processor must make a non-local reference. The processor requests control of the bus, and when control is granted, sends the target read address over the bus. The appropriate device (for example, memory or I/O interface) decodes that address and supplies the requested data back over the bus to the processor, as shown in Fig 4-26, where the communicated data is shown in gray.

   When two processors are communicating through global shared memory on the bus, one must acquire bus control to write the data; the other processor must later acquire bus control to read it. Each word transferred in this fashion requires two bus transactions. This approach requires modest hardware and maintains high flexibility, because the global memories and I/O interfaces are accessible over a common bus. However, the use of

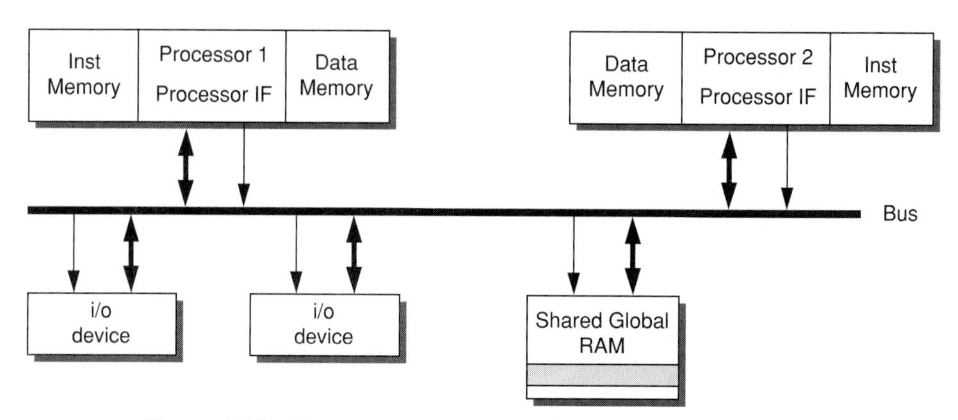

Figure 4-26 Two processors access shared memory over bus.

global memory does not scale well with the number of processors and devices, because bus traffic leads to long and unpredictable contention latency.

2. **Local processor memory accessed over a general processor bus:**

Configurable processors may allow local data memories to participate in general-purpose bus transactions. These data memories are primarily used by the processor to which they are closely coupled. However, the processor controlling the local data memory can serve as a bus slave and respond to requests on the general-purpose bus, as shown in Fig 4-27.

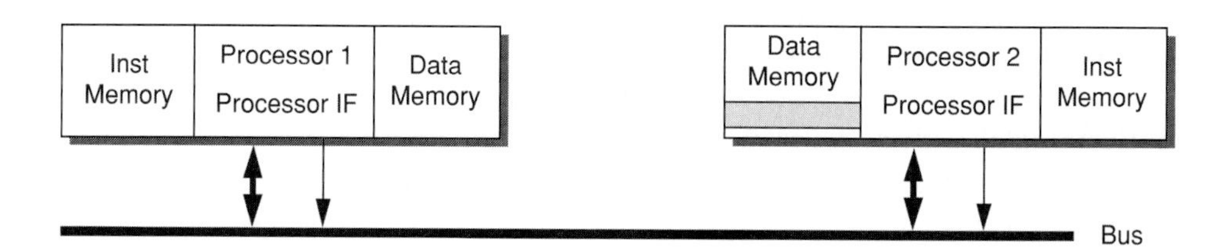

Figure 4-27 One processor accesses local data memory of a second processor over a bus.

In this case, the read by Processor 1 may require access arbitration at two levels: first when Processor 1 requests access to the general-purpose bus, and second, when the read request reaches Processor 2. The read request from Processor 1 arrives over Processor 2's processor interface and may contend with other requests for local data-memory access from tasks running on Processor 2. Two arbitration levels may increase the access latency seen by Processor 1, but Processor 2 avoids access latency almost entirely, because latency to local data memory is short (usually one or two cycles).

This latency asymmetry between Processor 1 and Processor 2 encourages *push communication*: when Processor 1 sends data to Processor 2, it writes the data over the bus into Processor 2's local data memory. If the write is buffered, Processor 1 can continue execution without waiting for the write to complete. Thus the long latency of data transfer to Processor 2 is hidden. Processor 2 sees minimal latency when it reads the data, because the data is local. Similarly, when Processor 2 wants to send data back to Processor 1, it writes the data into Processor 1's local data memory.

3. **Multiported local memory accessed over local bus:**

   When data flows in both directions between processors and latency is critical, a locally shared data memory is often the best choice for intertask communications. Each processor uses its local data memory interface to access a shared memory. This memory could have two physical access ports (two memory references satisfied each cycle) or could be controlled by a simple arbiter, where one processor's access is held off for a cycle if the other processor is using the single physical access port. See Fig 4-28.

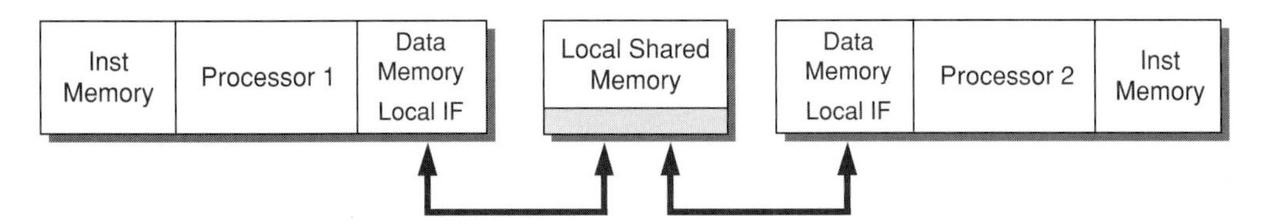

Figure 4-28 Two processors share access to local data memory.

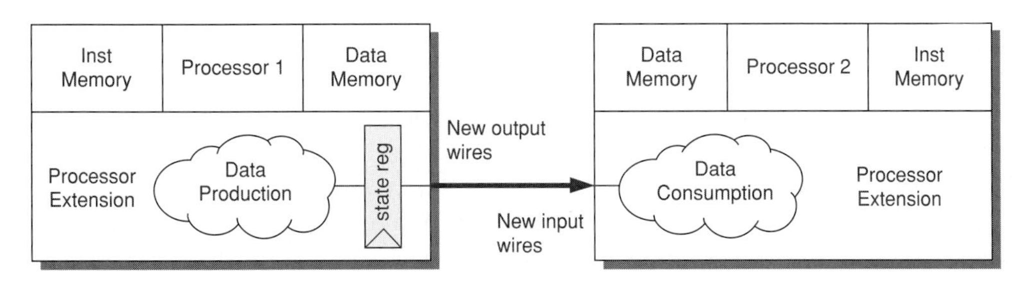

Figure 4-29 Direct processor-to-processor ports.

Arbitration for a single port is preferred in area- and cost-sensitive applications, especially when shared-memory utilization is modest, because a true dual-ported memory is about twice as big per bit when compared to single-ported RAM. However, a true dual-ported memory may be the better choice when the shared memory is very small or when absolute determinism of access latency is required.

4.4.2 Direct Connect Ports

Direct processor-to-processor connections reduce cost and latency for communication. They allow data to move directly from one processor's registers to the registers and execution units of another processor. A simple example of direct connection is shown in Fig 4-29. This example takes advantage of exportation of state registers and importation of wire values (features found in some extensible processors) to create an additional dedicated interface within each processor and to directly connect them.

Whenever the Processor 1 writes a value to the output register, usually as part of some computation, that value automatically appears on the output pins of the processor. That same value is immediately available as input value to operations in Processor 2. Wire connections can be arbitrarily wide, allowing large and non-power-of-two-sized operands to be transferred easily and quickly.

Some versions of Tensilica's Xtensa processor allow you to create registers with exported state, operations that write these states, and other operations that use these new input values from other exported states. All of these features are expressed in the TIE language, and this capability is described in more detail in Chapter 6.

The operation that produces the data for the output state register may be as simple as a register-to-register transfer, or it may be a complex logic function based on many other processor state values. Similarly, the input value can simply be transferred to another processor state within Processor 2 (register or memory), or it could be used as one input to a complex logic function.

This form of direct connection still requires some handshake between the two processors. The consumer of data may need to signal to the producer that the data in the register has been used so that the producer can write the next data value. The producer may need to signal the consumer that new data is available. This signaling can be done in several ways, including the following:

Consumer-to-producer port: An architect can make two additional port connections, each just one bit wide, one from the consumer processor back to the producer processor, and one from producer to consumer. The consumer asserts its acknowledge output when the data has been used. The producer uses this signal as part of the decision in the code to generate the next output value. The producer asserts its "data-ready" handshake output when the next data value is available. The consumer should negate its "acknowledge" signal, in preparation for the next assertion when the next data word has been processed. The handshake is shown in the timing diagram in Fig 4-30. Because this transaction requires at least one full instruction execution per signal transition, this method consumes at least a dozen cycles per data word transferred.

A variant of data queues, described in Section 4.4.3 Data Queues, creates producer—consumer handshake signals automatically. The data ready signal is equivalent to the push into the tail of a queue, and the acknowledge signal is equivalent to the pop from the head of a queue. A flag bit, set by data ready and cleared by acknowledge coordinates the two tasks.

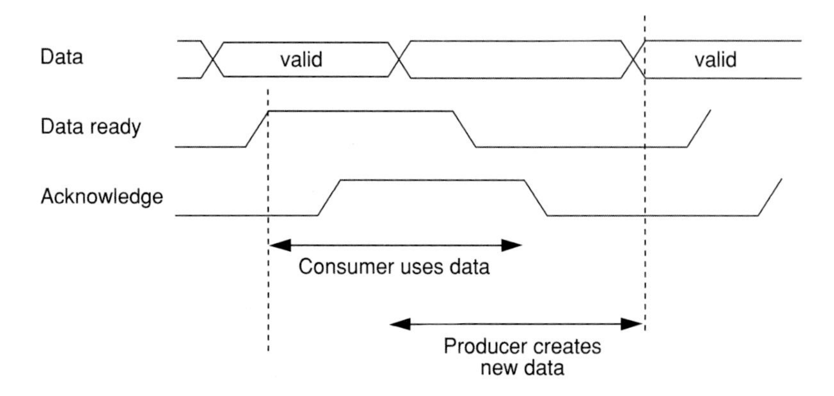

Figure 4-30 Two-wire handshake.

Interrupt-driven handshake: The data transfer can also be controlled by interrupts between the two processors. When the producer processor has created the data and placed it on its output port, it also asserts a signal on an output wire connected to an interrupt input of the consumer processor. The consumer processor handles the interrupt as soon as it can (after any higher priority interrupts are handled), and accepts the data from the input port within the interrupt handler. The consumer's interrupt handler then asserts its own output signal, which is connected to an interrupt input on the producer processor. The producer interrupt handler can then drive new data to the consumer. The basic structure of the interrupt-driven handshake is shown in Fig 4-31.

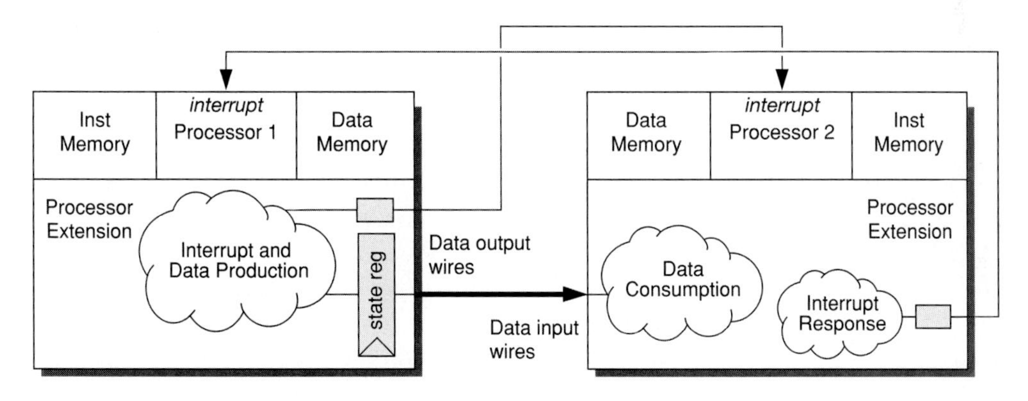

Figure 4-31 Interrupt-driven handshake.

4.4.3 Data Queues

The highest-bandwidth mechanism for task-to-task communication is hardware implementation of data queues. One data queue can sustain data rates as high as one transfer every cycle or more than 10 GB per second for wide operands (tens of bytes per operand at a clock rate of hundreds of MHz) because queue widths need not be tied to a processor's bus width or general-register width. The handshake between producer and consumer is implicit in the interfaces between the processors and the queue's head and tail.

When the data producer has created the data, it pushes it into the tail of the queue, assuming the queue is not full. If the queue is full, the producer stalls. When the data consumer is ready for new data, it pops it from the head of the queue, assuming the queue is not empty. If the queue is empty, the consumer stalls.

Queues can also be configured to provide nonblocking push and pop operations, where the producer can explicitly check for a full queue before attempting a push and the consumer can explicit check for an empty queue before attempting a pop. This mechanism allows the producer or consumer task to move to other work in lieu of stalling.

Application-specific processors allow direct implementation of queues as part of their instruction-set extensions. An instruction can specify a queue as one of the destinations for result values or use an incoming queue value as one source. This form of queue interface, shown in Fig 4-32, allows a new data value to be created or used each cycle on each queue interface. A complex processor extension could perform multiple queue operations per cycle, perhaps combining inputs from two input queues with local data and sending values to two output queues. The high aggregate bandwidth and low control overhead of queues allows application-specific processors to be used for applications with very high data rates where processors with conventional bus or memory interfaces are not appropriate because they cannot handle the required high data rates.

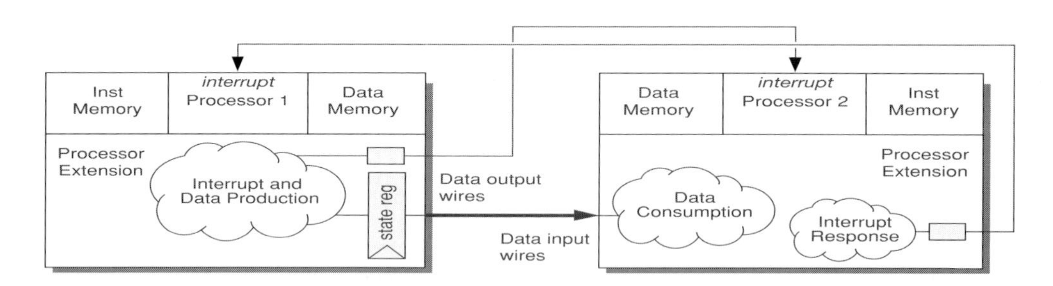

Figure 4-32 Hardware data queue mechanism.

Queues decouple the performance of one task from another. If the rate of data production and data consumption are quite uniform, the queue can be shallow. If either production or consumption rates are highly variable, a deep queue can mask this mismatch and ensure throughput at the average rate of producer and consumer rather than at the minimum rate of the producer or the minimum rate of the consumer. Sizing the queues is an important optimization driven by good system-level simulation. If the queue is too shallow, the processor at one end of the communication channel may stall when the other processor slows for some reason. If the queue is too deep, the silicon cost will be excessive.

One processor can employ queue communications with multiple partners. When the queue operations are directly incorporated into the instruction set, the code sequence entirely determines which queue is written or read. Sometimes, less direct mapping is desirable, so the code sequence that produces or consumes data can be separated from the selection of the source or destination queue.

Two methods for flexible queue selection are possible. First, the ultimate data destination can be included in the data transfer. This destination information is pushed into a common queue with the data. This queue feeds other queues, where simple logic pops the destination identifier and uses it to choose the correct destination-specific queue in which to push the corresponding data. Full flexibility of queue width makes this approach economical. For example, a 2-bit destination specifier and a 32-bit data word would be combined in a 34-bit common queue, perhaps feeding a set of four 32-bit queues, as shown in Fig 4-33.

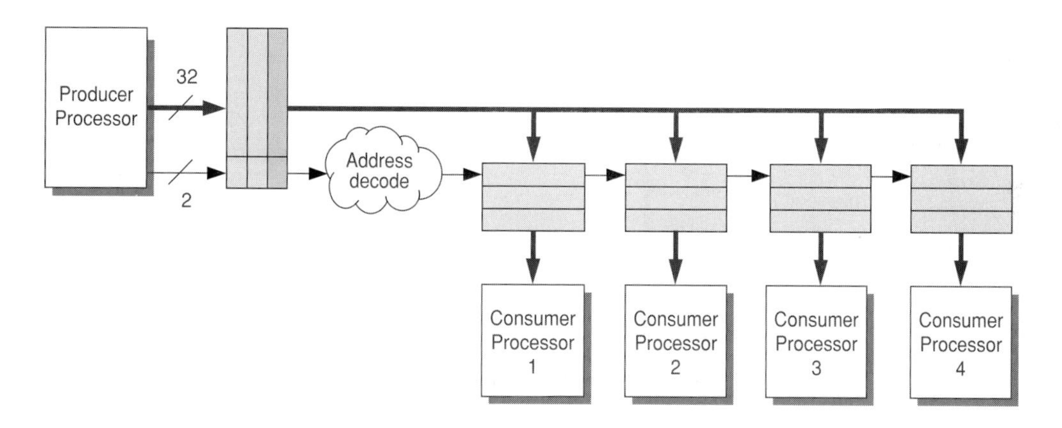

Figure 4-33 Producer enqueues destination with data.

Second, the queue head and tail can be mapped into memory so that a processor store is used to push a value and a processor load is used to pop a value. These operations can be blocking (producing a stall if the queue is full or empty) or nonblocking (processor may test the state of the queue before attempting the push or pop). Fig 4-34 shows a simple system with one producer and two consumers. The queues are mapped into the address spaces of the processors (here shown using the local-memory space with a 1-cycle access time), so any store to the address of the queue tail causes a push, and any load from the address of the queue head causes a pop.

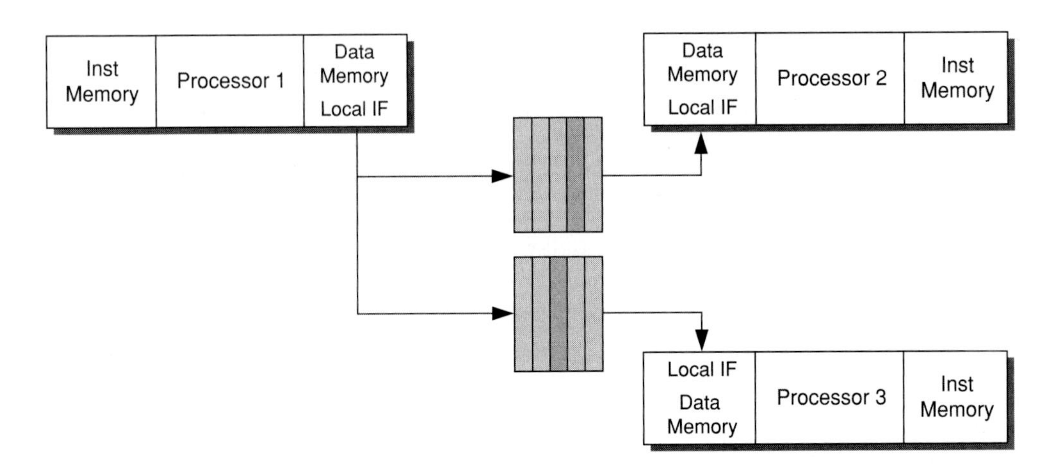

Figure 4-34 One producer serves two consumers through memory-mapped queues.

When the data rate is relatively low, the queue depth can be reduced, even to a single entry—a register that is written by the producer and read by the consumer. This *mailbox register* serves as a simple and convenient path between producer and consumer. A memory-mapped set of mailbox registers is shown in Fig 4-35. When the two tasks pass data back and forth, the same register can be used for transfers in either direction.

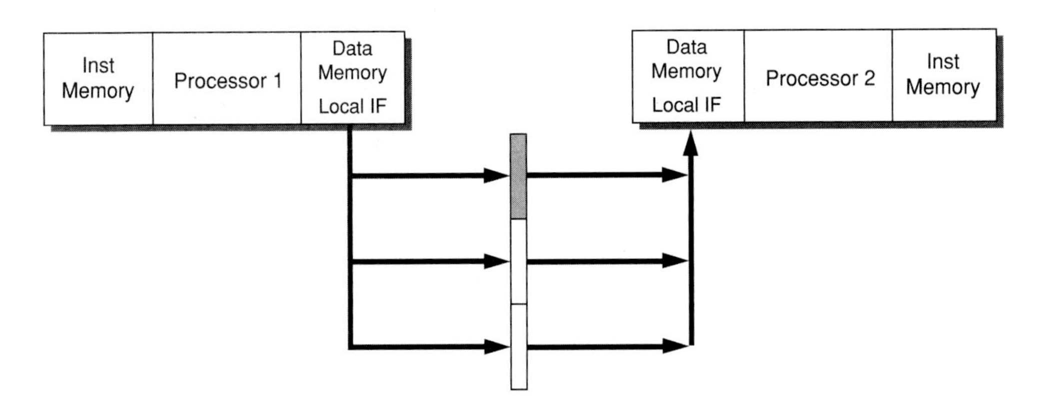

Figure 4-35 Memory-mapped mailbox registers.

Memory-mapped and instruction-mapped queues serve a wide range of processor communication uses. They work especially well at high data rates with relatively shallow buffering. At lower data rates, buses provide ample communications bandwidth. For applications with very deep buffering requirements, queues must be implemented in RAM or replaced with a shared-memory communication mechanism.

4.4.4 Time-Multiplexed Processor

Tasks can also communicate without any connection between different hardware blocks. Several tasks can share a single processor and communicate without any special hardware mechanism. An RTOS provides the capability to allocate time slices to each task so that all tasks appear to proceed in parallel. The RTOS also permits tasks to communicate through shared memory. The memory may not be physically shared with any other processor, but time-slicing of the processor creates the effect of shared memory. The RTOS must allocate a region of a processor's real memory for use as shared memory. Some operating systems may allow restrictions that prevent tasks from accessing that address range to avoid unexpected data corruption.

The RTOS typically provides programming abstraction for message passing, allowing the semantics of data queues to be implemented purely in software. Time-multiplexing one processor to interleave tasks is primarily a software mode, but requires basic hardware support. The hardware must implement at least one timer that generates periodic interrupts, which are used by the RTOS to ensure that no task hogs more than its allocated share of the processor. While different real-time operating systems provide different variants of task scheduling and intertask communication, the principles are largely common. Some simple operating systems do not implement preemptive time-slicing but rely instead on cooperative scheduling among tasks, where each task must explicitly relinquish control of the processor to allow other tasks to execute.

The above list of hardware mechanisms is not exhaustive. For example, some multiprocessor systems are built with dedicated message-passing buses, where short messages use one mechanism—a dedicated low-wire-count bus for low-contention distribution—while long messages or shared-memory communications use shared wide buses. This combination of mechanisms reduces contention latency for urgent messages, yet adds little to hardware cost.

4.5 Performance-Driven Communication Design

The central premise of the advanced processor-centric SOC design is this: parallelism occurs naturally in SOC design and designers can exploit that parallelism through both computation and communication means. In fact, the more the designer understands about the task interaction and the internal demands of each task, the more efficient the final design.

4.5.1.1 The Role of System Modeling in Complex SOC Design

Using system simulation throughout the design process serves two crucial roles in this understanding. First, an early start to simulation, even when the architecture is still quite unsettled, gives the design team advanced insight into bottlenecks.

Second, as design decisions are nailed down through the course of design refinement, these decisions are incorporated into the system model. The model's role as a performance predictor gradually evolves into a role as verification test bench. One of the greatest challenges for testing complex subsystems is generation of realistic stimulus and accurate lists of expected outputs. An accurate system model allows the design team to replace almost any subsystem model with a more detailed trial implementation and to test the implementation as part of the complete system model.

These different roles for modeling impose a diverse—and sometimes incompatible—set of requirements on the modeling. The ideal modeling environment would provide all of the following:

- Simplicity and conciseness of early descriptions of communicating tasks, plus system I/O interfaces.
- Speedy simulation performance to allow complete systems to execute on large data sets.
- High accuracy, especially in subsystems and interfaces that must model precise hardware behavior.
- Seamless support for piecewise refinement of subsystems, moving from pure behavioral descriptions down into RTL and gate-level simulations of subsystem logic functions.
- Unified environment for specification, description, debug, and performance monitoring for functions described as abstract algorithms; as binary software running on processors, including extensible processors; and as hardwired logic functions.
- Flexibility to model system-level communications flows working either from initial task code or from rough estimates of task input and output behavior.

Even experienced architects cannot always guess the traffic patterns of data movement among a set of tasks. Sometimes the obvious data movement among input, output, and key tasks is obscured by other transfers including instruction cache misses, data cache misses, extra data copying between local memories, and other tasks with unexpected data needs. The more complex the set of tasks, the greater the likelihood that initial back-of-the-envelope traffic estimates are wrong.

Sometimes the design team knows roughly where the busiest links exist, but not the amount of traffic. Sometimes the busy links are not at all where simple analysis suggests.

Building an early system model lets the design team understand the traffic profile. This model can be built even before any task assignment and communications structure development is attempted, as described in Fig 4-9, or as soon as an initial computation and communication structure is available. When model development is simple, it is also easy to update as the design team explores different system organizations.

Even when no executable task code is yet available, a simple traffic generator approximating the expected size and frequency of messages or shared memory references can serve as a useful stand-in. The generated traffic will load down shared buses in a way that mimics the probable behavior of the final task. A proposed system topology can be entirely prototype tested using just traffic generators.

The investment in a system model can be leveraged further by using it in subsystem verification. One of the biggest system-verification challenges is generation of a comprehensive set of realistic inputs to stimulate all important data paths and state transitions. If the model is useful enough to verify functionality and performance of the system hardware and software, then the model may serve well to provide stimulus and check expected results for each subsystem.

The profile of system activity may evolve significantly as the system is tuned and refined. For example, when individual tasks are untuned, the memory and processor-to-processor traffic may appear to be modest because overall system performance is itself modest. As the algorithms and code are accelerated and system performance improves, memory and intertask-communication transfer rates may also increase.

To test a subsystem, a designer replaces the subsystem's high-level model with a lower-level implementation model. For example, an abstract model that just implements an algorithm using generic transactions at the boundary may be replaced with a cycle-accurate model with detailed pin-level interfaces.

Alternatively, a cycle-accurate model might be replaced with the RTL implementation or even the gate-level net list of the block. Flexibility in the interface of each subsystem encourages quick substitution of a more detailed model to replace the abstract model. The substitution relies on mixed-mode simulator capability in which high-level models such as a fast processor instruction-set simulator are embedded in a simulation environment implementing lower-level RTL or gate-level logic simulation.

4.5.2 System Modeling Languages

In theory, systems can be modeled in almost any programming environment. In practice, the choice of modeling method is based on language familiarity, modeling support, model availability, and simplicity. The rationale for standardization or automation in modeling environments is twofold:

1. **Component model libraries:** Just as block libraries and block reuse reduce the effort of developing and verifying hardware blocks, model libraries reduce the effort of developing and verifying models. In fact, effective reuse of complex blocks such as processors often depends on availability of corresponding simulation models. Widely adopted standards for system modeling tend to create a model marketplace, especially for commonly used and complex models. The attractiveness of a model depends on execution speed, conformance to real hardware behavior, and suitability of interface modeling.
2. **Model construction and analysis tools:** The purpose of modeling systems is to develop insight into the system behavior. The ease of accurate system description and the richness of analysis help determine the value of system modeling. Analysis views include an enormous range of system and component-level information, such as time-series display of events, visual display of cumulative event statistics, and debug of processor code.

Four system-modeling languages are discussed next: ordinary C, System Verilog, SystemC, and XTMP.

Ordinary C: Historically, system developers have often created ad hoc models in C. C has the advantage of familiarity and cost, but C lacks of any standardization in mechanisms for communication and concurrency of execution. C does not include natural support for arbitrary-width bit vectors, a common requirement for modeling hardware blocks. Moreover, the models of common components—processors, memories, buses and standard I/O devices—must be developed or interfaced from scratch.

System Verilog: Hardware designers familiar with hardware description languages such as Verilog or VHDL, might move up to a modeling environment such as System Verilog. This superset of Verilog adds a number of capabilities for object-oriented programming and system support similar to C and C++. System Verilog does not yet support a wide range of models and system-level analysis tools. Its sweet spot is likely to be test benches used for verification of hardware blocks.

SystemC: SystemC, a class library built with C++, is an increasingly attractive choice for system-level modeling and analysis. The evolution of the SystemC language and environment is coordinated by a consortium of EDA and intellectual property companies, the Open SystemC Initiative (OSCI). The SystemC 1.0 release in 1999 provided a standard threads package for modeling concurrent execution of components and basic signal modeling capability. The SystemC 2.0 release in 2002 added general-purpose abstractions important to transaction-level modeling, including *channels*, *interfaces*, and *events*, implemented as C++ class libraries. These features enable a very broad range of communication primitives and modeling levels, including communication primitives such as mutual-exclusion locks, semaphores, and FIFOs, and abstract models of parallel computation such as Kahn Process Networks, static dataflow, and discrete event simulation of transactions.

A basic description of SystemC nomenclature helps explain the SystemC modeling:

A SystemC **model** is a collection of **modules**, each of which typically corresponds to a system component or interface to the external environment (e.g., network connections, sources of system stimulus, monitors of system behavior). A module contains **processes**, **ports**, and **channels**. A **process** implements the modeled behavior of a module. Processes are inherently concurrent with one another. A **channel** implements interfaces. An **interface** is a set of publicly-visible functions (**methods** in C++ terminology) for access to the behavior of a **channel**. A process in one module accesses the interface on a channel through a C++ object, a **port**.

SystemC has become well established in system modeling research, and is starting to gain traction in actual SOC development. SystemC's strength comes from its generality—a very broad range of model types can be built with the common class libraries. Moreover, standard SystemC-based models and SystemC interfaces for accurate models are becoming more widely available. Processors from ARM, Tensilica and MIPS, for example, all support SystemC. The generality of SystemC also somewhat impedes rapid adoption. The SystemC language presents many choices, for example in the representation of transaction interfaces, so that interoperability among independently-developed SystemC models requires considerable attention. The power and generality of SystemC somewhat increases the overhead for adoption, and may concentrate its near-term use more in system verification tasks than in early architectural exploration. Nevertheless, SystemC is likely to become pervasive over time. Section 4.9 provides a glimpse of the many good texts on SystemC and SystemC-based model development.

XTMP: Tensilica's Xtensa Modeling Protocol adopts much of the same technology as full SystemC, but focuses on the tasks of greatest interest to the system architect. Like SystemC, XTMP includes links to standard thread-execution packages for easy modeling of concurrency. Like SystemC, XTMP it provides a standardized programming interface for modeling of arbitrary devices at the transaction level. Unlike SystemC, the environment provides the built-in component models of greatest initial interest to the architect doing processor-centric SOC design—the processor models, memory models and address maps. XTMP also enables quick C-based construction of the model of the high-bandwidth heart of the SOC—the connections among processors and memories. Finally, XTMP provides simple interfaces for initialization and booting of processors, and connection to source-level debuggers for code running on the simulated processors.

The simplicity and conciseness of XTMP models make it attractive for rapid model development and early architecture exploration of complex SOCs. XTMP models are easily encapsulated in SystemC modules if and when the system-modeling effort requires the additional formalism of the SystemC language. The encapsulation may be done either at the level of individual XTMP components, so each XTMP component becomes a separate SystemC model, or at the level of an entire computational subsystem, where a large collection of processors and their memories operates as a unified module within a SystemC simulation model. Moreover, the simulation models for all processor types are derived from a single extensible family, so estimates of size, power, performance, and interface behavior for all modeled processors are directly comparable. The following section describes introduces system model construction using XTMP because of its simplicity. The same modeling task could be done in SystemC.

4.5.3 System Modeling Example: XTMP

Early system modeling is one key to improved system architecture. A unified method to describe system organization, communications mechanisms, and application software boot helps the design team move from back-of-the-envelope estimates to accurate simulation more quickly. One such simulation environment is Tensilica's Xtensa Modeling Protocol (XTMP), an API and runtime environment for rapid multi-processor description and analysis. XTMP employs its own lightweight simulation engine, and it also generates SystemC-compatible models. SystemC is an emerging hardware/software co-design standard, built as a set of C++ methods. XTMP can be used for simulating homogeneous and heterogeneous multiple processor subsystems as well as complex uniprocessor architectures. The XTMP API allows system designers to write customized, multithreaded simulators to model complicated systems. Designers can instantiate multiple similar and dissimilar Xtensa processor cores and use the XTMP API to quickly connect these simulated processor cores to memories, peripherals, interconnects, and designer-defined subsystems (which are called XTMP devices). XTMP encapsulates Tensilica's automatically generated, cycle-accurate instruction-set simulators, and through them supports multiple-processor source-level debug for the tasks running on those processors.

Because the Xtensa instruction-set simulator (a model of the application-specific processor created by the Xtensa processor generator) is written in C, it runs much faster than does simulation of a processor simulation model coded in an HDL such as Verilog or VHDL. It is also easier to modify and update a C simulation model than a hardware prototype. Thus, XTMP allows designers to create, debug, profile, and verify their combined hardware and firmware systems early in the design process, which encourages a more thorough investigation of the system's solution space. Like a standalone instruction set simulator for single processors, XTMP is a tool to help the designer quickly check the overall performance and correctness of firmware, targeting multiprocessor system design.

4.5.3.1 Developing an XTMP Model

A system simulation consists of a connected set of simulation components or models. The designer instantiates simulation components, connects them together to form a system, loads a program into each processor's memory, initializes a source debugger for each processor, and then starts the simulation.

In `XTMP_main()`, the top level of the simulation program, the designer performs the following steps:

1. **Create the components to be simulated:** The designer can create Xtensa processors with `XTMP_coreNew()`, basic memory devices with `XTMP_memoryNew()`, and some other simple utility devices. The designer can also write new models of more complicated devices and allow the rest of the simulator to access them with *callback functions*. A custom device can also communicate with a processor using the `XTMP_setInterrupt()` function.
2. **Connect the components together:** After creating the Xtensa processors and the devices that will be a part of the system simulation, the designer connects them to each other. Each Xtensa core can have a main processor interface (PIF) and up to five other interface ports (XLMI—the high-speed Xtensa Local Memory Interface, data ROM and RAM, and instruction ROM and RAM). The designer can connect a device directly to an Xtensa processor interface (PIF) or to an internal port using the `XTMP_connectToCore()` function. However, the designer may find it more convenient to create a *connector*, attach it to a processor's PIF or another interface port, and then connect multiple devices to the connector using the `XTMP_connect()` function. *(Note XTMP can also simulate connections built from direct ports and queues, but these XTMP features are not illustrated in this example.)* A connector acts as a simple memory map that can route transactions from multiple cores to multiple devices.

3. **Control the simulation:** After creating and connecting all the simulation components, the designer can

   - Load the program for each Xtensa processor core in the system by calling `XTMP_loadProgram()`.
   - Call `XTMP_enableDebug()` for any processor cores that the designer would like to debug with an Xtensa debugger.
   - Call the various API functions that control the simulation of the entire system.
   - Start or step the execution of all the simulated cores simultaneously.
   - Disable or reset individual processor cores while the others continue to run.
   - Access the internal registers and memory spaces of any of the processor cores (allowing instrumentation of the simulation).

XTMP Components

XTMP has a set of predefined components. Multiple independent instantiations of each processor type are fully supported. Some of the most common are

`XTMP_core:` an Xtensa processor created by the Xtensa processor generator

`XTMP_memory:` a simple memory device (XTMP_memory uses pages to avoid allocating regions of memory that are not used).

`XTMP_connector:` a device for connecting cores to other devices via the PIF or other ports. There can be a single global address map or distinct maps for each processor.

`XTMP_device:` a designer-defined device that interacts with the rest of the simulator via callback functions.

`XTMP_lock:` a lock device to help with multiprocessor synchronization.

4.5.3.2 XTMP Simulation Commands

The most common XTMP simulation commands include the following:

`XTMP_loadProgram():` loads target program into memories.

`XTMP_start():` runs the simulator until finished or for a fixed number of clock cycles.

`XTMP_disable()`, **`XTMP_enable()`** and **`XTMP_reset()`**: disables, enables, and resets a particular core.

`XTMP_step()`: steps the program on one processor core, while other cores are disabled.

4.5.3.3 A Simple XTMP Example

The code fragment in Fig 4-36 shows XTMP code for a simple system simulation, including a system RAM, and system ROM and one processor, with the configuration `my_xtensa_config`, running the application binary `sieve.out`. The system address map is not explicit in the system description, but inherited from the processor's address configuration, which specifies default location for system RAM and ROM.

```
1: XTMP_params p;
2: XTMP_core core;
3: XTMP_connector connector;
4: XTMP_memory rom, ram;
5: p = XTMP_paramsNew("my_xtensa_config", NULL);
6: core = XTMP_coreNew("cpu", p, NULL);
7: rom = XTMP_sysRomNew("rom", p);
8: ram = XTMP_sysRamNew("ram", p);
9. 0: connector = XTMP_connectorNew("connector", XTMP_byteWidth(core), 0);
10: XTMP_connectToCore(core, XTMP_PT_PIF, 0, connector, 0);
11: XTMP_addMapEntry(connector, rom, romStart, romStart);
12: XTMP_addMapEntry(connector, ram, ramStart, ramStart);
13: XTMP_loadProgram(core, "sieve.out", NULL);
14: XTMP_start(-1);
```

Figure 4-36 XTMP code for single-processor system description

The example in Fig 4-36 follows this flow:

1. **Set up the simulation (Lines 1-4):** The example sets up a simulation within the function `XTMP_main()`, analogous to the main function of a regular C program. First, objects needed for the simulation are declared: a parameter object, an Xtensa processor core, a `connector`, and two memories. Then, each object is created and given a name (for example: *cpu* for the `XTMP_core` object), which is used by the simulator in any error messages that refer to that object.
2. **Read the parameter file (Lines 5 and 6):** To create a processor, the designer must first use `XTMP_paramsNew()` to load a parameter file that describes the processor configuration. Reading in a parameter file creates an *XTMP_params* object that can be used to create other objects. In the example, the parameter file for the `my_xtensa_config` configuration is read and one processor instantiated. The parameter object is then used to create the core and to simplify the creation of the memories.
3. **Create the memories (Lines 7 and 8):** The memories are built by providing the size of the memory block. The convenience functions `XTMP_sysRamNew()` and `XTMP_sysRomNew()` can be used to automatically call `XTMP_memoryNew()` with the sizes of the system RAM and ROM that are stored in the `XTMP_params` object. XTMP allows for the creation of additional memories that do not overlap with the default base system RAM and ROM to be connected as well.
4. **Connect the core and memories (Lines 9-12):** The next task is to connect the core's PIF to the memories using the `XTMP_connector` object.
5. **Load and run code (Line 13 and 14):** Finally, the code is loaded into memory, and XTMP can start and run the simulation to completion with a call to `XTMP_start(-1)`.

4.5.3.4 A Multiple-Processor XTMP Example

The potential leverage of rapid system description is shown more clearly when the system has multiple processors, a more complex address map, and a wider variety of devices and shared memories.

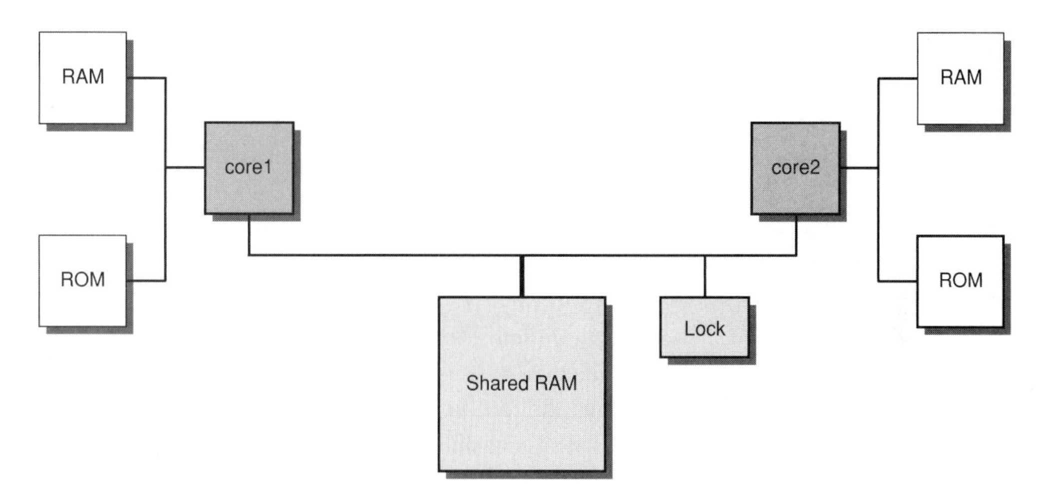

Figure 4-37 Block diagram for multiple processor XTMP example.

The impact of rapid, accurate multiple-processor modeling on early architecture development is potentially dramatic. System models can often be developed in a matter of hours in contrast to the months typically required to stitch together large-scale RTL-based SOC simulations. C-based XTMP models run significantly faster than do simulations of RTL models—generally at least 100 times faster when compared to standard RTL logic simulators for the same level of system complexity. The integration of XTMP simulation models with the software development environment is seamless because source-level debugging and rapid firmware modification through C programming is intrinsic and efficient. Ease of system simulation at the earliest design stages is one of the significant benefits of processor-centric SOC design.

An XTMP example with two processors that are instances of the same Xtensa configuration, five memories, and a lock to synchronize both processors is shown in Fig 4-37 and the code that defines this system appears in Fig 4-38.

Lines 1–8 declare local variables used in system configuration, including the declaration of arguments to the simulators that assign unique process identifiers to each processor. Unique processor identifiers can be important if multiple processors are running identical code. The identifier allows the code on each processor to select the unique subset of the total task for which it is responsible.

Line 10 identifies one processor configuration, `s1`

Lines 11–14 and lines 16–19 respectively instantiate two processors of that configuration, together with their private RAM and ROM areas. The memories are sized, but no address is assigned to them.

Lines 21 and 22 define a lock device, which supports memory sharing.

Lines 23 and 24 allocate two connectors, which will have assigned addresses and various devices—especially memories—connected to them.

Line 25 defines a 64KB memory area that will be used for shared-memory operations controlled by the lock.

Lines 27 and 29 connect each of the connectors to a corresponding processor's processor interface (PIF).

Lines 31–38 assign addresses for each memory or device as seen from each processor. XTMP supports different address maps for each processor. No address map entry is necessary for devices that are not accessible from a particular processor.

Lines 40 and 41 load the task software into each of the two processors. The same task could be loaded into both processors if the task code itself used the processor ID to distinguish the role of each processor in the distributed application.

Finally, line 42 starts the execution of system simulation.

```
1: XTMP_params p;
2: XTMP_core core1, core2;
3: XTMP connector router1, router2;
4: XTMP_memory rom1, rom2, ram1, ram2;
5: XTMP_memory shared_mem;
6: XTMP_lock lock;
7: char *sim_arg1[] = {"--proc_id=1", NULL};
8: char *sim_arg2[] = {"--proc_id=2", NULL};
9:
10: p = XTMP_paramsNew("s1", NULL);
11: core1 = XTMP_coreNew("cpu1", p, sim_arg1);
12: rom1 = XTMP_memoryNew("rom1", p, 0x00200000); /*2MB space*/
13: XTMP_memorySetReadOnly(rom1, true);
14: ram1 = XTMP_memoryNew("ram1", p, 0x00800000); /*8MB space*/
15:
16: core2 = XTMP_coreNew("cpu2", p, sim_arg2);
17: rom2 = XTMP_memoryNew("rom2", p, 0x00200000); /*2MB space*/
18: XTMP_memorySetReadOnly(rom2, true);
19: ram2 = XTMP_memoryNew("ram2", p, 0x00800000); /*8MB space*/
20:
21: lock = XTMP_lockNew("lock", XTMP_byteWidth(core1),
22: XTMP_isBigEndianFromParams(p), 1 );
23: router1 = XTMP_connectorNew("router1", XTMP_byteWidth(core1), 0);
24: router2 = XTMP_connectorNew("router2", XTMP_byteWidth(core2), 0);
25: shared_mem = XTMP_memoryNew("shared", p, 0x10000);/*64K mem*/
26:
27: XTMP_connectToCore(core1, XTMP_PT_PIF, 0, router1, 0);
28:
```

```
29: XTMP_connectToCore(core2, XTMP_PT_PIF, 0, router2, 0);
30:
31: XTMP_addMapEntry(router1, rom1, 0x80000000, 0x80000000);
32: XTMP_addMapEntry(router1, ram1, 0x80100000, 0x80100000);
33: XTMP_addMapEntry(router1, lock, 0x21000020, 0x21000020);
34: XTMP_addMapEntry(router1, shared_mem, 0x21010000, 0x21010000);
35: XTMP_addMapEntry(router2, rom2, 0x80000000, 0x80000000);
36: XTMP_addMapEntry(router2, ram2, 0x80100000, 0x80100000);
37: XTMP_addMapEntry(router2, lock, 0x21000020, 0x21000020);
38: XTMP_addMapEntry(router1, shared_mem, 0x21020000, 0x21020000);
39:
40: XTMP_loadProgram(core1, "proc1.out", NULL);
41: XTMP_loadProgram(core2, "proc2.out", NULL);
42: XTMP_start(-1);
```

Figure 4-38 XTMP code for dual-processor system.

4.5.4 Balancing Computation and Communications

The daunting gate-counts and available concurrency of deep-submicron silicon opens the door to an enormous variety of possible system organizations. The number of processors and other hardware blocks; the configuration of the processors, the structure, number, type, and width of communication paths; and the selection of algorithms are all potential variables in optimizing a system. In theory, the designer could define some formal optimization problem, but in practice, the diversity of goals makes this impractical. Instead, a processor-centric SOC design seeks to satisfy seven conditions:

1. Correctly implement the set of tasks that transform input data to output data.
2. Assign tasks to processors so that the computational load of the tasks never exceeds the capacity of the corresponding processor.
3. Provide enough processors (and other hardware blocks) in the right configuration to execute the complete set of tasks with the required computational throughput.
4. Set a structure of communication links (buses, queues, ports) with adequate width, latency, buffering depth, and arbitration priorities to ensure adequate performance for all data transfers.
5. Provide enough memory to handle the instruction memory, private data memory, and shared-data memory to accommodate the needs for all processors and buffering of task-to-task communication.
6. Provide sufficient excess programmability, computational capacity, and communications bandwidth to support all expected changes in product requirements.
7. Do all this at the lowest possible terms of development cost, cost in silicon size, chip packaging, fabrication technology, and mandatory support chips.

Goals 2, 3, and 4 capture the heart of the optimization process. The number of processors, the configuration of processors, and the communication mechanisms between processors all interact with one another and with the set of software tasks. The architect may need to explore several iterations of system structure to achieve a good balance.

For example, the designer may discover that an instruction-set extension can significantly accelerate some task. That extension creates extra capacity on that processor, particularly for tasks that can exploit the same instruction-set extension. Moving more tasks onto that processor adds the memory-configuration and communications-interface requirements of the new tasks to the existing workload. Recombining these tasks might team a pair of communicating tasks together on one processor such that the two tasks can no longer operate in parallel to achieve low end-to-end latency. Conversely, the combination of tasks together on one processor might eliminate the need for any shared memory accessible from multiple processors, simplifying the on-chip communications structure or reducing requirements for off-chip memory.

The details of iterative optimization vary from system to system. The ability to rapidly model and evaluate possible alternatives in system structure, communications mechanisms, processor and memory configuration, and task software contributes to the design team's effectiveness in the face of complex goals.

4.6 The SOC Design Flow

This chapter's introduction to the basic issues of concurrency and communications is the right backdrop for proposing a preferred design flow for complex SOC design. This design flow is not a mechanical sequence of steps, but a recommended framework for translating high-level requirements into efficient hardware and software implementations that exploit the availability of application-specific processors used in large numbers, as the essential fabric (or atoms) of SOC design.

4.6.1 Recommended Design Flow

The recommended design-flow process is sketched in Fig 4-39 and described in this section.

In Fig 4-39, the block arrows show the basic flow of refinement from high-level application requirements down to detailed hardware and software implementation. Within the flow are four major subflows:

1. Optimization of computation by mapping tasks to tuned processors
2. Optimization of communication by selection and tuning of interconnect among blocks, especially processors
3. Detailed software development, including operating systems and I/O code
4. Detailed VLSI implementation, including logic design, verification, physical design, and silicon fabrication

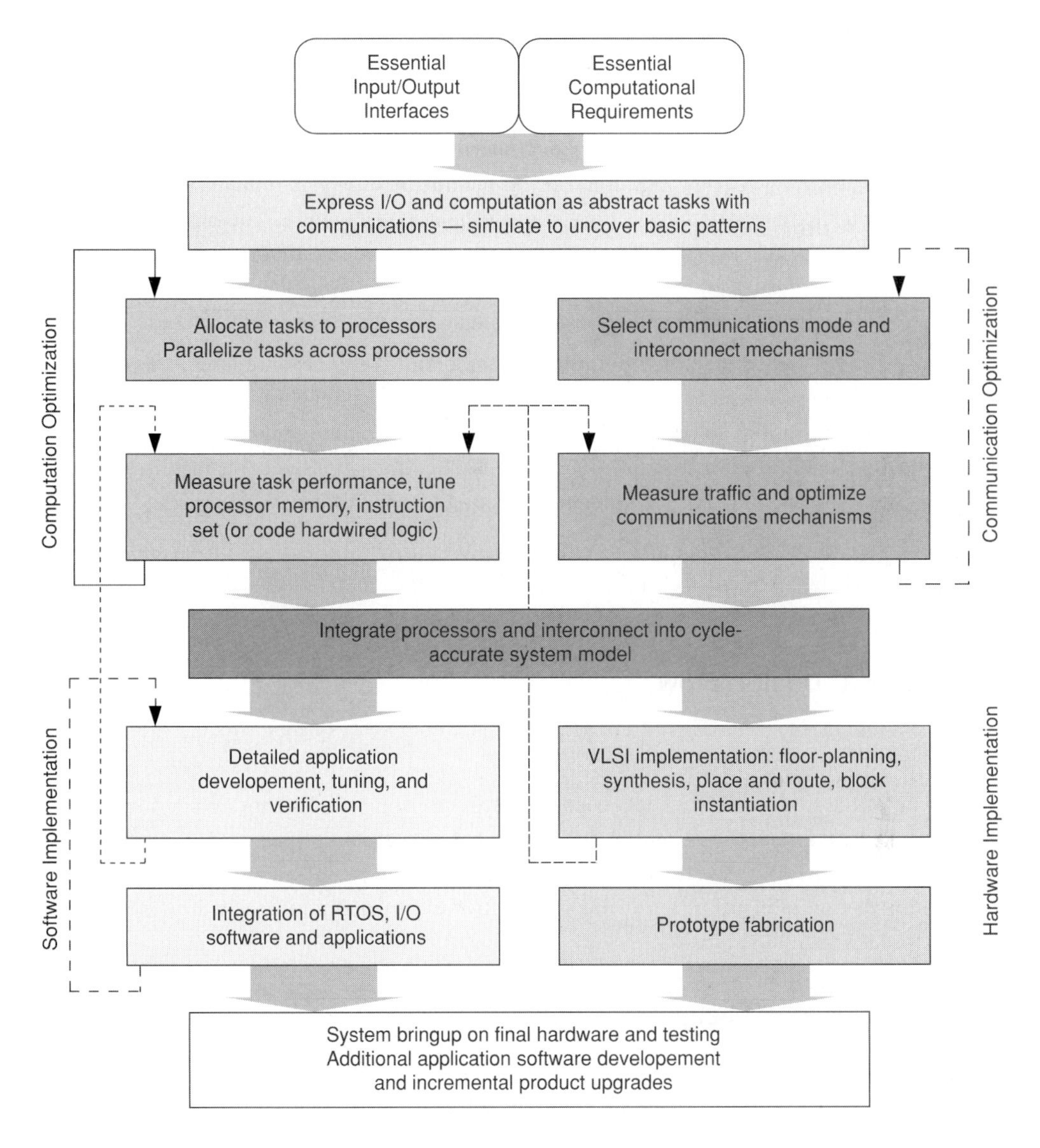

Figure 4-39 Advanced SOC design flow.

Any process of incremental refinement can lead to important design insights, particularly if early choices dictate implementations that are too slow, too expensive, or too complex. Those insights should trigger revision of earlier decisions. Some of these important feedback loops are shown:

- Discoveries within computation optimization
- Tradeoffs in communication tuning
- Cost and performance results in hardware design feeding back into changes in computation and communication structure
- Unanticipated results from detailed software development requiring changes in processor configuration

The earlier stages of the design flow, especially the development of an abstract task model of the system and exploration of parallel communication and computation structures, have been discussed in this chapter. Detailed configuration and tuning of processors and their corresponding communications structures is a key aspect of this method and the focus of the next three chapters. These chapters show how use of configurable processors gives the design team earlier insight, delivers more performance headroom, and reduces design iterations for each block and for the connected system of blocks.

The last five stages of the design flow—detailed hardware and software implementation and bring-up—are relatively conventional. If good decisions are made early in the design process, including the building of an accurate and complete system model, the details of implementation should yield few surprises.

Some of the traditional conflicts between hardware and software are avoided because the team shares a common reference model. In particular, hardware prototype development is off the critical path for much of the software development. Final hardware/software integration should go smoothly because, in truth, hardware and software are integrated early in the processor-centric approach to SOC design.

4.6.2 Shifts in SOC Design Methodology

The suggested SOC design flow implies a number of innovations relative to SOC design techniques commonly employed today. Key recommended methodology transitions include the following:

- **Prefer simulation over hardware prototypes.** Prototypes still play a role, especially for modeling external interfaces or for systems that are built from pre-existing SOCs. However, the sheer complexity of leading-edge SOCs makes hardware prototyping (even with large FPGAs) expensive and time consuming. Because FPGAs implement logic somewhat differently than do ASICs, FPGA-based prototypes can also be inaccurate. The same effort invested in developing good system-simulation models can result in virtual platforms that are available earlier, are easier to debug, and require no hardware maintenance.

Moreover, simulation models can be reproduced and distributed to development teams almost for free, allowing greater parallelism within the development process.

- **Defer hardware/software partitioning.** The availability of application-specific processors increases the range of tasks that can be implemented in a software-centric manner. Functions that once mandated early assumption of major RTL-based hardware design now can remain in software form, assuming extended processors as implementation vehicles. Of course, the design team may still decide to move particularly simple and efficiency-sensitive tasks into a hardware-only implementation, but the fraction of the design that follows this path is significantly reduced.

  Moreover, attempted configuration of an application-specific processor to meet task performance and cost goals may provide a good starting point for a more completely hardwired data-path-plus-state-machine logic implementation. Much of the effort once focused on hardware/software partitioning can be devoted to "software/software partitioning"—the parallelization of complex system applications into the best number and arrangement of sub-tasks executing on tuned processors. Blocks built from "software plus processors" are intrinsically easier to tune, debug, and reprogram. These attributes allow design teams to complete the system design more quickly.

- **Leverage a greater range of communications primitives.** Both the greater intrinsic bandwidth of deep-submicron silicon and the greater interface flexibility of configurable processors create new system architecture options. Instruction-mapped and memory-mapped data queues, low-overhead shared memories, and wide crossbar and bus interconnects all provide potential order-of-magnitude reductions in communication latency and order-of-magnitude bandwidth improvements relative to conventional 32-bit processor bus communications.

- **Anticipate design reuse.** Compared to a purely hardwired design approach, the advanced SOC design flow promises easier chip reuse. To build a new but related system, don't build a new chip with block reuse. Reprogram the processor fabric in the existing chip instead!

  Design teams quickly realize that application-specific processors offer great leverage for design reuse as well. Once a processor has been configured to fit an application domain, its architecture, logic implementation, and software environment are likely to be useful for many related tasks. Design teams may also reuse applications code, either in source form, using application-specific data-types and operators, or in binary form, mapped to the base instruction-set architecture plus application-specific instructions. If the new design needs new functions, they can be added either purely in software, using the existing extended instruction set, or with new instruction features. Significantly, new features are added without breaking binary code compatibility, so old code runs on both old and new processor configurations, and new code can take advantage of the new features. This opportunity to seamlessly upgrade processors in either hardware or software increases the attractiveness of design reuse.

- **Develop hardware and software in parallel.** The traditional model of SOC development forced most software development to wait until silicon prototypes were complete. This limitation could add an extra year to the overall development schedule and prevent the timely incorporation of software insights into the chip architecture. The new SOC design methodology makes the entire development process more software-centric, allows the software team to start meaningful development sooner, and encourages tighter collaboration between the hardware and software teams. Mismatches in assumptions are identified and fixed more quickly. New cost targets, performance goals, and the system impact of new features can all be assessed more easily. The focal point for this collaboration is the system model, especially the cycle-accurate system model that serves as a rendezvous point in the design flow after initial computation and communication design and before detailed hardware and software implementation.

This design flow is not a precise prescription. The process of large SOC design is so complex that no simple formula will suffice, particularly because design rarely starts with a "clean sheet of paper." Prior experiences of the design team and pre-existing software, hardware blocks, and tools all influence the design flow. Nevertheless, this flow can serve as a guide to discovery of new and more efficient system solutions, particularly ones that take fuller advantage of extensible processors.

4.7 Non-Processor Building Blocks in Complex SOC

This book puts processors on center stage, but processors are not the only actors in the system. Some types of building blocks—especially memories and I/O interfaces—play principal roles. Other blocks are optional, especially where designers have a choice between implementing a function as hardwired logic and implementing the function as software on a configurable processor. This section outlines the issues with these supporting blocks.

4.7.1 Memories

Processors require memories. Off-chip memories—often nonvolatile—hold boot code that allows the processors in the system to initialize themselves and the other system functions. On-chip memories, closely coupled to the processor pipeline, hold the instruction and data sets most immediately required by the processor's application code. Other RAMs hold data shared across tasks or serve as input and output buffers to handle nonuniform rates of data production or consumption. Even on-chip ROMS make economic sense in some SOCs.

Multiple local RAMs may be used in each complex function block in an SOC. As the number of function blocks grows, the chip as a whole may incorporate scores of small RAM blocks, and the total RAM area may approach or even exceed half the total chip area. Moreover, the designer must provide a means to initialize and test each RAM. This proliferation of RAMs dictates a more systematic approach to RAM use in SOCs. The following guidelines may help designers take better advantage of RAMs:

1. *Off-chip RAM is much cheaper than on-chip RAM, at least for large memories.* The cost per bit of commodity DRAM is roughly 2 microcents per bit, while the price of fast on-chip SRAM is on the order of 30 to 40 microcents per bit (as of 2003). Off-chip DRAM access does require additional pins on the chip, consumes more power, and takes many cycles to access, but applications that require megabytes of temporary storage almost always use off-chip DRAM.
2. *Use caches and shared memories when on-chip RAM requirements are uncertain.* Dedicating memories to a particular hardwired logic block or deeply embedded processor gives the highest bandwidth and lowest latency but may hamper flexibility. The designer sometimes faces a delicate tradeoff between making a RAM too small—risking the correct functioning of the system—and making a RAM unnecessarily large—unduly increasing the silicon area and cost. If the processor is configured with instruction and/or data caches, the design becomes more tolerant of growth in application-code or data-set size. The processor's cache-management hardware automatically holds the most commonly used portions of the instruction code and data sets in fast local memory, while other less frequently accessed instructions and data reside only in shared memory, often in off-chip DRAM. Organizing the system so that several processors can share one memory both aids communication among the processors and allows the processors to allocate memory more flexibly. The maximum utilization of the shared RAM may be significantly less than the sum of the maximum usage of each of the processors. Therefore, the shared memory can often be smaller than several separate memories.
3. *Perform a system-performance sanity check by assessing memory bandwidth.* Many data-intensive applications do not face computation-rate bounds but memory-bandwidth bounds, particularly after processor configuration has compressed the necessary computation into fewer cycles. Look at the memory-transfer requirements of each task and ensure that the processor's local memories can handle the local traffic and that the total data traffic in shared memories stays well below the memories' capacities. If no behavioral code is yet available for a task, implement a traffic generator to make contention for shared communications resources (buses, memories) more realistic.

Watch out for contention latency in memory access. Increase memory width or increase the number of memories that can be active to overcome these bottlenecks. Detailed simulation may reveal additional, subtler contention issues, but when the back-of-the-envelope calculation suggests a bandwidth shortfall, serious performance issues are likely. Pay particular attention to tasks that must move data from off-chip memory, through the processor, and back to off-chip memory. Such flow-through tasks can quickly consume all available bandwidth.

4.7.2 I/O Peripherals

I/O peripherals vary widely in type, cost, and bandwidth requirements. Many peripherals implement industry-standard communications connections and are included in the SOC for the convenience of system design.

These standard interfaces often have quite modest bandwidth demands and put no special burden on interconnect structure or software device drivers. These interfaces often include simple low-speed connections to USB, CAN, and I2C standard serial buses; audio DACs and A-to-D converters; and system-debug devices. Eight- and 16-bit data interfaces are often adequate for these peripherals and the designer can safely connect low-bandwidth peripherals together on one narrow peripheral bus. In many cases, these peripheral devices are available as inexpensive and well-tested intellectual property blocks, implemented with RTL—though the electrical specifications of the interface port may create a need for specialized I/O pin drivers and receivers on the SOC.

High-bandwidth interfaces—especially interfaces requiring tens or hundreds of megabytes per second of data bandwidth—present more of a design challenge. These interfaces include LAN and WAN network connections and video ports. The processor-centric SOC design methodology offers two approaches to efficient integration of high-bandwidth I/O:

1. *Use an autonomous DMA engine:* The control registers of the I/O interface logic are directly read and written by processor operations, but data to and from the interface are moved by the interface logic using DMA (direct memory access). The DMA engine may move the data directly into the local memory of the appropriate processor, sharply cutting the access latency seen by the task that processes the incoming data. Similarly, the DMA engine can pull the data from the processor's memory once the processor has signaled that the data is ready. DMA engines are often built into the logic of the I/O interface, together with logic that performs flow control, basic error checking, and data formatting. All this hardwired logic, however, can make the peripheral large, hard to model, and subject to obsolescence as new protocols appear in the market.
2. *Couple the I/O interface tightly to a processor:* The small size and rich interface flexibility of configurable processors enables a new role for these small task engines as software-driven peripheral controllers. Hardwired circuits are still required for the lowest level data conversion—for clock recovery and serialization/deserialization—but many of the other interface functions can be moved efficiently into a processor. The state machine controlling the peripheral is replaced with firmware, typically implemented in C. Many of the functions once residing in the device driver as part of the RTOS can also be moved down into the I/O processor. Error checking and recovery, data formatting and buffering, routing of the data to and from multiple destinations, and interrupt service can all be handled by the dedicated processors rather than in specialized hardware or RTOS code. The raw data-transfer interface of the I/O device can be tied directly to input and output data queues of the configurable processor. In the limit, such an interface can sustain transfer rates of more

than 50GB per second between local memory and the I/O interface, including on-the-fly computation of checksums or basic data formatting. The tightly coupled processor approach increases the overall parallelism of the system and reduces the fragility of the design due to excessive reliance on hardwired logic.

4.7.3 Hardwired Logic Blocks

Application-specific processors are a potent alternative to hardwired logic blocks, but RTL-based logic design still makes good sense under certain conditions. If an RTL block has already been designed and verified, then simply reusing that block is appropriate, particularly if requirements are deemed unlikely to change. Similarly, small blocks (less than 10–20K logic gates) are less likely to be buggy and are less subject to late changes, even if designed from scratch. These blocks may play an important role in moving or transforming data somewhere in the SOC architecture. Interfacing processors with hardwired blocks presents some special opportunities to reduce cost and increase flexibility. Two interface mechanisms bear mention:

1. *Map hardware registers into local memory space:* When the operations of the hardware block are controlled through built-in registers, those registers can be mapped into the system's global address space or into a processor's local address space. The processor performs writes to send commands and input data to the hardware function and performs reads to receive status and output data from the hardware. Alternatively, the designer can tie hardware status signals such as "operation done" to processor interrupts to enable greater concurrency between hardware computation and unrelated processor operations. This approach makes the hardware block look much like an I/O device, and it makes the controlling software look much like a standard device driver.
2. *Extend the instruction set to directly stimulate hardware functions:* The input signals of the hardware block are directly controlled by output states and queues of the processor, as described in Section 4.4.3 Data Queues. The output signal of the hardware block directly controls the processor's input ports and queues. The designer can then specify arbitrary new processor instructions that take hardware block outputs as instruction-source operands and use hardware block inputs as instruction-result destinations (thus avoiding the use of intermediate registers and greatly accelerating the task). The entire flow of data from the processor through the hardware block and back into processor registers and memory can be implemented within a pair of processor instructions. An example of this tight coupling between processor and external data-path logic is shown in Fig 4-40, where the pipelines associated with the two instructions are shown in yellow and red.

This tightly coupled approach effectively absorbs the functionality of the existing hardwired logic into the processor and puts it entirely under the control of software. The flexibility of the block can be further enhanced by replacing any internal control registers in the hardwired block with corresponding processor registers and new processor outputs from these

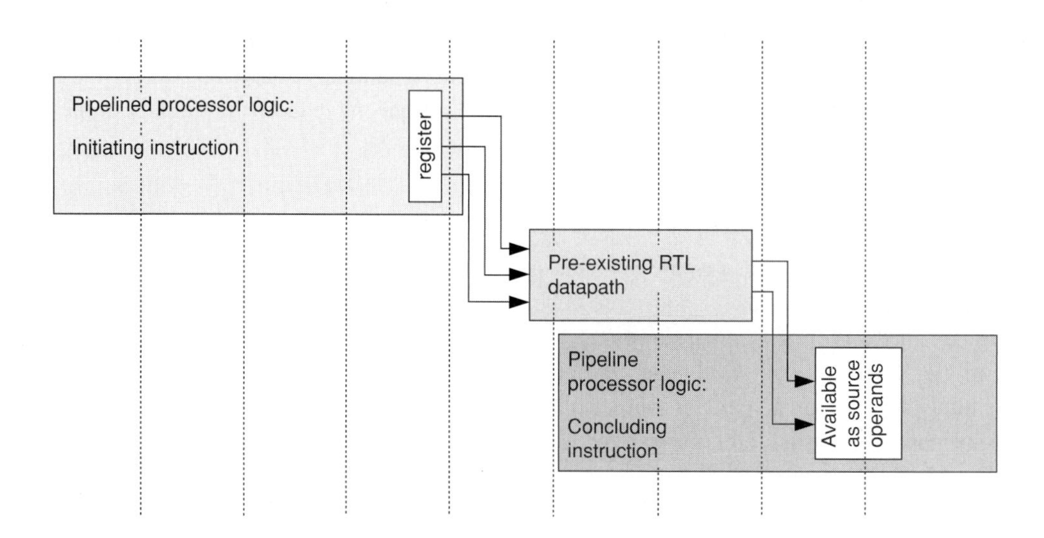

Figure 4-40 Direct attachment of RTL to processor pipeline.

processor registers to the external logic they control. The boundary between hardwired block and processor almost disappears. Moving registers from hardwired logic to the processor makes the registers automatically visible to the compiler, debugger, and simulator. It also reduces the overhead for any processor operations on those registers. The improvement in debuggability due to full register visibility alone may make processor-centric design compelling for some SOC projects.

These methods for handling memory, I/O, and hardwired logic offer the designer meaningful improvements in bandwidth, software control, and ease of integration. The remarkable flexibility of configurable processor memory and logic interfaces will no doubt enable even more SOC design creativity as the processor-centric methodology matures and proliferates.

4.8 Implications of Processor-Centric SOC Architecture

The notion of putting many processors on one SOC has been around for a decade, but limitations in the processors, in the existing design methodologies, and in the target applications themselves have limited the actual deployment of processor-centric SOC designs. The combination of greater silicon density, the demand for higher performance and better power efficiency, and the emergence of smaller and more flexible processor cores is changing the landscape of SOC design. This chapter has sketched one approach to systematic design, and succeeding chapters work further down into the details, especially for more thorough use of processors in SOC designs. These methods promise both to shorten and simplify the design process and to improve the flexibility and reusability of the chips and systems that use the new SOC design approach. While the overall trends toward electronics complexity, density, and efficiency seem clear, two threads are worth further exploration—the impact of external technology trends on SOC architecture and the impact of processor-centric SOC designs on end products.

Transistors are now incredibly cheap—even in a high-performance ASIC flow, a logic transistor costs about 10 microcents—and the cost will undoubtably continue to fall. Designers can and should use more of these cheap transistors if it helps them achieve their other project goals: performance, form-factor, time-to-market, and battery life. On the other hand, design complexity is increasingly expensive. The effort and cost to create and verify a complex function increases much more than linearly with transistor count.

We can expect, then, that design teams will gravitate to design methods that allow them to create complex, high-efficiency designs more quickly, even if they consume more transistors. Using more general-purpose processor cores may not be an acceptable solution because these processors are both larger and less efficient than application-specific processors, and they provide no productivity benefits compared to automatically configured processors cores and software. The concept of design reuse—redeploying existing proven hardwired logic blocks in new chips—will certainly continue. However, the implementation of new functions will be in "software and processor" form. In fact, we expect processor- and software-based functions to gradually dominate libraries of reusable function blocks.

Nanometer silicon technology is raising increasing concerns about power dissipation, signal integrity, and physical design. Designers are apt to gravitate to methods that provide structured approaches to power management and physical design. A more modular, systematic, "paint-by-numbers," physical implementation style with strong guarantees on optimal power management and signal integrity is likely to be preferred to a "blank-canvas," physical-implementation style. Using processors as the key logic building block, rather than arbitrary RTL, enables this more systematic design style, because all configurations of a base processor share common physical design flows, common clock-gating mechanisms, and similar floor planning constraints. The processor flow is optimized once, so the design team for a single SOC need not worry so much about power, signal, and layout issues.

Instead, the design team can focus its efforts on architectural optimizations that can yield much larger benefits. In 1998, Sandborn and Vertal of University of Maryland argued that 80% of the features controlling overall performance and cost were determined during the system-level design phase, which occurs during the first 20% of the overall development. The economics of designing and building big silicon chips bears directly on the preferred architecture. Designers will inexorably bias their design strategies toward approaches that enhance the return on their SOC design investments. They work to increase the benefits of the chip by

- *Increasing volume*—Making the chip compelling and flexible enough to be used in multiple systems
- *Increasing selling price*—Making the chip or system features more compelling
- *Reducing manufacturing cost*—Trimming die size and packaging requirements

The design team also works to cut development cost by

- *Reducing design time*—Configuring and programming processors rather than designing and verifying RTL by hand
- *Streamlining the design team*—Leveraging software expertise more fully compared to hardware coding and verification
- *Eliminating much of the risk of silicon respins*—Accommodating most functional changes in software updates

Together, the fall of transistor costs, the rise of nanometer implementation effects, and the basic economics of complex SOC design all appear to favor this software-centric, processor-based SOC approach. Just as underlying technology and business forces seem to favor this new SOC design methodology, the impact of this approach is percolating up into application architectures. This approach uses the critical application tasks as the starting point and builds more flexible computing architectures to manage complexity and maximize chip volume. Algorithm researchers will feel fewer constraints to restrict themselves to architectures that are simple enough to implement in RTL. If an algorithm is expressible in C or C++, it is directly optimizable on a configurable processor. Algorithm designers will gravitate to approaches that can expose adequate fine-grained and coarse-grained concurrency to design tools, especially to high-level communications profiling and compiler-based processor-generation tools. The bounds on algorithm complexity will be set less by human RTL developers and more by software analysis technology.

The drive for concurrency is also likely to be reflected in more regular computation and communication structures, particularly for problems with high degrees of concurrency (image processing, high-channel-count communications processing) or more modest volumes (where the unique requirements of a single system cannot justify a complete SOC development). In these cases, standard SOC platforms are likely to emerge, using large uniform arrays of processors, where the processors are neither completely generic nor narrowly application-specific, but tuned for an entire application domain (e.g., communication routing, video coding, graphics rendering, image processing, and signal recovery). We are likely to see a broad range of SOC devices with widely varying degrees of specialization and widely different target applications. In fact, we are just at the start of the advanced SOC era. The implications of SOC design based on large numbers of application-specific processors are just starting to appear through the fog.

4.9 Further Reading

System-level design is a vast field, and methods vary widely across different application domains. Nevertheless, some texts and articles provide useful background.

- One key dimension of system-level design is modeling and analysis, and the SystemC language is emerging as a popular vehicle for model development. The following texts provide a useful introduction: Thorsten Grotker (Editor), Stan Liao, Grant Martin, and Stuart Swan, *System Design with SystemC*, Kluwer Academic Publishers, 2002; and Wolfgang Muller, Wolfgang Rosenstiel, and Jurgen Ruf (Editors), *SystemC: Methodologies and Applications*, Kluwer Academic Publishers, 2003.
- While SOCs represent just a subset of all embedded electronics systems, many of the design issues are common. A good introduction to embedded system design issues can be found in Arnold S. Berger, *Embedded Systems Design: An Introduction to Processes, Tools and Techniques*, CMP Books, 2001.
- A good introduction to system design in networking, especially with application-specific processors, can be found in Patrick Crowley, Mark A. Franklin, Haldun Hadimioglu, and Peter Z. Onufryk, *Network Processor Design: Issues and Practices*, Volume 1, Morgan Kaufmann, 2002.
- A good introduction to multimedia applications, especially content encoding, can be found in Jerry Gibson, Toby Berger, Tom Lookabaugh, Rich Baker, and David Lindbergh, *Digital Compression for Multimedia: Principles & Standards*, Morgan Kaufmann, 1998. Gonzalez's and Woods's popular reference for image processing may also give important insight into SOC systems design and algorithms: Rafael C. Gonzalez and Richard E. Woods, *Digital Image Processing* (2nd Edition), Addison-Wesley, 2002.
- For those interested in queuing theory, see this popular text: Donald Gross and Carl M. Harris, *Fundamentals of Queueing Theory, (*3rd edition) Wiley-Interscience, 1998.
- Many mesh architectures have been explored. One interesting approach is the MIT RAW architecture: Elliot Waingold et. al., "Baring it all to software: Raw machines." In *Computer*, September 1997.
- In 1998, Sandborn and Vertal of University of Maryland argued that 80% of the features controlling overall performance and cost were determined during the system-level design phase, which occurs during the first 20% of the overall development. Peter A. Sandborn, and Mike Vertal. "Analyzing packaging trade-offs during system design." In *IEEE Design and Test of Computers, 15 (3), July–September 1998,*10-19.
- The characteristics of on-chip interconnect scaling are thoroughly discussed in Ho, Mai, and Horowitz, "The Future of Wires." In *Proceedings of the IEEE*, 2001.
- The argument for multiple processors for design simplification is ably made by Jack Ganssle, "The Art of Designing Embedded Systems," *Newnes*, 1999.

More technical information on Wind River VxWorks functions and general insight into communications-based programming can be found at the Wind River Web site: *http://www.windriver.com.*

CHAPTER 5

Configurable Processors: A Software View

Chapters 5 and 6 give complementary perspectives on a central issue for processor-centric SOC design—the relationships between individual tasks and the optimized processor designs on which they run. Chapter 5 looks at optimized processor design through the eyes of the software developer, including the processes of compiling, simulating, profiling, and updating applications and the impact of processor configuration on the programmer's processor model. Along the way, the chapter introduces some of the mandatory elements of automatic processor generation, including the nature of processor descriptions, the variety and purpose of processor-generator outputs, and the basic forms of optimized instruction sets. Chapter 6, in contrast, looks at optimized processor design through the eyes of the hardware developer, including the processes of converting an RTL data path into a processor instruction set and developing a C or C++ algorithm for task sequencing, in place of hardwired state machines.

Why should the software developer care about application-specific processor optimization? After all, a general-purpose processor can execute any program written in a high-level language. In some system domains, the application performance, power dissipation, and manufacturing cost of a generic processor may meet all of the software developer's goals. In many embedded systems, however, the software developer can and should take an active role in achieving demanding system-performance and efficiency goals. An application-specific processor may be valuable, even essential, whenever the software developer faces one or more of the following challenges:

- The system requires application performance beyond the reach of any available processor running at any realistic clock rate.

- There may be adequate processors available, but the cost, power dissipation, silicon integratability, or limited supply of chips implementing those processors make them incompatible with system design and volume manufacturing goals.
- The necessary performance can be reached by a combination of software on a general-purpose processor and a hardware accelerator for key functions, but the inflexibility of the acceleration hardware, the communication overhead between the software and the accelerator, or the inadequacy of tools available for hardware/software co-development make the "CPU plus accelerator" architecture unattractive or unmanageable.
- The target application consists of a set of tightly communicating tasks, each of which may require an entire processor to reach adequate performance. Available general-purpose processors lack the high-bandwidth and low-latency communication capabilities to support the tightly coupled communication needed to achieve necessary throughput.
- The SOC has a tight area budget. Available general-purpose processors with good performance tend to be larger than needed for task-specific applications because their general-purpose nature requires that these processors have features that suit a wide variety of applications.

In any of these cases, the software team should consider using application-specific processors. Application-specific processors use the development flow, tools, and languages most familiar to software developers but extend the reach of a software-centric approach to SOC design into high-performance, data-intensive problems previously only solvable with hardware accelerators. Using multiple application-specific processors better leverages the software team's efforts, multiplies the team's efficiency, and makes the software team more valuable within the entire system-design effort.

This chapter outlines the core issues of performance-oriented embedded-processor architecture and software development. It walks through a detailed method for analyzing applications and for creating processor instruction-set architecture extensions with improved performance. It provides detailed guidelines for creating efficient, reusable instruction sets for a range of common computational patterns and defines new software techniques to exploit these new, tailored instruction sets.

The chapter also examines the thorny issues of memory-system configuration and programming for memory-bandwidth-intensive data-streaming applications. It closes with a detailed introduction of advanced development methods, including use of long-instruction-word architectures for higher concurrency and reusability and fully automatic instruction-set generation within the compiler. Altogether, this chapter highlights a significant transformation of the embedded-application development process, boosting available software performance while reducing the effort needed to achieve application throughput requirements using software.

5.1 Processor Hardware/Software Cogeneration

We commonly think of a processor as a hardware device, or perhaps as an instruction-set architecture, but this view is a simplification. In fact, a useful processor brings together a number of interacting components. Some of these components, such as the instruction-set definition are abstract; some, such as the integrated circuit that implements the instruction-set architecture, are more concrete. This ensemble includes the following basic elements:

- The definition of the processor's capabilities, especially the fit of the processor's instruction set, interface, and system-programming functions to the system architecture and target applications.
- The hardware circuit implementation.
- The software-development environment (including compilers, assembler, debugger, instruction-set simulator, and source-management tools).
- The runtime software environment available to applications (real-time operating systems and libraries).
- The documentation for efficient use of all of the components listed above.
- In some cases, the body of available applications is a key attraction of particular processor architectures, especially for general-purpose processor platforms that must run a wide range of tasks chosen by the system's end user.

Consequently, a processor's architectural value cannot be measured by a simple quantitative assessment of the performance characteristics of circuit. Instead, its value derives from the richness, interoperability, and appropriateness of all of the associated components. The core idea behind extensible processors is this: increase the processor's utility by improving the fit between application and processor design; by reducing the circuit cost; and by retaining the rich set of software tools, operating systems, and other software infrastructure associated with complete processor environments.

This section explores the software aspect of a processor through three perspectives: the programmer's view of a processor architecture, the major components of the software environment, and how these components are generated for new extended processor architectures.

5.1.1 Applications, Programming Languages, and Processor Architecture

The primitive data types and operations of a high-level language are an imperfect match to the natural computational needs of many embedded applications. At the most basic level, the core of application development is a multistep process of translation—first from a human-language product specification into an algorithm written in a formal programming language, then from a programming language into a sequence of processor instructions and/or a collection of transistors implementing the intended function. The quality of the fit between the product specification and the programming language, and between programming language and underlying hardware, deeply influences the cost, performance, battery life, and flexibility of the resulting electronic system. A programming language like C is widely used because it can express any algorithm. Unfortunately, the nature of the language creates a computational bottleneck that exposes three fundamental issues:

- **Operations that are innately parallel must be expressed serially.** The C language is inherently data-serial and instruction-serial (though not truly task-serial). Improvements in semiconductor technology increase the opportunities for concurrency at all levels, but C programs may struggle to directly exploit these opportunities. Chapter 4 discusses system architecture and task-level concurrency in some depth. Concurrency within a task and implemented with one processor is the first major theme of this chapter.
- **Memory is represented as a single, unified address space.** To exploit concurrency, different threads and different parallel operations must often access distinct memories. The ability to arbitrarily assign global pointers in C makes it difficult for compilers to detect concurrency inherent in algorithms.
- **Applications are complicated by their expression as operations on integers.** For some applications, 32-bit integers are the natural data type in the design of the algorithms. For other applications, the real data type is more idiosyncratic to the task. It may be a stream of short, fixed-point, complex numbers; an array of pixels each of which has multiple color components; or a set of fields making up a protocol header, where each bit field has a different interpretation. Each of these data types can be emulated with conventional integer operations, but processors that can directly handle application-specific data types can sidestep the lowest common denominator of generic data types and integer operations. Building processors that more directly implement the application's native data types and intrinsic operations constitutes the second major theme of this chapter.

C is a fine programming language and will likely remain the mainstay of embedded-application programming for a long time. But the C language only provides built-in support only for the most basic set of (primarily) integer data-types and operations, as shown in Fig 5-1. Note that most C implementations perform all integer arithmetic as 32-bit operations.

Because the natural operations and data types encountered in embedded-system design are more complex, programmers must synthesize or emulate more complex data types and operations from the language's primitive, built-in data types and operations to express their application-level intent. Sometimes, optimizing compilers are smart enough to efficiently map complex operations into short sequences of primitive integer operations. In other cases, the mapped instruction sequences are much longer and run more slowly than what might be achieved if the processor primitives more closely matched those of the target applications.

Data Type	Typical definition on a 32-bit processor	Typical supported operations
`signed int, long`	32-bit signed integer	add, subtract, multiply, divide, shift, logical-AND, logical-OR, logical-XOR, logical-NOT, bitwise-AND, bitwise-OR, bitwise-XOR, compare, array indexing, memory referencing, conversions
`unsigned int, long`	32-bit positive integer	add, subtract, multiply, divide, shift, logical-AND, logical-OR, logical-NOT, bitwise-AND, bitwise-OR, bitwise-XOR, compare, array indexing, memory referencing, conversions
`signed short`	16-bit signed integer	memory referencing and conversions
`unsigned short`	16-bit positive integer	memory referencing and conversions
`signed char`	8-bit signed integer	memory referencing and conversions
`unsigned char`	8-bit positive integer	memory referencing and conversions
`float`	32-bit floating-point number	Add, subtract, multiply, divide, memory referencing, compare, convert to/from integer
`double`	64-bit floating-point number	Add, subtract, multiply, divide, memory referencing, compare, convert to/from integer

Figure 5-1 Standard C language data types and operations.

5.1.2 A Quick Example: Pixel Blending

A short example highlights this "semantic gap" between an application's natural operations and the actual application code as expressed in C and implemented on an integer processor optimized for running generic, integer-based C programs. Consider an application that manipulates pixels. A basic and natural primitive pixel operation is *blend*—to lay one pixel on top of the other, using the top pixel's alpha value (pixel transparency). For each color component (red, green, blue) of the pixel, the blended value is a function of the color component value and the "alpha" value (the degree of transparency) of each of the two pixels. The equation for combining the red color components, for example, of two source pixels, “Over” and “Back,” to a result pixel called “New” is

$$New_{red} \leftarrow Over_{red} \times Over_{alpha} + Back_{red} \times Back_{alpha}$$

When mapped into integer C, the pixel’s component values might be expressed as four 8-bit values packed together in a 32-bit word, where the four values [a, r, g, b] correspond to alpha, red, green, and blue. The C code for the blend function might look like the code in Fig 5-2 (using standard integer data types), where the foreground pixel is `over` and the background pixel is `back`. Note: The programmer could choose to implement this algorithm using C structs containing C fields to represent the pixel [a, r, g, b], but the compiled machine code would be similar:

```
unsigned int blend(unsigned int over, unsigned int back) {
  unsigned char overAlpha = ((over & 0xff000000) >> 24);
  unsigned char oldBackAlpha = ((back & 0xff000000) >> 24);
  unsigned char revAlpha = 0xffóoverAlpha;
  unsigned char newAlpha = ((((oldBackAlpha)) * revAlpha) +
   (((overAlpha)) * overAlpha) >> 8);
  unsigned char Red = ((((back & 0x00ff0000) >> 16) * revAlpha) +
   (((over & 0x00ff0000) >> 16) * overAlpha) >> 8);
  unsigned char Grn = ((((back & 0x0000ff00) >> 8) * revAlpha) +
   (((over & 0x0000ff00) >> 8) * overAlpha) >> 8);
  unsigned char Blu = ((((back & 0x000000ff)) * revAlpha) +
   (((over & 0x000000ff) ) * overAlpha) >> 8);
  return ((newAlpha << 24) | (Red << 16) | (Grn << 8 ) | (Blu ));
}
```

Figure 5-2 Pixel-blend function in C.

Compared to the verbal description, the resulting code is fairly large and difficult to understand. The *blend* operation requires about 33 distinct integer operations to compute the intended pixel value. The implementation of this version of the blend operation on a RISC processor will be slow and the machine code will be fairly large—perhaps as much as a thousand instruction bits.

An application-specific processor can implement the *blend* algorithm as a single instruction and can easily operate at the rate of one *blend* operation per cycle. The pixel data type becomes native to the source language, and the software requires only a few tens of bits to express this operation in the processor’s machine code. See Section 5.3.4 for a detailed example of how the pixel-blending operation might be implemented with an application-specific processor.

Application-specific data types and operators bridge the semantic gap between generic integer C and the specifics of a target embedded application. In fact, many software engineers develop good libraries of new data types and operator primitives, and they write core algorithms in terms of these primitives as a data abstraction.

Similarly, object-oriented programming languages such as C++ are particularly appropriate vehicles for implementing new application-specific data types and operators, though on non-extensible processors the mapping of these new operators to the processor instruction set must use long sequences of integer operations instead of simpler, faster operations that operate directly on the application's native data types.

New programmer-defined data types and application-specific operators may simply be good programming style on traditional processors, but on extensible processors this approach also optimizes applications more quickly, produces code that executes faster, and reduces code size. These are the three key benefits that application-specific processors offer to embedded software developers.

5.2 The Process of Instruction Definition and Application Tuning

The most complex tasks in SOC development are almost always expressed as software algorithms written in some programming language. Regardless of the quality of the match between the application intent and the programming language, the developer must implement these tasks as machine code running on some processor within the SOC. Moreover, the developer must pursue a complex set of objectives, including system cost, application performance, power efficiency, design reusability, and development effort. When the SOC developer starts from an application already expressed in a programming language, the obvious process is this:

1. Make simple assumptions about the target hardware.
2. Compile and simulate the application.
3. Profile the code to find where and by how much objectives are missed.
4. Change the hardware and/or the application code and repeat steps 2 through 4 until all objectives are met.

This application-tuning process has long been an important element of electronic-system development, even without meaningful developer control over the target-processor design. With the advent of automated processor generation, processor configuration becomes an essential element of SOC design.

This section teaches the basic technique of software-driven processor configuration: application profiling and development of processor extensions. The section includes recommended methods for designing simple, fast, and reusable application-specific instruction sets.

5.2.1 Profiling and Performance

Software developers profile code to help them reduce the number of cycles required to perform an application task. Reducing the number of execution cycles isn't just important for improving overall system throughput. If the task can be accomplished in fewer cycles, then a given performance level can be achieved at lower clock frequency and consequently at lower power. Alternatively, reducing the number of execution cycles creates performance headroom so developers can add new system features during the initial product development. They can also add new features over time as product upgrades because the processor has the available bandwidth to implement these upgrades in addition to the tasks it's already executing. As a result, the focus of task profiling is to determine the performance level for a task and processor, to identify the "low-hanging fruit" for improvement, and to assess the impact of trial improvements to the processor's instruction set and architecture.

Before the advent of configurable processors, profiling identified candidate functions and loops for C-level tuning or for conversion to assembly code. The availability of configurable processors means that the software developer now has even more ability to refine and improve application performance through profiling. Code profilers report a wide range of performance characteristics, such as the number of cycles spent executing each instruction, each line of source code, or each function. They also report the number of events, including exceptions, cache misses, or communications operations. The opportunity to measure a wide range of dynamic system behavior through one approach and tool environment makes profile-based tuning a powerful technique.

The software developer controls a number of important variables that may affect a task's performance profile:

- The application or workload may consist of a number of algorithms working together. The relevant goal for workload performance will be some aggregate of the performance goal for each workload component.
- There may be several potential versions of the application source code available for a task. Each version may have different performance characteristics or may make different tradeoffs among code size, data-bandwidth requirements, hardware cost, and execution speed.
- The software developer can choose to tune portions of the application either in C or in assembly code. Tuning can improve performance but it requires effort and can result in trading off code size, execution speed, generality, portability, and readability.
- Most algorithms have characteristics that vary with the input data set. Performance for different input sets will vary, sometimes significantly. Once the developer establishes the sensitivity to different data sets through profiling, the developer can choose the appropriate input data set for performance tuning.
- The compiler may allow the programmer to control a variety of different compilation goals and options, including the optimization level and choice of libraries, which allows the developer to make tradeoffs between execution speed and code size.
- A wide range of memory and instruction-set options may be candidates for the initial processor configuration. Each option offers a unique mix of tradeoffs.

Together, this set of choices forms the initial design space for the software developer. The developer uses the profiling environment, usually based on simulated execution, to assess the baseline application performance across the target workload and input data.

Exploration of simple choices of compiler options and basic processor configurations may achieve quick improvements to the application cycle count before concentrating on application-specific instruction-set extension.

Application-tuning often involves looking both at the global picture of application behavior and performance—looking at the forest—and at the details of the inner loops—looking at the trees. Sometimes the issues are completely independent: poorly chosen algorithms can be too slow even with the finest instruction-set tuning.

Sometimes the issues interact. For example, many application-specific processor tuning optimizations only work across multiple loop iterations or code sections. The following section provides a quick overview of high-level performance tuning and then takes a more detailed look at tuning of inner-loop code.

5.2.1.1 High-Level Performance Tuning Tips

Maximizing application performance starts with insight into the application's overall performance profile. Once the performance-critical regions are identified, the first question to ask is this: What is the upper bound on overall application performance, assuming these critical regions were optimized to take zero time? If the upper bound on performance doesn't satisfy the system's essential performance requirements, some fundamental restructuring of the application or reassessment of requirements may be necessary. On the other hand, if the performance requirements are potentially within reach, the developer can start "peeling the onion" to get to an appropriate implementation of the processor and the code.

The developer should consider the following questions in starting the tuning process:

- Have I chosen a good algorithm and decomposition as a starting point? Is there a faster basic algorithm for the size of my data set?
- Have I profiled the code with an appropriate basic processor configuration? Is there an available predefined instruction-set extension—floating-point extensions, for example—that directly addresses my application's core computation needs?
- Do the assessments of initial performance suggest that memory-system design, especially access to data, dominates all other performance issues? (Memory-system tuning is covered in some detail later in this chapter.)
- Have I chosen appropriate compiler flags? Is the appropriate level of code-performance or code-size optimization turned on?
- Does my code have pathological characteristics that limit performance, such as
    - aliased parameters in critical functions?
    - frequent and unintended type conversions among 8-bit, 16-bit and 32-bit integers?

- unnecessary use of global variables instead of local variables?
- use of pointers where arrays are natural?

Any of these characteristics may inhibit important compiler optimizations or generate inefficient code on many processors. Once the basic code has been globally tuned, a more detailed inner-loop profiling becomes the starting point for focused optimization and processor tuning.

5.2.1.2 Low-Level Performance Tuning

The baseline profile highlights hot spots in the application code: the functions, the loops, and even the individual lines of code that consume the most execution cycles. A detailed profile can also reveal the contribution of each processor feature to the number of cycles in a section of code. Many factors may cause the processor's real performance to diverge from the theoretical peak performance. Common sources of performance degradation include

1. Instructions that take multiple cycles to execute.
2. Instruction sequences that stall when data passes from one instruction to the next in the processor's pipeline.
3. Fetching instructions from memory may require several cycles.
4. When required operands are not found in the data cache, the processor may stall until the operand can be fetched from main memory.

Selecting an appropriate processor can reduce both the number of instructions and the number of execution cycles consumed by critical instruction sequences.

A short profiling example for the byte-swap function underscores the basic issues. This is the same simple C function introduced in Chapter 1, Section 1.4, a function that commonly occurs in networking applications:

```
int i,a[],b[]
for (i=0; i<4096; i++) {
 a[i] = (((b[i] & 0x000000ff) << 24) | ((b[i] & 0x0000ff00) << 8) |
 ((b[i] & 0x00ff0000) >> 8) | ((b[i] & 0xff000000) >> 24));}
```

Fig 5-3 profiles the core function of the simple byte-swap task prior to optimization. The profile shows that another procedure calls this function 100 times. The compiler has performed a number of advanced optimizations to improve performance as much as possible. It has unrolled the inner loop to save address-calculation cycles. The compiler has also selected a zero-overhead loop instruction to eliminate loop count and branch overhead.

Code

```
for (i=0; i<4096; i++) {
a [i] = (((b[i] & 0x000000ff) << 24) |
(((b[i] & 0x0000ff00) << 8) |
(((b[i] & 0x00ff0000) >> 8) |
(((b[i] & 0xff000000) >> 24));|
```

Execution Cycles	Data Cache Stalls	Total Cycles	Assembly Code	
100	0	100	`loopgtz`	`a8,.LBB6_byteswap`
204,800	717,654	922,454	`l32i.n`	`a89, a9, 4`
204,800	0	204,800	`l32i.n`	`a4, a9, 0`
204,800	0	204,800	`addi.n`	`a9, a9, 8`
204,800	0	204,800	`srai`	`a11, a8, 8`
204,800	0	204,800	`and`	`a11, a11, a12`
204,800	0	204,800	`srai`	`a13, a4, 8`
204,800	0	204,800	`and`	`a13, a13, a12`
204,800	0	204,800	`srli`	`a14, a4, 24`
204,800	0	204,800	`slli`	`a3, a4, 24`
204,800	0	204,800	`srli`	`a15, a8, 24`
204,800	0	204,800	`slli`	`a2, a8, 24`
204,800	0	204,800	`and`	`a8, a8, a12`
204,800	0	204,800	`and`	`a4, a4, a12`
204,800	0	204,800	`slli`	`a4, a4, 8`
204,800	0	204,800	`slli`	`a8, a8, 8`
204,800	0	204,800	`or`	`a2, a2, a8`
204,800	0	204,800	`or`	`a3, a3, a4`
204,800	0	204,800	`or`	`a13, a13, a3`
204,800	0	204,800	`or`	`a11, a11, a2`
204,800	0	204,800	`or`	`a11, a11, a15`
204,800	0	204,800	`or`	`a13, a13, a14`
204,800	0	204,800	`s32i.n`	`a13, a10, 0`
204,800	0	204,800	`s32i.n`	`a11, a10, 4`
204,800	0	204,800	`addi.n`	`a10, a10, 8`
			`.LBB6_byteswap:`	
4,915,300	717,654	5,632,954		

Figure 5-3 Execution profile for swap (before optimization).

The figure shows the source code for the core function and the resulting assembly code using the Xtensa ISA. It also tabulates the number of instruction execution cycles and stalls for each assembly instruction. For this example, only data-cache misses (accessing `b[i]`) are significant because the compiler has already made optimal use of the processor's base instruction set. Hand-coded assembly language will not result in faster code.

Even in this base case example, the processor's memory system has been optimized to refill 16 bytes per cycle into the cache every second iteration of the unrolled loop to reduce the number of cycles spent fetching data when a cache miss occurs. Configurability of the bus width, transfer size, and other memory-interface characteristics allows the designer a wide range of tradeoffs between minimizing chip cost and maximizing application throughput

Optimizing the processor instruction set drastically reduces the number of cycles in the inner loop. As Fig 5-4 shows, one new instruction called `swap` replaces a longer sequence of ordinary integer instructions in the original code. The subsequent compilation using the `swap` extension performs all the same optimizations—loop unrolling and zero-overhead-loop code generation—as in the original code.

For the optimized case, two effects reduce total number of execution cycles. First, the number of executed instructions drops by about three times with the addition of just the one new instruction, byteswap. Second, increasing the cache-line size from 16 bytes to 64 bytes reduces the number of cache misses by four times and the number of miss stall cycles by three times, to miss only one for every eight iterations of the unrolled loop. The processor bus interface is no wider, but the memory system transfers large blocks more efficiently than small blocks, so the net result is a further savings in execution cycles. Alternatively, the system designer could set up a direct memory access (DMA) transfer, so an external agent pushes the data into the processor's local memory in anticipation of use. This approach would eliminate the cache misses and the associated wasted cycles.

Execution Cycles	Data Cache Stalls	Total Cycles	Assembly Code	
100	0	100	`loopgtz`	`a8,.LBB6_byteswa1`
204,803	258,140	307,325	`132i.n`	`2, a9, 0`
204,803	0	204,803	`132i.n`	`a11, a9, 4`
204,803	0	204,803	`addi.n`	`a9, a9, 8`
204,803	0	204,803	`swap`	`a11, a11`
204,803	0	204,803	`swap`	`a12, a12`
204,803	0	204,803	`s32i.n`	`a12, a10, 0`
204,803	0	204,803	`s32i.n`	`a11, a10, 4`
204,803	0	204,803	`addi.n`	`a10, a10, 8`
			`.LBB6_byteswap:`	
4,915,300	717,654	5,632,954		

Figure 5-4 Execution profile for byteswap (after optimization).

5.2.2 New Instructions for Performance and Efficiency

The basic process of instruction-set extension has three phases:

1. Identifying the application sections that need acceleration.
2. Defining new instructions that are faster or more efficient than existing general-purpose instructions.
3. Incorporating the new instructions into the application code.

This section focuses on the definition of new instructions. Application-specific instructions can be faster and more efficient than the processor's baseline instructions because each new instruction usually combines a number of basic operations together. An instruction's set of operations can be dependent on one another (where the output of one operation serves as an input to others) or these operations can be independent. The combining of operations follows three forms:

- **Fused operations:** The combination of a set of dependent operations into a single complex operation with the same input operands, output operands, and function. This combination creates opportunities to share input operands, eliminates the need to store and fetch intermediate operands, and executes operations in parallel.
- **Compound operations:** The combination of different, independent, parallel operations into one instruction. Combining operations into compound instructions is effective even when no dependencies exist between the operations and is especially effective when operations share input operands.
- **Single-instruction/multiple data (SIMD) operations:** The duplication of an operation so a series of identical operations on different data inputs can be performed in parallel.

5.3 The Basics of Instruction Extension

Description of processor instruction-set extensions is central to design of application-specific processors. Sometimes the performance target can be reached with a simple selection of predefined processor features (for example, changing the parameters of the processor's caches or including optional instruction-set packages such as floating-point or DSP instructions). More often, big improvements in efficiency come only with new instructions tailored specifically to the target application.

Any processor instruction must specify two forms of essential information:

Operands: The set of processor registers, immediate values, and memory locations used to designate the source values and destination locations for the computations.

Operation: The transformation function from source operands to destination operands.

Other elements of the instruction definition include

- Instruction encoding
- Mapping of processor state to high-level-language data types
- Mapping of instructions into high-level-language functions
- Microarchitectural details, such as pipelining and hardware sharing
- Verification support, such as test vectors or assertions about instruction behavior

A trivial TIE example appeared in Chapter 1, but a very short introduction to TIE in this chapter will help explain basic issues of instruction-set extension. A basic TIE description on an instruction-set extension consists of two main parts: the declaration of programmer state (in the form of register files and state registers) and the description of new instructions that use the programmer state as operands. In the simplest case, programmer state takes the form of register files:

```
regfile <name> <width> <depth> <short_name>
```

and state registers:

```
state <name> <width>
```

Register files are usually indexed by a field within the instruction word, so the register entry is an *explicit* operand in the instructions that read or write that register file. On the other hand, state registers are used *implicitly* by instructions that read or write that register.

In the simplest case, new instructions take the form of operand declarations:

```
operation <name> {<explicit operands>} {<implicit operands>} {<assignment
    statements>}
```

TIE assignment statements use the same syntax as that of combinational logic in the Verilog hardware description language, where each statement is either an assignment to a result operand of an expression (a function of input operands):

```
assign <name> = <expression>
```

or an assignment to a temporary variable used elsewhere in the operation:

```
wire <name> = <expression>
```

where `wire` is a one-bit-wide signal by default (but can be declared and computed with any width as required by the application).

The TIE instruction-set-extension language includes range of primitive functions that can be used as operators in expressions. This book develops a number of instruction-set-extension examples from these basic TIE primitives, and other TIE language features are introduced along the way. TIE code describes extensions to be made to an existing Xtensa processor architecture. The Xtensa processor generator uses application-specific TIE descriptions to tailor the design of an Xtensa 32-bit RISC processor for different applications. The baseline Xtensa processor provides basic RISC processor capabilities including a general-purpose register-oriented instruction set with compact (16- and 24-bit) instructions, a 32-bit register file (named AR), and one read/write port to data memory (which can be as wide as 128 bits).

NOTE The advantage of extending the instruction set of an existing processor is to speed development. Design teams can and have developed processors from scratch, but it takes a lot of time to develop and debug a processor and its associated software-development tools. Extending an existing processor takes far less time and is far more efficient.

The point of introducing TIE syntax here is not to focus on the language details but to highlight the major opportunities and decisions faced by the engineer optimizing a processor for specific applications. Processor optimization requires the balancing of three goals (which occasionally conflict):

1. **Application performance:** Performance is often tied to the number of basic operations executed by one instruction. Wider and deeper register files also help to improve operational parallelism. Parallel operations generally increase hardware cost and instruction specialization. Wider and deeper register files generally increase hardware cost, but they also increase instruction flexibility (they do not increase specialization).
2. **Instruction length:** Instructions with many operands or wide operands (immediate fields or operand specifiers for deep register files) may fully consume the available instruction width (as many as 24 bits for the Xtensa processor and as many as 64 bits for the Xtensa LX processor, described in Section 5.8). The opcode that specifies each operation also consumes instruction encoding space, independent of the number and size of operations. For example, no more than 224 instructions can be encoded in a 24-bit instruction word, even if all the instructions have no operand specifiers. Instruction sets with many operations and operations with many independent operands are more flexible, but may exhaust the available encoding space. Complex instruction sets with large register files imply considerable data-path logic to implement the functions and storage. Interestingly, the logic for instruction decoding is usually small compared to the logic required to implement the

data-path functions controlled by the instructions so decoding-logic size is *not* a major consideration in hardware cost.

3. **Instruction-set flexibility:** The ideal instruction set is both closely tailored to provide all the features and parallel operations required by the target applications, and fully flexible to be equally efficient across all applications that may run on the processor. In practice, the ultimate set of target applications is not knowable at the time of design, so the designer must make some guesses. Providing more operations—applicable to a greater range of operands and available in freer combinations—makes a more general-purpose instruction set but also increases hardware cost and puts additional pressure on instruction length. Building highly tuned instructions that combine all the basic computation of a key algorithm in a single instruction reduces instruction encoding difficulty (and often hardware cost), but sharply restricts the flexibility and use of the new instruction.

Understanding and balancing these three factors is central to the effective use of configurable processors.

5.3.1 Instruction Extension Methods

Focusing instruction-set-extension efforts on performance-critical sections of code revealed by profiling provides the greatest benefit for the effort expended. New instructions can also improve instruction-memory storage requirements. Processor improvements that reduce the number of execution cycles also tend to reduce code size and memory-bandwidth requirements. To develop new instructions, the software developer works from the overall performance profile and goes through the following process:

- Survey the overall performance profile, identify the performance-critical sections of code, and determine if accelerating those sections can achieve the overall target application performance. For example, if only 20% of the total execution time is spent in all of the application's inner loops, getting a 3x improvement in application performance just by tuning those loops (even by driving the execution time of those loops to zero) will be impossible.
- Examine the C source and generated assembly code and assess whether a better source algorithm or better assembly-code generation can significantly improve performance. Accelerating a poor algorithm may yield an impressive relative performance gain but still not achieve application performance goals.
- Look at the profile using an initial processor configuration, including the stalls associated with the memory system. Configure features such as basic arithmetic capabilities, cache sizes, and bus widths until the time spent executing instruction sequences and loading necessary input data dominates overall performance. Caches and interfaces should be sized to minimize time consumed by instruction- and data-cache misses, or should enable DMA transfers to eliminate cache misses altogether.

For each performance-critical section, the designer can use the following process to define new, more appropriate processor instructions:

1. **Look at the C or assembly code and identify the critical *sequence of dependent operations* in each section of code.** A sequence of dependent operations is a set of instructions where the result of each instruction in the set is directly used as input to the next. The sequence can include control-flow operations (branches), but some optimizations, such as fusing operations, may not be useful across different basic blocks within the section of code. Note the number and types of operands for the sequence. For example, the following fragment of C code has a sequence of five dependent operations (a load, a multiply, an add, a shift, and a store):

```
int i;
unsigned short *x,temp1,temp2,temp3,temp4,*y, loopconstant=5;
temp1 = x[i];
temp2 = temp1 * loopconstant;
temp3 = temp2 + temp2;
temp4 = temp3>>1;
y[i] = temp4;
```

2. **For critical code sequences, identify opportunities to replace memory operands with operands held in registers.** This change saves address calculations, data loads, and stores.
3. **Identify opportunities to replace variable operands with constants.** This change saves memory and register reads and reduces the number of operands that must be specified, thus saving instructions.
4. **For memory operands, determine the simplest form of operand addressing** that generates the correct sequence of load and store addresses. The number of operands and operand specifiers can be reduced if successive addresses can be created as fixed offsets from a base register value or by simple, automatic register increments.
5. **Identify the minimum necessary computational bit width,** especially where the width of standard C integer data-types (8-bit, 16-bit, 32-bit) may be overkill for the actual computation needs. To do this, the programmer must fully understand the incoming operand precision required by the application. Successive arithmetic operations generally increase the dynamic range of successive results, so overflow or underflow handling may be necessary. Techniques such as extended precision for some results, or saturation on underflow and overflow may be implemented to meet precision requirements.
6. **Fuse dependent operations into a set of new instruction descriptions** constrained by the number of operands that can be encoded in the available instruction-word width. This step is so common that it is discussed in more detail in Section 5.3.4. Fusion saves both instruction bits and cycles. Execution cycles are saved if fusing operations either reduces

the number of instructions required to encode the sequence or if fusing two operations reduces the total latency along the sequence of dependent operations.

7. **Combine independent operations that share operands** into new instruction descriptions constrained by the number of operands that can be encoded in the available instruction-word width. This approach saves operand specifier bits, which can allow more operations to be executed per cycle and reduces the number of register ports. Fig 5-5 compares the separate and combined versions of instructions as expression trees. The input registers outlined in bold specify the same register and are therefore eliminated by fusion.

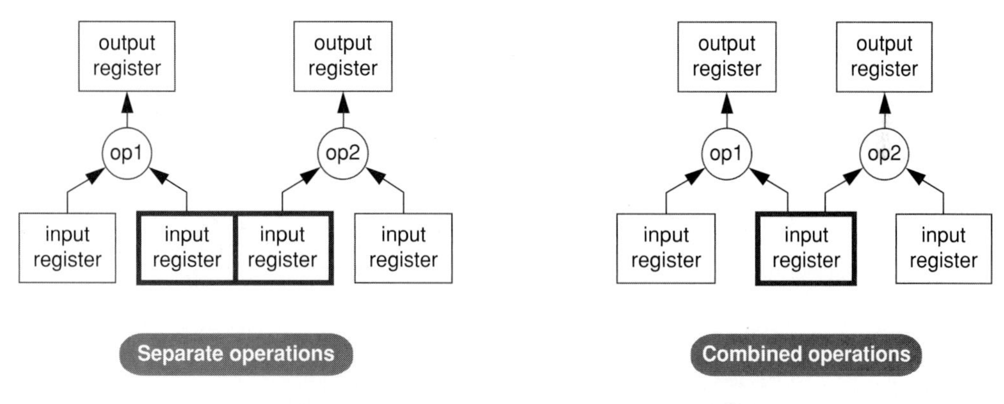

Figure 5-5 Combining of common input specifiers.

8. **Simplify load-and-store instructions by restricting address offsets or source- and destination-register specifiers.** This version of fusion saves instruction bits and may allow more operations per cycle. For example, it might be possible to reduce an 8-bit address offset to 2 bits, so that only 4 words near the base address register can be reached by the new instruction. Four words may be adequate for certain application-specific instructions. Similarly, the load-destination or store-source operands might be specified as part of a much smaller, 2-entry register file, so a 4-bit register specifier in a more general-purpose instruction could be reduced to one bit. If these two techniques are combined, a load instruction that originally required 16 bits of operand and offset specifier (8-bit address offset, 4-bit base-address-register specifier, 4-bit destination-register specifier) could be reduced to 7 bits of operand and offset specifier (2-bit offset, 4-bit base-address specifier, 1-bit destination-register specifier), as shown in Fig 5-6.

   In fact, creating an instruction that uses a dedicated state register as the base address could eliminate even the base-address register specifier. It's important to note that architects of general-purpose processors cannot be this restrictive with operand specifiers because they must design processors with broad applicability across a wide range of applications. But software developers tailoring an application-specific processor for a target application can define far more limited specifier ranges precisely because they know the application code requirements in advance. When address calculations are made with separate add-immediate or subtract-immediate operations, the immediate range can be reduced to just the necessary increments, saving bits from the immediate specifier.

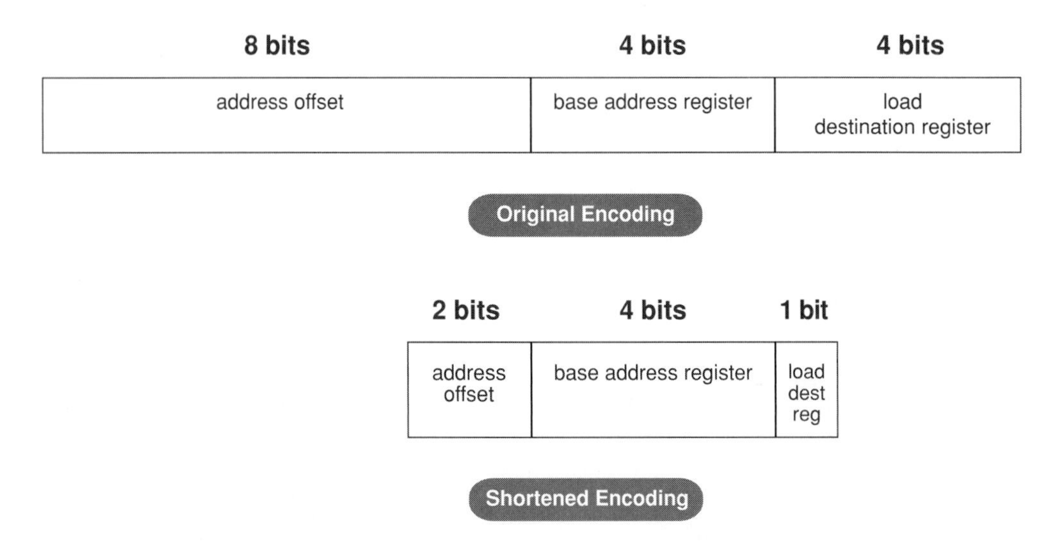

Figure 5-6 Reduced operand specifier size.

9. **Combine address updates (e.g., auto-increment or auto-decrement) into memory operations.** This fusion both reduces the number of operand specifiers (the increment or decrement amount may be implicit in the operation) and the number of operations (a separate add or subtract instruction may be eliminated from the critical code sequence). Consider, for example, the Xtensa assembly code sequence shown in Fig 5-7. Inside the loop, a 64-bit extended-register-file load instruction (with a conventional offset field) is preceded by an addition to the base register of the address stride (8 bytes) to advance a pointer to the next data value in memory.

```
            loop        a0, LoopEnd

            l64i        x0, a1, 0
            add         a1, 8
LoopEnd:
```

Original

```
                loop           a0, LoopEnd

                l64.inc        x0, a1
LoopEnd:
```

Combined load/increment

Figure 5-7 Combine address increment with load instruction.

The original pair of instructions can be replaced by an instruction that directly and implicitly incorporates the 8-byte increment operation. This fusion reduces the number of instructions from two to one and the total number of specifiers from five to two

10. **Combine loads and stores with computational operations.** This fusion eliminates the operand specifier for the load result and the computation operation input, or the computation result plus the operand specifier for the store source. For example, the load from the previous example may write an operand that is later used as the source for a byte rotate, as shown in Fig 5-8. The load, including the address increment, can be further combined with the rotate. This fusion saves instructions and operand specifiers. It also achieves higher performance through greater parallelism among operations. If the unrotated data value is also required elsewhere in the calculation, it could be specified as a second destination of the `l64.inc.rotb` instruction.

```
                loop           a0, LoopEnd

                l64.inc        x0, a1
                rotb           x1, a1, 7
LoopEnd:
```

Separate load/increment and rotate

```
                loop           a0, LoopEnd

                l64.inc.rotb x1, a1, 7
LoopEnd:
```

Combined load/increment/rotate

Figure 5-8 Combine load with compute instruction.

11. **Combine operations from different iterations.** Unroll loops to create more opportunities to combine operations into new compound instructions. Newly uncovered long sequences

of dependent operations can be used to create deeply pipelined instructions. New uncovered combinations of non-dependent operations can be used to create new wide SIMD instructions. SIMD or vector methods are outlined in Section 5.3.6, and Chapter 7 discusses SIMD methods and pipelining in even more detail.

12. **Estimate timing and area.** For a processor with a set of newly defined instructions, work with the processor design tools and the hardware team to generate initial estimates of area and critical timing paths. Instruction definitions may be refined by specifying an *instruction schedule* to spread the execution-unit logic across several pipeline stages. Code can be restructured to overlap execution of long-latency operations from more than one loop iteration (*software pipelining*).
13. **Share register files and operators.** Look across all the instruction definitions and identify large register files and data-path operations that are similar. Organize and generalize those functions so they are shared as much as possible.
14. **Assign C types.** Where appropriate, assign C-language type names to new register files to allow the compiler to manage that storage class. Add data-type conversion descriptions where appropriate.
15. **Build tests.** Develop test vectors so that the function of each instruction can be tested outside the context of the end application.
16. **Recheck timing and area.** Recheck the configured processor's silicon area and critical-path timing and iterate through steps 1 to 8 above until performance goals are achieved

Instruction-description format details and the supporting processor profiling and configuration tools shape this process, but the basic sequence follows the principles above.

5.3.2 Upgrading the Application

Defining new processor instructions does not necessarily make those instructions available to the application. Compilers generally cannot identify opportunities to optimally use any arbitrary new instruction, though at the end of this chapter we describe an important set of circumstances in which completely automated instruction-set design and use is not only possible, but also highly efficient.

In many cases, new instruction definitions rely on application developer insights that are unavailable to the compiler, so parts of the application (the critical inner loops) will need to be rewritten to exploit the new instructions. As discussed at the beginning of this chapter, the primitive integer data types and operations available in C are an imperfect match to the intrinsic computational structure of many embedded applications. Re-expressing key algorithms in terms of new application-specific data types and operations may, as a secondary benefit, improve the readability and maintainability of the application.

New application-specific instructions define new operators, and those functions become available to the high-level language programmer in three ways:

1. **As intrinsic functions:** The operation is instantiated explicitly as a function, so a new operation—for example, `my_func`—with two sources and one result, is used through a C function call as `result = my_func(source1,source2);`.
   - Direct instantiation through intrinsic functions is simple, though complications arise if the function can only be applied to certain data types (for example, only to 32-bit integers) or storage classes (for example, only to values known to be held in general registers). Without automatic, application-specific data-type conversions, the programmer must perform manual data-type conversion from application-specific data types to the default (integer) data types used in the original source code.
   - Instantiation of intrinsic functions fits naturally with automatic support of new data types and storage classes (for example, additional register files). New data types, if defined along with new register files, are available to the programmer via an advanced compiler specifically developed to support configurable, application-specific processors. Advanced compilers can automatically allocate variables of the new type to the new register file. They generate code to dereference pointers and to manage arrays so loads and stores of values to and from the register file are supported for this new data type. With this level of compiler support, the programmer's primary task is to identify where intrinsic functions corresponding to the new instructions should be used to replace existing sequences of generic, high-level-language code. Because the new instructions were developed precisely for this purpose, replacement should be easy.
2. **Through inference by the compiler:** If the compiler is aware of the functional definition of a new operation, it can automatically instantiate the operation without any extra work by the software developer. The semantic definition of a new operation called `my_func` may be expressed as a template—an expression in terms of primitive C operations on its built-in data types such as char, short, and int. A sophisticated compiler may be able to identify all occurrences of that expression, or its equivalents, and translate those occurrences into instantiations of the operation `my_func`. Automatic inference by the compiler is an important by-product of the automatic generation of new instruction extensions. Section 5.9 describes Tensilica's XPRES technology, a methodology and toolset for fully automatic generation of instruction-set extensions with full inference of generated instruction by the compiler.
3. **Via C++ operator overloading:** The new operation can be invoked explicitly through use of a specially defined operator for that source type. A new operation called `my_func` operating on data type `my_type` might be associated with the operator symbol "+", so the function `my_func` is instantiated as `source1+source2` when the two sources are of type `my_type`. Note that the semantics overloaded onto the operator + may bear no resemblance to the originally designated arithmetic operation of addition. The existing character + is simply convenient shorthand for the binary operation implemented by "my_func". A good rule-of-thumb is to use overloading where the new operator is already

commonly used in the application domain. Using "+" to add two vectors or for Galois field arithmetic is natural. Using "+" for accumulating a checksum may not be natural—the intrinsic name of the function that generates the checksum may cause less confusion to other developers reading the code.

The programmer may directly reassign appropriate variables to the newly defined type and modify the source code to apply the new operators.

Sometimes, the programmer has already structured the code by using important application-specific data types and implementing a library of operators on those data types. In these cases, only the data-type definition and library implementation need be modified to use the new intrinsic support. If the application is developed in C++, operator overloading may also be used.

As an example, suppose the developer defines a new 8-entry-by-16-bit register file named `Vec8x16_Reg`. This register file is used as the basis for a C++ class called `Vec8x16`. The developer provides a function to increment each of the eight elements of the vector as a new instruction named `IncVec8x16`. This Vector class is defined and the ++ operator is associated with the instruction by using the following code:

```
class Vec8x16
{Vec8x16_Reg _data;
 public: Vec8x16& operator++(int) {IncVec8x16(this->_data); return *this;} };
```

This C++ code identifies that the operation ++ applied to a variable of type `Vec8x16` should be implemented via a call of the function `IncVec8x16`. With this shorthand, a program can increment all eight vector entries in a variable `vec` of type `Vec8x16` simply by using the syntax: `vec++`. For some developers and some applications, use of this shorthand may improve both efficiency (due to comprehensive use of accelerated instructions) and understandability of the code (because application-specific operators raise the abstraction of the coding style to a level appropriate to the application). Of course, not all application-specific data types and operators have such intuitive mappings for common C operators onto the new functions.

Different processor-generation systems differ in the use of these three methods. The availability of these mechanisms for creating instruction extensions and the nature of the extensions affect the ways that software developer exploits the extensions in the application code.

Regardless of the coding style, the new source code incorporating application-specific data types and functions is often simpler than the original code and generates fewer bytes of machine instructions. Smaller code size may, in turn, improve cache-hit rates and reduce bus traffic, improving overall system performance still further.

More aggressive compound instructions that perform several unrelated operations per cycle or operate in SIMD fashion on several operands in parallel may require more sophisticated source modifications. If the instruction operates on *N* data operands per instruction, SIMD extensions require that loops be restructured so that only 1/*N* iterations are performed. If the iteration count is not known to be evenly divisible by N, then fix-up code for the remaining few iterations may be needed.

Similarly, if the instructions combine unrelated operations to enhance pipelining—for example, if the operand load for one iteration is merged with the computation of a previous iteration—*preamble* and *postamble* code may be needed to deal with loop start-up and termination.

Often, instruction-set enhancement is just one of several tasks driving SOC development. Other activities such as application development and porting, algorithm tuning, development of system interconnect, running of large test cases, integration with other subsystems, and chip floor planning may all be happening in parallel. Each of these activities may either change the design requirements or open up new opportunities for tuning processor instruction sets. In practice, incremental enhancement of instruction sets may continue from the earliest architectural studies to finalization of processor RTL. The ability to rapidly deploy modified data types and intrinsic functions into application source code and to rapidly evaluate the hardware impact of these instruction-set extensions make such architectural explorations and modifications very suitable for SOC development.

5.3.3 The Tradeoff between Instruction-Set Performance and Generality

Instruction-set enhancement is based on combining previously separate operations into one instruction. This process inevitably leads to tradeoffs between the performance of programs that use the new instructions and the generality of the extended instruction set. Combining operations increases concurrency and performance only if the application actually contains that particular combination of operations. Similarly, restricting operand ranges to reduce hardware cost or to permit the combining of operations in one instruction only helps if the application code works within those restrictions.

Processor performance generally improves if a large number of operations and operands can be freely combined into instructions. The freedom to combine instructions can significantly increase instruction-memory size, the amount of instruction-decoding logic, the number of register file ports, and data-path complexity. The resulting growth in silicon area increases chip cost and power dissipation. In cost- and power-conscious SOC designs, instruction width is constrained to reduce the number and width of the operand specifiers.

Each of the following instruction-design techniques reduces the number of bits required for operand specification and the overall register and execution-unit complexity:

- **Fuse dependent operations** to eliminate the destination specifier of one operation and the source specifier of the succeeding dependent operation, especially if no other operation in the program uses the result of the first operation. This fusion eliminates two identical specifiers from the instruction stream.
- **Combine operations that use the same sources.** This fusion reduces both the encoding redundancy and the number of operand fetches.

- **Restrict the range of immediate values or address offsets.** This reduction eliminates unused zero values from the instructions.
- **Replace variables with constants.** This replacement eliminates an operand specifier altogether. It also has the important side effect of simplifying and speeding up the implementation logic for the operation.
- **Convert explicit use of a register-file entry to implicit use of a dedicated register.** If only one value of a specific data type is needed in the critical code sequence, a dedicated register can be implicitly specified in the operation without consuming any instruction bits for an operand specifier.
- **Minimize register-file depth to shrink the width of the register specifier.** Register files need only to be deep enough to handle the degree of concurrency required by the target algorithm. Reducing the number of entries saves instruction bits and may reduce the cost of large register-file structures.

Even with these techniques, the generality of the resulting instruction set strongly correlates to the processor generator's capacity to accommodate instruction definitions with thorough support for implicit operands, flexible operand encoding, and many small operand specifiers per instruction word. The processor's ability to handle long instruction words (larger than 32 bits) may dramatically improve instruction-set generality. The methods and impact of flexible-length instruction encoding are described in Chapter 4, Section 4.4.

A closer examination of techniques for application-specific design will make the issues more concrete. We now look at three basic instruction-set enhancement primitives: operator fusion, combining of independent operations, and SIMD. In each case, we provide simple examples using TIE.

5.3.4 Operation Fusion

Operation fusion is the process of combining a set of simple dependent operations into an equivalent complex operation with the same overall input and output operands and function. This process eliminates the need to store and fetch intermediate operands, creates opportunities to share input operands, and executes some operations in parallel.

For example, operands held in a general 16-entry register file each require a 4-bit specifier. A three-operand instruction (two inputs, one output) can be combined with another three-operand instruction (also two inputs, one output) if the result of the first instruction serves as an input to the second. The new fused operation has four operands (three inputs, one output). Fig 5-9 compares the unfused and fused version of an instruction sequence as expression trees, where the input and output registers outlined in bold specify the same register and are eliminated by fusion. The fusion of two operations, each requiring 12 operand-specifier bits, becomes one operation with 16 operand-specifier bits.

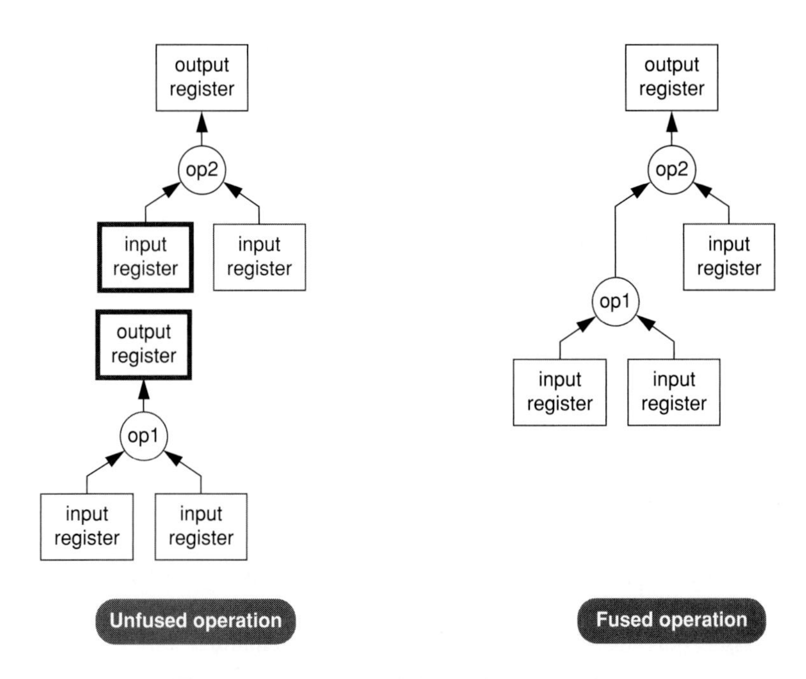

Figure 5-9 Fusion of dependent operations.

Consider the computation of just one color component from the pixel blend function in Fig 5-2.

```
unsigned char overAlpha = ((over & 0xff000000) >> 24);
unsigned char Red = ((((back & 0x00ff0000) >> 16) * (0xff-overAlpha) +
  ((( over & 0x00ff0000) >> 16) * overAlpha) >> 8);
```

This operation might be implemented by the sequence of primitive integer operations shown in Fig 510, where back and over are inputs, and Red is the output. The variables `t0` to `t9` are intermediate results.

This set of operations might be expected to require 10 distinct RISC instructions, assuming all variables are in registers and all the constants can be encoded in the instructions or are preloaded into registers. This simple example demonstrates some of the methods used for instruction-set description. The 10 operations could be fused into just one instruction with two input reads and one output write. Fig 5-11 shows a complete TIE description of this fusion, assuming the input values are already stored in the processor's general-purpose (AR) registers.

```
t0 ← over & 0xff000000
t1 ← t0 >> 24
t2 ← back & 0x00ff0000
t3 ← t2 >> 16
t4 ← 0xff-t1
t5 ← t3 * t4
t6 ← over & 0x00ff0000
t7 ← t6 >> 16
t8 ← t7 * t1
t9 ← t5 + t8
Red ← t9 >> 8
```

Figure 5-10 Simple operation sequence example.

```
1: operation fusedred {in AR back, in AR over, out AR Red} {} {
2: wire [31:0] t0 = over & 32'hff000000;
3: wire [7:0] t1 = t0 >>24;
4: wire [31:0 t2 = back & 32'h00ff0000;
5: wire [7:0] t3 = t2 >> 16;
6: wire [7:0] t4 = 8'hff6 t1;
7: wire [15:0] t5 = t3 * t4;
8: wire [31:0] t6 = over & 32'h00ff0000;
9: wire [7:0] t7 = t6 >> 16;
10: wire [15:0] t8 = t7 * t1;
11: wire [15:0] t9 = t5 + t8;
12: assign Red = {24'h000000,t9 >> 8};}
```

Figure 5-11 Basic operation sequence TIE example.

The same temporary names are retained for clarity, though the output could be computed in a single expression just like the original C code, yielding the same instruction function and logic implementation:

```
operation fusedred {in AR back, in AR over, out AR Red} {} {
 assign Red = {24'h000000,
  (((back & 32'h00ff0000)>>16) * (8'hff6((over & 32'hff000000)>>24)
  + ((over & 32'h00ff0000)>>16) * ((over & 32'hff000000)>>24)) >> 8};}
```

A very basic introduction to TIE is given in the appendix, but a short explanation of this example gives you the gist. The *operation* statement defines a new instruction, `fusedred`, with two explicit source operands (`back` and `over`) and one result operand (Red) that are all entries in the base Xtensa AR register file. There are no implicit operands (uses of state registers or built-in interface signals), so the second pair of brackets in the operation statement is empty. Implicit operands do not appear in the instruction word.

The body of the operation definition is a series of assignments that all occur whenever the instruction is executed. The left-hand side of an assignment is either an output operand or a wire that is internal to the execution unit. By default, wires are one bit wide, but an arbitrary width can be explicitly declared. (Though this “wide wire” is more properly called a bus, the “wire” name applies no matter the number of wires.)

The right-hand side of an assignment can be any expression using TIE operators and the names of wires and source operands. The TIE operators are essentially identical to the operators of combinational Verilog and include arithmetic, logical, comparison, shift, and bit-field-extraction and bit-field-concatenation operators. TIE integer operations are unsigned.

In this example, the specific constants for data masking and shift amounts are pulled into the instruction definition. This reduces the mask and shift operations to simple bit selections, making them almost free in terms of gates and circuit delay. Nevertheless, fused operations often have longer propagation delay from input to output than simple RISC operations. In some cases, such a complex instruction could reduce the maximum operating frequency of the processor clock if the entire execution of the instruction is constrained to one cycle. If the instruction execution is pipelined to produce results after two or three cycles, the processor clock is less likely to degrade.

Pipelining means that the processor can start a new computation of this type every cycle because the intermediate results of the operation are stored in temporary registers. However, any instruction that depends on the result `Red` must wait an extra cycle or two to start. Fig 5-12 compares non-pipelined and pipelined implementation of this simple function, where the source operands come directly from a clocked register and the destination results go into a clocked register in both cases. In the version with pipelined logic, two additional pipeline registers are shown. Consequently, less logic must execute in each clock cycle.

Depending on the nature of the computation, even a complex-looking series of dependent operations can still fit into one processor cycle. In other cases, complex operations require longer latency, though rarely as much as the total latency of the original series of primitive operations. In TIE, result pipelining is simply expressed by including a schedule statement as follows:

```
schedule red_schedule {fusedred} (def Red 3;)
```

This statement adds two extra cycles of latency (for a total of three cycles) for the result operand. The processor generator automatically implements appropriate pipeline interlocks and operand-forwarding paths for any instruction that might depend on these operands.

If the number of input operands is large, or if the operation needs to compute multiple results, state registers are quite useful. For example, suppose the designer wanted to incorporate

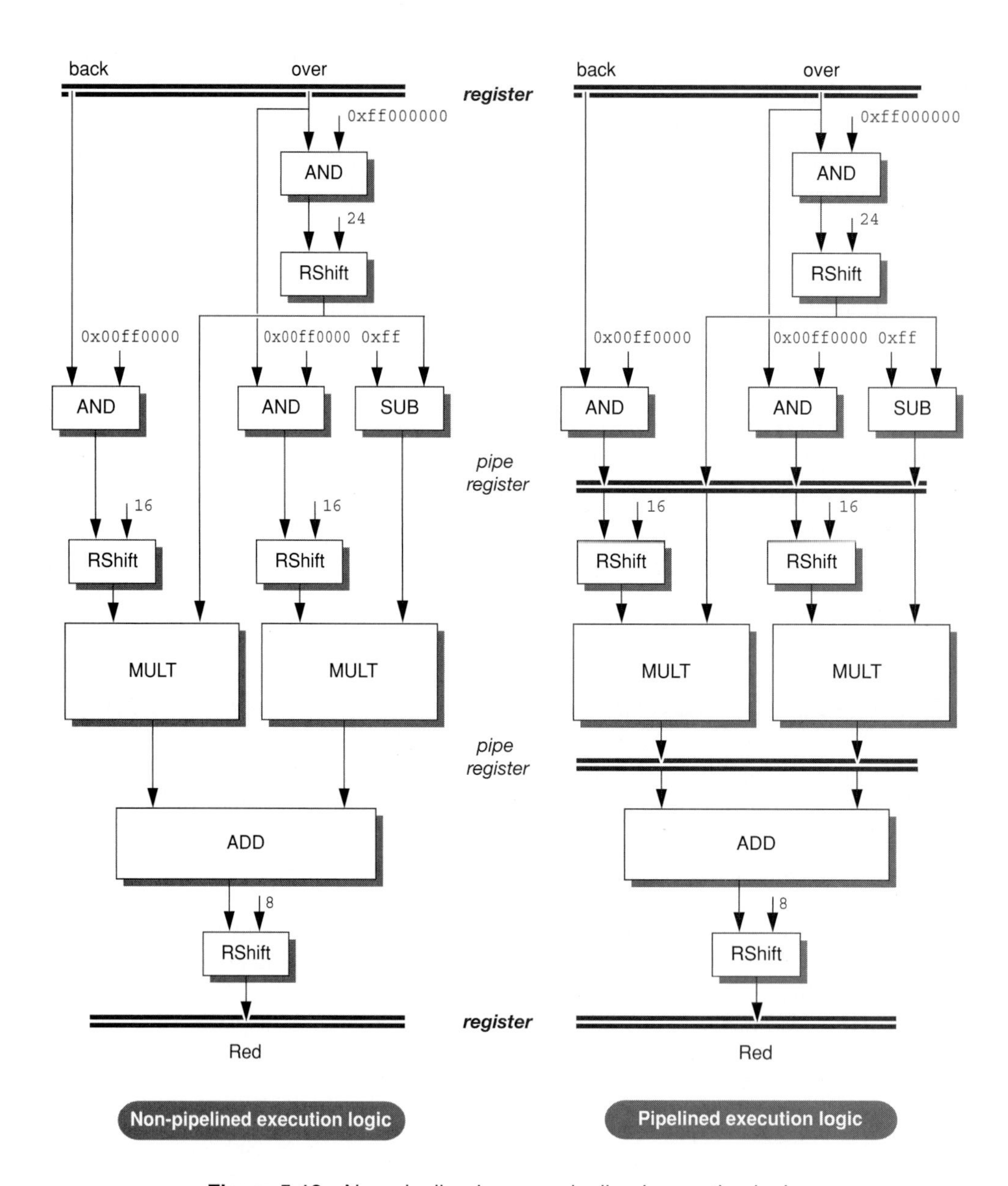

Figure 5-12 Non-pipelined versus pipelined execution logic.

an 8-bit color bias in the blending operation by adding a offset to every result. The TIE implementation, including the bias value in a state register, is shown in Fig 5-13.

```
1: state redbias 8 add_read_write
2: operation fusedred {in AR back, in AR over, out AR Red} {in redbias} {
3:   assign Red = {24'h000000, redbias +
4:   (((back & 32'h00ff0000)>>16) * (8'hffó((over & 32'hff000000)>>24) +
5:   ((over & 32'h00ff0000)>>16) * ((over & 32'hff000000)>>24)) >> 8)};}
6: schedule red_schedule {fusedred} {def Red 3;}
```

Figure 5-13 Pixel blend with color bias in state register TIE example.

Note that all four color components of the pixel blend function in Fig 5-2 could be computed in a single instruction. Success in fusing operations relies on three factors:

1. Reducing the number of instructions to encode the functions,
2. Reducing the number of operands to fit the encoding of the new instructions, and
3. Reducing the number of register-file ports to supply all the operators.

Reducing the number of instructions often shrinks the total latency of the set of operations because more than one primitive operation can be performed in one cycle. Even when the latency of the operation is not reduced, the new operation will be more fully pipelined, so that the instruction can be repeated with new operand values as rapidly as every cycle. The ability to rapidly repeat an instruction's execution using different operands increases overall throughput.

Pipeline stages hold intermediate results in hidden registers within the execution unit. These hidden pipeline registers are not visible to software and serve to increase the maximum operating frequency of the processor implementing the operations.

The primary goal of fusion is to increase concurrency among basic operations. To achieve good concurrency, however, the number of operands must be reduced so that

- The operations can be packed into as few instructions as possible.
- The hardware requirements for those operations, especially register-file read and write ports, are kept to a minimum.

The basic methods summarized here also tend to reduce the total complexity and latency of the set of operations:

1. **Share input operands:** Look for parts of the computation where several computations share the same input operand or the output operand of one operation is solely used as the

input operand of another. If all the operations requiring that operand can be combined, the number of fetches and number of instructions will be reduced.

2. **Substitute constants for variables:** In many applications, the values of constants are intrinsic to the application type and are not subject to change. For example, a multiplication by π will not require redefinition. In other cases, operand values are sufficiently stable so that there's a vanishingly small risk of obsolescing an instruction because the value has changed. For example, constants used to define image color spaces are industry standard and unchanging. Substituting constants for stable values held in registers or register files eliminates operand fetches and simplifies computational logic. The reduction in logic complexity can be significant. Even when the application cannot restrict itself to one constant value, an immediate field in the instruction can encode a small number of possible operand values. This approach may achieve much of the same reduction in logic area and operand-specification bits as restriction to a single constant value. For example, the shift of one variable by another requires a full barrel-shifter unit. However, the shift of a variable by a constant is just a selection of bits, which requires no logic gates and incurs no circuit delay. A fixed-function barrel shifter can be implemented with wires alone.
3. **Replace memory tables with logic:** Some algorithms use lookup tables as an intrinsic part of the computation. When that table is not subject to change, it can often be efficiently replaced by logic that performs the equivalent function without a memory reference. This transformation reduces the instruction latency and the number and size of operand specifiers. The resulting instruction is often more power efficient than a memory reference, particularly for small tables. It also frees the memory system to perform other load and store operations.
4. **Merge simple operations with others:** In general-purpose processors, many computation cycles are often consumed performing simple bit selection and logical operations. When those operations are subject to operation fusion, they can often be implemented with only a few logic gates per bit and minimal delay. Simple operations—shifts by constants, masking and selection of bits, and additions with just a few bits of precision—can be combined with more complex operations such as wider additions, multiplications, comparisons, and variable shifts at low circuit cost and with little delay.
5. **Put operands in special registers:** Many inner-loop computations load a series of source operands from memory, perform some complex operations, and then store a corresponding stream of result operands back to memory. There are additional individual scalar variables that are set or used within the loop many times, but these variables are not stored back to memory inside the loop. These variables may be set before the loop and merely used in the loop. Alternatively, these variables may be used and set during each loop iteration, accumulating the results for use after the loop terminates. In either case, heavily used operands can be held in added processor registers dedicated to these variables and used only by the new fused instructions created specifically for the code in this inner loop. A special register of this type can be referenced implicitly by the instruction and therefore requires no

operand-specifier bits in the instruction. Even when a special multi-entry register file is used, it can be shallower than a general-purpose register file (because of its special-use nature), and therefore, fewer operand-specifier bits are needed to specify a location within that special, dedicated register file. The overall effect is a reduction in the number of operand specifiers required in the inner loop and a reduction of the number of register-file ports or memory references within the inner loop.

Together, these techniques significantly reduce the number of instructions required to execute a program. The reduced number of instructions usually reduces computation latency and increases the throughput of the computation's inner loop. Instruction fusions may increase the degree of instruction-set specialization, but designers can often anticipate likely areas of change within an application and can therefore develop operation fusions that remain applicable even as the application evolves.

5.3.5 Compound Operations

The second basic instruction-enhancement technique is the construction of compound instructions. Compound instructions combine two or more independent operations into one instruction. Combining operations is frequently effective even when no dependencies exist between the operations. This technique is especially effective when the independent operations happen to share common input operands, but it is still effective even when the technique does not reduce the number of operand specifiers. The principal goal of this technique is reduction of the total number of instructions. Even with a narrow instruction word (32 bits or less), an instruction can specify two (and sometimes even three) independent operations. Naturally, a narrow instruction word restricts the number and size of specifiers, but two simple techniques commonly apply:

Reduce the number of combinations: The target application determines the mix of basic operations and their frequency of occurrence. If only the more commonly occurring operations are combined, the number of distinct combinations may be sharply reduced. Rarer operations are left as individual instructions. They are not combined.

Reduce the size of operand specifiers: As with operation fusion, reducing the number of operands in compound instructions eases encoding constraints (and reduces logic cost, logic delay, and power dissipation). Many of the same techniques that work for fused encodings apply here, especially sharing common input operands, using special registers or register files, and simplifying some operations by substituting constants for variable operands.

A simple example of encoding two independent operations into a set of compound instructions is described in the following fragment of TIE code, shown in Fig 5-14. Each compound instruction has two independent opcode fields. The first field holds one of two possible operations (shift bytes left, shift bytes right) on the general-purpose 32-bit (AR) registers. The second field holds one of two possible operations (or, and) on an added 128-bit, 4-entry register file.

```
1: regfile XR 128 4 x
2: function [127:0] sbl ([127:0] a, [5:0] sa) shared {assign sbl = a<<sa;}
3: function [127:0] sbr ([127:0] a, [5:0] sa) shared {assign sbr = a>>sa;}
4: function [127:0] and ([127:0] a, [127:0] b) shared {assign and = a & b;}
5: function [127:0] or ([127:0] a, [127:0] b) shared {assign or = a | b;}
6: operation sll.or {in AR a,inout AR b,in XR c,inout XR d} {} {
7: assign b=sbl(b,a[5:0]); assign d=and(d,c);}
8: operation srl.or { in AR a,inout AR b,in XR c,inout XR d} {} {
9:  assign b= sbr(b,a[5:0]); assign d=or(d,c);}
10: operation sll.and { in AR a,inout AR b,in XR c,inout XR d} {} {
11:  assign b=sbl(b,a[5:0]);assign d=and(d,c);}
12: operation srl.and { in AR a,inout AR b,in XR c,inout XR d} {} {
13: assign b= sbr(b,a[5:0]); assign d= or(d,c);}
```

Figure 5-14 Compound instruction TIE example.

This example starts with the declaration of a new register file—the XR register file—with four entries, each 128 bits wide. The short name of the register file is x, so assembly coders can use the register names x0, x1, x2, and x3 to refer to the new entries. Assembly code, however, is not necessary to use the register file. The **regfile** declaration creates a new C data type called XR, and the compiler will allocate any C variable declared with type XR to the XR register file. The processor generator also creates 128-bit load-and-store operations for the XR register file to enable pointer dereferencing and to allow the processor to save and restore the XR register file's state.

The TIE code defines four new operations: `sll`.or, `sll.and`, `srl.or`, and `srl.and`. These operations combine either a left or right bit shift on the base AR registers (sll, srl) with a bit-wise logical OR or AND on the XR registers. These are defined as shared functions (see Chapter 7 for details) to ensure that all the operations use the same hardware. The four compound operations each use two operand sources from the AR register file (with the second operand source also serving as the destination for the result of the operation) and two operand sources from the XR register file (with the second operand source also serving as the destination for the result of the operation).

Fig 5-15 shows one encoding of the four instructions that could be generated from the TIE code in Fig 5-14 within a 24-bit instruction format.

For monolithic instructions, the description of independent operations is simple and concise. However, if the number of combinations becomes large—say, for example, all combinations of two sets of eight operations (64 instructions)—the instruction description can become ungainly because each combination is separately enumerated. Very-long-instruction-word (VLIW) or flexible-length-instruction methods, such as described in Chapter 4, Section 4.5, make large-scale instruction combination simpler because the description of the building-block

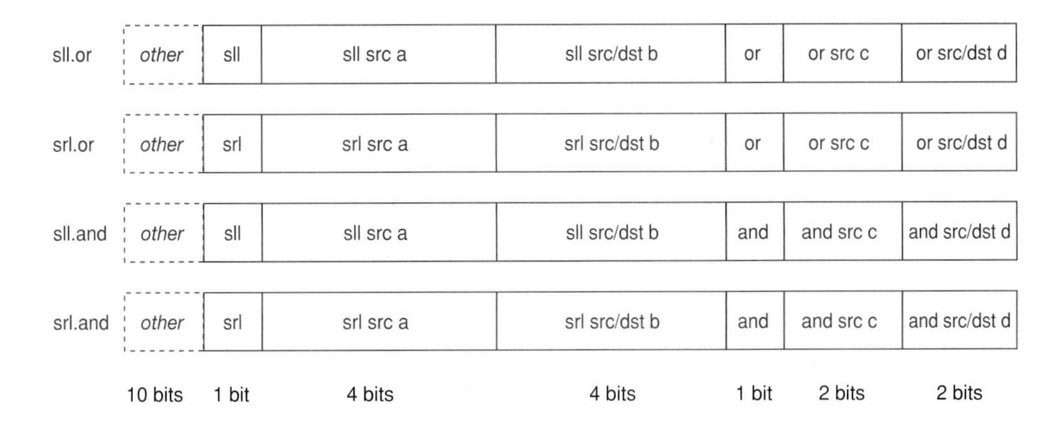

Figure 5-15 An encoding of four compound instructions.

operations is separate from the description of the allowed operation combinations.

5.3.6 SIMD Instructions

Many of the most important and performance-demanding applications share a common characteristic: the same sequence of operations is performed repetitively on a stream of data operands. Many error-coding, video-compression, audio-processing, and digital-signal-recovery tasks fit this model. This trait enables a class of instruction sets where a single instruction sequence of can be applied in parallel to a closely related group of operands. Several operands can be packed into a wide compound operand word or vector. The set of instructions operates on the entire vector using atomic operations, defining an SIMD architecture. SIMD architectures often provide large performance benefits for data-intensive applications.

The overall bit width of the operand vector is the product of the width of the individual primitive operands M and the number of operands packed into a vector N, resulting in an operand-vector bit width of M*N. For simple cases, this bit width applies to the operation units, the registers, and the interface to data memory (when needed). Fig 5-16 shows a memory, register file, and a set of parallel arithmetic logic units (ALUs) that use four 32-bit operands packed into each 128-bit SIMD data word or vector.

When an *N*-wide vector operation executes every cycle, the SIMD architecture delivers N times the performance of an architecture that performs the same operation on just one simple operand per cycle. In combination with compound instructions—for example to combine the SIMD load, address update, and SIMD ALU operation—this approach may reduce 3*N* cycles to one.

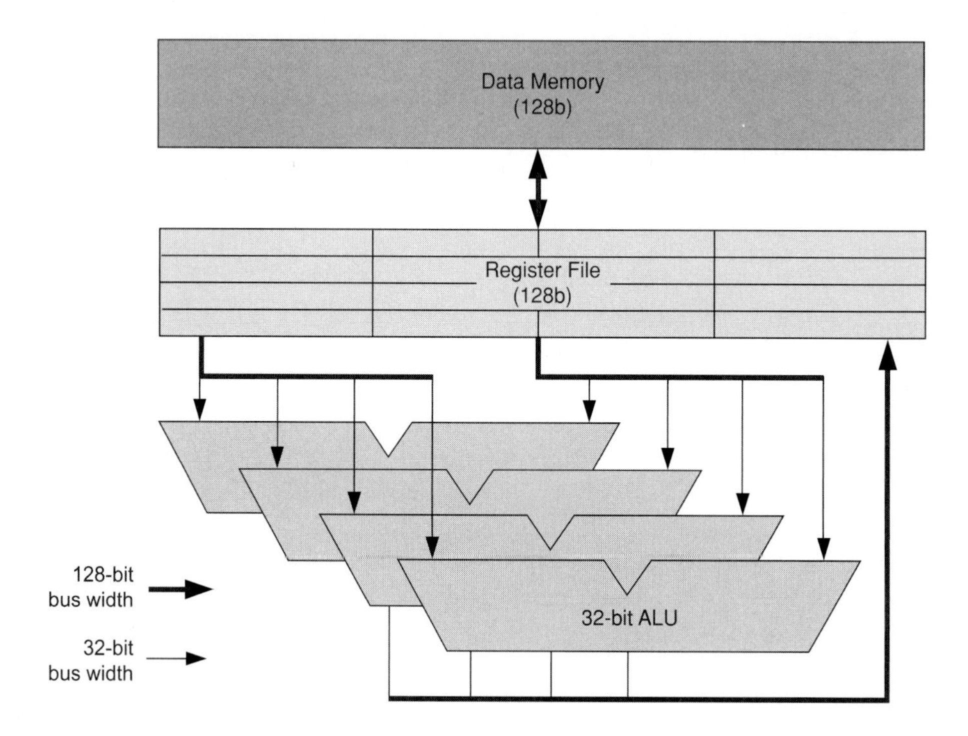

Figure 5-16 Simple SIMD data path with four 32-bit operands.

Three considerations affect the choice of *N*:

1. The data-path and register widths are set by *M*N*. If the data path becomes too wide, heavy wire loading may degrade the implementation's maximum clock frequency. As a result, data-path widths of more than a few hundred bits may be impractical.
2. If the application always operates on a sequences of operands whose length is an integer multiple of *N*, no special handling of short vectors is needed. Unfortunately, sequence lengths often vary or are not known at the time the processor is generated. The SIMD instruction set may include features to handle cases when some number of operands less than *N* remains at the end of a vector, or the excess operands can be manipulated with ordinary scalar (non-SIMD) instructions. In many cases, the bulk of the SIMD instructions can assume that all vectors contain exactly N operands, with just a few instructions—for loading, storing, and accumulating results across vector elements—that deal with the short-vector situation.
3. If the SIMD instruction set loads and stores full vectors, the vector length is limited to the maximum configuration width of local memory, typically a power of two. (Note: traditional vector supercomputers implement vector lengths that are longer, sometimes much longer, than the memory-word width, and transfer the vector sequentially among memory,

register files, and execution units.) In some cases, it is appropriate to make vector registers and operation unit data paths wider than memory, either for extended precision (guard bits) or to reduce the cost of the memory and the memory interface by allowing multiple memory operations to load or store a full vector. More complex load/store or bus-interface units can handle some of the mismatch between the vector register's width and local-memory or global-bus width, but basic SIMD performance is often bound by the bandwidth of the interface between local memory and the vector register.

SIMD instructions are easily described. Fig 5-17 shows a very short example of an instruction set that operates on 32-bit operands that are packed as four operands to a 128-bit vector, similar to the structure shown in Fig 5-16. For simplicity, just two SIMD vector instructions are defined here: a 32-bit unsigned `add` and a 32-bit exclusive `or`. These SIMD instructions operate on an eight-entry register file.

```
1: regfile vec 128 8 v
2: operation vadd {in vec a, in vec b, out vec c} {} { assign c =
3: {a[127:96]+b[127:96],a[95:64]+b[95:64],a[63:32]+b[63:32],a[31:0]+b[31:0]};}
4: operation vxor { in vec a, in vec b, out vec c} {} { assign c =
5: {a[127:96]^b[127:96],a[95:64]^b[95:64],a[63:32]^b[63:32],a[31:0]^b[31:0]};}
```

Figure 5-17 SIMD instruction TIE example.

This TIE example defines a new eight-entry register file, 128 bits wide. The two SIMD instructions vadd and vxor treat each register-file entry as a vector of four 32-bit operands. The two instructions operate on each of the 128-bit source vector's four 32-bit operand values separately and then concatenate the four 32-bit results into a 128-bit result vector that is written back to the destination vector register. An inner loop that once employed a conventional add can now use the vector add instruction, but with only one fourth the iterations. So, the original C code,

```
int i,*x,*y;
for (i=0;i<N;i++) {y[i] = y[i] + x[i];}
```

becomes this equally simple, but faster C code (assuming that the data arrays x and y can be aligned on a four `int` boundary):

```
int i; vec *vx,*vy;
for (i=0;i<N/4;i++) {vy[i] = vadd(vy[i],vx[i]);}
```

The big improvements in overall application performance, discussed in Chapter 3, come from the combination of these three basic instruction-extension techniques: fusion, compound operations, and SIMD.

5.4 The Programmer's Model

Creating the ideal programmer's model for a target application is the crux of application-specific processor generation. If discovery and implementation of the ideal model is easy and yields a good SOC cost and good application performance, the automatic processor generator inevitably becomes one of the software developer's primary design tools.

The programmer's model captures the essential processor features available to software. This model has two basic dimensions:

1. Storage elements including registers, register files, and memories.
2. Instructions that transform the state of the storage elements.

Further, we can include machine encoding in the programmer's model as a third dimension, because the programmers may use fully compiled and assembled programs—binary programs—as a natural form of program reuse:

3. Instruction encoding, including both operand specifiers and opcodes for each instruction.

The programmer's model constitutes an agreement between the hardware and the software. Every correct hardware implementation of the programmer's model should produce correct results for any valid program on any set of input data, although different architecture implementations may have very different characteristics. The number of execution cycles, maximum clock frequency, and cost may vary widely with the implementation. The programmer's model is explicitly visible to the assembly language programmer, and it is also implicitly encoded in the compiler, the debugger, the simulator, and in the precompiled runtime software infrastructure, which includes libraries and real-time operating systems.

Most software development has shifted from assembly code to high-level languages, so the suitability of an architecture for programming from C and C++, for example, is increasingly important. Assembly code use lingers in only two roles: where shortcomings in the compiler prevent efficient mapping from high-level language code to machine instructions and where special operations or registers, often associated with processor initialization, testing, and exception handling, must be specified—operations that may not be accessible via a particular high-level language compiler.

5.4.1 The Base User Instruction Set

A typical programmer's processor model is described in the next set of figures. Fig 5-18 describes typical Xtensa processor state. Fig 5-19 describes typical Xtensa instruction format. Fig 5-20 describes many typical Xtensa instructions. The Xtensa processor's base user-level instruction set is used as an example. The model consists of the storage description, including main memory (addressable as bytes, half-words, and words) and the instruction-set description.

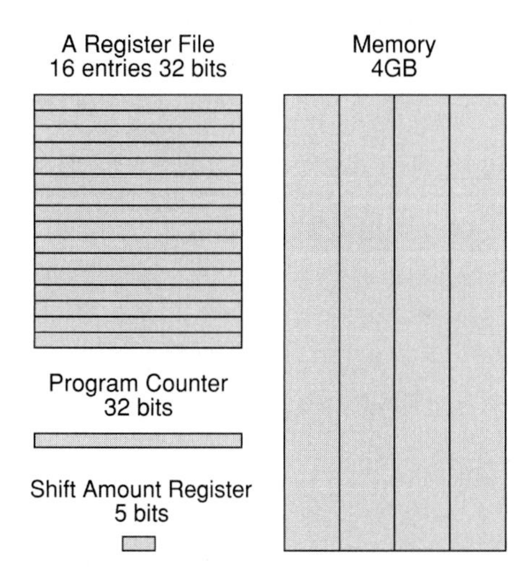

Figure 5-18 Typical processor state (Xtensa).

The Xtensa processor's baseline instruction set is very similar to those of most 32-bit RISC processors. Like other RISC processors, the Xtensa processor uses a load/store architecture and has a 32-bit, general-purpose register file that serves as the source and destination for most of its instructions. There are some differences, however, because the Xtensa processor was designed specifically for embedded applications. One major difference is that the Xtensa processor employs instruction formats with both 16- and 24-bit lengths instead of the typical RISC processor, which has 32-bit instructions. The Xtensa processor's more compact instruction for-

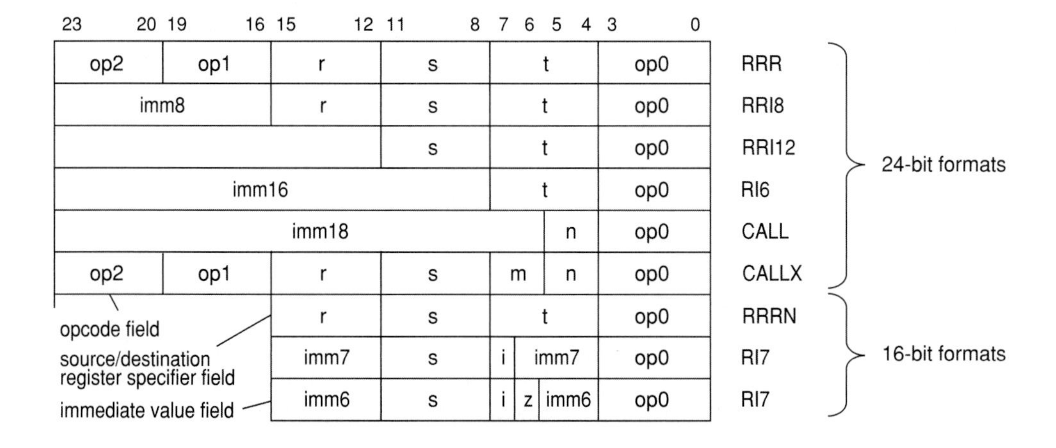

Figure 5-19 Typical instruction formats (Xtensa).

mats allow for a smaller software footprint in memory, which is often important for memory-strapped embedded-system designs. (For simplicity, details of encoding and formal definition of instruction semantics are omitted in the figures.)

Group	Menomic	Format	Brief Description
Load	L32R	RI16	2-bit load PC relative
	L32I	RRI8	32-bit load
	L16UI	RRI8	16-bit unsigned load
	L16SI	RRI8	16-bit signed load
	L8UI	RRI8	8-bit unsigned load
Store	S32I	RRI8	32-bit store
	S16I	RRI8	16-bit store
	S8I	RRI8	8-bit store
Jump/Call	J	RI16	Unconditional jump, PC relative
	JX	RRR	Unconditional jump, address in register
	CALL0	CALL	Call subroutine, PC relative
	CALLX0	CALLX	Call subroutine, address in register
	RET	RRR	Subroutine return
Branch	BEQZ	RRI12	Branch if equal to zero
	BNEZ	RRI12	Branch if not equal to zero
	BGEZ	RRI12	Branch if greater than or equal to zero
	BLTZ	RRI12	Branch if less than zero
	BEQI	RRI8	Branch if equal immediate
	BNEI	RRI8	Branch if not equal immediate
	BGEI	RRI8	Branch if greater than or equal immediate
	BLTI	RRI8	Branch if less than immediate
	BGEUI	RRI8	Branch if greater than or equal unsigned immediate
	BLTUI	RRI8	Branch if less than unsigned immediate
	BBCI	RRI8	Branch if bit clear immediate
	BBSI	RRI8	Branch if bit set immediate
	BEQ	RRI8	Branch if equal
	BNE	RRI8	Branch if not equal
	BGE	RRI8	Branch if greater than or equal
	BLT	RRI8	Branch if less than
	BGEU	RRI8	Branch if greater than or equal unsigned
	BLTU	RRI8	Branch if less than unsigned
	BANY	RRI8	Branch if any of masked bits set
	BNONE	RRI8	Branch if none of masked bits set (all clear).

Group	Menomic	Format	Brief Description
Branch (Continued)	BALL	RRI8	Branch if all of masked bits set
	BNALL	RRI8	Branch if not all of masked bits set
	BBC	RRI8	Branch if bit clear
	BBS	RRI8	Branch if bit set
Arithmetic/ Logical	ABS	RRR	Absolute value
	ADD	RRR	Add source registers
	ADDX2	RRR	Add with multiply by 2
	ADDX4	RRR	Add with multiply by 4
	ADDX8	RRR	Add with multiply by 8
	SUB	RRR	Subtract source registers
	SUBX2	RRR	Subtract with multiply by 2
	SUBX4	RRR	Subtract with multiply by 4
	SUBX8	RRR	Subtract with multiply by 8
	NEG	RRR	Negate
	ADDI	RRI8	Add source and immediate
	ADDMI	RRR	Add source and immediate x 28
	AND	RRR	Bitwise logical AND of source registers
	OR	RRR	Bitwise logical AND of source registers
	XOR	RRR	Bitwise logical XOR of source registers
Shift	SLLI	RRR	Shift left logical immediate
	EXTUI	RRR	Extract unsigned field immediate
	SRLI	RRR	Shift right logical immediate
	SRAI	RRR	Shift right arithmetic immediate
	SRC	RRR	Shift right combined (from SAR)
	SRA	RRR	Shift right arithmetic (from SAR)
	SLL	RRR	Shift left logical (from SAR)
	SRL	RRR	Shift right logical (from SAR)
	SSA8x	RRR	Set shift amount register for little/big-endian byte align
	SSR	RRR	Set shift amount register (SAR) for shift right logical
	SSL	RRR	Set shift amount register for shift left logical
	SSAI	RRR	Set shift amount register immediate

Group	Menomic	Format	Brief Description
Moves	MOV	RRR	Move (instruction idiom)
	MOVEQZ	RRR	Move if equal to zero
	MOVGEZ	RRR	Move if greater than or equal to zero
	MOVI	RRI8	Move immediate
	MOVLTZ	RRR	Move if less than zero
	MOVNEZ	RRR	Move if not equal to zero
Processor Control	RSR	RRR	Read special instruction
	WSR	RRR	Write special instruction
	XSR	RRR	Exchange Special register
	ISYNC	RRR	Instruction fetch synchronize
	RSYNC	RRR	Instruction register synchronize
	ESYNC	RRR	Register value synchronize
	DSYNC	RRR	Load/store synchronize
	MEMW	RRR	Order memory accesses before with memory access after

Figure 5-20 Typical instruction description (Xtesna).

In principle, the programmer's model provides everything a software developer needs to know. In reality, there are a number of other details that may be relevant, including the timing of each instruction or interactions between instructions that affect timing. This additional information helps the programmer make choices when several different instruction sequences might have the same effect. The details of instruction timing depend on the processor's pipeline. Chapter 7 discusses the important details of processor pipelines in more depth.

5.4.2 The Application-Specific Instruction Set

Section 5.2 suggested a number of plausible processor-instruction extensions, but we can characterize application-specific instruction sets more generally. An algorithm produces output data from input data through a series of transformations on the processor's state. Each transformation takes input data and/or internal state and produces output data and/or new values for internal state. State-to-state transformations define the instructions. Each instruction performs a set of basic transformations, such as reading state from memory locations and registers, performing a combinational function on that data, and writing computation results back to memory locations and registers. RISC instructions perform just one simple data transformation per instruction, either transferring data between memory and the central register file or performing a combinational function on two values obtained from two registers and writing the result to one register.

Application-specific instructions that perform the same transformations using much shorter instruction sequences outperform generic instruction sets as implemented by RISC processors. This transformation efficiency stems from two characteristics of application-specific instructions:

1. Application-specific combinational functions are not limited to simple arithmetic, logical, and bit-manipulation operations. The functions can be essentially any combination and series of operations. In the limit, one application-specific instruction could compute any function on any number of inputs. The practical limit on the number of inputs is hundreds or thousands of bits. Again, because application-specific instructions target specific applications and algorithms, they need not be universal or general-purpose. Consequently, application-specific instruction-set extensions are often decidedly nonorthogonal. This design philosophy is completely contrary to orthodox RISC processor design, but it is very efficient and entirely appropriate for the design of application-specific processors that target specific embedded applications.
2. Application-specific instructions can read inputs from several different memory locations and registers in parallel and they can write the transformed results into several different memory locations and registers simultaneously. The number and width of the source and destination locations aids instruction parallelism.

From this perspective, new instructions are actually easy to describe. The developer simply describes the processor's registers and memories and the input-state-to-output-state function for each new instruction. However, effective use of application-specific instructions depends on understanding the answers to three big challenges:

1. How does the developer exploit application-specific instructions in the course of software development? What tools and coding methods can be used?
2. How does the developer choose among different instruction sets, all capable of performing the necessary transformations of the algorithm? Which instruction sets are low-cost; which reduce power; which permit high processor clock frequency; which are most flexible; and which reduce program size?
3. How does use of an application-specific instruction interact with other system functions and development tasks? What effect do these new instructions have on system initialization, interrupt and exception handling, switching among multiple tasks, code debugging, and software upgrades?

5.4.3 The System-Programming Instruction Set

A processor's instruction set almost always includes registers and instructions not normally used by applications programs. These features control the exception-handling conditions (including interrupts and timers), aid code debugging, provide memory protection, and perform other special functions. We call this class of features *system-programming support*. These additional features help the processor switch more rapidly among tasks, and they prevent the operations of one application from interacting—either inadvertently or maliciously—with the operations of another application. In many cases, system-programming issues are independent of the issues of application-specific processor configuration, but there are notable exceptions. Here are a few examples where configuration and system-programming support interact:

Configuration of system features: Different system environments have very different requirements for system-control features. For example, the number of interrupt signals may determine the processor's fit to system requirements. Configurable features such as the number of interrupts can be changed, with the side effect of changing the system-programming support model for the software developer. Moreover, the interrupt priority levels and the interrupt vectors might all be configurable. As a result of this variability, the number, format, and function of interrupt-control registers depend on the configuration. The programmer's model must also reflect changes in the number of timers or in the type of memory protection. If the system software needs full memory protection with virtual-to-physical address translation for each memory page, the developer will specify a memory management unit. The number and organization of control registers and translation-buffer entries become part of the programmer's model too.

Context switching: Often, one processor runs several different application tasks, switching among the tasks according to time slices, user demands, or availability of task inputs and resources. To switch between tasks, the entire computational state, especially the state of the processor registers for Task A, must be saved and the state of Task B substituted. Most often, the state of Task A's registers are stored in memory, in a data structure managed by the operating system, for later restoration when Task A is allowed to restart. The code for saving and restoring the task state is usually part of the operating-system kernel code. In other cases, the processor uses special hardware to switch register contexts, so that Task A's registers are hidden and Task B's registers become visible to the programmer. This scheme reduces the time to switch between tasks. If a processor extension adds new registers to the programmer's model, the kernel code must track these changes for context-switching purposes. Automatic generation of context-switching code allows the operating system to easily accommodate the needs of tasks using extended instruction sets.

Application-specific exceptions: Occasionally, system functions play a vital role in task design and processor optimization. For example, the application may intermittently encounter unusual circumstances or error conditions in the course of execution. The application code can directly test for these rare conditions at every point at which they might occur, but constant testing significantly slows program execution. Instead, the processor might establish error conditions that are implicitly checked during every appropriate operation. If the error condition occurs, processor execution switches to an exception-handling routine. This routine can take whatever error recovery or reporting action is required, and then resume task execution. Even if the overhead for taking the exception, handling it, and returning to normal execution is large, the error should be so rare that the performance impact is minimal.

A good example of this sort of exception is divide-by-zero detection for floating-point arithmetic. Occurrences are rare, but could occur during any divide operation and must be checked. It would be unreasonably expensive to add application code that checks every divide, but establishing floating-point exceptions as part of the programmer's model is quite reasonable. The configuration of the system-programming support becomes an important part of the application developer's capabilities.

5.5 Processor Performance Factors

Many factors contribute to processor performance and embedded-software developers should understand the essentials of effective processor tuning. A processor's instruction-set architecture has one direct and two indirect influences on processor performance. The instruction set directly determines the number of instructions required to implement an algorithm. The instruction set indirectly affects the clock frequency that can be achieved in a target semiconductor implementation technology. The longest path through the processor's circuits limits its maximum clock frequency. If the architectural implementation requires a number of sequential operations to be packed together into one clock cycle, that fused operation may create the longest circuit path in the processor and will limit the processor's maximum operating frequency.

The instruction set also indirectly affects the average number of clocks per instruction. If the instruction set must perform a long series of operations over several clock cycles, the combination of long-latency instructions and dependencies between instructions (as one instruction must wait for an earlier instruction to complete) increases the average number of clock cycles per instruction.

Code size may have an important indirect influence on performance through the behavior of instruction caches. Architectures with denser code will experience lower miss-rates and will consequently achieve higher performance for a given cache configuration than architectures with "loose" code. Alternatively, architectures with denser code can achieve the same performance as architectures with looser code using significantly smaller instruction caches. Instruction- and data-cache configurations, even if not formally part of the instruction-set architecture, contribute significantly to application performance as well as to SOC cost and power dissipation.

The overall goal of processor optimization can often be summarized as the combination of the following goals:

1. Select or create an instruction set that increases algorithm encoding efficiency. That is, reduce the number of instructions executed.
2. Reduce the number of cycles required to execute those instructions, including the stall cycles when the processor is waiting for some data or a hardware resource.
3. Increase the processor's clock frequency in the available semiconductor implementation technology and design method to reduce total execution time.
4. Reduce the energy required to perform the intended computation by dissipating less power per cycle and by requiring fewer cycles. Reducing the number of external bits or memory references may be particularly important to power reduction.

Of these four goals, reducing of the number of instructions per program (the dynamic instruction execution count) usually has the most direct influence on performance. The central role of instruction count in improving performance encourages adding and using application-specific instructions that do more useful work for a specific application or set of applications than the traditional general-purpose instructions. This increase in effectiveness may come from performing a significant number of primitive operations with one instruction, or it may come from providing new operations and appropriately sized storage that better match the real computation demanded by the algorithm. The process of discovering and designing new, better-fitting instructions is this chapter's central topic. However, reducing instruction count is not enough by itself. Other considerations stand out.

First, new instructions that combine a number of operations pipelined over several cycles are not intrinsically better than sequences of more primitive instructions. Performance improves only if the new instruction executes in fewer cycles than the instruction sequence it replaces. Elegant does not necessarily mean fast.

RISC architecture emerged, in part, as a reaction to increasingly complex instruction sets in CISC processors, where a single instruction could perform complex calculations but took many cycles to execute because complex instructions were often implemented as a sequence of micro-instructions. RISC architectures implement only simple operations, but a sequence of simple operations is sometimes faster than the same sequence encoded into a complex instruction via microcode.

Compilers often exploit the context of the computation, reuse subexpression calculations, and cache common values in local registers for faster access. If new instructions are added to improve performance, they should execute in fewer cycles than the equivalent sequence of primitive instructions or allow greater concurrency between the new instructions and other processor operations. Otherwise, there's no real reason to add the new instructions.

Second, the new instructions must not unduly decrease the processor's maximum clock frequency. For a synchronous-logic processor implementation, the longest signal-propagation path through any instruction's implementation logic sets the maximum clock frequency for all processor instructions. If new, complex instructions have long logic paths, the decrease in processor operating frequency may degrade application performance more than cycle-count reduction improves it.

Third, execution stalls may degrade task performance. A basic understanding of the causes of stalls helps tune the processor configuration. Three broad types of stalls are important:

1. **Memory stalls:** Modern processor pipelines are usually designed with the assumption that instructions and data are available in memories closely tied to the processor logic and accessible within one, or at most a few, processor cycles. The processor will stall for several cycles if a needed instruction or data word resides in a remote, slow memory. The processor's microarchitecture may try to hide this latency using architectural features such as caches and *nonblocking loads* (the processor continues to execute instructions that do

not depend on the data value being fetched from remote memory). Nonblocking loads are rare in low-cost embedded-processor cores because of their implementation complexity and cost. Caches were once rarely seen in embedded processors but are now quite common. A number of processor and system-configuration choices may influence the number of stalls associated with cache misses:

- *Cache size:* Larger instruction and data caches increase the chances that a desired word is in the cache. For embedded applications, there is often a pronounced drop in cache misses once the cache reaches a certain size because all frequently used instructions or all frequently referenced data can reside in the cache simultaneously once a size threshold is reached. Using caches of at least this threshold capacity significantly reduces stalls and performance sensitivity to main-memory latency.
- *Write-back vs. write-through data caches:* Data caches can deal with store operations using two basic policies to ensure consistency between the values in the cache and the values in main memory: write-through and write-back. A write-through cache policy means that writes store data to the cache and to main memory simultaneously. The write-through operation causes no stalls if the bus and main memory can handle the processor's write traffic. If the main memory or bus is slow and the processor executes many store operations (or if several processors contend for access to the bus or main memory), the processor will stall while waiting for previous write-through store operations to complete. (Stores to nonlocal memory are typically buffered to decouple the fast processor from the slow memory transfer, but processor and system logic must ensure that these pending values are not ignored by subsequent reads to the same location and that all memory accesses actually occur in the expected order).

  A store to a cache using a write-back policy initially stores data only in the cache, not to main memory. When that cache entry is evicted from the cache to make room for another, cache-control logic finally writes the data to main memory. This approach generally reduces write traffic to main memory, but it may increase the number of stalls associated with a cache miss for two reasons. First, a "dirty" cache line may have to be evicted from the cache before the new data is brought in. Second, write-back caches generally require an entire cache line to be present in the cache. Therefore, on a write miss, the processor may stall waiting for the rest of a new cache line to be brought into the cache.
- *Cache set associativity:* Set associativity is a popular cache-design technique that allows the processor to look in several possible locations in the cache to find a desired instruction or data word. Due to the specifics of cache addressing and the nature of data and instruction locality, having several ways that cache can store information with related addresses increases the chances that a desired piece of data or instruction will be in the cache. In general, the more places the processor can

look, the greater the chance that several commonly used entries can coexist in the cache. Increasing cache set associativity reduces the cache-miss rate, though it may increase the size of the cache-control logic.

- Cache-line size: When the processor fetches instructions or data from memory to fill the cache, it generally does not fetch just the desired instruction or operand but a whole block of code or data around the desired instruction or operand. This scheme exploits the fact that memory references are commonly clustered in adjacent addresses (the principle of spatial memory locality). Depending on the application's pattern of memory references and the time required to fetch a cache line from memory, processor stalls may be reduced either by increasing the cache-line size (reducing the number of fetches) or by decreasing the cache-line size (reducing the transfer time of the blocks).
- *Main bus width:* The processor's main bus must be wide enough to support the widest possible single-cycle data transfer and wide enough to meet cache-line transfer-rate requirements. A wider main bus reduces the number of cycles required to transfer a block of data or instructions over the bus, thus reducing the number of stall cycles, especially for large cache line sizes and fast memories. When the memory's transfer rate is lower than the bus transfer rate, a narrower bus will save silicon area yet deliver the same performance.
- *Memory latency:* Fast, modern processor cores almost always have fast, closely coupled memories for local data storage and caches. The longer access latency of slower main memories accessed over the processor's main bus can still significantly degrade application performance. Caches mitigate the effect of slow main memory, but faster main memories can noticeably reduce memory stalls. One way to speed up main memory is to move it from off-chip memory onto the SOC with the processor. However, it's very hard to build large, fast-on-chip memories, so these memories must not be closely coupled (attached to a single-cycle local-memory bus) if they aren't fast enough to keep up with the processor clock. If a significant fraction of a task's instruction or data references are made to slow memory, the impact of reducing memory latency can be significant. A large number of references to loosely coupled memory might occur when a data structure is shared between two tasks running on separate processors and must therefore reside in main memory to maintain memory coherency between the tasks. The impact of high memory latency may be mitigated by longer cache-line sizes, second-level caches, or the use of DMA to push the data into a processor's local memories.

2. **Data dependency stalls:** In a simple world, all processor operations complete in one cycle. Unfortunately, this constraint either precludes many important but complex operations from appearing as processor instructions or it forces the processor to run at a much lower clock frequency and throughput. Often, the results of one instruction are not imme-

diately used by others, so pipelined processors usually outperform non-pipelined processors because compilers have become adept at squeezing most of the data dependencies out of instruction sequences. Nevertheless, when the result of one long-latency instruction is used by a succeeding instruction, the succeeding instruction (and possibly others) must wait for the data. Common instructions that incur additional latency include loads (especially loads that experience cache misses) and complex arithmetic operations such as multiplication. The processor hardware hides all the details of pipelining from the application software developer, but developers who want to understand the subtler impact of data dependencies and pipelines should look at the detailed discussion of pipeline effects in Chapter 7.

3. **Resource stalls:** Often, a processor shares a resource among several instructions or pipeline execution stages. In theory, the shared resource could be duplicated to permit simultaneous use of the resource by more than one instruction. In practice, too much hardware may be required to provide duplicate resources for all possible usage conflicts. For example, many small processors implement multiplication without a fully pipelined multiplier. If a second multiplication instruction attempts to execute before an earlier multiplication operation has finished, the second instruction must wait (stall) until the multiplier is available. Especially when resource conflicts are rare, providing arbitration for a single resource is more efficient than providing multiple copies of the resource. If the processor does incorporate multiple equivalent computation units, it may still need to arbitrate access to those units during peak demand.

4. **Branch delays:** Processor pipelining also affects changes in control flow. By default, a processor executes instructions stored sequentially in memory. When a branch instruction executes, more than one cycle is required to fetch the branch instruction from memory, decode it, compute the branch-target address, and decide whether the branch is taken. Meanwhile, a pipelined processor will fetch instructions subsequent to the branch, decode them, and start to execute them. If the branch, in fact, turns out to be taken, those subsequent instructions should not execute, but they will already be in various stages of execution within the processor pipeline.

Most pipelined processor architectures choose to abandon partially executed instructions following a branch if the branch is taken. Abandonment creates execution bubbles in the pipeline, where no useful work is performed during the one or two cycles following a branch. Fig 5-21 shows a simple sequence of Xtensa processor instructions, where instructions in the light gray area are executed before the branch instruction. Instructions in the unshaded area are fetched after the branch instruction is fetched (but before the branch condition is known) and are abandoned before they complete if the branch is taken. The instructions at the label "Target" (in dark gray) are fetched and executed only after the target address is known and the branch is taken. No other instruction can be fetched or executed in lieu of the two abandoned instructions, so the taken branch wastes two execution cycles in this example.

```
         l32i    a2, a3, 16
         add     a5, a6, a7
         beq     a2, a5, Target
         s32i    a5, a3, 16
         sub     a5, a8
         ...
Target:  l16i    a6, a7, 8
         sll     a6, 1
```

Figure 5-21 Branch delay bubbles.

Software developers and processor designers adopt a number of strategies to minimize branch bubbles. First, profile-based code generation can determine the most heavily used paths through the code and organize the control flow so that branches are taken on less frequently used paths. Second, the processor architecture can introduce instructions that hide branch latency in many circumstances. The Xtensa architecture's zero-overhead loop option fits this category. Alternatively, conditional and unconditional branches can be constructed to take a branch without bubbles, with a penalty only if the branch is not taken. Third, high-end desktop processors have evolved quite elaborate dynamic branch-prediction mechanisms to make intelligent guesses about the location and direction of branches with a recovery mechanism for those cases when the guess is wrong. Simple versions of these mechanisms are starting to appear in high-end embedded cores. Complex branch-prediction mechanisms, however, can degrade a processor's operating frequency, increase its silicon area and power dissipation, or both. The benefit from spending precious silicon resources on branch prediction should be weighed against the benefit of other, perhaps highly leveraged, extensions directed at application performance.

5.5.1 The Software Development Environment

Large embedded applications present two engineering challenges to the software developer: complexity and optimality. Sophisticated software tools play a crucial role in addressing both needs.

5.5.1.1 Compiler

The central components of the software development environment are the compiler and the debugger. As more software development shifts from assembly code to high-level languages, the assembler's role diminishes, serving primarily as a back-end translator for compiled code and as a manual tool for developing specialized system-software routines. C and C++ are the most common high-level languages for embedded-software development, although Java appears to be gaining favor, especially were easy application portability is more important than performance.

Compilers vary quite widely in the quality of the code they produce. A number of code-optimization techniques developed for high-performance computer systems are completely applicable to embedded software, but they have percolated down into embedded-processor compilers slowly because a big slice of the embedded-software market has only modest performance requirements. Some general-purpose optimization techniques, such as global optimization, interprocedural analysis, scheduling, software pipelining, and feedback-directed optimization are found in more advanced embedded compilers. Some optimization techniques, such as code vectorization and VLIW instruction scheduling may be irrelevant for some embedded-processor architectures but very effective with others, especially for processors with SIMD execution units and configurable processors that can freely combine a number of independent operations in a single instruction.

5.5.1.2 Debugger

The debugger allows the programmer to walk through the code as it executes—function by function, line by line, or instruction by instruction. The programmer can set breakpoints or watch points, examine the processor state, and display it in a variety of formats. A modern debugger has good facilities for navigating through a large number of source files compiled together into a complete program. The debugger works in conjunction with a number of interfaces to the real or simulated processor. It may interact with a hardware processor over a network connection; it may connect to a software-based system simulation based on a single-processor instruction-set simulator (ISS); or it may execute on the target hardware.

The debugger must display and permit the updating of all of the processor state and decode all machine-instruction encodings into assembly code. Historically, configuring a debugger for a new instruction set has been a process of manually coding sequences to read and write each element of programmer state and to decipher each machine instruction into appropriate text strings. However, automatic generation of these state-access sequences is essential to efficient and effective debugger use whenever instruction-set extensibility includes addition of new programmer state.

Extension of the debugger has particular benefits when a processor is used as an alternative to RTL design, the topic of Chapter 6. All states and execution units added to the processor become visible in the source-level debugger. Debugging with this unified view of software and hardware states and functions improves engineering productivity for both subsystem-level verification and system integration. When hardware and software teams must work together on complex tasks with tight interaction between the application software and low-level hardware-centric tasks, the unified view provided by the debugger reduces confusion about the interaction of system hardware and software.

Debugging of real-time code or multiprocessing code presents special issues. The mechanisms to increase visibility—breakpoints, single stepping through code, examination of processor state—all change the timing behavior of the code and may interfere with the handling of timing-critical events, including interrupts and external-device interactions. The most difficult type of problems to debug in multiple-processor systems relate to race conditions that fail to occur if a debugger slightly perturbs the system's timing behavior. Real-time trace hardware can capture relevant information about program behavior over an interval without interfering with program behavior. The level of detail in the captured trace can range from all state changes to just a few important events. The trace can then be replayed through the debugger software so the developer can step through the interval and examine program state as if the program were running live. Additionally, the use of simulation allows full programmer visibility via the debugger without perturbing the system.

5.5.1.3 Profiler

The software-development environment almost always includes some form of code profiler. Good code profiling becomes even more important in the world of performance-driven designs. The profiler runs the application with typical input data and gathers detailed statistics on where and how the processor spends its time executing code within the application. This data may come from the actual execution of the code on a physical processor using the technique of program counter (PC) sampling.

For PC sampling, the profiler interrupts the normal application execution at regular intervals during code execution and records the state of the program counter and the execution stack for later analysis. With enough interrupts, the profiler accumulates a statistically meaningful picture of the application software's hot spots. For more accurate and complete statistics, the ISS can capture profile data. This data can include information on the branch behavior, on the precise caller of each function, and on the type and duration of stalls. Profile information may be as simple as a table of functions ordered by estimated execution counts. The profile gives the programmer a quick insight into where to find application hot spots. Profiling allows the programmer to focus optimization efforts on the most important places in the code, driving both algorithmic tuning and code optimization. Profiling aids optimization of code written in either assembly language or a high-level language such as C or C++.

Good code profiling is valuable for any software project where performance is key, but it takes on special importance in the development of configurable processors. Profiling focuses attention on the right routines for algorithm optimization, it quickly highlights the appropriate targets for instruction-set extensions, and it gives rapid feedback on whether changes in processor configuration or instruction set have produced meaningful improvements in overall software performance.

Each hot spot identified by the code profile reflects a region in the code where much time is spent. Many embedded applications confirm the Pareto principle (the "80/20 rule"): 80% of the time is spent executing 20% of the code. Some particularly computationally intensive tasks even follow a "99/1 rule." Consequently, attacking the hot spots in the order of time spent to reduce execution time is a common practice because it is so eminently sensible and produces excellent results. The developer first addresses the inner loop with the highest cycle count, selecting or describing new instructions to simplify and accelerate that loop's computation until the time spent in that loop drops below that of other loops. Optimization of the most time-consuming loop causes a different loop to become the slowest. The developer continues to optimize the "slowest" loop until the overall performance goal is achieved.

Sophisticated embedded-system developers working on performance-critical software have long been aware of the importance of tuning hot-spot code, but prior to the availability of application-specific processors, they lacked tools powerful enough to fix tougher performance problems. Once obvious improvements in basic software algorithms are made (for example, replacing simplistic sorting algorithms that take on the order of n^2 iterations substituting with more sophisticated sorting algorithms that require on the order of $n \log n$ iterations), the developer must delve into low-level tuning or "micro-optimizations" of C code or even drop down to assembly code to squeeze more performance out of the hot spot if the code is running on a general-purpose processor.

Using assembly coding has three problems. First, assembly coding is time consuming and error-prone because the programmer must work at a very low abstraction level and must therefore write code by combining a detailed knowledge of the bare-metal processor architecture and each instruction's syntax and semantics with an equally detailed knowledge of the application. Second, the more effort expended on assembly coding, the less portable code becomes. Third, the benefit of assembly coding is limited because the coder has the same set of generic instructions that are available to the compiler, and compilers have become very good at generating efficient code. Opportunities to cleverly choose instructions to beat the compiler's code generation effort may be modest because compilers have improved to the point where they leave human programmers little room to improve. As optimizing compiler technology continues to improve, the efficiency gap between handwritten assembly code and compiled code continually shrinks.

5.5.1.4 Simulator

When competitive, high-volume, embedded electronics systems could be built using off-the-shelf components, the simulator's role was insignificant. Designers selected the components (including the processor chips), built printed circuit boards, and debugged both the hardware design and the software on prototype hardware. As long as prototyping was fairly easy, this approach to system design, development, and integration worked well.

However, SOCs have radically altered the development landscape. As system cost, power, and performance pressure has grown, SOC integration has become mandatory across a wide range of products. This shift makes it nearly impossible to build early hardware prototypes. As a result, simulation has become far more important to embedded software development.

Three additional factors help drive this trend. First, the complexity of individual subsystems, the number of subsystems, and the unusual communication topologies possible with on-chip subsystems all make it extremely difficult or impossible to emulate a complex SOC architecture by building a board using processors, FPGAs, and other off-the-shelf chips. For example, some system architectures depend fundamentally on very wide buses between major subsystems. Closely coupled architectures with buses hundreds of bits wide are very difficult to prototype at a reasonable cost with real-time performance and good accuracy.

Second, microscopic (but noninvasive) debugging insight into detailed processor behavior, including cycle-by-cycle interactions between instructions and between processors and memories, is generally unavailable from hardware prototypes using off-the-shelf chips because full visibility into such systems might require tens of thousands of additional pins running at hundreds of megahertz. That's clearly impractical for any but the most exotic and expensive prototype.

Third, even when hardware prototypes for complex SOCs can be built, they are often physically large and expensive and are therefore difficult to maintain in significant numbers. The needs of SOC development, especially software development, set today's simulation requirements. Four desired characteristics stand out:

1. **Speed:** Developer productivity and verification coverage depends heavily on the time required to run and test the code. So, fast simulation, up to millions of instructions per second on a fast PC, makes software simulation an appropriate vehicle for many verification jobs.
2. **Low cost:** As the software content of electronics systems grows, the number of software or firmware developers working on a project also grows. Provisioning every developer with a low-cost simulation environment that runs on commodity desktop PCs becomes increasingly important to system-development cost. Moreover, simulators (and the other software tools) must suit two audiences: the initial SOC developer, who makes decisions about system architecture and processor configuration, and the SOC user, who develops systems and software that run on that SOC architecture.
3. **Accuracy:** The simulation model must not only model all functions of the processor correctly; it must also model pipeline and interface performance with cycle accuracy. High accuracy allows the designer to make detailed choices in software algorithms, coding style, and processor extensions. The simulator should both accurately report cycle counts and give accurate insight into exactly where the cycles are spent.
4. **System-capable:** To support an expanded role for processors in SOC design, the simulator must be part of an environment that provides fast and accurate modeling of multiple processors, memories, interconnect, and other subsystems. Rapid system-model development and full multiple-processor debug support are essential.

Two caveats apply. First, there is no universal modeling standard for arbitrary function blocks (though SystemC may eventually emerge as a standard for exchanging high-level models), so the system designer must either manually develop fast models for non-processor subsystems or drop down into a lowest-common-denominator description medium such as RTL for non-processor blocks. Second, software simulation does not run in real time, so simulation interfaces to external real-time devices and networks are awkward. More expensive emulation systems or full-speed hardware prototypes may still be required for some especially demanding system-development and testing tasks.

5.5.1.5 Support Tools

The software development environment may include a wide set of support utilities for preparing, manipulating, and examining binary files. These utilities usually include the linker, archive manager, symbol-table viewers, and other useful tools for managing binary code. These tools serve the same roles in SOC design as in any embedded software effort.

An important exception is the linker. For single-processor systems, assigning a final address to each block of code and data is relatively simple. Application code, library code, and local data are assigned to addresses in memories that are local to the processor or cacheable by the processor. Data sections that correspond to I/O buffers and devices are mapped to uncached or global memories. The developer writes simple linker scripts to control these mappings.

For multiple-processor systems, address assignment is harder. Some application code is specific to one processor. Some applications have copies of the code running on several processors. Sometimes the application code for each processor is unique and private, but the library code can be safely shared to save memory space. Some data sections are private, while other memory areas are shared by various processor combinations to facilitate interprocessor communication. Some data sections must be visible to I/O devices. The linker must resolve all these constraints.

Development of linker scripts is particularly complex for heterogeneous multiple-processor (MP) systems where different processors map shared-memory blocks to different addresses in their individual memory maps. Moreover, processors with particular instruction-set extensions may be able to run code that is unsuitable for more basic processor configurations.

Multiple-processor linker automation is an important adjunct to easy development of applications for sophisticated MP architectures. Automated MP linking makes the complex, manual development of linker scripts unnecessary. MP linker automation must account for all the constraints on the code that may run on each processor configuration by determining the addresses of all code and data sections for all software running on each processor.

Fig 5-22 shows how two processors, each with private and shared code and private and shared data (including I/O buffers, stack, and semaphores), might map their code and data sections into SOC memory. Note that the two processors have different address maps. That is, the hardware may map the same physical memory location to different addresses from the individual viewpoints of the two processors. The mapping must put each section in the appropriate type of memory (e.g., off-chip RAM, local processor RAM) with appropriate cache attributes (e.g., uncached, cached with write-back policy). Automatic requirements analysis and creation of link scripts for each processor makes resolution of these complex address assignment requirements easy.

5.5.1.6 IDE

The complex process of developing embedded software is often managed through an integrated development environment (IDE). The IDE provides a host of visually oriented tools to manage the complex set of source files and libraries, to view and edit source code, to build the source

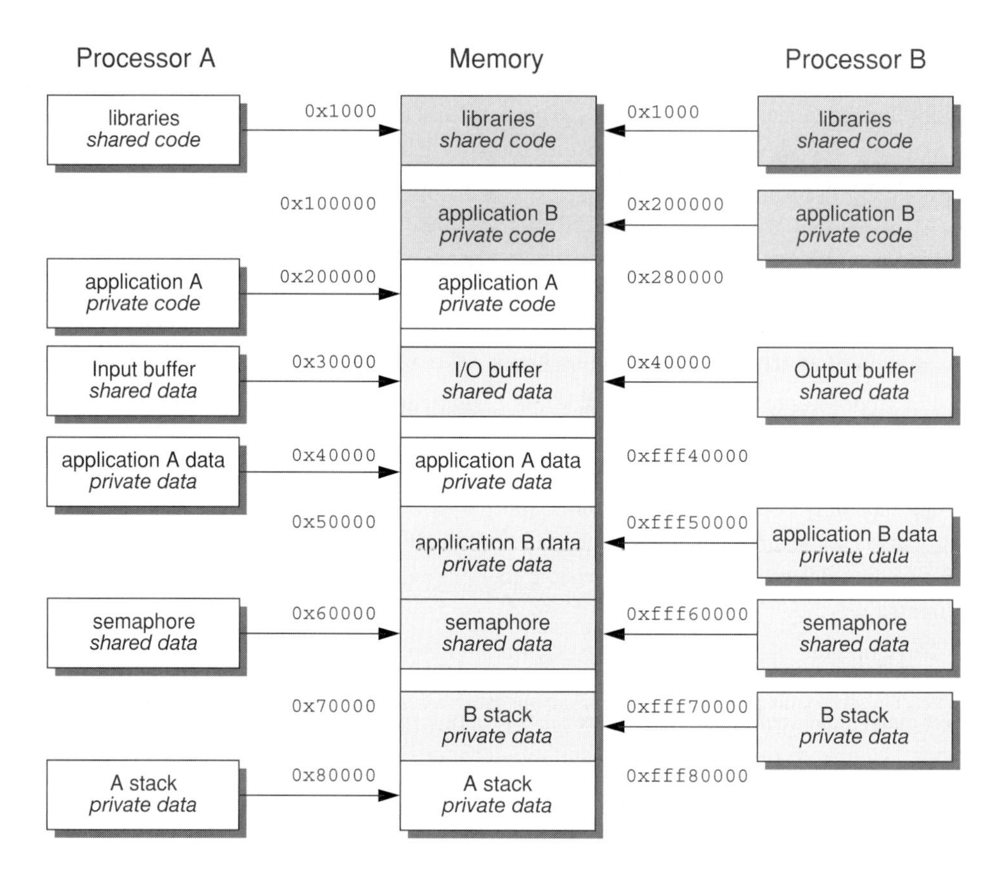

Figure 5-22 An MP (multiple-processor) linker resolves addresses for code and data with processor-specific address maps.

files into properly configured and linked binary code files, and to debug the application. The IDE may also provide an interface to code profiling.

For software tasks that are particularly sensitive to performance or efficiency, further analysis tools may be crucial. This is particularly true for processor-centric SOC software development for two reasons. First, when processors are applied to data-intensive computations formerly implemented with hardwired logic, maximizing throughput becomes even more important. The programmer may need to visualize complex interactions to squeeze out every last cycle to match or beat hardwired logic speeds. A graphical IDE provides appropriate infrastructure to connect C/C++ source code, assembly code, cycle-by-cycle performance simulation, and the state of variables in the computation. Second, the IDE provides a suitable environment for visualizing complex interactions among tasks running on one or several processor cores. Understanding the complex interactions among applications contending for access to shared variables and understanding the characteristics of data as it streams from producer to consumer are both critical to application correctness and application performance. The advantages of a good IDE might appear to be independent of SOC design methodology, but the presence of an IDE for processor-centric SOC software development verges on mandatory.

Key IDE capabilities include the following:

Detailed visual profiling—Point-and-click navigation among views of the execution profile—by function, by source-line, and by instruction—allow the software developer to drill down to the deepest layer necessary to gain insight into exact code behavior and the instruction sequences that consume the most time.

Interactive configuration—Rapid creation and modification of processor configuration and instruction-set-extension descriptions, including rapid feedback on estimated size, power, and clock frequency

Pipeline visualization—Detailed views of instruction execution, including visual presentation of pipeline stages, processor-state changes, execution stalls, and instruction overlap. These views provide insight into performance issues and can aid both source-level coding and extended-instruction design.

Multiple-processor system description—High-level capture of the architectural relationship of a set of processors, memories, interconnects, and other functions for the purpose of quickly building multiple-processor simulations.

MP debug and profiling—Interactive execution of multiple processor models for functional verification and performance estimation. Both overall performance and specific issues of processor-to-processor communication can be rapidly evaluated.

RTOS visualization—Interactive visualization of the interaction among multiple software tasks running under the control of a real-time operating system, sharing data structures, competing for processor timeslots, and waiting for system resources such as network and storage devices.

As the software content of SOC-based projects increases, and as programmability percolates deeper into data-intensive tasks, the software-development environment becomes increasingly central to the engineering process. In fact, it is reasonable to expect convergence between the software-oriented (interactions of the state of tasks) and hardware-oriented (cycle-by-cycle signal values) views, especially with the cycle-accurate performance capabilities outlined here. With convergence, the IDE becomes the central view of the development project and can be used by all of the system architects, software developers, and hardware engineers working on an SOC project.

5.5.2 The Software Runtime Environment

Runtime environments for SOC applications vary widely. In some cases, the runtime environment enveloping the application code is minimal when the processor directly replaces hardwired logic. In other cases, the runtime environment might include complex operating systems, networking stacks, resident debug monitors, and a suite of applications communicating with file systems, databases, and remote servers. The choice of an appropriate software environment can be a major architectural decision, so the ability of the processor environment to support a fluid range of options may aid rapid convergence toward the best system design.

5.5.2.1 Operating Systems

An operating system is a complex collection of runtime software components. It includes a kernel, the core software that manages the system's resources and facilitates communication among software tasks. The kernel supports multitasking; interrupt handling; exception signals; and intertask communication via shared memory, message queues, and mailboxes. The kernel may also be responsible for protecting one task from accidental or malicious interference from others. More sophisticated real-time operating systems may provide kernel-controlled memory protection and restricted access to system resources in conjunction with memory-management hardware. Real-time operating systems are usually configurable, in the sense that optional modules such as specific communications-support functions can be omitted or scaled back to limit the instruction- and data-memory requirements of the software.

The OS kernel is particularly important if tasks are dynamically started and terminated in response to changing system actions or when complex sharing of hardware resources (memory, disk space, and network connections) is required. Some systems with a single application that starts with the booting of the system and runs forever require no task switching, resource management, multithreading within a task, or task-to-task communication. For these systems, the kernel reduces to a simple set of initialization and exception-handling libraries. Running a single-threaded task can vastly simplify programming and significantly reduce the memory footprint of the runtime software.

Despite the importance of the kernel's system services, it may not be a major consumer of processor cycles. As a result, the kernel is not necessarily a key target for application-specific instruction-set extension. However, fast interrupt handling, context switching, and memory copying may be important performance bottlenecks in the kernel, and application-specific extensions can significantly improve the performance of these functions.

The OS often includes facilities for initialization, exception handling, general-purpose runtime libraries, debug support, various application-oriented protocol stacks, general-purpose applications, and other middleware (defined below). These same components may also be used without a multitasking OS kernel and can serve as a static library of useful runtime functions. Major functions outside the basic kernel operations include the following:

Initialization and exception handlers: Initialization code executes as soon as the processor exits the reset state. Initialization code sets memory and registers to a known state and prepares the processor to run applications written in high-level languages and to handle interrupts and exceptions. It also prepares the processor to run the OS kernel. During initialization, interrupts are disabled. The initialization code prevents and avoids exceptions during the initialization process. For a configurable processor, where system-programming support functions may vary depending on the configuration, the initialization code itself must also configurable. Interrupt handling, on-chip debug hardware, memory-protection and memory-management hardware, timers, coprocessors, caches, local memories, and other configuration-dependent functions are all initialized according to the configuration requirements. Consequently, when a designer changes a processor configuration during SOC development, the initialization code must also change to accommodate those processor modifications.

Runtime libraries: A wide variety of runtime libraries may be available to support application development. Many of these libraries are general-purpose and serve to raise the abstraction level for general C and C++ programming.

The ANSI C library includes

- `stdlib.h`: For basic arithmetic, string manipulation, data-type conversion, sorting functions, and memory allocation.
- `ctype.h`: For character data-type manipulation.
- `stdio.h`: For RTOS-independent file and buffer management.
- `string.h`: For string and memory copies, initialization, and comparison.
- `signal.h`: For exception management by high-level-language programs.
- `time.h`: For RTOS-independent real-time clock access.
- `stdarg.h`: For management of application argument lists.

Various other libraries may also be available to support system programming and application development.

Application-oriented middleware: Different application domains require different application-programming environments. These sets of basic services for networking, file-system access, security, and basic multimedia operations allow much more sophisticated software systems to be developed using pretested software components for common, domain-specific protocols. For example, networking middleware might include the Internet Protocol router stack—Point-to-Point Protocol (PPP), Open Shortest Path First (OSPF), and Routing Information Protocol (RIP) components—and standard management and server applications such as the Simple Network Management Protocol (SNMP) or a Web server. In a consumer-device environment, middleware might include device drivers for popular connectivity interfaces (Ethernet, USB, 802.11 Wireless LAN), graphics and multimedia packages, flash-memory file systems, and standard media encoding and decoding packages.

Debug monitors: The scale and complexity of embedded-software development has encouraged the emergence of a range of embedded debug mechanisms. In some cases, no debug-related software runs on the target processor—special on-chip debug hardware allows full access from a host debug system. In other cases, a small runtime debug monitor increases debugger capabilities and performance and reduces hardware cost. The monitor enables placement of software breakpoints in the code, eases buffering of debug information for the host system, and permits extraction of profile information from running hardware.

The runtime environment is usually structured so that all of the system hardware dependencies outside the processor are packaged together as a board-support package (BSP). The BSP includes the relevant I/O device drivers, address maps, boot code, and memory-system initialization routines for a particular combination of processor and system environment.

5.5.3 Processor Generation Flow

The technical benefits of an application-tuned processor are clear: more performance, lower power dissipation, and the opportunity for adding product features in software. All the work of creating an application-specific architecture could be done by hand: manual specification of new instruction set, memories, and interfaces; manual design of processor hardware descriptions in RTL; manual verification of the hardware against the architecture specification; manual development and validation of new compilers; manual porting of real-time operating systems; and so forth. The time, the cost, the bug risk, and the scarcity of specialized design skills make manual development of application-specific processors thoroughly impractical. When application-specific processor creation is automated, more designers and a greater range of system designs can exploit the benefits of processor-centric SOC design.

Understanding the basic mechanisms for processor-hardware and processor-software generation from a common specification helps the designer comprehend the capabilities—and the limitations—of application-specific processors. Fig 5-23 shows the basic inputs and outputs of a processor generator. The input to the generator is a configuration description, which may contain both descriptions of new instructions and the selected parameters for other processor features. The generator's output includes all the major development tools and runtime software, the RTL description of the processor hardware, and associated scripts and tools needed to streamline the implementation and verification of the hardware. Fig 5-23 lists the contents and attributes for each major block produced by the processor generator.

5.6 Example: Tuning a Large Task

The task of accelerating a large software application illustrates the process of profiling code and extending a processor's instruction set to reduce the number of execution cycles and increase throughput. As an example, consider the RSA (Rivest-Shamir-Adelman) cryptographic algorithm. The RSA algorithm is crucially important to many diverse security applications, especially for e-commerce. The algorithm is computationally difficult, even with superior implementation techniques. High performance is required if complex transactions utilizing RSA digital signatures and key exchanges are to become pervasive in applications such as e-commerce.

The RSA algorithm is an example of a public-key cryptography system. A public-key system employs two separate but related keys: a public key and a private key. Data encrypted with the public key can be decrypted only with the private key, and vice versa. The public key can be freely distributed; without the matching private key, it is of no value in decrypting other data. The RSA algorithm is based on the mathematical operation of modular exponentiation. Logically, a large integer is raised to a large integer power and the result divided by a large integer, retaining only the remainder. These integers may be thousands of bits long. In practice, 1024 bits is a common size, although larger sizes have been specified for keys that will be used for long time periods. Modular exponentiation requires lots of multiplication. Performing one 1024-bit private-key operation requires more than 20,000 32-by-32-bit multiplications.

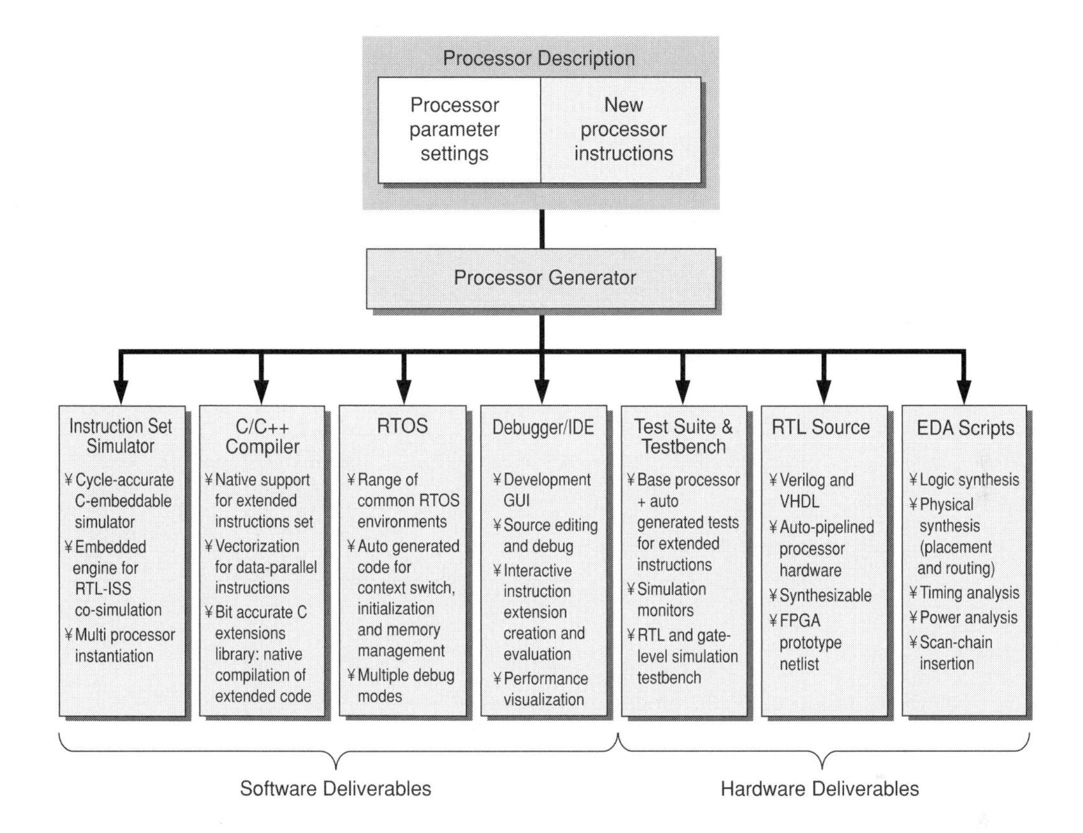

Figure 5-23 Application-specific processor generator outputs.

This example is based on the open source library for the Secure Socket Layer standard, which uses RSA encryption at its core. The OpenSSL code is a good test case for a complex software system. The source code consists of 738 C source and include files, so sophisticated profiling tools are crucial to uncovering the relevant hot spots. This code has already been ported to a number of different processor architectures, and the initial compilation of the code for the base processor is simple and quick. Fast multiplication is essential for RSA calculations, so a fast 32-bit multiplier is included in this example's baseline processor configuration. This multiplier also permits direct access to the upper 32 bits of the 64-bit multiplication result, and this feature is also used in the baseline code. The test case used for this example performs two private-key encryptions and one public-key encryption. The computation structure is similar for private and public encryption, though private-key encryption requires more than 10 times the total computation operations of the public-key encryption.

The initial port of this algorithm resulted in the profile shown in Fig 5-23. The figure lists the six functions that consume the most cycles—each of the six functions consume at least 2% each of the total 27 million cycles required to execute the OpenSSL code. All six of the most time-consuming functions involve arithmetic for RSA's extended-precision computations (the function prefix "bn" refers to "big-number" extended-precision calculations).

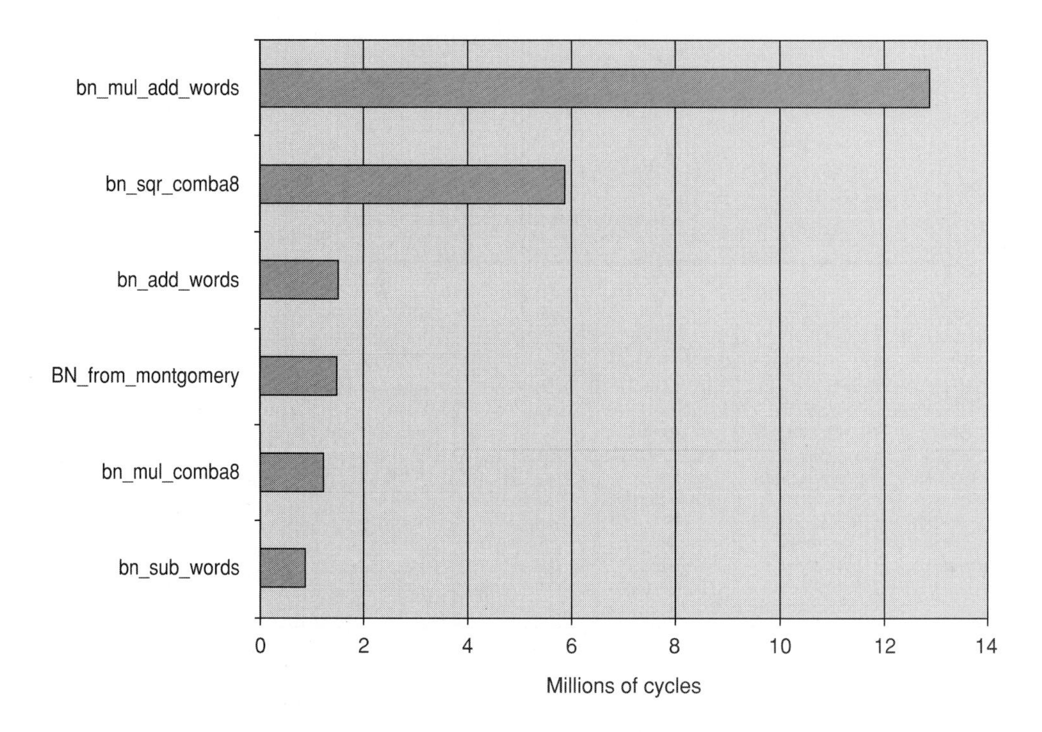

Figure 5-24 SSL out-of-box code profile (total: 27M cycles).

We focus on the dominant routine, `bn_mul_add_words()`. One new instruction to perform a 32-bit multiplication, with accumulation of carries, serves as an efficient building block for arbitrary-precision multiplication. The new instruction, `bnmac32`, is described in just a few lines of TIE code (see the appendix on SSL). The addition of the instruction also creates a corresponding new C function, `bnmac32()`. The incorporation of this intrinsic function in the OpenSSL code simplifies carry handling in the inner loop of `bn_mul_add_words()` and eliminates two branch instructions. The new function's impact on performance is shown in the profile in Fig 5-24. The addition of just this one instruction reduces the key routine's cycle cost from 13 to 5 million cycles and reduces the total cycle count of the entire algorithm to 19 million cycles, down from 27 million.

The function `bn_mul_add_words()` is the single biggest user of multiplication operations in the RSA code, but big-number multiplication is used throughout the other functions as well. Changing the other functions to exploit the fast `bn_mul_add_words()` function and adding two new instructions (`bnmul32` and `bnadd32`) actually shifts more of the computation burden to the function and increases the total time spent in `bn_mul_add_words()` but sig-

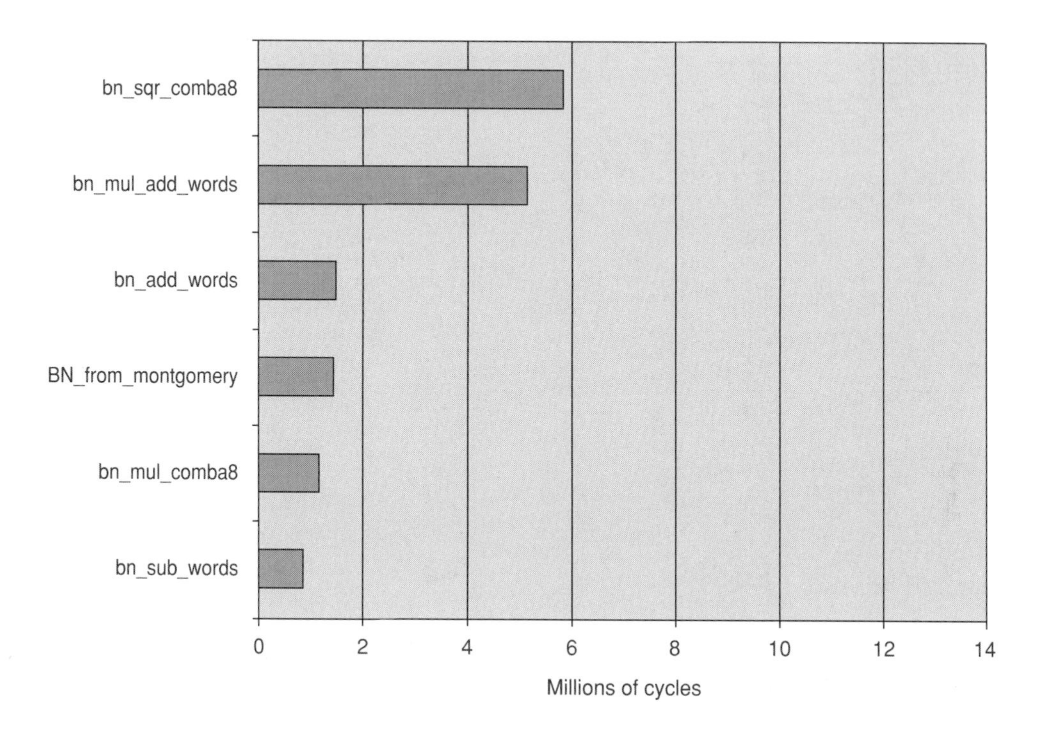

Figure 5-25 SSL code profile after initial optimization (total 19M cycles).

nificantly reduces the time spent in other routines that use multiplication. The resulting profile of time-consuming functions is shown in Fig 5-25. This version of the RSA code, which uses the optimized 32-bit multiplication implemented in three new instructions, is nearly twice as fast as the original implementation (14M cycles versus 27M cycles).

However, extensible processors are not limited to 32-bit operations. In general, doubling the multiplier width cuts the total number of multiplications in big-number arithmetic by a factor of four.

We can take advantage of this potential by defining a new set of 64-bit multiplication-with-carry-accumulation instructions. The input and output values of the new 64-bit multiplication instructions are held in newly defined 64-bit registers, and these registers are served by new 64-bit load-and-store instructions. Moreover, these new load-and-store instructions automatically increment the target address to further reduce the number of instructions in the inner loops of key functions.

Using these new 64-bit instructions, the most important OpenSSL functions were modified to reduce the number of loop iterations by a factor of two at each level of the loop nest. The rewritten functions accommodate all possible encryption sizes from 512 to 4096 bits. Fig 5-26 shows the final optimized code profile using the 64-bit instructions. Performance has again dou-

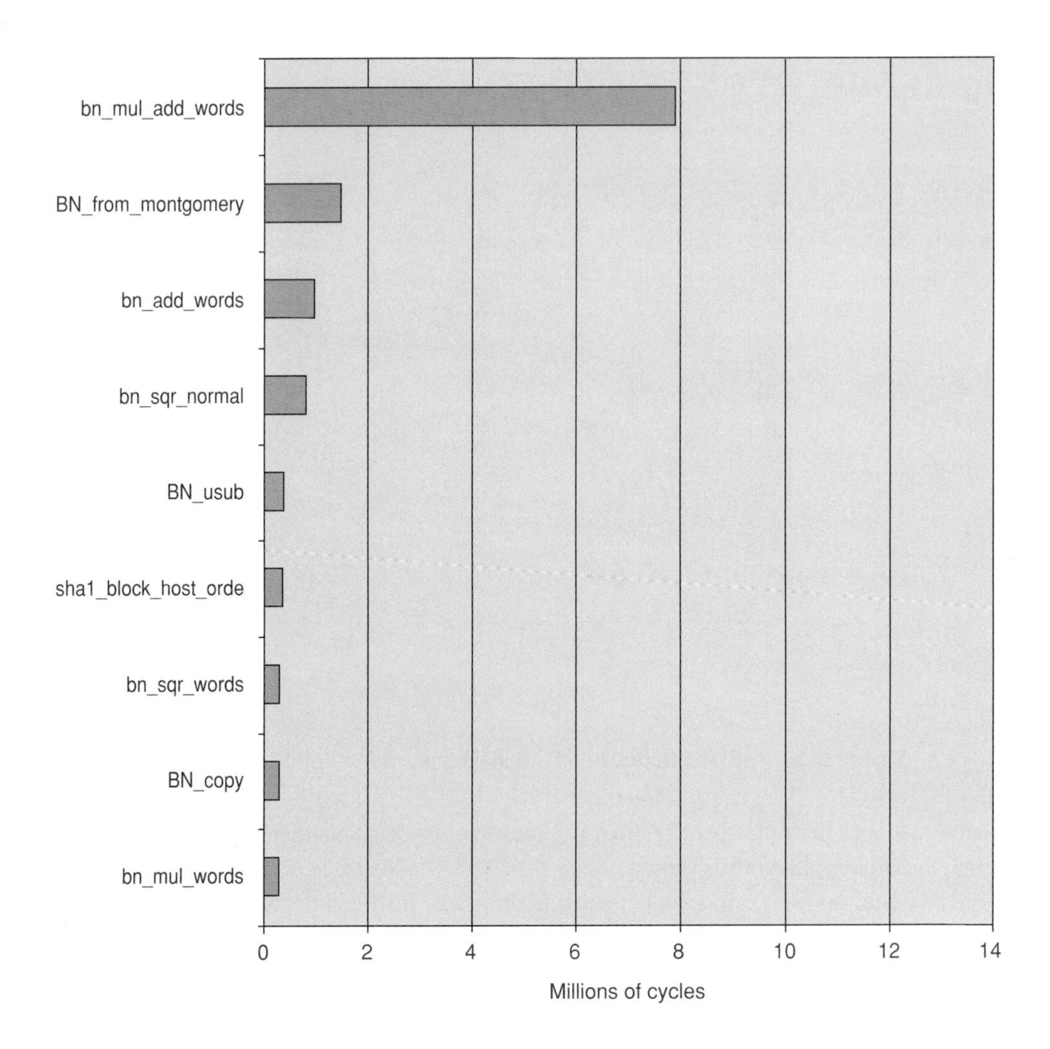

Figure 5-26 SSL code profile with full 32-bit optimization (total 14M cycles).

bled (from 14M to 7M cycles), which is four times faster than the original out-of-the-box reference code.

Further acceleration could easily be achieved by expanding from a 64- to a 128-bit multiplier. However, diminishing returns set in at this point. Doubling the multiplier width again—with a resulting increase of four times the chip area—only boosts overall performance by another 10%, so this further optimization might not represent a good engineering tradeoff, depending on the system's design goals.

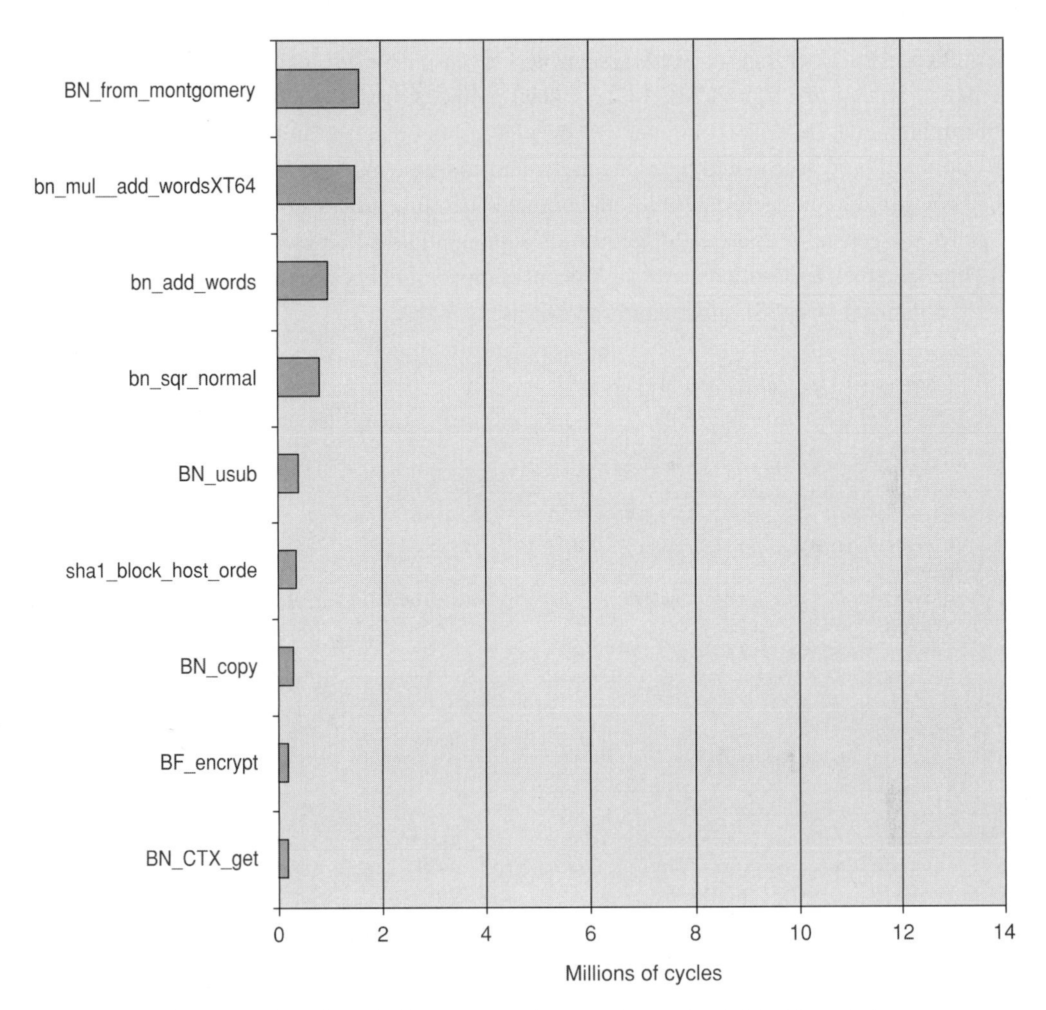

Figure 5-27 SSL final 64-bit optimized code profile (total 7M cycles).

This example demonstrates the utility of profiling tools to identify hot spots, which leads to the creation of instructions that improve hot-spot performance. With only eight new instructions (two of which are included just to accelerate the saving and restoring of the states of the added 64-bit registers), performance improvements of as much as 12.5x for specific functions were achieved. In a test involving RSA public- and private-key operations on modulii of 512, 1024, 2048, and 4096 bits, the final version of the RSA instruction set used just 12% of the cycles of the original version, an 8.18x improvement in total cycle count.

The critical inner loop of the big-number multiplication routine shrank from 18 instructions (including two branches) to just three instructions (with no branches), and the routine was invoked half as often. Overall, the cycle count drops to about 7M cycles, down from the original 27 million. At a clock rate of 300MHz—which is reasonable for automated synthesis, placement, and routing of 130nm CMOS logic—the optimized processor core would generally outperform high-end, PC-class computer system processors operating at a frequency of GHz, and the optimized processor would be a lot smaller and consume much less power. The 4x speed-up when compared to the performance of an optimized baseline processor—and more than a factor of 10 from a general-purpose RISC processor without a fast 32 bit multiplier—is fairly typical for large, complex applications with significant computational content. The complete TIE code for the optimized OpenSSL processor is shown in Fig 5-28.

```
1: state carry 32 add_read_write
2: state w 32 add_read_write
3: state w64 64 add_read_write
4: state carry64 64 add_read_write
5: state r 64 add_read_write
6: state statea 64 add_read_write
7: coprocessor BN 1 {carry,w,w64,carry64,r,statea}
8:
9: operation STR64I {inout AR ars} {in r, out VAddr, out MemDataOut64} {
10: assign VAddr = ars;
11: assign MemDataOut64 = r;
12: assign ars = ars + 8; // increment for next fetch of r
13:     }
14: operation STA64 {in AR ars} {in statea, out VAddr, out MemDataOut64} {
15:  assign VAddr = ars;
16: assign MemDataOut64 = statea;}
17:
18: operation LDR64 {in AR ars} {out r, out VAddr, in MemDataIn64} {
19:  assign VAddr = ars; //don't incr, must store back same r[i], then incr
20: assign r = MemDataIn64;}
21:
22: operation LDA64I {inout AR ars} {out statea, out VAddr, in MemDataIn64}
23: {
24: assign VAddr = ars;
25: assign ars = ars + 8; // increment for next fetch of a
26: assign statea = MemDataIn64;}
27: operation BNMAC64 {inout AR ars} {inout carry64, in w64, in statea, in r ,
28: out VAddr, out MemDataOut64} {
29: wire[127:0] temp1 = TIEmac(w64,statea,carry64,1'b0,1'b0);
30: wire[127:0] temp2 = TIEadd(temp1,r,1'b0);
31: assign carry64 = temp2[127:64];
32: assign VAddr = ars;
33: assign MemDataOut64 = temp2[63:0];
34: assign ars = ars + 8;}
35: operation BNMAC32 {inout AR ars, in AR art} {inout carry, in w} {
36: wire[63:0] temp1 = TIEmac(w,art,ars,1'b0,1'b0);
37: wire[63:0] temp2 = TIEadd(temp1,carry,1'b0);
38: assign carry = temp2[63:32];
39: assign ars = temp2[31:0];}
```

```
40:
41: operation BNMUL32 {out AR arr, in AR ars, in AR art} {inout carry}{
42: wire[63:0] temp1 = TIEmul(ars,art,1'b0);
43: wire[63:0] temp2 = TIEadd(temp1,carry,1'b0);
44: assign carry = temp2[63:32];
45: assign arr = temp2[31:0];}
46:
47: operation BNADD32 {out AR arr, in AR ars, in AR art} {inout carry} {
48: wire [32:0] temp1 = TIEadd(ars,art,carry,1'b0);
49: assign arr = temp1[31:0];
50: assign carry = {31'h0,temp1[32];}
51:
52: schedule BNMAC64 {BNMAC64} {
53: use ars 1; use w64 1; use statea 1; use carry64 1; use r 2;
54: def ars 2; def carry64 2;}
55: schedule LDR64 {LDR64} {use ars 1;def r 2;}
56: schedule LDA64I {LDA64I} {use ars 1; def statea 2; def ars 2;}
57: schedule STR64I {STR64I} {use ars 1;use r 1; def ars 2;}
58: schedule STA64 {STA64} {use ars 1; use statea 1;}
59: schedule bnarith {BNMUL32} {
60: use ars 1; use art 1; use carry 2; def carry 2; def arr 2;}
61: schedule BNMAC32 {BNMAC32} {
62: use w 1; use art 1; use carry 2; use ars 1;
63: def carry 2; def ars 2;}
```

Figure 5-28 TIE source for OpenSSL acceleration.

5.7 Memory-System Tuning

The bulk of this chapter has focused on changing the processor's programming model to improve application performance and efficiency. The interaction of the application's memory references and the memory-system design can dramatically affect performance, so the software developer should understand the basic issues and techniques of memory-system tuning. The major parameters for processor memory-system design were already discussed in Chapter 4, Section 4.1. The purpose of memory tuning is to explore the target application's sensitivity to memory-system parameters and to choose parameters for each processor that balance performance, processor cost, and size appropriately. The processor simulator plays a central role in this assessment, because the simulator models and reports the expected performance with a breakdown of memory-related stalls.

5.7.1 Basic Memory-System Strategy

Memory-system configuration consists of two phases. First, the designer must establish the strategy for each major segment of instruction and data storage, answering basic questions such as the following:

- Will the processor execute its initialization code from off-chip, read-only, or flash memory; or will that code be preloaded into a local on-chip instruction RAM? This question is important because the boot-code location helps determine the kind of memory-bus interface that is required. If the processor never needs access to remote memory except for initialization code, a simpler and smaller memory interface may be appropriate.
- Is the performance-sensitive application code small enough to fit entirely in instruction RAM local to the processor, or is an instruction cache necessary?
- How does the processor load application input data? Is it loaded from remote input buffers or memories, is it pushed into the processor's local data RAM by an outside agent such as a DMA controller or another processor, or does the processor directly access the data using input, ports and queues? These questions are important because data I/O references also help determine the kind of memory-bus interface that is required.
- How does the application result data exit the processor? Is it sent to remote output buffers or memories, is it pulled from the processor's local data RAM by an outside agent such as a DMA controller or another processor, or does the processor transfer the data through direct output ports and queues? These questions are important (like the previous set) because data I/O references also help determine the kind of memory-bus interface that is required.
- Can all of the performance-sensitive data (including application data, maximum-sized stack, and other variable data storage) fit entirely in data RAM local to the processor, or will it be necessary to emulate a large, local data RAM using a local data cache and a large external memory?

Sometimes the instruction or data segments are so large that instruction and data caches are necessary for good overall performance. Unfortunately, the complex interactions among memory references may make cache behavior appear nondeterministic. This trait presents a significant problem for some embedded applications, where certain instruction sequences or data regions must be accessible with a small, constant latency.

This need can be addressed either by closely coupling both cache and local RAM to the processor (with time-critical instructions or data allocated to addresses mapped to the local RAM) or by using cache locking to prevent certain data or instruction lines from being evicted from the cache once the correct contents are loaded. Cache locking temporarily makes selected cache regions (on a line-by-line basis) act as local memory.

5.7.2 Detailed Memory-System Tuning

Once the basic organization of instruction, data, and cache memory has been established, the second phase—detailed memory-system tuning—can proceed. Analyzing memory stalls associated with the processor's memory-system configurations drives this process. Fig 5-28 shows a typical application-performance profile for an MPEG4 encoding application, including instruction-cache-miss stalls, data-cache-miss stalls, store-buffer stalls (for a write-through cache), and other instruction-execution delays (exceptions, source interlocks, and branch-taken delays).

	Number Executed	Event Type	Number of Events	Cycles	Cycles per Instruction (CPI)	Cumm CPI
Instructions	74,579,289	Committed Instructions	74,579,289	74,579,289	1.000	1.000
		Uncached	1,452	13,277	0.000	1.000
		ICache misses	428,244	6,126,010	0.082	1.082
		Source interlocks	870,990	870,990	0.012	1.094
		Taken branches	2,884,870	6,565,980	0.088	1.182
		Exceptions	72	370	0.000	1.182
Data Reads	14,974,107	DCache bad misses	374,164	5,637,396	0.076	1.258
Data Writes	6,357,577	Store buffer stalls	—	—	0.000	1.258
		Total		**93,783,312**		**1.258**

Figure 5-29 Memory system profile and parameters.

Two perspectives guide detailed memory-system tuning:

- The macro view of the memory system's aggregate performance across all instruction and data references in a complete application.
- The microview of the memory system's behavior, especially data references, in the key application hot spots or inner loops.

5.7.3 Aggregate Memory System Performance

The macroview is driven by the accumulated dynamic statistics of all program-memory references. The cumulative statistics often give little insight into why, for example, cache misses occur, but they serve as the foundation for tuning overall application throughput. Simulating the application with a range of different memory-system parameters establishes the tradeoffs between application performance and memory-system implementation cost.

Typically, the behavioral patterns of an application's instruction references and its data references are uncorrelated—the statistics of each should be examined to determine the appropriate configuration. The impact of various parameters is summarized below:

- **Memory-access latency (first word):** Reducing the memory latency for the first word of a cache-refill block always reduces the penalty for cache misses and increases application performance.
- **Memory-access time (each additional word of the block):** Reducing the incremental delay for each subsequent word in a block reduces the penalty for cache misses (when the cache-line size is larger than one memory word) and increases application performance.
- **Cache size:** Increasing the size of the cache almost always reduces the number of cache misses and thus increases application performance.
- **Cache set associativity:** Increasing the number of ways in the cache almost always reduces the number of cache misses and increases application performance.
- **Cache-line size:** Increasing the cache-line size may increase or decrease application performance, depending on the pattern of memory references, because longer cache lines take more time to load from or write back to main memory and increase the risk of interference among lines (longer cache lines mean fewer total lines in a given size cache). When all the words in the cache line are used before a line is evicted from the cache, a longer line size generally improves application performance, particularly when the latency to access the first memory word is long and the incremental time for additional words is short. Conversely, when the application uses little of each cache, a shorter cache line may improve performance, particularly if the memory latency is short or the cost of each incremental word of the cache line is large. This uncertainty is why accurate simulation is so important.
- **Refill width:** Increasing the number of bits that can be transferred into or out of the cache on each cycle generally reduces the delay for reading or writing a cache line and improves application performance, especially for long cache lines. The refill size often corresponds to the width of the bus connecting the processor to main memory, so instruction- and data-refill widths are almost always the same, though this is not mandatory.
- **Write-back vs. write-through data cache:** Choosing a write-back cache generally reduces the number of write operations to the remote (on-chip or off-chip) memory associated with the data cache, because several processor stores to locations within one cache line result in just one write operation on the processor-to-main-memory bus. If the memory write bandwidth or the bus bandwidth to the memory is relatively narrow, choosing a write-back policy may increase application performance. If write bandwidth is not a problem, a write-through cache can reduce the delay incurred by data-cache misses and increase application performance. The data-cache miss delay is often lower for write-through caches because making room for a new cache line (by evicting a victim line) never requires the write-back of a dirty cache line. In addition, write-back caches may stall the processor during a write operation to allow the cache to load all of the other words in the target cache line before the write occurs. This type of stall doesn't happen with write-through caches.

Each of the memory-system configuration choices affects the silicon cost of the processor implementation. Larger caches, wider refill width, shorter cache-line size, and greater set associativity all increase the silicon area for RAM, logic, or both. Of these choices, increasing cache capacity and decreasing the cache-line size generally cause the biggest increases in silicon area.

Remote-memory latency is often a central concern in processor configuration. Caches reduce that sensitivity, sometimes dramatically. In many cases, however, optimizing the system's main memory, independent of the processor's local-memory system, may be critically important to improving overall performance. Accessing off-chip dynamic RAM, for example, may require 50 to 100 processor cycles per access, particularly if the path from processor to memory winds through several different on-chip buses or if the RAM interface is not optimized for rapid access times.

Fig 5-29 shows a memory hierarchy with caches and RAM local to the processor, global on-chip RAM, off-chip RAM, and off-chip flash memories. The access latency from the processor may increase tenfold for each level in the hierarchy:

- local memory access latency ≈ 1 cycle
- global on-chip RAM access latency ≈ 10 cycles
- off-chip RAM access latency ≈ 100 cycles

Configurable processors allow the designer to tweak many different memory-system parameters. Understanding the performance tradeoffs and choosing the optimal configuration may be a complex process. Charts of application performance as a function of key parameters can help the designer visualize those tradeoffs and finalize the memory system more quickly. Fig 5-30 shows simulated data-cache performance results for an optimized JPEG encoding application. It plots the total number of execution cycles as a function of cache size, cache-line size, and set associativity.

For this application, data-cache behavior is an important design consideration. The simplest cache (4 KB, direct-mapped, 16-byte line) has a load-miss rate of 13.4%, while the most complex (32 KB, 4-way set associative, 64-byte line) has a load-miss rate of 1.9%. This chart clearly shows that larger cache sizes are better, but it also suggests diminishing returns above 16 KB.

Line size—the number of bytes brought in on each cache miss—is also a significant factor. Moving from 16- to 64-byte cache lines creates more performance benefit than doubling the cache size. This result is notable because longer cache lines actually reduce silicon cost, while doubling cache size can be expensive (in 0.13μ technology, a 16Kbyte RAM array requires roughly $1mm^2$). The figure also shows that two-way set associativity is clearly better than direct-mapped (one-way) cache, but going from two-way to four-way set associativity has less dramatic benefits.

This simulation is based on a single-processor design without consideration of other processors that may be contending for access to shared, nonlocal memory. The single-processor

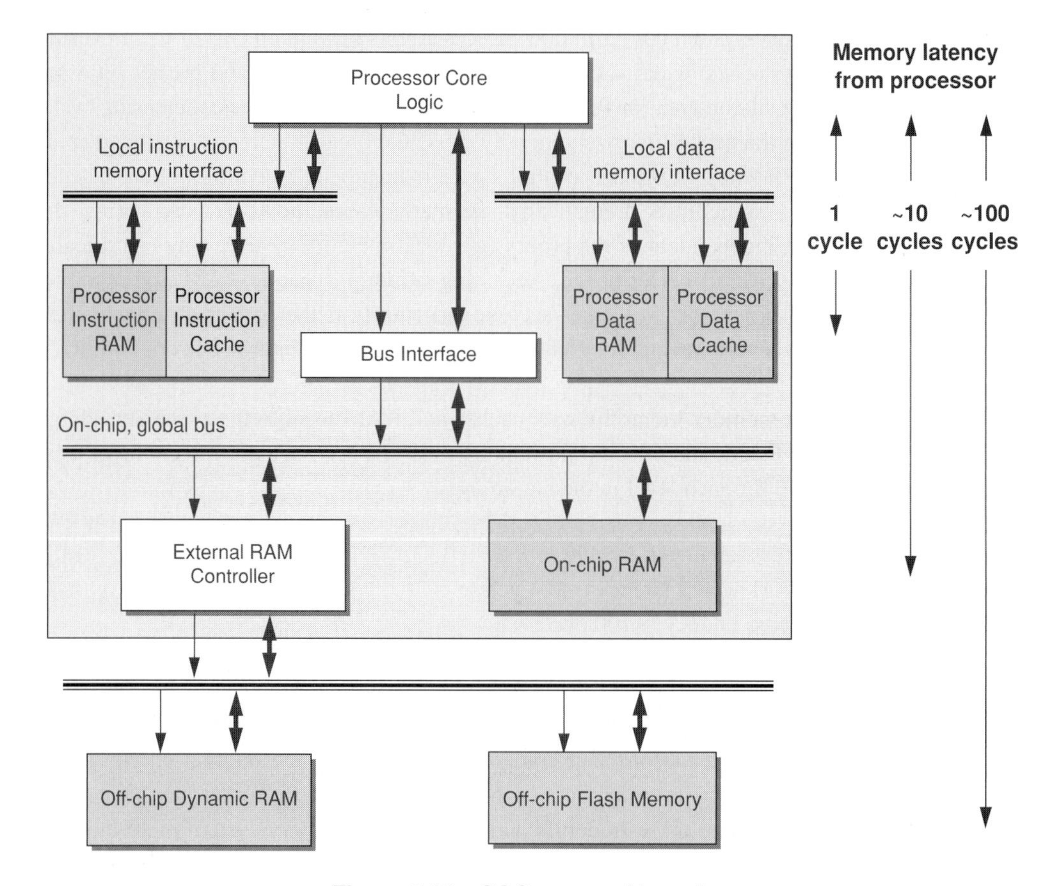

Figure 5-30 SOC memory hierarchy.

behavior may change depending on the memory design, the interconnect structure between processors and memory, and the pattern of memory references by other processors. In particular, the effective memory-access latency may increase due to memory or bus contention. When the set of processors is known, resimulation with more accurate modeling of processor-to-memory contention may give more exact predictions of final system performance. This refined simulation may suggest further improvements to the memory-system configuration of some processors, perhaps including increased cache size or set associativity, a change to write-back policy, or an increase in the cache-refill block size.

5.7.4 Inner-Loop Data-Reference Tuning

For some data-intensive applications, the data-reference pattern is simple. But memory performance is often so critical to overall system performance, that additional analysis is worthwhile. When the application's inner loop must load a long sequence of data, perform some computa-

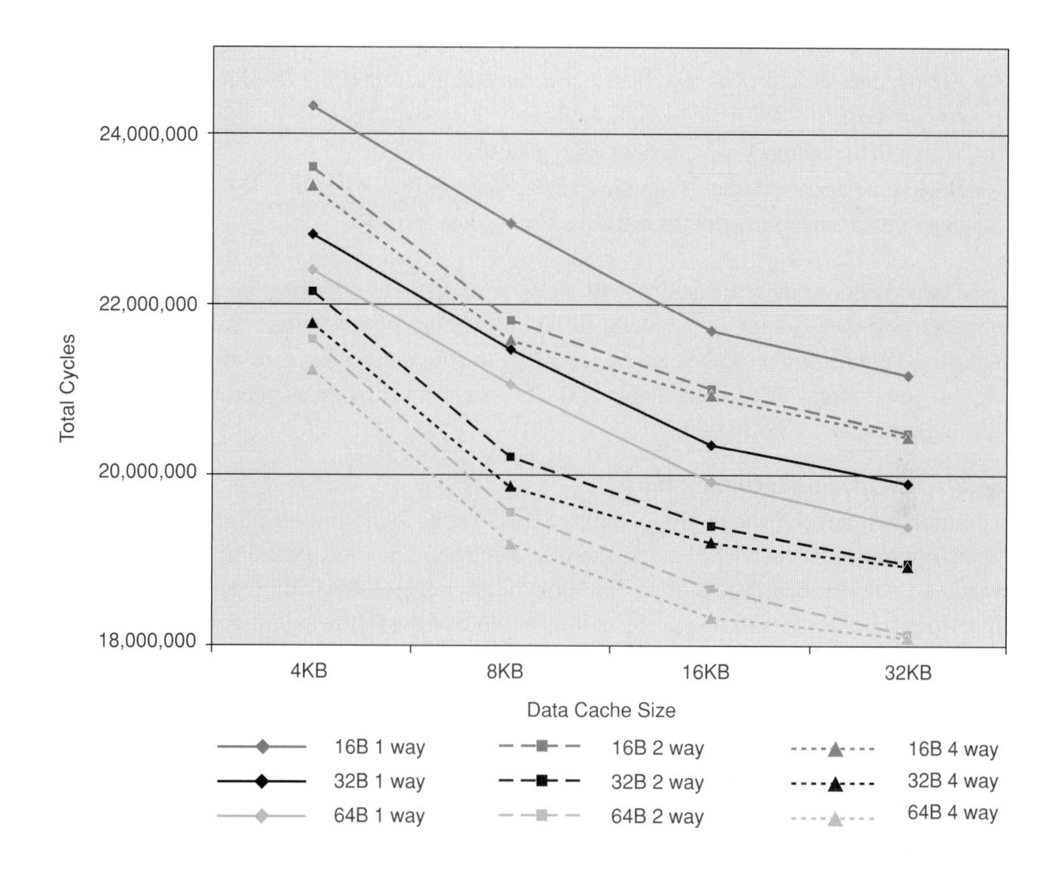

Figure 5-31 Data memory stalls graph.

tion, and send out the result sequence, this is termed a *data-streaming* application. The data is not used again outside that computation, so the desired characteristics of the data-memory system may be largely dictated by the needs of this data streaming. Usually, this use-once data-reference pattern is known to the application developer from the outset and becomes a central concern when setting the basic memory system strategy. Occasionally, the significance of the inner-loop data-reference pattern is only noted during detailed cache tuning. The data-reference pattern often appears in the form of excessive cache misses, with no performance sensitivity to cache size.

Configurable processors offer a wide range of options than can achieve good application performance with data-streaming applications. It is essential that the designer understands the size of data references and the sequence of data addresses in the inner loop, and is able to answer questions such as

- What are the sizes of the data structures used in the loop?

- Is each data structure completely used within the loop, or are only selected fields used?
- Are the data structures accessed at sequential addresses, at some fixed increment, or in random order?
- Where in the memory system does each data stream reside?
- Does some agent outside the application's processor (an I/O device, DMA controller or other processor) overwrite the old data as new data arrives?

The answers to these questions will allow developers to optimize the processor and the application together. Sometimes, several different memory optimizations have the same basic effect, so the optimization can be chosen according to cost, convenience, or availability. In other cases, no good optimization is available and the system must be redesigned to break the data-streaming performance bottleneck.

5.7.4.1 Managing Incoming Data

A common case involves a single incoming data stream, found initially in nonlocal memory. Without optimization, each load suffers the full latency of a non-local memory reference, which may run from fewer than 10 to as many as 100 processor cycles. Such long latencies cripple system performance. To combat this problem, one or more of the following improvements might be considered:

Use DMA to move the data into local memory: DMA comprises a range of techniques that relieve the processor from the responsibility of moving data between remote I/O interfaces and the processor's local memory. The DMA approach employs a separate data-movement engine to read data from an input device and write it into the application processor's memory, or to read data from processor memory and write it to an output device. DMA may also take place between memories, usually between a large, slow, remote memory and the processor's smaller-but-faster local memory. The DMA engine is controlled by a set of registers used to set the source and destination address or I/O device, the size of each DMA transfer, and the total number of transfers. Sophisticated DMA engines may manage several different simultaneous transfers and reload their own control registers from a chain of DMA command blocks stored in memory.

DMA exploits the potential concurrency between memory movement and processor computation. DMA use is built on the assumption that processors are expensive and hardwired data-movement engines are cheap. In practice, a second embedded-processor core may serve as a very effective DMA engine while offering much easier programmability and unified system simulation plus on-the-fly data manipulation in exchange for a possibly slightly higher silicon cost compared to a fixed-function DMA engine.

Cache the data buffer: Map the incoming data memory to a cached address. The cache will naturally load the data a full cache line at a time. If the line is fully used, because the individual data elements are close to the size of the line or because the data accesses are sequential, the number of cache misses will shrink and performance will improve by amortizing the memory latency across all the loads to that cache line. The benefit is especially large if the cache-refill block's first-word access latency is large but the latency for accessing additional words is small.

Longer cache-line sizes and wider refill sizes are generally advantageous for applications that employ a sequential access pattern. Moreover, additional performance will be gained if the processor implements a non-blocking cache-miss policy or provides nonblocking prefetch instructions (allowing some other instructions to execute while the cache refill occurs).

If an external agent, such as a DMA controller, changes data behind the processor's back, cache coherency issues arise. Large-scale, multiprocessor computer systems use hardware to enforce the consistency of caches (cache coherency) and other memories among processors. For cost reasons, software coherency is more common than hardware-based cache coherency in embedded systems.

Explicitly manage cache lines: RISC processors commonly include instructions to explicitly manage the state of cache lines. The software developer can use these instructions to ensure that the necessary data is in the cache when needed and that other useful data in the cache avoids unnecessary eviction due to cache collisions. Before reading the data, the application invalidates the corresponding entry in the cache to force a data fetch into the data cache from main memory. After the application has written new data, the application can explicitly force the cache line containing the new data back to main memory with a write-back cache operation. Using this method, an entire cache line can be brought into cache with a single block-read transfer or written out to memory or to an output device with a single block-write transfer. A typical set of data-cache instructions is listed below (including the typical conditions for use). Each of these instructions operates on a specific address (where a number of addresses all refer to the same index in the cache). The following typical operations use the Xtensa processor instruction names:

- **`Data-Cache Hit Invalidate:`** If the addressed line is present in the cache and not locked, then invalidate that line. This instruction ensures that subsequent reads of this address will cause a cache miss and the entire cache line containing that address will be fetched from main memory.
- **`Data-Cache Hit Write-back:`** If the addressed line is present in the cache, and the entry is dirty (i.e., it is a write-back cache and the most recently written data values exists only in cache), then write the entire cache line containing that address back to memory. This operation ensures that data is visible to other processors and I/O interfaces by forcing writes to memory shared with an I/O controller or another processor.
- **`Data-Cache Hit Write-back Invalidate:`** If the addressed line is present in the cache, invalidate the line after first writing back any dirty data. This instruction is used after the last write to a location, which will later be used, and perhaps modified, by some external agent.
- Cache operations also include instructions that specify a cache-line index rather than an address. These instructions operate on an entry in the cache, regardless of what memory

address range is currently present in that entry. Operations such as `Data-Cache Index Write-back` and `Data-Cache Index Write-back Invalidate` are useful when a whole region of memory must be forced back to main memory before a subsequent use by an output device or another processor. For example, execution of a `Data-Cache Index Write-back` operation performed on index 0 of a 4Kbyte cache assures that no dirty data appears in that cache line and that the most current copy of the line appears in main memory, for all addresses that map to index 0 of the cache—that is, addresses 0x0, 0x1000, 0x2000, 0x3000, and so forth, for a cacheable address range starting at address 0x0.

Sometimes, the supported cache operations include prefetch instructions that move data from memory to the cache if it is not already there. The prefetch instructions may also include hints about expected use of the line, such as hinting that the line is likely to be used only once (especially in a data-streaming situation) so it can be designated as a preferred candidate for eviction from the cache. The processor hardware can use these hints to improve cache behavior. For example, if the prefetch hint indicates that this data will be used only once, then the set-associative cache hardware can mark this set as "least recently used," so it will be preferentially evicted after use instead of other sets that may contain more reusable data. Note: These cache instructions are also useful for data- and instruction-cache RAM testing and power-on initialization.

Map the I/O device into local memory: Configurable processors are not constrained to communicate with I/O devices over a general-purpose bus. Key I/O interface controllers can be mapped into the processor's local-memory space and directly accessed over a low-latency, high-bandwidth interface just like local RAM. This organization allows incoming data to be moved into processor registers exactly as if it had already been placed in local data RAM, without extra bus cycles or DMA. This approach does make the I/O interface private to that one processor. If other processors or logic need access to the same I/O resources, either the processor to which the I/O controller is attached must act like an I/O controller and move the data into a shared-address space or the processor must be configured to expose some of its local memory space as shared memory to other processor bus masters. Both approaches work well for high-bandwidth I/O interfaces where systemwide traffic reduction may reduce overall bus contention and power dissipation.

5.7.4.2 Managing Outgoing Data

Sometimes the sequence of writes for outgoing result data presents challenges similar to those of a sequence of reads for incoming source data. Using an address range or cache configuration that allows outgoing data writes to ignore the cache is often useful. Each word written by the application then passes through to the bus or memory system without stalling the processor as long as the processor's write buffering and the average memory-write performance are sufficient to keep pace with the write traffic.

Write-back caches may work better than write-through caches because they potentially allow an entire line to be written in one memory transaction, but they suffer two potential liabilities. First, if the processor allocates a cache line upon the first write to a word in that line, the processor must read the entire line from memory, only to immediately overwrite it with new values. This property of write-back caches may unnecessarily double the memory traffic and increase latency if the processor stalls waiting for the entire line to enter the cache. Second, this allocation of lines to outgoing data may pollute the cache with data that is not needed for subsequent computations.

Explicit management of cache lines is often the simplest and most predictable method to manage outgoing data, especially if the bus or memory system is optimized for block transfers. Cache operations such as **`Data-Cache Hit Write-back`** and **`Data-Cache Index Write-back`** are particularly useful.

More complex cases frequently arise when processing multiple data streams. For example, two data streams might be small enough that each could individually reside in cache, but the addresses of the two streams conflict so that the benefit of caching is lost because of *cache thrashing*, which occurs when actively referenced portions of the two data streams fall into the same range of cache indices. Simulation can reveal these complex interactions and aid the selection of low-cost, high-performance alternatives. In general, the problem is reduced if the degree of associativity of the cache is at least as great as the number of incoming and outgoing streams.

5.8 Long Instruction Words

Instruction-set performance relates to the number of useful operations than can be executed per unit of time or per clock. High performance does not guarantee good flexibility, however. Instruction-set flexibility relates to the wider diversity of different applications whose computations can be efficiently encoded in the instruction stream. A longer instruction word generally allows a greater number and diversity of operations and operand specifiers to be encoded in each word.

RISC architectures generally encode one primitive operation per instruction. Long-instruction-word architectures encode a number of independent subinstructions per instruction, with operation and operand specifiers for each subinstruction. The subinstructions may be primitive generic operations similar to RISC instructions or they may each be more sophisticated, application-specific operations such as those described previously in this chapter as processor extensions. Making the instruction word longer, for any given number of operands and operations makes instruction encoding simpler and more orthogonal.

Long-instruction-word processors are not always faster than RISC processors. Sometimes, the benefit of RISC execution-unit simplicity boosts maximum clock frequency, and the execution of several distinct RISC instructions per cycle can compensate for the relative austerity of RISC instruction sets. Nevertheless, when RISC instruction sets are found in the most demanding, data-intensive tasks, they are implemented with super-scalar implementations that attempt to execute multiple instructions per cycle, mimicking the greater intrinsic operational parallelism of long instruction words.

Fig 5-32 shows an example of a basic long-instruction operation encoding example. The figure lays out a 64-bit instruction word with three independent subinstruction slots, each of which specifies an operation and operands. The first subinstruction (subinstruction 0) has an opcode and four operand specifiers—two source registers, an immediate field, and one destination register. The second and third subinstructions (subinstructions 1 and 2) have an opcode and three operand specifiers—two source registers and one source/destination register. The 2-bit format field on the left designates this particular grouping of subinstructions. It may also designate the overall length of the instruction if the processor supports variable-length encoding.

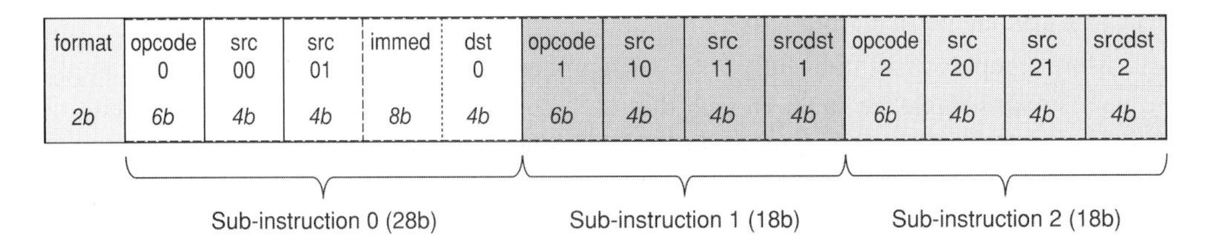

Figure 5-32 Example of long-instruction word encoding.

Clearly, there is a hardware cost associated with long instruction words. Instruction memory is wider, decode logic is bigger, and a larger number of execution units and register files (or register file ports) must be implemented to deliver instruction parallelism. Larger numbers of bigger logic blocks are incrementally harder to optimize, so maximum clock frequency can drop compared to simpler, narrower instruction encodings such as RISC. Nevertheless, the performance and flexibility benefits can be substantial, particularly for data-intensive applications with high inherent parallelism.

In some long-instruction-word architectures, each subinstruction has almost completely independent resources: dedicated execution units, dedicated register files, and dedicated data memories. In other architectures, the subinstructions share common register files and data memories and require a number of ports into common storage structures to allow effective and efficient data sharing.

Long-instruction-word architectures also vary widely on the question, How "long" is a long instruction? For high-end computer-system processors such as Intel's Itanium family and for high-end embedded processors such as Texas Instruments' TMS320C6400 DSP family, the instruction word is very "long" indeed—hundreds of bits. For more cost- and power-sensitive embedded applications, "long" may be just 64 bits. The essential processor architecture principles are largely the same, however, once multiple, independent subinstructions are packed into each instruction word.

5.8.1 Code Size and Long Instructions

One common liability of long-instruction-word architectures is large code size compared to architectures that encode one independent operation per instruction. This is a common problem for VLIW architectures, but it is an especially important one for SOC designs where instruction memories may consume a significant fraction of total silicon area. Compared to code compiled for code-efficient architectures, VLIW code can often require two to five times more code storage. Fig 5-32 compares the total code size of a VLIW DSP (TI TMS320C6203) with Tensilica's Xtensa processor for the EEMBC Telecom (discussed in Chapter 3) suite, with both straight compilation from unmodified C and with optimized C code. No assembly code was used.

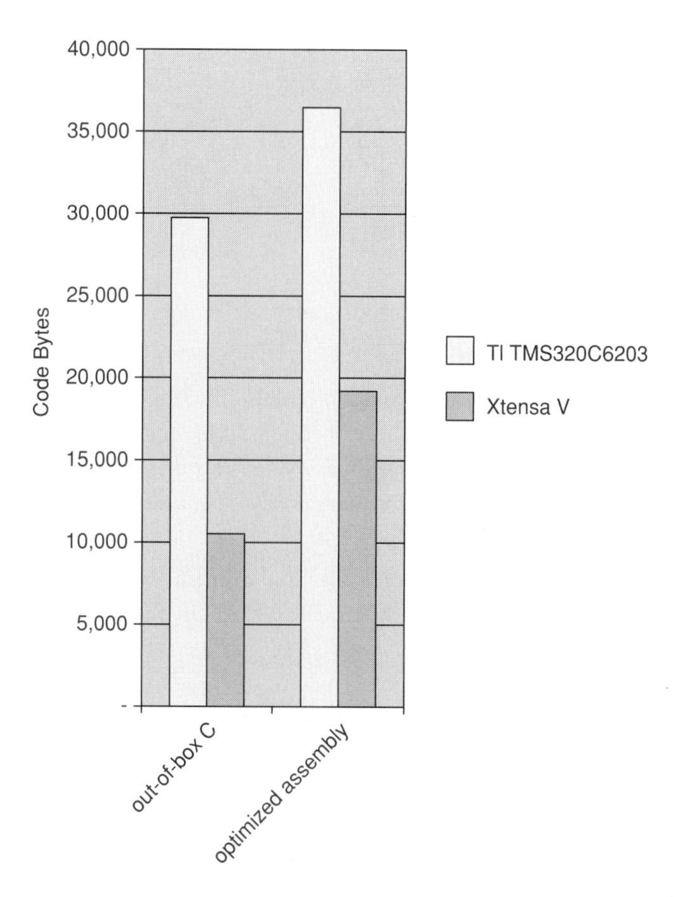

Figure 5-33 EEMBC telecom code size comparison.

Similarly, Fig 5-33 compares the total code size of a VLIW media processor (Philips Trimedia TM1300) with Tensilica's Xtensa processor for the EEMBC Consumer suite, both with straight compilation from unmodified C and with full optimization of the C. No handwritten assembly code was created for the optimized Tensilica processor.

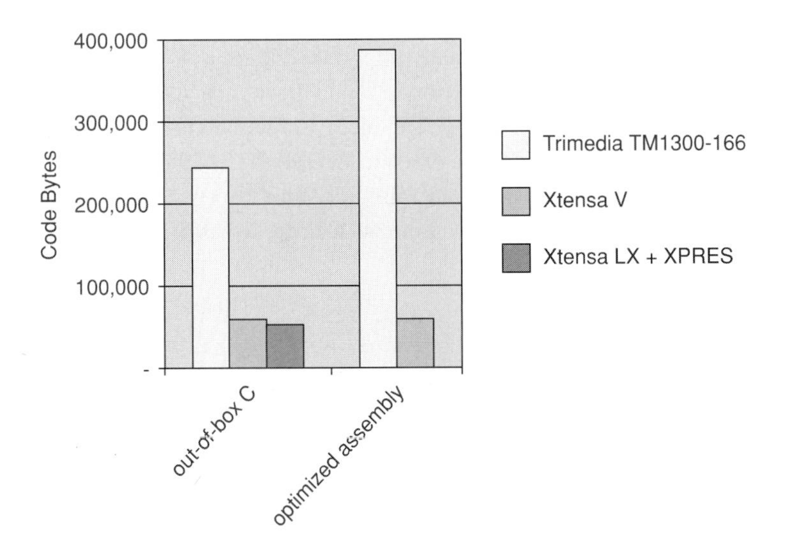

Figure 5-34 EEMBC consumer code size comparison.

Code bloat stems, in part, from instruction-length inflexibility. If, for example, the compiler can find only one operation whose source operands and execution units are ready, it may be forced to encode several subinstruction fields as NOPs (no operation). Instruction storage is already a major portion of embedded SOC silicon area, so code expansion translates into higher cost, poorer instruction-cache performance, or both.

A second source of VLIW code bloat is the loose encoding of frequent operations commonly found in VLIW processors. The TI TMS320C6203 DSP, for example, requires 32 bits of instruction to specify a 16-bit multiplication and 32 bits to specify a 16-bit add, so the common multiply/accumulate (MAC) combination takes at least 64 bits. If a loop containing many MACs is unrolled four times (to amortize the cost of branch and address calculations), the resulting eight MAC operations require 512 bits of instruction storage, not counting the additional bits for any loads, stores, branches, or address-calculation instructions.

However, long instructions do not necessarily lead to VLIW code bloat. A long-instruction-word implementation of Tensilica's Vectra LX DSP architecture needs about 20 bits within the instruction stream to specify eight 16-bit MACs executing in SIMD fashion, not counting the additional bits for any loads, stores, branches, or address-calculation instructions.

One attractive solution for long-instruction-word code bloat is to use a more flexible range of instruction lengths. If the processor allows multiple instruction lengths, including short instructions that encode a single operation, the compiler can achieve significantly better code size and instruction storage efficiency compared to traditional VLIW processor designs with fixed-length instruction words. Reducing code size for long-instruction-word processors also tends to decrease bus-bandwidth requirements and reduces the power dissipation associated with instruction fetches. Tensilica's Xtensa LX processor, for example, incorporates flexible-length instruction extensions (FLIX). This architectural approach addresses the code size challenge by offering 16-bit, 24-bit, and a choice of either 32- or 64-bit instruction lengths. Designer-defined instructions can use the 24-, 32-, and 64-bit instruction formats.

Long instructions allow more encoding freedom, where a large number of subinstruction or operation slots can be defined (although three to six independent slots are typical) depending on the operational richness required in each slot. The operation slots need not be equally sized. Big slots (20–30 bits) accommodate a wide variety of opcodes, relatively deep register files (16–32 entries), and three or four register-operand specifiers. Developers should consider creating processors with big operation slots for applications with modest degrees of concurrency but a strong need for flexibility and generality within the application domain.

Small slots (8–16 bits) lend themselves to direct specification of movement among small register sets and allow a large number of independent slots to be packed into a long instruction word. Each of the larger number of slots offers a more limited range of operations, fewer specifiers (or more implied operands), and shallower register files. Developers should consider creating processors with many small slots for applications with a high degree concurrency among many specialized function units. These architectures often look like microcoded engines, as described in Chapter 6.

5.8.2 Long Instruction Words and Automatic Processor Generation

Long-instruction-word architectures fit very well with automatic generation of processor hardware and software. High-level instruction descriptions can specify the set of subinstructions that fit into each slot. From these descriptions, the processor generator determines the encoding requirements for each field in each slot, assigns opcodes, and creates instruction-decoding hardware for all necessary instruction formats. The processor generator can also create the corresponding compiler and assembler for the long-word processor. For long-instruction-word architectures, packing subinstructions into long instructions is a very complex task. The assembler can handle this packing, so assembly source-code programs written by programmers need only specify the operations or subinstructions, giving less attention to packing constraints. The compiler generates code with instruction-slot availability in mind to maximize performance and minimize code size, so it generally does its own packing of operations into long instructions.

Fig 5-35 shows a short but complete example of a very simple long-instruction-word processor described in TIE with FLIX technology. It relies entirely on built-in definitions of 32-bit integer operations and defines no new operations. It creates a processor with a high degree of potential concurrency even for applications written purely in terms of standard C integer operations and data types. The first of three slots supports all the commonly used integer operations, including ALU operations, loads, stores, jumps, and branches. The second slot offers loads and stores, plus the most common ALU operations. The third slot offers a full complement of ALU operations, but no loads and stores.

```
1: format format1 64 {base_slot, ldst_slot, alu_slot}
2: slot_opcodes base_slot {ADD.N, ADDX2, ADDX4, SUB, SUBX2, SUBX4, ADDI.N,
3:    AND, OR, XOR, BEQZ.N, BNEZ.N, BGEZ, BEQI, BNEI, BGEI, BNEI, BLTI, BEQ,
4:    BNE, BGE, BLT, BGEU, BLTU, L32I.N, L32R, L16UI, L16SI, L8UI, S32I.N,
5:   S16I,S8I, SLLI, SRLI, SRAI, J, JX, MOVI.N }
6: slot_opcodes ldst_slot { ADD.N, SUB, ADDI.N, L32I.N, L32R, L16UI, L16SI,
7:   L8UI, S32I.N, S16I, S8I, MOVI.N }
8: slot_opcodes alu_slot {ADD.N, ADDX2, ADDX4, SUB, SUBX2, SUBX4, ADDI.N,
9:   AND, OR, XOR, SLLI, SRLI, SRAI, MOVI.N }
```

Figure 5-35 Simple 32-bit multislot architecture description.

The first line declares a 64-bit format, `format1`, containing three slots, `base_slot`, `ldst_slot`, and `alu_slot`. Lines 2 to 5 list all the TIE instructions that can be packed into the first of those slots: `base_slot`. In this case, all the instructions happen to be predefined Xtensa LX instructions, but new instruction could also be included in this slot. The processor generator also creates a NOP for each slot, so the software tools can always create complete instruction, even when no other operations for that slot are available for packing into a long instruction. Lines 6 to 9 designate the subset of instructions that can go into the other two slots.

Fig 5-36 defines a long-instruction-word architecture with a mix of built-in 32-bit operations and new 128-bit operations. It defines one 64-bit instruction format with three subinstruction slots (`base_slot`, `ldst_slot`, and `alu_slot`). The description takes advantage of the Xtensa processor's predefined RISC instructions, but also defines a large new register file and three new ALU operations on the new register file.

The first five lines are identical to those of Fig 5-35. The sixth line declares a new register file 128-bits wide and 32 entries deep. The seventh line lists the two load and store instructions for the new wide register file, which can be found in the second slot of the long instruction word. The eighth line defines a new immediate range, an 8-bit signed value, to be used as the offset range for the new 128-bit load-and-store instructions. Lines 9-14 fully define the new load-and-store instructions in terms of basic interface signals **`Vaddr`** (the address used to access local data memory), **`MemDataIn128`** (the data being returned from local data memory), and **`MemDataOut128`** (the data to be sent to the local data memory). The use of 128-bit memory data signals also guarantees that the local data memory will be at least 128 bits wide. Line 10 lists the three new ALU operations that can be put in the third slot of the long instruction word.

Lines 15-19 fully define those operations on the 128-bit wide register file: add, bitwise-AND, and bitwise-OR.

```
1: format format1 64 {base_slot, ldst_slot, alu_slot}
2: slot_opcodes base_slot {ADD.N, ADDX2, ADDX4, SUB, SUBX2, SUBX4, ADDI.N,
3:   AND, OR, XOR, BEQZ.N, BNEZ.N, BGEZ, BEQI, BNEI, BGEI, BNEI, BLTI, BEQ,
4:   BNE, BGE, BLT, BGEU, BLTU, L32I.N, L32R, L16UI, L16SI, L8UI, S32I.N,
5:  S16I, S8I, SLLI, SRLI, SRAI, J, JX, MOVI.N }
6: regfile x 128 32 x
7: slot_opcodes ldst_slot {loadx, storex} /* slot does 128b load/store*/
8:  immediate_range sim8 -128 127 1 /*8 bit signed offset field */
9: operation loadx {in x *a, in sim8 off, out x d} {out VAddr, in
10: MemDataIn128}{
11: assign VAddr = a + off; assign d = MemDataIn128;}
12: operation storex {in x *a, in sim8 off, in x s} {out VAddr,out
13:  MemDataOut128}{
14: assign VAddr = a + off; assign MemDataOut128 = s;}
15: slot_opcodes alu_slot {addx, andx, orx} /* two new ALU operations on x
16:  regs */
17: operation addx {in x a, in x b, out x c} {} {assign c = a + b;}
18: operation andx {in x a, in x b, out x c} {} { assign c = a & b;}
19: operation orx {in x a, in x b, out x c} {} { assign c = a | b;}
```

Figure 5-36 Mixed 32-bit/128-bit multislot architecture description.

With this example, any combination of the 39 instructions (including NOP) in the first slot, three instructions in the second slot (loadx, storex, and NOP), and four instruction in the third slot can be combined to form legal instructions—a total of 468 combinations. This simplified example specifies almost enough instructions to densely populate a long instruction word. The first slot needs about 21 bits, the second slot only needs about 19 bits, the third slot needs about 17 bits, and the format/length field required four bits—for a total of roughly 62 bits. This example shows the potential to independently specify operations to enable instruction-level parallelism. Moreover, all of the techniques for improving the performance of individual instructions—especially fusion and SIMD—are readily applied to the operations encoded in each subinstruction.

```
1: format pair 32 {shift, logic}
2: regfile X 128 4 x
3: slot_opcodes shift {xr_srl, xr_sll }
4: operation xr_sll {in AR a,inout AR b} {} {assign b=b<<{a[3:0],3'h0};}
5: operation xr_srl {in AR a,inout AR b} {} {assign b=b>>{a[3:0],3'h0};}
6: slot_opcodes logic { xr_or, xr_and }
7: operation xr_and {in X c,inout X d} {} {assign d=d & c;}
8: operation xr_or {in X c,inout X d} {} {assign d=d | c;}
```

Figure 5-37 Compound operation TIE example revisited.

The compound operation technique, as described in Section 5.3.5, can be applied within subinstructions, but long instruction words also encourage the encoding of independent operations in different slots. Fig 5-37, for example, reimplements the simple compound instruction example of Section 5.3.5 using long-instruction-word TIE with operation slots.

The first line defines a 32-bit wide instruction, a new format, and the two slots within that format. As in Fig 5-14, the next line declares a new wide register file. Lines 3 to 5 define the instructions (byte shifts) that can occupy the first slot. Lines 6 to 8 define the instructions (bitwise-AND and bitwise-OR). Altogether, this TIE example defines four instructions, representing the four combinations. If these were the only instructions, the processor generator would discover that this format requires only 16 bits to encode: 10 bits for the "shift" slot (two four-bit specifiers for the two AR register entries, plus one bit to differentiate shift left from shift right) and 6 bits for the "logic" slot (two 2-bit specifiers for the two X register entries, plus one bit to differentiate AND from OR).

This example also underscores an important side benefit of independent specification of the operations for compound instructions—easier software use. In the first version of compound instruction use, in Fig 5-14, the compiler may have a difficult time automatically recognizing when the combined byte-shift/logical operation can apply. The programmer must make the use of this combination explicitly through a C-intrinsic function or through direct assembly coding. When the two operations are described independently, as in Fig 5-37, the underlying primitive shift or logical operation can be more directly expressed by the programmer or simply inferred by the compiler from unmodified C code. Section 5.9 explores compiler inference more fully. This approach allows the compiler and assembler to create and exploit concurrency more easily.

5.9 Fully Automatic Instruction-Set Extension

The basic algorithm analysis and instruction-design process described in Chapter 4, Section 4.2.2 is systematic but complex. Compiler-like tools that design application-specific instruction sets can automate much of this process. In fact, the more advanced compiler algorithms for code selection, software pipelining, register allocation, and long-instruction-word operation scheduling can also be applied to discovery and implementation of new instruction definitions and code generation using those instructions.

Automatic processor generation builds on the basic flow for processor generation shown in Chapter 1, Fig 1.1, but adds the creation of automatic processor architecture, as shown in Fig 5-37.

However, automatic processor generation offers benefits beyond simple discovery of improved architectures. Compared to human-designed instruction-extension methods, automation tools eliminate any need to manually incorporate new data types and intrinsic functions into the application source code and provide effective acceleration of applications that may be too large or complex for a human programmer to assess. As a result, this technology holds tremendous promise for transforming the development of processor architectures for embedded applications.

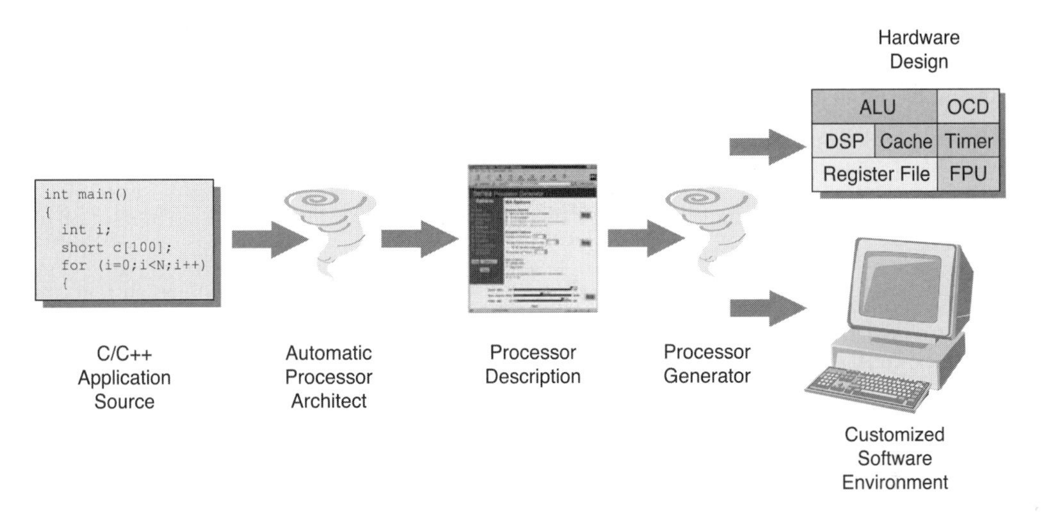

Figure 5-38 Automatic processor generation concept.

The essential goals for automatic processor generation are:

1. A software developer of average experience should be able to easily use the tool and achieve consistently good results.
2. No source-code modification should be required to take advantage of generated instruction sets. (Note that some ways of expressing algorithms are better than others in exposing the latent parallelism, especially for SIMD optimization, so source code tuning can help. The automatic processor generator should highlight opportunities to the developer to improve the source code.)
3. The generated instruction sets should be sufficiently general-purpose and robust so that small changes to the application code do not degrade application performance.
4. The architecture design-automation environment should provide guidance so that advanced developers can further enhance automatically generated instruction-set extensions to achieve better performance.
5. The development tool must be sufficiently fast so that a large range of potential instruction-set extensions can be assessed—on the order of thousands of variant architectures per minute.

The requirement for generality and reprogrammability mandates two related use models for the system:

- Initial SOC development: C/C++ in, instruction-set description out.
- Software development for an existing SOC: C/C++ and generated instruction-set description in, binary code out.

Tensilica's XPRES (Xtensa Processor Extension System) compiler implements automated processor instruction-set generation. A more detailed explanation of the XPRES flow will help explain the use and capability of this further level of processor automation. Fig 5-39 shows the four steps implemented by XPRES compiler. All of these steps are machine-automated except for optional manual steps, as noted.

The generation of a tailored C/C++ compiler adds significantly to the usefulness of the automatically generated processor. Even when the source application evolves, the generated compiler looks aggressively for opportunities to use the extended instruction set. In fact, this method can even be effective for generating fairly general-purpose architectures. As long as the basic set of operations is appropriate to another application, even if that application is unrelated to the first, the generated compiler will often use the extended architecture effectively.

The automatic processor generator internally enumerates the estimated hardware cost and application-performance benefit of each of thousands of configurations, effectively building a *Pareto* curve, such as that shown in Fig 5-40. Each point on the curve represents the best performance level achieved at each level of added gate count. This image is a screen capture from Tensilica's Xplorer development environment, for XPRES results on a simple video motion-estimation routine (sum-of-absolute-differences).

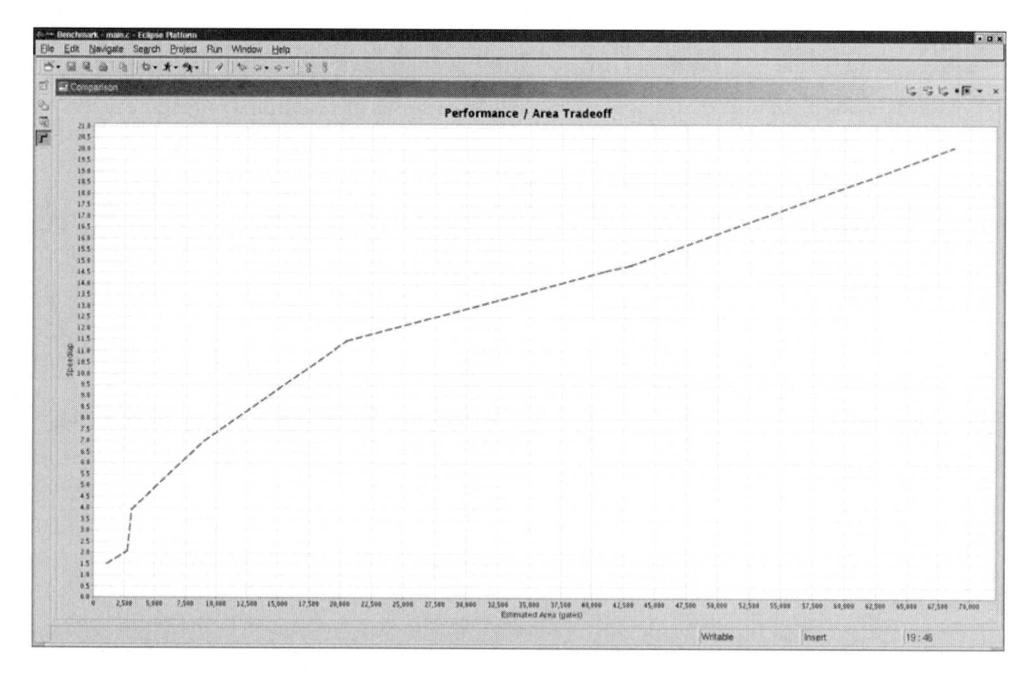

Figure 5-40 Automatic generation of architectures for sum-of-absolute-differences.

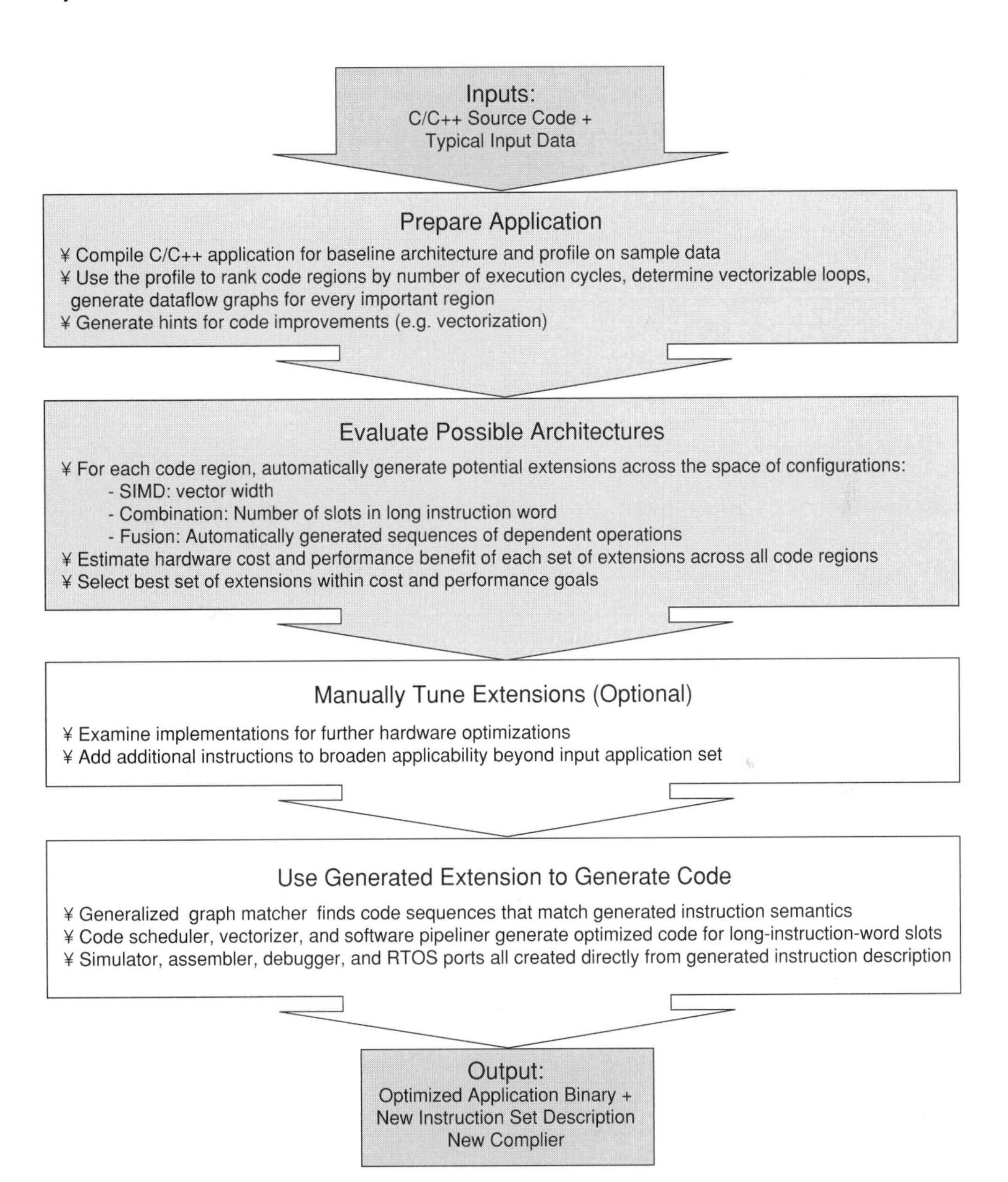

Figure 5-39 XPRES automatic processor generation flow.

Automatic generation of instruction-set extensions applies to a very wide range of potential problems. It yields the most dramatic benefits for data-intensive tasks where much of the processor-execution time is spent in a few hot spots and where SIMD, wide-instruction, and operation-fusion techniques can sharply reduce the number of instructions per loop iteration. Media- and signal-processing tasks often fall squarely in the sweet spot of automatic architecture generation. The automatic generator also handles applications where the developer has already identified key applications-specific functions, implemented those functions in TIE, and used those functions in the C source code. Fig 5-41 shows the results of automatic processor generation for three applications using the XPRES compiler, including one fairly large application: an MPEG4 video encoder.

Application	Speed-up	Baseline Code Size	Code Size with Acceleration	MIPS32 Code Size (gcc –O2)	Configurations Evaluated	Generator Run Time (minutes)
MPEG4 Encoder	3.0x	111KB	136KB	356KB	**1,830,796**	30
Radix-4 FFT	10.6x	1.5KB	3.6KB	4.4KB	**175,796**	3
GSM Encoder	3.9x	17KB	20KB	38KB	**576,722**	15
GSM Encoder (FFT ISA)	1.8x	17KB	19KB	38KB	-	-

Figure 5-41 Automatic architecture generation results for DSP and media applications.

The figure includes code-size results for the baseline Xtensa processor architecture and the automatically optimized Xtensa processor architecture for each application. Using aggressively optimized instruction sets generally increases code size slightly, but in all cases, the optimized code remains significantly smaller than that for conventional 32-bit RISC architectures. The figure also shows the number of configurations evaluated, which increases with the size of the application. The automatic processor generator run time also increases along with the size of the application, but averages about 50,000 evaluated configurations per minute on a 2GHz PC running Linux.

The figure also shows one example of generated-architecture generality. The GSM Encoder source code was compiled and run, not using an architecture optimized for the GSM Encoder, but for the architecture optimized for the FFT. While both are DSP-style applications, they have no source code in common. Nevertheless, the compiler automatically generated for the FFT-optimized processor could recognize ways to use the processor's FFT-optimized instruction set to accelerate the GSM Encoder by 80% when compared to the performance of code compiled for the baseline Xtensa processor instruction set.

Completely automatic instruction-set extension carries two important caveats:

1. Programmers may know certain facts about the behavior of their application, which are not made explicit in the C or C++ code. For example, the programmer may know that a variable can take on only a certain range of values or that two indirectly referenced data

structures can never overlap. The absence of that information from the source code may inhibit automatic optimizations in the machine code and instruction extensions. Guidelines for using the automatic instruction-set generator should give useful hints on how to better incorporate that application-specific information into the source code. The human creator of instruction extensions may know this information and be able to exploit this additional information to create instruction sets and corresponding code modifications.

2. Expert architects and programmers can sometimes develop dramatically different and novel alternative algorithms for a task. A different inner-loop algorithm may lend itself much better to accelerated instructions than the original algorithm captured in the C or C++ source code. Very probably, there will always be a class of problems where the expert human will outperform the automatic generator, though the human will take longer (sometimes much longer) to develop an optimized architecture.

The implications of automatic instruction-set generation are wide-ranging. First, this technology opens up the creation of application-specific processors to a broad range of designers. It is not even necessary to have a basic understanding of instruction-set architectures. The basic skill to run a compiler is sufficient to take advantage of the mechanisms of automatic instruction-set extension.

Second, automatic instruction-set generation deals effectively with complex problems where the application's performance bottleneck is spread across many loops or sections of code. An automated, compiler-based method is easily able to track the potential for sharing instructions among loops, the relative importance of different code sections based on dynamic execution profiles, and the cumulative hardware-cost estimate. Global optimization is more difficult for the human designer to track.

Third, automatic instruction-set generation ensures that newly created instructions can be used by the application without source-code modification. The compiler-based tool knows exactly what combination of primitive C operations corresponds to each new instruction, so it is able to instantiate that new instruction wherever it benefits performance or code density. Moreover, once the instruction set is frozen and the SOC is built, the compiler retains knowledge of the correspondence between the C source code and the instructions. The compiler can utilize the same extended instructions even as the C source is changed.

Fourth, the automatic generator may make better instruction-set-extension decisions than human architects. The generator is not affected by the architect's prejudice against creating new instructions (design inertia) or influenced by architectural folklore on rumored benefits of certain instructions. It has complete and quite accurate estimates of gate count and execution cycles and can perform comprehensive and systematic cost/benefit analysis. This combination of benefits therefore fulfills both of the central promises of application-specific processors: cheaper and more rapid development of optimized chips and easier reprogramming of that chip once it's built to accommodate the evolving system requirements.

5.10 Further Reading

Anyone interested in modern processor architecture would be well-served by the most important textbook and standard reference on computer architecture: John L. Hennessy, David A. Patterson, and David Goldberg, *Computer Architecture: A Quantitative Approach*, (3rd edition) Morgan Kaufmann; 2002.

A number of books focus on the architecture of specific processors. An overview can be found in Patrick Staken, *A Practitioner's Guide to RISC Microprocessor Architecture*, Wiley-Interscience, 1996. Useful information on ARM, MIPS and PowerPC processors can be found in the following:

- David Seal, *ARM Architecture Reference Manual*, Pearson Educational, 200.
- Dominic Sweetman, *See MIPS Run*, Morgan Kaufmann, 1999.
- Tom Shanley, *PowerPC System Architecture*, Addison-Wesley Pub Co, 1996

There is a wide variety of books on embedded-system programming, but Michael Bahr's book focuses on programming embedded systems with high-level languages, a critical productivity breakthrough necessary for modern embedded system development: Michael Bahr, *Programming Embedded Systems in C and C ++*, O'Reilly & Associates, 1999.

Compiler design is a specialized domain, but increasingly important to the architecture of processors and processor-based systems. Often, the key question is not "What is the right instruction set for the programmer?" but "What is the right instruction set for the compiler?" See Steven Muchnick, *Advanced Compiler Design and Implementation,* Morgan Kaufmann, 1997.

More details on the Secure Socket Layer and the popular open source implementation, Open SSL, can be found at *http:/www.openssl.org*.

CHAPTER 6

Configurable Processors: A Hardware View

SOCs solve complex, data-intensive application problems—problems of performance, power-efficiency, and tight integration of diverse functions. Tuned processors provide a potent combination of performance and programmability for solving these problems, but the process of finding the right solution to these problems often looks different to software developers and hardware developers. Chapter 5 focused on using an existing algorithm written in a high-level software language to develop an optimized processor to run that and similar applications. This is the natural perspective of a software developer.

This chapter, by contrast, takes the perspective of the hardware developer. The developer may be asked to implement a complete application using hardware methods, or perhaps to use hardware to boost the performance or efficiency of key portions of the application because the application software running on a general-purpose processor is too slow. In either case, significant new hardware must be developed, and the challenges of rapid design, thorough verification, and fast response to changing requirements should be foremost in the hardware developer's mind. As a rule, no software implementation exists or that the software reference code is so inefficient that simply running it on a processor has already been dismissed as impractical.

Problems for which a hardware-centric approach represents a suitable solution are easy to recognize. The designer may see that the application needs more data operations per cycle than a general-purpose processor can efficiently provide. The designer may note that the data-flow bandwidth through the function exceeds the interface capacity of available processors. Even very early in the development process, the developer may have a hardware block diagram in mind or may have already developed an RTL-based hardware design for a related task. The developer may know that substantial hardware parallelism not only is possible within a task, but is also essential to meeting the performance requirements of the block's specifications.

Just as a hardware-based approach has significant benefits relative to software running on traditional processors, there are also significant liabilities. Development complexity, the risk of hardware bugs, and the shifting sands of product requirements all make a purely hardwired design approach dangerous for complex tasks. Application-specific processors can achieve high parallelism and offer complete programmability, extending the power of hardware-intensive designs into applications with complex controls. To exploit the unique capabilities of application-specific processors, though, the developer must be able to translate between traditional hardware primitives—both data-path and control structures—and corresponding processor primitives. This chapter compares these primitives and outlines a number of methods for migrating hardware designs and design concepts into processor-based designs.

6.1 Application Acceleration: A Common Problem

One standard method used to address a complex, performance-intensive application is to partition the application work between software running on a generic processor and a hardware block—an *application accelerator*—that performs key data-intensive computations under the direction of software running on the processor. A simple block diagram of this partitioned approach is shown in Fig 6-1. The application code running on the processor may send data and commands directly to accelerator registers, or it may write data and commands to a shared buffer memory or queue so the accelerator can fetch them and perform the requested operations. Similarly, the accelerator may hold the results in its own registers, waiting for the application code running on the processor to retrieve them, or the accelerator may store the results in a buffer memory pending retrieval by the application code.

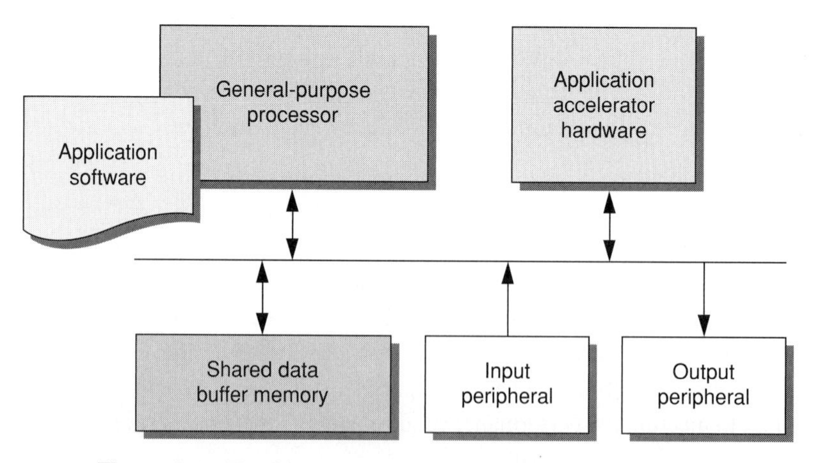

Figure 6-1 Traditional processor + accelerator partitioning.

The fundamental limitations of this design approach are fourfold:

1. The methods for designing and verifying large hardware blocks are labor-intensive and slow.
2. The requirements for the portion of the application running on the accelerator may change either late in the design or even after the SOC is built, mandating new silicon design.
3. The adaptation of the application software and verification of the combined hardware/software solution may be awkward and/or difficult.
4. The requirement to move data back and forth among the processor, accelerator, and memory may slow total application throughput, offsetting much or all of the benefit derived from hardware acceleration.

Ironically, the promise of concurrency between the processor and the accelerator is also often unrealized because the application, by nature of the way it is written, may force the processor to sit idle while the accelerator performs necessary work. In addition, the accelerator will be idle during application phases that cannot exploit it.

Application-specific processors offer two alternatives to accelerator-based design. Fig 6-2 shows the incorporation of the accelerator function into the processor through configuration and extension of the processor to match the performance needs of the accelerator. The higher level application software and the lower level application acceleration functions (coded as firmware for the extended processor) both run on the one processor. This approach eliminates the processor—accelerator communication overhead and often reduces the total silicon cost. It makes the accelerator functions far more programmable and significantly simplifies integration and testing of the total application.

It also allows the acceleration hardware to have intimate access to all of the processor's resources. Many acceleration operations can make good use of a general-purpose processor's resources. In addition, merging of the acceleration extensions with the processor repurposes the processor's gates as the acceleration controller so that a separate acceleration controller is not needed.

This design approach is particularly attractive when then application's software is otherwise idle during accelerator operations or when the data or command communication between the application software and accelerator hardware presents a significant performance bottleneck. The data need not move between processor and accelerator when they are one and the same block.

Fig 6-3 shows an alternative approach—conversion of the accelerator to a separate processor configured for application acceleration. This second processor runs in parallel with the general-purpose processor. It receives commands and communicates status and data either through registers or through shared data memory (which may be located within either processor or external to both processors). This design approach is particularly attractive when there is substantial potential parallelism between the portion of the application running on the general-

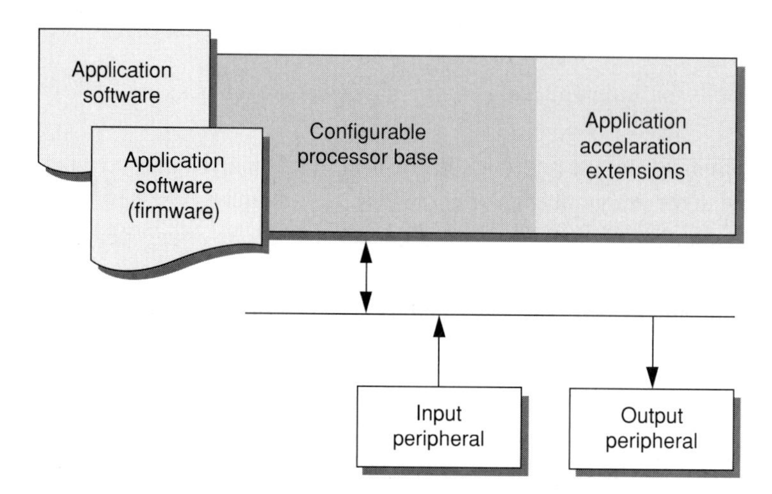

Figure 6-2 Incorporation of an accelerator into a processor.

purpose processor and the accelerated portion of the application running on the application-specific processor.

While data and commands must still flow between the two processors, application-specific processors can offer optimized data interfaces that may eliminate any bottlenecks between two appropriately configured processors. Optimizations include fast, shared memory-mapped RAM buffers; memory-mapped queues; and direct connection via processor ports mapped into the instruction sets of both the general-purpose and application-acceleration processors. The bandwidth achieved through these methods may be more than 10 times that of conventional processor—bus transfers, and both the sending and receiving tasks may incur essentially zero overhead for each transfer. Optimized interface options and issues are discussed in greater depth in Section 6.5.

This chapter focuses on the key design decisions and basic techniques for implementing hardware-based functions, such as accelerators, using application-specific processors. Moving from an RTL-based design or a hardware-style block diagram to an application-specific processor configuration plus software is a straightforward design transition. This chapter discusses the three major dimensions of that design transition:

1. **Data flow:** The core data-path hardware becomes a set of instructions.
2. **Control-flow:** The set of finite-state machines or microprograms becomes a high-level-language program.
3. **Interface:** The signals that comprise the hardware interface to surrounding blocks become memory-mapped and instruction-mapped buses and I/O ports of the processor.

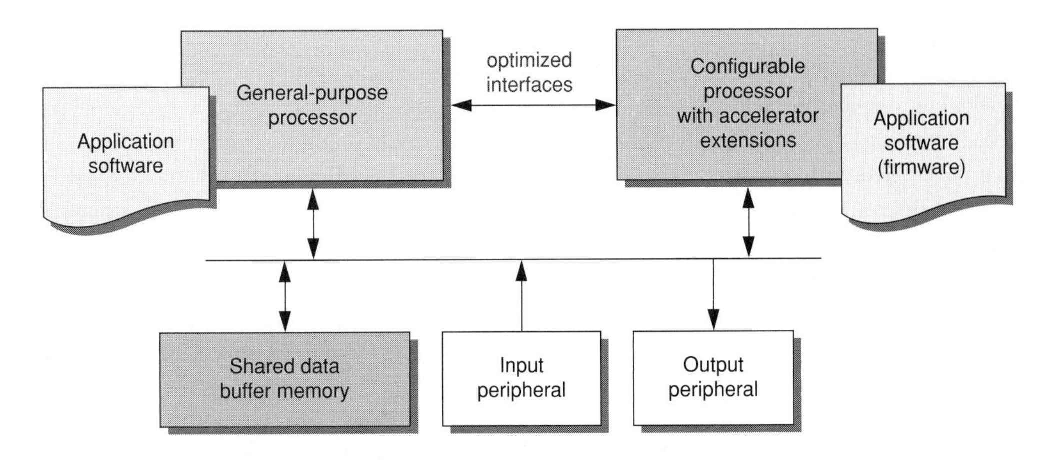

Figure 6-3 Implementation of an accelerator as an application-specific processor.

Throughout the chapter, a key consideration recurs: How should the designer make choices to achieve a good combination of performance and reprogrammability? A number of basic techniques achieve both high degrees of useful parallelism within the processor and easy reuse even after silicon is built. These techniques improve the versatility and value of the design over time. These methods also help the designer to take narrowly applicable hardware features and generalize them into instruction sets and interfaces with a wider range of usefulness across many possible future application needs.

The last section of this chapter presents methods for verifying the correctness of an application-specific processor's instruction set, control program, hardware signal interface, and VLSI implementation. These methods will also aid implementation engineers responsible for verifying processors used for software-intensive tasks, as described in Chapter 4.

6.2 Introduction to Pipelines and Processors

Pipelining plays a central role in efficient hardware design. It can significantly increase a function unit's clock rate with only modest impact on silicon area, power dissipation, and latency. Whenever data throughput and area efficiency are priorities, designers should consider the use of pipelining. Unfortunately, standard logic-design methods do not always make pipelined implementations easy to develop or debug, so the promises of aggressive pipelining sometimes go unfulfilled in real-word designs.

Processors, by contrast, are routinely pipelined to achieve high throughput. The incorporation of new function units into a pipelined processor takes full advantage of the long history of pipelined processor design and makes pipelining both easier to understand and easier to exploit. This section introduces processor pipelining and the basic issues of incorporating new application-specific function units into a processor pipeline. Some of the subtler and more advanced ramifications of pipelines are deferred to Chapter 7.

6.2.1 Pipelining Fundamentals

Pipelining is a fundamental design technique for boosting performance and efficiency in any logic circuit, but data dependencies between operations in the pipeline introduce potential delays and complexities that can prevent pipelined operations from attaining their maximum potential in terms of performance and efficiency. Pipelined processors in particular must deal with latency in control flow, data-memory references, and execution units. A quick picture of the workings of configurable-processor execution pipelines will help hardware designers exploit the useful correspondence between hardware design and processor configuration.

Automated processor design is feasible, in part, because processors maintain such a simple, robust abstraction of instruction-sequence execution. In particular, most processor architectures enforce two properties of instruction execution:

Instruction execution is atomic: If an instruction executes at all, it executes completely—either all of the instruction's destination storage locations (registers and memory) change, or none change.

Instruction execution is sequential: Instructions are executed in their entirety in the order indicated by the sequence of program-counter values. All of the state changes for one instruction occur before any of the state changes for the next. In practice, clever processor implementers may find ways to substantially overlap the execution of operations within the processor, so that the actual state changes (particularly memory loads and stores) may not be strictly sequential. However, the processor's microarchitectural implementation tracks all of these state-change reorderings and ensures that any changes in state *visible to the programmer* appear to occur in strict sequence.

Implementation techniques such as pipelining involve overlapping the execution of a number of instructions, starting one instruction long before the previous instruction has finished. This overlap might appear to challenge the principles of atomicity and sequentiality, but there are well-known implementation techniques that enforce these principles even when the processor hardware has high concurrency and multiple deep pipelines that write to many registers at different pipeline stages. Some of these advanced implementation techniques, such as operand bypassing, source-destination register interlocks, and write cancellation for exceptions, are presented in Chapter 7.

6.2.2 RISC Pipeline Basics

Fig 6-4 shows a simple picture of a five-stage RISC (Reduced Instruction Set Computing) pipeline. Normally, one instruction starts every cycle. The current instruction state passes from pipe stage to pipe stage, with specific steps in the instruction's execution taking place in each stage. Normally, one instruction completes every cycle.

Spreading instruction execution across several pipe stages typically improves overall clock rate by reducing the amount of logic, hence delay, in each stage. Pipelining improves performance with a modest increase in cost and power compared to performing all the work of the instruction during one clock cycle.

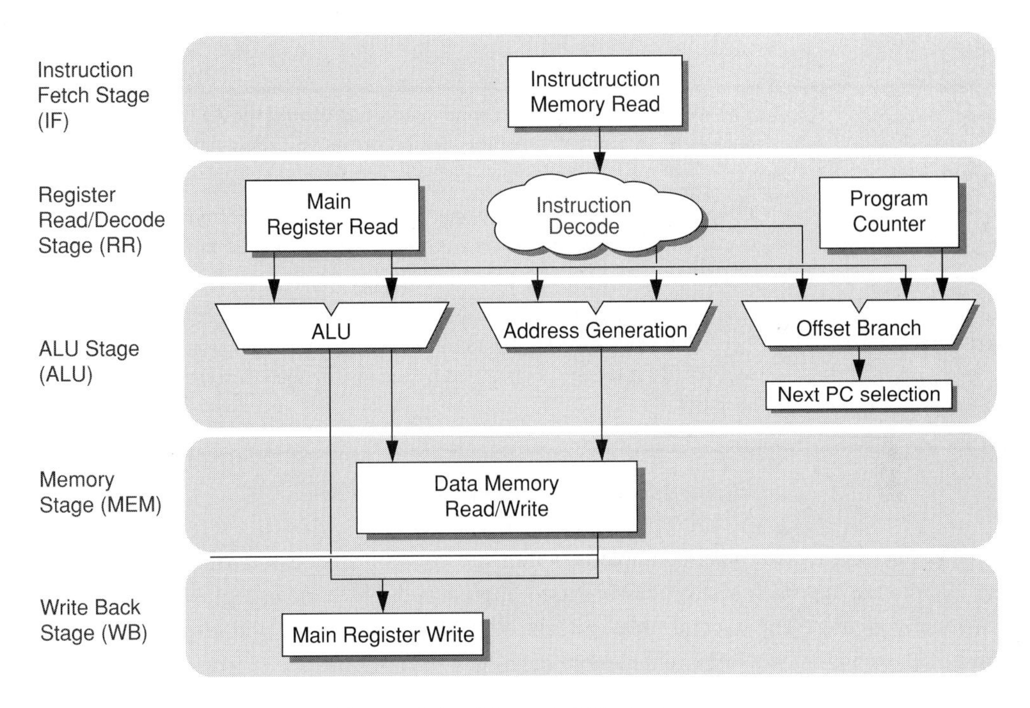

Figure 6-4 A basic RISC pipeline.

This form of pipeline reflects a balance of goals: good clock rate; modest hardware cost; low design complexity; easy code generation for high-level language programs; and broad applicability to a range of control-intensive, integer-based applications. A basic understanding of processor pipelining may help the hardware developer more fully appreciate the opportunities and tradeoffs in the use of application-specific processors for traditional hardware tasks. The function of each of the five stages is shown in Fig 6-5.

Stage	Function
Instruction Fetch	Fetches an instruction from memory using the current program counter. The program counter then advances to point to the next instruction's memory location.
Register Read/Decode	The instruction's source-register specifier fields serve as indices into the register file. The opcode fields are decoded to determine the instruction type and other operation controls.

Stage	Function
ALU	The source registers and decoded instruction control the ALU calculation, data-memory address generation, or branch-target computation.
Memory	If the instruction is a memory operation, the data is read from or written to the data memory. Read data is aligned and sign-extended as specified by the instruction.
Write-Back	The result from the ALU operation or memory load is written back into the register file at the index specified by the instruction's destination-register specifier.

Figure 6-5 Pipe stages in five-stage RISC.

Details of processor pipeline implementation are not typically central to successfully mapping a hardware approach to a processor-based approached. Nevertheless, pipeline details will occasionally matter. The second-order effects of pipeline interlocks and bypassing, branch delays, and exceptions are discussed in more detail in Chapter 7.

6.2.3 Pipelines for Extended Instruction-Set Implementation

The addition of new register files and execution units to a processor's hardware is conceptually simple, but the impact on the processor pipeline is important. New instructions fit the basic model of the RISC pipeline—instructions are fetched and decoded, source operands are read from registers (and memory), execution units transform these operands into result values, which are then written back into registers (and memory). All this is familiar territory for RISC processor designers. However, the addition of new instructions modifies the structure of the processor in four important ways:

1. The extended processor can include new registers, register files, and ports to data memory. New processor state may be of any width and format appropriate to the application.
2. Instructions are no longer so simple. A single instruction can take sources from an unlimited number of source registers, register files, and memory locations; create an unlimited number of results; and then store these results in registers, register files, and memory locations. In this manner, one new instruction can perform potentially dozens of basic operations.
3. The pipeline may be extended to allow instructions to continue for many cycles to compute results, if necessary. These additional cycles allow deeply pipelined and iterative execution units (units that repeat a basic operation many times to compute a result) to fit comfortably within the pipeline.

4. New external interfaces to outside logic and other processors can be built into the pipeline, with their operations mapped into either memory-access operations or into instruction semantics. In this way, new coprocessor interfaces enable high-bandwidth communications with very low overhead.

Fig 6-6 shows the addition of a new register file and a complex execution unit to a basic five-stage RISC pipeline. In this example, a new register file with three read ports (provides three source operands) and two write ports (stores two destination values) is added. An extended execution unit uses the three new register values plus one main register source operand at the beginning of the ALU stage. The new execution unit also uses the memory-data result at the beginning of the write-back stage. The new unit produces one result, to be used for memory addressing at the end of the ALU stage, which is written back into the base register file using the write port of the existing main register file. Finally, the new execution unit produces two more values at the end of Extended Stage 1, which are stored in the extended register file.

Even if the set of extensions includes instructions with a wide variety of latencies, sources, and destinations, an automated processor generator can readily determine all possible source-destination conflicts and include processor hardware that dynamically checks for register-number conflicts on every cycle. If any source operand of a newly decoded instruction matches a destination operand of an instruction still flowing anywhere in the extended pipeline, the processor stalls the new instruction until all necessary source values are available. Automatic conflict detection simplifies programming and minimizes effective operation latency. The instruction-extension mechanism even recognizes when an instruction can iteratively use a single function unit across a number of cycles to save silicon area. This ability is especially useful for complex functions such as high-precision multiplication.

6.2.4 Guarantee of Correctness in Processor Hardware Extension

Fig 6-6 highlights another important characteristic of extended-processor hardware: minimal disruption to the processor's basic structure and logic. Adding new state registers and execution units touches few parts of the base processor design. All logic associated with instruction fetches, branch operations, loads, stores, base integer operations, and exceptions is untouched by extensions. Even as the instruction-decode unit is extended to recognize new instruction formats and operations, the decoding logic associated with baseline instructions does not change.

Automatic processor generation also satisfies a fundamental tenet of correctness: ***All code that runs on the base processor architecture runs with identical results and performance on the extended processor.*** Extended instructions affect only opcodes that were previously undefined or reserved. Note that various optional instruction packages may assume use of overlapping parts of the processor's instruction-encoding space, so only one of these optional packages can be implemented in any given processor configuration.

A well-behaved instruction-extension system must support a second crucial tenet: ***It must be structurally impossible for any instruction extension to specify anything that could disrupt***

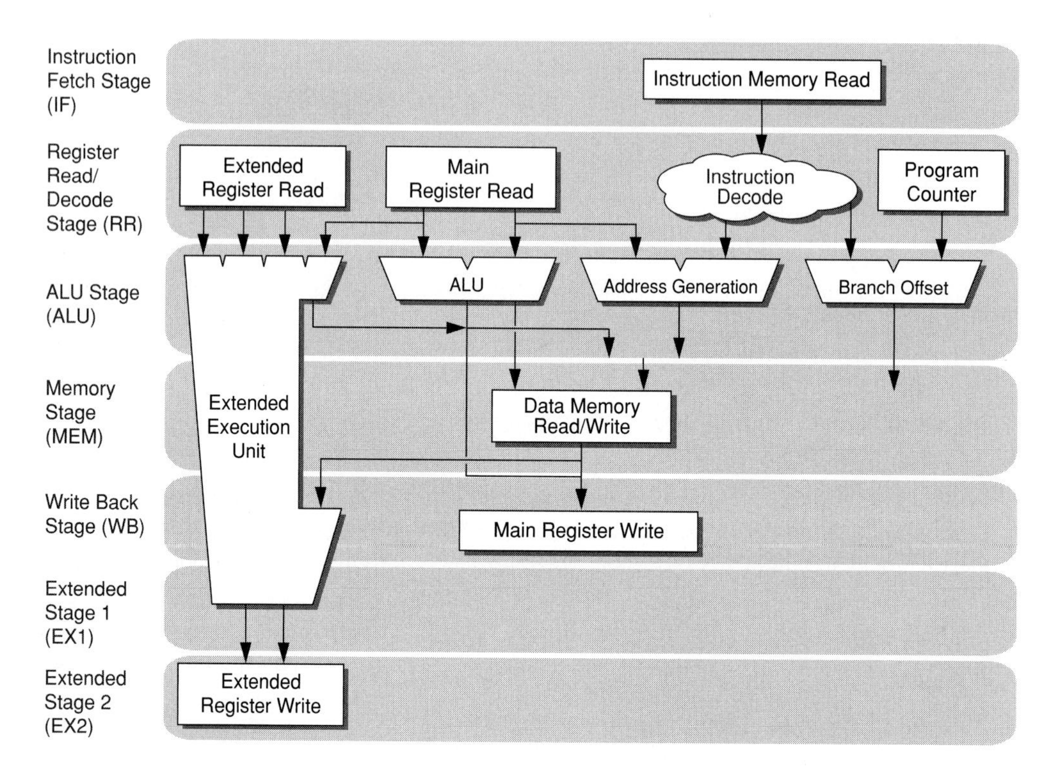

Figure 6-6 Pipeline with extended register file and complex execution unit.

the behavior of any base instruction or cause the machine to enter an unknown execution state. With these two guarantees, the designer's only worry is specifying instructions that are actually useful for the target application. Designers need not worry that new instructions might "break" a processor, because the processor generator will not allow this to happen.

6.3 Hardware Blocks to Processors

The application-specific processor is an alternative to a hardwired block for implementing a task, but what kind of hardwired logic does a processor replace? Three characteristics typify functions, traditionally implemented as RTL, which migrate naturally into processor-based implementations:

1. **High computational content:** A significant number of logical, arithmetic, and data-arrangement operations must be performed on each data element passing through the

block. The "significant number" may vary by application from one per cycle to tens of operations per cycle.

2. **Moderate to high control complexity:** The computation may be controlled by a single finite-state machine, a set of interacting finite-state machines, or some form of microcode sequencer. In any case, the number of combined states or microinstructions ranges from as few as ten to hundreds.
3. **Wide, fast data interfaces:** Tasks that require high computational rates often require a wide-bandwidth interface as well. The source and destination of the data may be external memory or interface ports. Interface data rates may be as low as a few megabytes per second or as high as gigabytes per second.

Tasks lacking these three attributes, especially the first two, may not be good candidates for processor-based implementation. Low control complexity generally indicates a simple, stable task with little risk of change. Very simple state machines implemented in RTL are generally easy to design and verify and rarely change after a chip is prototyped. Tasks with low computation content generally offer a designer less opportunity to take advantage of tuned data paths. At the other extreme, some computational pipelines are so long that they make a poor match to a processor pipeline. Some high-end graphics pipelines, for example, are tens or hundreds of cycles deep and are so regular that software programmability is less necessary.

6.3.1 The Basic Transformation of Hardware into Instructions

RTL designers commonly look at hardware-based computation as a set of register-to-register transformations. The registers include input and output registers that connect to external interfaces and state registers that hold settings or intermediate results. The hardware block may also contain addressable memories that hold arrays of intermediate results or tables of values used in the computation.

Often, the algorithm can be designed at the register- and memory-transfer level, as shown in Fig 6-7. This figure shows the computational data flow of a simple logic block that performs two multiplications, one addition, and one memory reference. The figure shows a flow from the block's input registers to its output registers. The block's data path may have control inputs that determine the selection of inputs, the data-path computation, and the destination of results. It may also have control outputs that feed the state of the data path to other blocks or into the control state machines. For simplicity, this figure ignores the data paths used to load the state registers and data memory. The figure also defers the question of how large hardware blocks that appear in a design more than once, such as a multiplier, might be shared to save silicon area.

This task could be directly translated into hardware as is, but it might not be efficient even with a direct RTL implementation. The function is not pipelined, so it can produce new results every clock cycle, one clock cycle after new inputs are available. The designer may save cost by recognizing that different computations can share a single hardware element, which is accomplished by multiplexing the appropriate input values into the shared function unit and steering

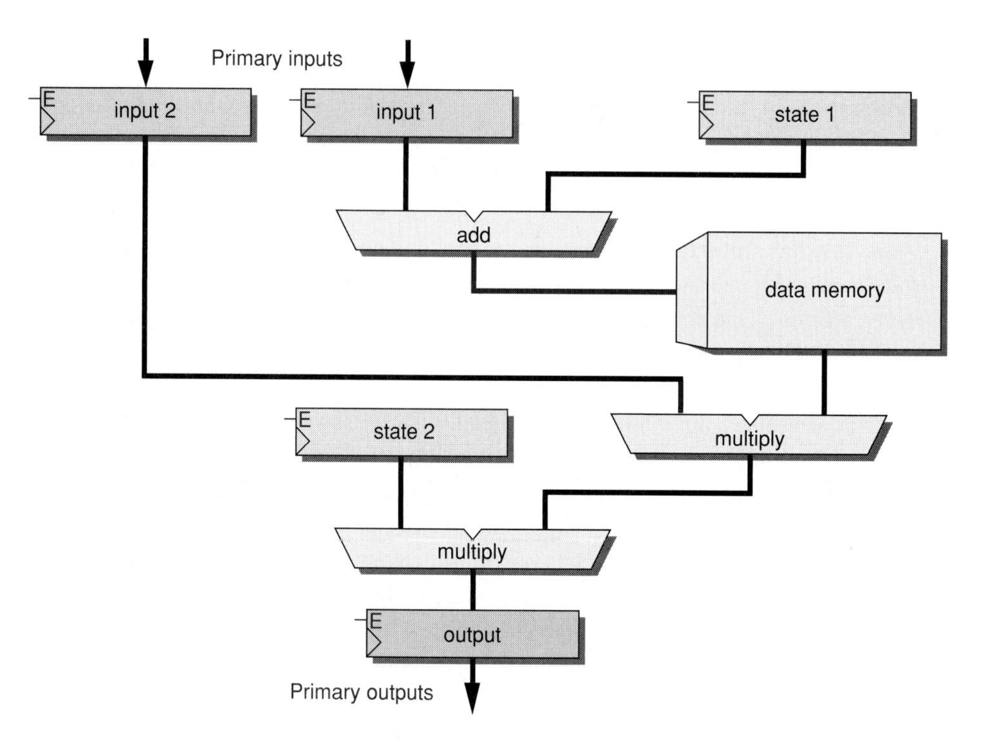

Figure 6-7 Simple hardware-function data flow.

the results to the right register. The maximum delay from input to output may also exceed the desired clock period, so intermediate registers should probably be inserted into long paths to pipeline the computation and distribute operations across multiple clocks. The insertion of these intermediate registers increases the number of latency cycles but improves the circuit's overall clock rate. The introduction of two intermediate registers and the reuse of a multiplier (reducing the number of multipliers used from two to one) produce the data path shown in Fig 6-8.

Two additional 2:1 multiplexers, one on each input of the multiplier, allow the two multiplications to share one multiplier. Three temporary registers are added: one to hold each of the two multiplication results to allow multiplier sharing, plus one more register to hold the address-calculation result. All of these intermediate registers help to improve clock frequency. This design is capable of computing the original function and other functions as well. The computation would be spread over four clock cycles, from the availability of data in the input registers to the availability of data in the output register, as shown in Fig 6-9.

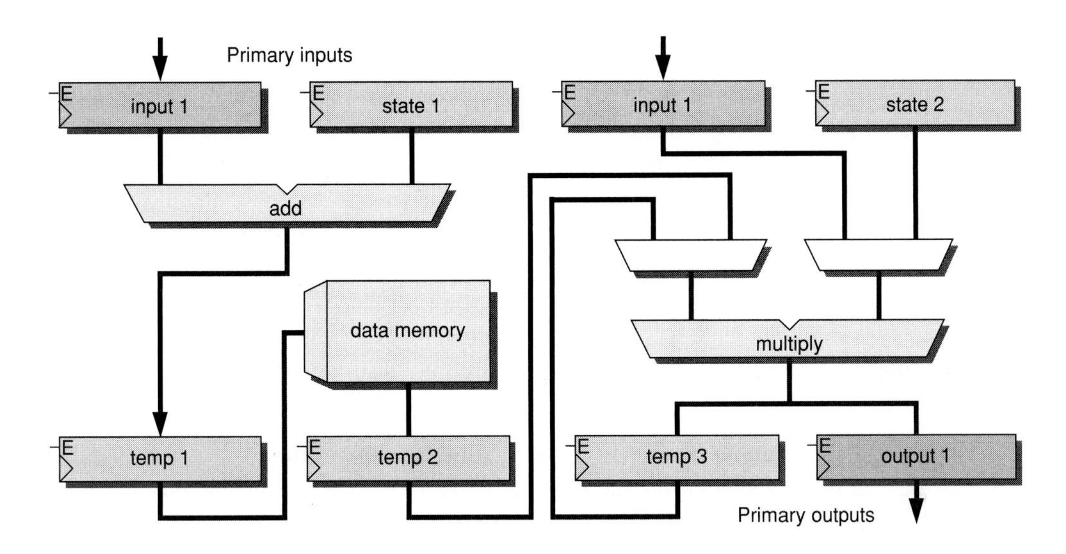

Figure 6-8 Optimized data path with operator reuse and temporary registers.

Cycle	Function
1	`temp1 = add(input1,state1)`
2	`temp2 = read(temp1)`
3	`temp3 = multiply(temp2,input2)`
4	`output1 = multiply(temp3,state2)`

Figure 6-9 Cycles for execution of pipelined function.

If this function were implemented in RTL, the data path would implement the functions above together with the two input registers, two state registers, three temporary registers, and the output register. Control of the function would probably be implemented using some finite-state machine whose outputs would drive the enable signals of the seven registers and the control inputs for the two multiplexers and the data memory. The mapping of this simple hardware function to a processor-based solution is quite simple and offers the designer a few basic choices. Three styles of implementation might be considered:

6.3.2 One Primitive Operation per Instruction

Each basic register-to-register operation in the above example could become a unique instruction. The developer might define four distinct instructions, one corresponding to each of the cycles shown above. In fact, processors often constrain the way data memory is used in the pipeline. The Xtensa processor, for example, pipelines all data-memory references and uses one cycle for an address calculation and one cycle for the actual memory operation, which results in total pipeline depth of two cycles between the inputs to the address calculation and result data becoming available from the memory. In using this data path as an example, we take advantage of the load operation's combination of an address-calculation cycle with a memory-access cycle to combine the operations of cycle 1 and cycle 2 into one instruction. Therefore, only three instructions are needed to express the four functions.

Sample TIE code for this function, assuming 24-bit data paths throughout, appears in Fig 6-10. The `state` declarations define the name and size of the two new temporary registers. Three new operations—`lookup`, `mul1`, and `mul2`—are defined using the registers shown—the architecture's input, output, and state registers plus the two added temporary registers. These registers serve as operand sources and destinations. In this Xtensa-based example, two predefined signals called **`Vaddr`** (virtual address) and **`MemDataIn32`** (32-bit read data returned from memory) are used to add a new data-load instruction. Note that the temporary register `temp1` is no longer explicitly required, but is replaced by an unnamed pipeline register placed between the address-calculation cycle and the memory-access cycle.

```
1: state input1 24 add_read_write
2: state input2 24 add_read_write
3: state output1 24 add_read_write
4: state state1 24 add_read_write
5: state state2 24 add_read_write
6: state temp2 24
7: state temp3 24
8: operation lookup {} {in state1, in input1, in MemDataIn32, out temp2, out
9:   VAddr} {
10: assign VAddr = {8'h0, input1 + state1};
11: assign temp2 = MemDataIn32[23:0];}
12: function [23:0] mult ([23:0]x,[23:0]y) shared {assign mult = x*y;}
13: operation mul1 {} {in temp2, in input2, out temp3} {
14: assign temp3 = mult(temp2,input2);}
15: operation mul2 {} {in temp3, in state2, out output1} {
16: assign output1 = mult(temp3,state2);}
```

Figure 6-10 Data-path function TIE example.

The first five lines of the TIE description declare the five state registers of the original function description of Fig 6-7. The next two lines declare the two temporary registers used to pipeline the execution. Lines 8 and 9 define the operands of the lookup operation, where the first pair of brackets surrounds the list of *explicit operands* (in this case, there are none), and the second pair of brackets surrounds the list of *implicit operands* (in this case there are five, including the built-in memory interface signals **Vaddr** and **MemDataIn32**).

Explicit operands correspond to C expressions and variables that will appear in the C code using this new function. These values are passed in and out of the instruction through entries in register files and the register index is specified in an instruction field. Any constant values appearing as immediate fields in the processor instruction are also included in the explicit operand list. Implicit operands correspond to state registers and special interface signals within the processor. However, implicit operands do not appear as corresponding variables or expressions in the C code that uses the TIE instruction or in fields within the assembly-language instruction. The implicit operands use no bits of the instruction word. Line 10 implements the address calculation. Line 11 completes the load operation by putting the result into the pipeline register (the implicit temp1). Line 12 declares a function to implement a single 24-bit-by-24-bit multiplier.

This example uses TIE's shared function capability, which automatically implements appropriate input multiplexers and output routing for shared-function units such as the 24-bit multiplier. The last four lines of the TIE code implement the operations corresponding to the two multiplications, making the connections to the source and destination registers.

The C code that implements this sequence using the new instructions is trivial because this instruction set refers implicitly to input, state, and output registers throughout the code. If the data-path function is used repeatedly, the function might be implemented by the following loop, where changing values of `input1` and `input2` ensure that a new computation takes place in each iteration:

```
for (i=0;i<N;i++) {
 <values for input1 and input2 change>
 lookup();
 mul1();
 mul2();
 <value of output1 is used>
}
```

The structure of this TIE implementation is shown in Fig 6-11. This structure is very similar to Fig 6-8 except for the replacement of the state register `temp1` by a pipeline register.

This version of the function is simple and area-efficient. It is also fairly flexible because the three operations are independent and can be reordered or reused as needed just by changing the C code. Other operations can be inserted as needed and can exploit other built-in processor resources. Moreover, these operations can be reordered or used in any combination to compute a wide range of related functions. The design could be made even more flexible by mapping all

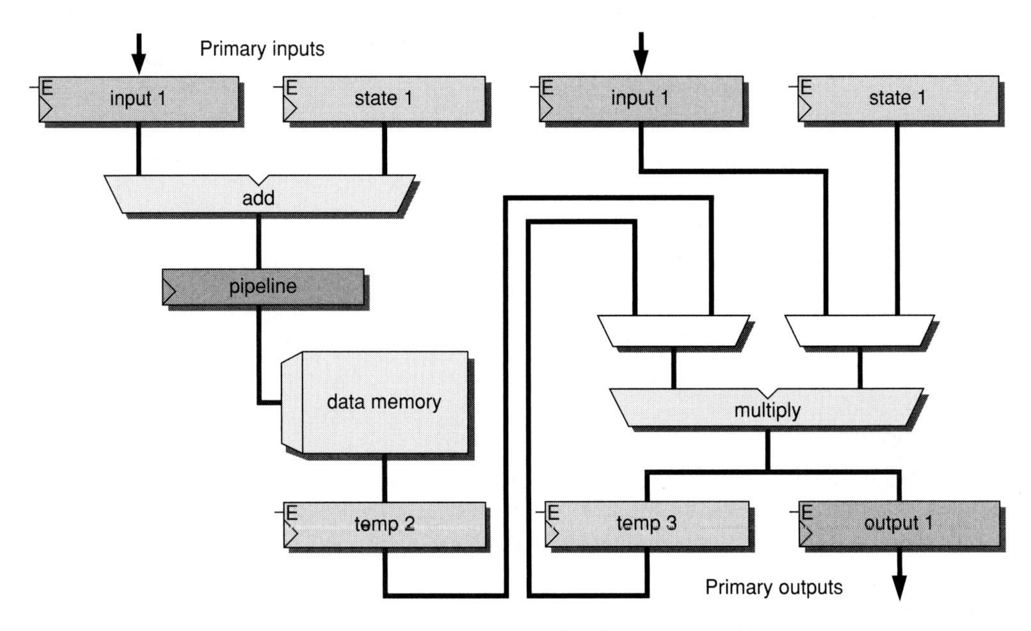

Figure 6-11 Data path with pipelined address and lookup.

the state and temporary registers into a unified register file, which would remove the strict connection of operations to dedicated source and destination states. By using one shared central register file—instead of separate input, output, state, and temporary registers—any of the new operations can be applied to any register in the register file for greater function flexibility. Fig 6-13 shows the structure of this alternative data-path implementation.

Fig 6-12 shows the relevant active logic for each cycle.

Cycle	Function	Implementation
1	`temp2 - read(add(input1, state1))` `[first cycle]`	input 1, state 1, add, pipeline
2	`temp2 = read() [second cycle]`	pipeline, data memory, temp 2

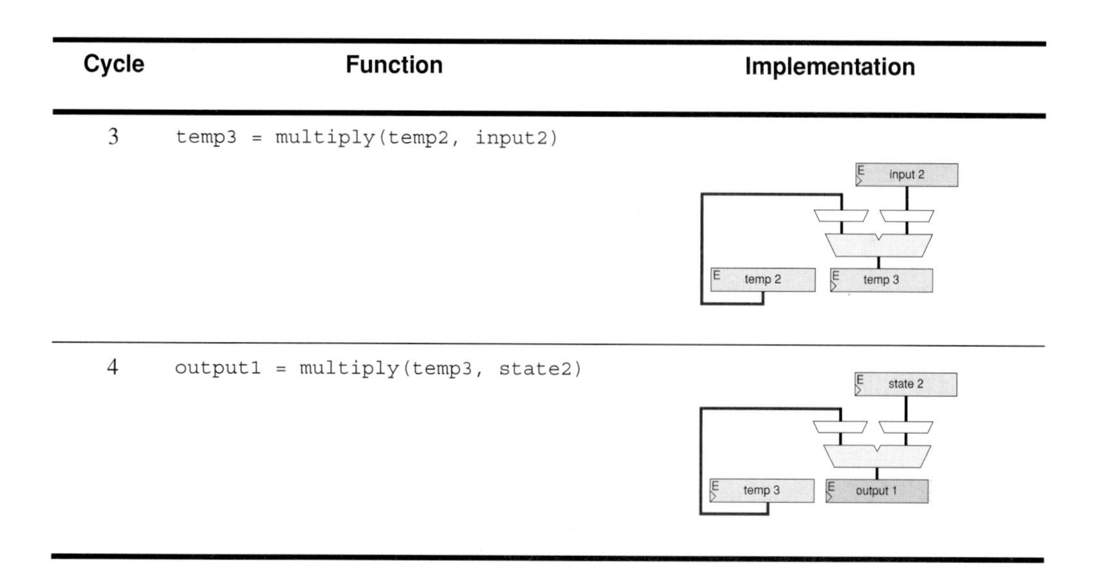

Figure 6-12 Cycles for execution of pipelined function, with combined address and load.

A by-product of a having central register file with only one write port is worth mentioning: all register file writes take place at the same point in the pipeline—in the pipe stage associated with the longest latency operation. In this example, the add-load sequence takes two cycles, so the register-write operation takes place two cycles after the operation is started. However, if the multiplication requires only one cycle, the register's write port may not be available. Therefore, the actual write to the register file would be delayed for one cycle. Without some special accommodation, this result would not be available to a subsequent instruction until that delay cycle has passed. The processor generator, however, can add bypass logic, as discussed in Chapter 7, to maintain the full benefit of a single-cycle multiplication while using a register file with one write port for all results.

A register file often requires less silicon area than the set of individual registers it replaces, especially if the number of registers is large but the number of register-file read and write ports is small. This area savings comes from two effects. First, the shared register file may need fewer total registers because intermediate results from different calculations can share temporary-register entries. Second, a VLSI circuit implementation of a register file is typically smaller than the implementation of the same number of discrete registers. Often, it's much smaller.

The major potential cost of the change from discrete registers to a unified register file is consumption of more instruction-encoding bits, because dedicated state registers are implicitly addressed (and therefore need no address bits in the instruction) while register files have explicit addressing and therefore require that an instruction have address fields that select registers in the file.

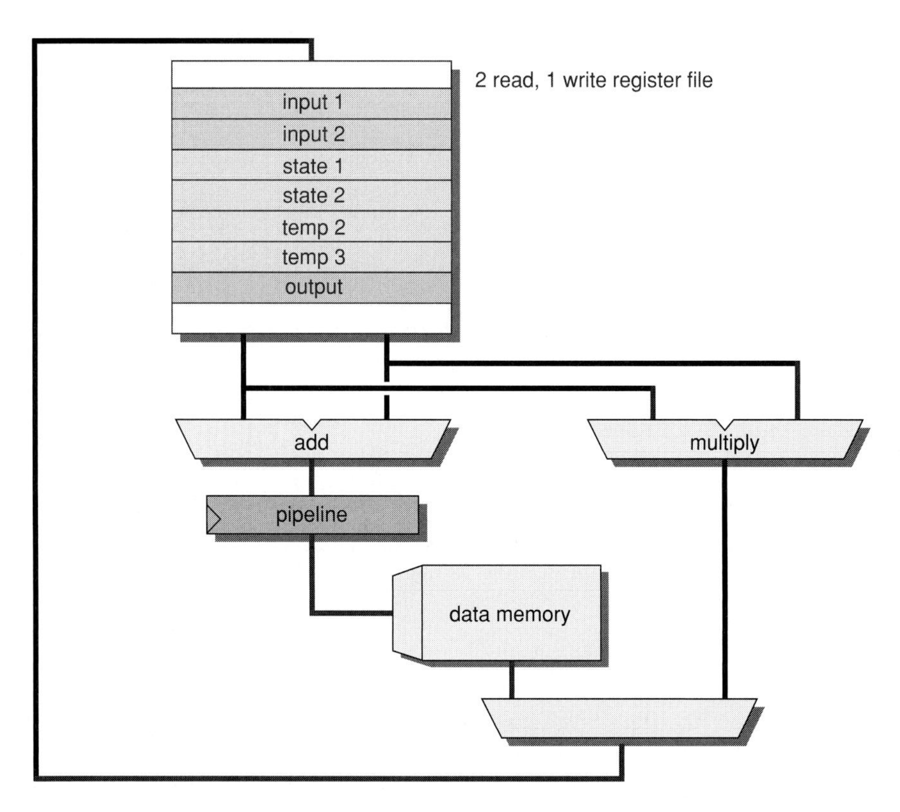

Figure 6-13 Data path with unified register file.

In addition, a unified register file may require additional operand-bypass logic if operations with different pipeline latencies feed the same register-file write port. Depending on the width of operands and the number of required source operands and results, either a new register file may be defined to address the different needs of the different operations or the existing base register file may be used or extended to support the new operations. A revised implementation in TIE, using a new eight-entry register file, is shown in Fig 6-14.

```
1: regfile XR 24 8 x
2: operation lookup {in XR in1,in XR in2,out XR out1} {in MemDataIn32,out
3:   VAddr} {
4: assign VAddr = {8'h0,in1 + in2};
5: assign out1 = MemDataIn32[23:0];}
6: operation mul24 {in XR m1, in XR m2, out XR mout} {} {assign mout = m1 *
7:    m2;}
```

Figure 6-14 Data-path function TIE example, with unified register file.

The first line in Fig 6-14 defines a new eight-entry, 24-bit-wide register file called `XR`. The next four lines define a new load operation that uses the sum of two registers as the memory address and returns a 24-bit value. In this simple example, the 24-bit operands are aligned on 32-bit boundaries. (Chapter 4 discusses how instruction-set extensions can handle data of arbitrary alignment.) The last two lines of this instruction-set extension define a general-purpose, 24-by-24-bit multiplication operation. Note that the multiplier no longer requires explicit sharing. A single multiplication instruction serves both multiplication operations. The explicit sharing via input multiplexing, found in Fig 6-11, is implicit in the use of the register file. Here is the corresponding C code, where the `input1`, `input2`, `state1`, `state2`, `temp2`, `temp3`, and `output` registers correspond to C variables of the same name, which are automatically allocated to entries in the XR register file:

```
for (i=0; i<N; i++) {
 XR input1, input2, temp2, temp3, output, state1, state2;
 temp2 = lookup(input1, state1);
 temp3 = mul24(temp2, input2);
 output = mul24(temp3, state2);
}
```

Note that the origin of the inputs and the final output destination are ignored.

Using a central register file further increases the instruction set's generality and reduces its cost. This simple operation encoding produces one output result every four cycles because the operations work in series on the data. When even higher throughput is required, greater parallelism can double or quadruple the output rate.

6.3.3 Multiple Independent Operations per Instruction

Additional operational parallelism can achieve higher throughput. *Long-instruction-word* techniques, such as those introduced in Chapter 4, Section 4.5, allow a number of *independent* operations to be encoded in one instruction word. Free combination of multiple operations during each execution cycle improves the tradeoff between flexibility and performance because the operations operate in parallel, yet the operations can be easily used in different combinations as computing requirements change.

Historically, long-instruction-word or VLIW methods have often been considered difficult to deploy because many VLIW software tools incompletely support the packing of operations together into useful instructions. When the environment of instruction-set-description tools and the corresponding assemblers and compilers can freely and automatically generate instruction sets and code that pack multiple operations per cycle, the designer's task becomes much more tractable.

For the simple data-path problem described in Section 6.3.2, sharing a single multiplier is good for area efficiency but it creates a performance bottleneck. Implementing two multipliers allows two multiplications to execute in parallel. The structure of the corresponding data path is shown in Fig 6-15.

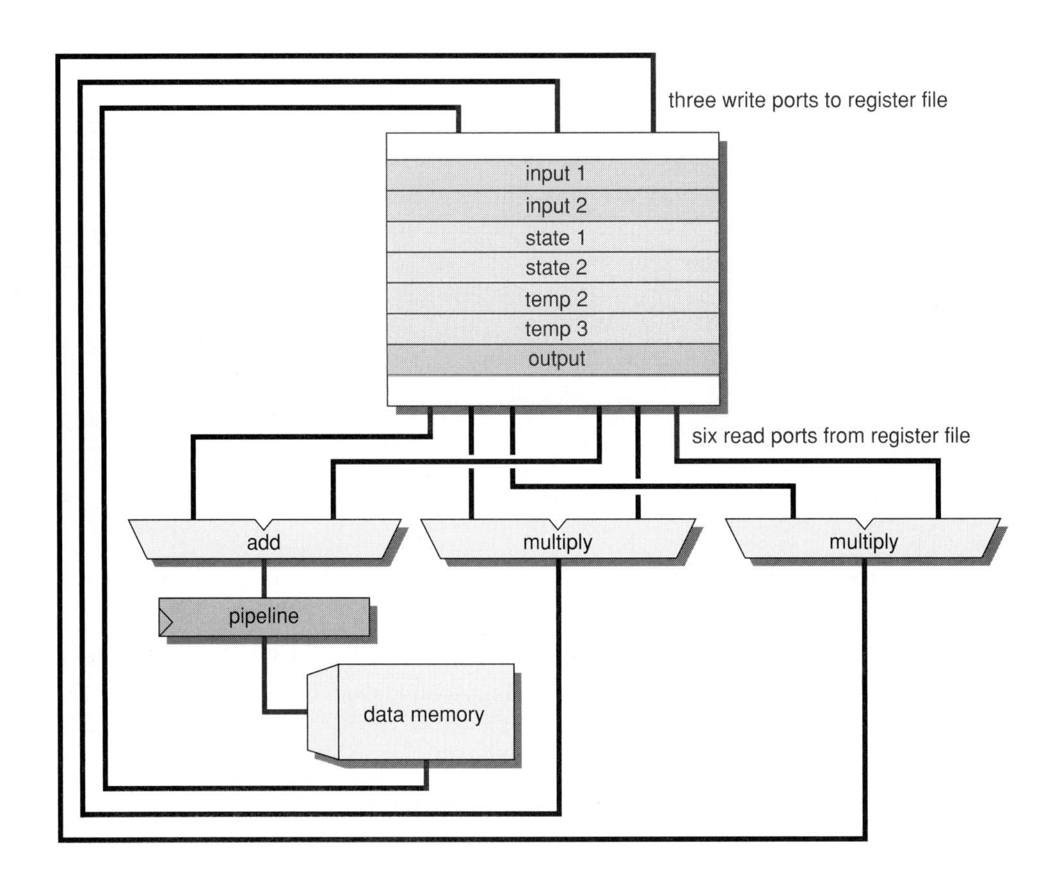

Figure 6-15 Data path with three independent operation pipelines.

The long instruction word that controls this data path can incorporate operation slots for each of the independent operations, using the FLIX technology described in Chapter 5 and expressed in TIE as shown in Fig 6-16.

The **`regfile`** declaration creates an extended eight-entry register file, 24 bits wide, with the name `XR`. The **`format`** declaration defines a new instruction format with three operation slots using 64-bits. The `slot0` operation slot can specify either the lookup operation or a `NOP`. The `slot1` operation slot can specify a `mul24` operation or a `NOP`. The `slot2` operation slot can specify a `mul24` operation or a `NOP`. From these definitions, the processor generator infers that the XR register file requires six read ports and three write ports to implement the three parallel operations.

This definition creates an instruction set in which any combination of the three operations can be executed in every cycle (resulting in eight different combinations). The instruction latency is still four cycles per result, but the compiler can take essentially the same C code shown above and start a new add-lookup-multiply-multiply sequence every cycle for a sustained throughput of one complete function execution per cycle.

```
1: regfile XR 24 8 xr
2: length l64 64 {InstBuf[3:0] == 14}
3: format format1 l64 {slot0, slot1, slot2}
4: slot_opcodes slot0 {lookup}
5: slot_opcodes slot1 {mul24}
6: slot_opcodes slot2 {mul24}
7: operation lookup {in XR in1, in XR in2, out XR out1} {in MemDataIn32, out
8:    VAddr} {
9: assign VAddr = {8'h0,in1 + in2};
10: assign out1 = MemDataIn32[23:0];}
11: operation mul24 {in XR m1, in XR m2, out XR mout} {} {assign mout = m1 *
12:     m2;}
```

Figure 6-16 Data-path function TIE example, with three operation slots per instruction.

This C code takes advantage of the compiler's automatic software-pipelining and instruction-scheduling capabilities, but may necessitate a larger register file to accommodate overlapping uses of the state values for `input1`, `input2`, `state1`, `state2`, `temp2`, `temp3`, and `output1` because different loop iterations may be active at the same time. In the example as shown, an eight-entry register file is still sufficient. Even if the processor initiates a loop iteration every cycle, only eight operand values are live at any given point in the loop. Fig 6-17 shows the possible scheduling of the code, with four iterations of the loop overlapping in execution.

Software-managed pipelining allows the code to use a complex combination of simple operations. The details of the code are given here to show the true overlap among operations. In practice, the compiler can generate such code automatically, placing only the one instruction of cycle 4 in the inner execution loop. The compiler would add *preamble* code (similar to the code in cycles 1–3) to get three iterations underway and *postamble* code (similar to the code in cycles 5–7) to wrap up the computation of three iterations once the requisite number of iterations is complete.

In the Xtensa assembly code shown here, the first instruction operand specifies the destination register, and the next two operands specify the source registers. The operations associated with the first iteration of the loop are highlighted in very light gray; the operations of the second iteration, in light gray; the operations of the third iteration, in medium gray; and the operations of the fourth iteration, in dark gray. Note, however, that buffering the values through one register file requires a register file with six read ports and three write ports, which may increase silicon area. During cycle 4, the cycle in which all four loop iterations are active simultaneously, the assignment of the eight register entries in active use are shown. To sustain a throughput of one iteration per cycle, the compiler could automatically unroll this loop so that each instruction in the inner loop looks like the combination in cycle 4.

This code sequence ignores, for the moment, the question of how incoming data enters the registers corresponding to `input1` and `input2`, and how outgoing data is removed from the register corresponding to the output. The important topic of I/O is handled in Section 6.5.

Cycle	Slot 0	Slot 1	Slot 2	Allocation of registers reg: variable[loop iteration]
1	lookup xr2,xr0,xr1	NOP	NOP	xr0: input1 [1],xr2:temp2 [1]
2	lookup xr7,xr0,xr1	NOP	NOP	xr0 : input1 [2],xr1:state1 [all],xr7:temp2 [2]
3	lookup xr2,xr0,xr1	mul24 xr4,xr3,xr2	NOP	xr0: input1 [3],xr1:state1 [all],xr2:temp2 [2], xr3:input2 [1],xr4:temp3 [1],xr7:temp2 [2]
4	lookup,xr7,xr0,xr1	mul24,xr4,xr3,xr7	mul24,xr6,xr5,xr4	xr0: input1 [4],xr1:state1 [all],xr2:temp2 [3], xr3:input2 [2],xr4:temp3 [2],xr5:state2 [all], xr6:output [1],xr7:temp2 [4]
5	NOP	mul24,xr4,xr3,xr2	mul24,xr6,xr5,xr4	xr1: state1 [all],xr3:input2 [3],xr4:temp3 [3], xr5:state2 [all],xr6:output [2],xr7:temp2 [4]
6	NOP	mul24,xr4,xr3,xr7	mul24,xr6,xr5,xr4	xr1: state1 [all],xr3:input2 [4],xr4:temp3 [4], xr5:state2 [all],xr6:output [3]
7	NOP	NOP	mul24,xr6,xr5,xr4	xr1: state1 [all],xr5:state2 [all],xr6:output [4]

Figure 6-17 Implementation of an accelerator as an application-specific processor.

6.3.4 Pipelined Instruction

Where high throughput is required but implementation cost is more important than flexibility, a single pipelined instruction can directly implement the target hardware function. The advanced issues of pipelined processor design are discussed in more detail in Chapter 7. The processor generator uses the TIE **`schedule`** directive to automatically insert pipeline registers at optimal points in the data-path logic to boost clock rate. The resulting structure is shown in Fig 6-18. This function unit implements the original data-path function of Fig 6-7 almost verbatim. The only change is the allocation of logic to four pipeline stages with automatic insertion of pipeline registers at each boundary instead of explicit designer specification of the three intermediate registers, `temp1`, `temp2`, and `temp3`. This approach is more concise in the TIE code and allows the processor synthesis tools to automatically balance delays across the pipeline register boundaries to maximize clock frequency.

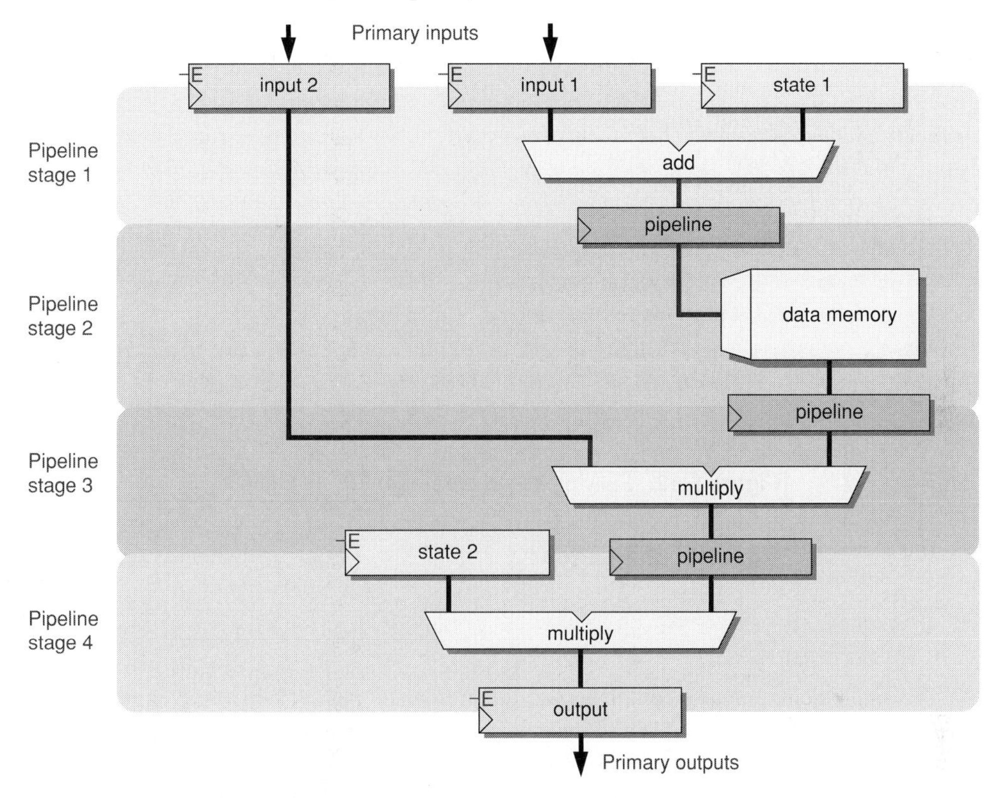

Figure 6-18 Fully pipelined instruction implementation.

Fig 6-19 shows the TIE hardware description for this design. The **`schedule`** statement triggers the automatic operation pipelining and indicates that the value of `input2` is not required until the beginning of the third pipeline stage, that the value of `state2` is not required until the beginning of the fourth pipeline stage, and that the output should be ready at the end of the fourth pipe stage.

The first five lines declare the input, state, and output registers. The single operation statement, spread over lines 6 to 10, defines the addition-load-multiply-multiply sequence. Lines 11 and 12 define the pipeline state in which inputs are used and outputs are available. Any source (**use**) or destination (**def**) left unspecified is used or defined in pipeline stage 1 except for values returned from memory, which are scheduled to arrive in a later pipe state.

```
1: state input1 24 add_read_write
2: state input2 24 add_read_write
3: state output 24 add_read_write
4: state state1 24 add_read_write
5: state state2 24 add_read_write
6: operation lookup.mul.mul {} {in input1,in state1,in input2,in state2, out
7:    output1, in MemDataIn32,out VAddr} {
8: assign VAddr = {8'h0, input1 + state1};
9: wire [23:0] mulout = MemDataIn32[23:0] * input2;
10: assign output = mulout * state2;}
11: schedule inst_sched {lookup.mul.mul} {use input2 3; use state2 4; def
12:    output 4;}
```

Figure 6-19 Fully pipelined data-path TIE example.

The corresponding C code is very simple:

```
for (i=0; i<N; i++) {lookup.mul.mul();}
```

This implementation also completes one function result per cycle in a fully pipelined fashion with no preamble or postamble code required. This method is simple, fast, and efficient. Its primary liability is relative inflexibility. The resulting compound instruction can be used only for one particular combination of addition, memory reference, and two multiplications. This instruction may still be attractive to the designer, however, if that four-operation sequence is considered very stable but the context of its use—the surrounding operations and interaction with other algorithms—is less stable. Consequently, this pipelined approach may give the best balance of performance, efficiency, and flexibility.

6.3.5 Tradeoffs in Mapping Hardware Functions to Processor Instructions

This section highlights the range of methods available to the hardware developer for mapping hardware functions to processor instructions. It also reveals key tradeoffs facing the developer, including the following:

- **Degree of pipelining:** Processor execution units can typically be pipelined to the same degree as traditional pipelines built with RTL.
- **Size (and power) of implementation, including register files:** Sharing execution units saves chip area but reduces throughput. Using register files instead of dedicated state registers may decrease chip area when relatively few parallel register reads and writes are

needed, but may increase the area when many reads and writes are needed each cycle. Use of common register files typically makes the resulting instruction set more flexible than when key values reside in dedicated state registers.

- **Instruction extension and C code simplification:** As shown by the TIE examples above, instruction-set descriptions are typically more concise than equivalent RTL descriptions. Even if the C language is relatively unfamiliar to the hardware developer, the expression simplicity makes the processor-centric design methodology easy to learn for most designers.
- **Generality of the resulting instruction set as function requirements change:** An essential benefit of processor-centric SOC design is flexibility after the chip is fabricated. Different instruction sets offer different degrees of flexibility to combine operations. In general, mapping hardware functions into instructions that use more instruction bits are more flexible but require more gates to implement. Mappings that make all operand selection and operand routing between function units implicit are less flexible, though somewhat smaller in silicon area.
- **I/O mechanisms:** Functional requirements often dictate whether input and output data travel through memory or over dedicated interfaces. Choices for input and output data transfer can affect throughput and cost, and are discussed in more detail in Section 6.5.

6.4 Moving from Hardwired Engines to Processors

This section outlines the simple sequence of steps developers can use to map complex functions, originally conceived as hardware blocks, into application-specific processors. The result is rapid implementation of blocks that rival the performance of hardwired logic yet remain fully programmable, so bugs are easier to fix and changes to requirements are accommodated just by changing the firmware running on the application-specific processor.

Most hardware developers are intimately familiar with the design of function blocks that combine data paths and finite-state-machine logic. Even when the theoretical benefits of application-specific processors are clear, the transition from RTL hardware design to processor-centric design may appear complex. The description of a simple set of techniques, however, demystifies the process and helps hardware developers exploit the potential of highly programmable processors that deliver the performance of hardwired logic.

Section 6.2 has already introduced the basic concepts of mapping a simple data-path function into a set of instructions and an instruction sequence. That section showed that a hardware data path can be mapped simply and directly into a set of processor instructions with comparable performance and efficiency, but greater reusability. The section detailed the trade-offs between instruction-set generality, silicon area, and data-path throughput. Thus, Section 6.2 dealt with the question of *why*.

This section deals with the larger question of *how* complete hardware engine designs—data path plus sequencing hardware—can be transformed into efficient, application-specific processors running firmware. It focuses on how data paths are controlled, especially how finite-state-machine sequencers are translated into code running on an application-specific processor and how specialized microcoded engines are generalized into processors.

One important caveat should be kept in mind—pipelined processors can have meaningful overhead for taken branches. Consequently, state-machine structures that expect to choose among multiple control arcs during almost every cycle may be difficult to translate into a processor with the same performance. Often, these issues can be mitigated by the advanced control-flow techniques described in Chapter 7. In other cases, the best answer is to retain the finite-state machine.

6.4.1 Translating Finite-State Machines to Software

Consider the finite-state machine shown in Fig 6-20. For clarity's sake, this state machine has just 19 states, but a real-world state machine might have many dozens of states. (Note: State machines can, unfortunately, be designed with hundreds of states. The complexity of managing such a design may quickly push the design team toward a microcoded engine or processor, as discussed in Section 6.4.3.)

In this simple example, each state corresponds to one operational cycle of the logic. Each arc from a state represents the conditions for transition to the next state. (For example: transition from state `idle` to state `S2` if `condition` is true.) If there is only one outgoing state-transition arc, the transition is unconditional. Transition conditions are based on status or data signals from the data path or logic being controlled. The state machine's current state controls the data path: selecting operations, steering data, and enabling register writes. The function block follows different paths through the state diagram depending on the data values generated by the data path and external inputs. Sometimes the state-to-state transition conditions represent rare but possible error conditions that force the state machine back into an initial state, such as the transition from state `S13` to state `idle` when `error` is true.

6.4.1.1 The State-Machine-to-Software-Program Conversion Flow

In many cases, the conversion of the state machine and data path into firmware running on an application-specific processor and a program is a simple task. Conversion follows this basic procedure:

1. For each data path variable in the function, decide whether it will be represented by a distinct state register or as a C variable. If it is a C variable, decide whether it should use a built-in C data type (int, short, char) or a new application-specific data type.

    a. Variables mapped to state registers are manipulated by application-specific instructions that implicitly operate on the corresponding state register.

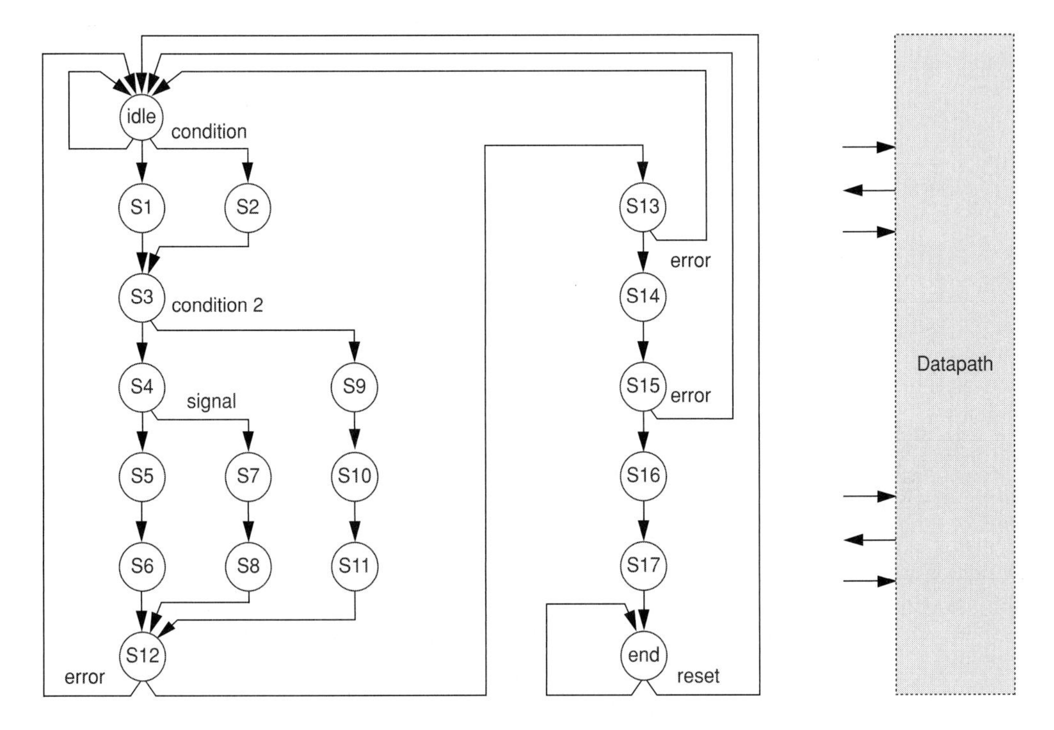

Figure 6-20 Simple finite-state machine.

b. Variables mapped to built-in C data types are usually held in general-purpose processor registers (or memory) and can be manipulated by base processor instructions and application-specific instructions.

c. Variables mapped to application-specific data types are generally held in new extended register files (or memory) and explicitly manipulated by application-specific instructions.

In general, mapping variables to state registers reduces the number of bits required to encode instructions and also reduces the flexibility of the implementation. Mapping variables to built-in and application-specific C data types allows the compiler to dynamically allocate available register-file entries to active variables and to hold other variables temporarily in memory. Using built-in C data types also allows the developer to use standard C operators to manipulate the data.

2. Take the following steps to map a hardware function into application-specific processor instructions:

   a. Identify the set of data-path operations associated with each state. Define several explicit independent operations, such as the one shown in Fig 6-15. Assigning one operation to each independent result is a natural partitioning.

b. Look for common data-path elements that may be shared among the operations of different states (such the multiplier in Fig 6-8). Shared data-path elements allow a smaller set of operations to cover all the needs in the block, save instruction-encoding space, and simplify the sharing of common logic blocks such as arithmetic units and registers.

c. Generalize the instructions so that one instruction can be used for several similar operations to minimize the number of new instructions. Guidelines for generalizing instructions are outlined in Section 6.4.2.

d. Use 32-bit arithmetic, logical, and shift operations whenever possible because compilers automatically map these operations to base processor instructions.

e. Use the long-instruction words with multiple independent operations per instruction (such FLIX, described in Chapter 5) to execute the independent operations in one cycle.

3. For each condition associated with a state-transition arc in the state machine, determine how the processor will test that branch condition. Two basic methods are commonly used:

   a. Arrange earlier operations so they leave the data value on which the condition is based in a register. Regular conditional-branch instructions can then use that value, or bits of that register, as a branch qualifier.

   b. Compute the branch condition within the data path and assign it to a new condition code in the configurable processor. Special branch or conditional instructions can then use these application-specific condition-code bits.

   Pipelined processor implementation introduces branch latency that can interfere with achieving a perfect one-to-one correspondence between a hardware finite-state machine and a software implementation of that finite-state machine running on the processor. Chapter 7 introduces a number of methods to mitigate branch latency.

4. Write the C program that implements the state machine. Each state-machine state represents a set of operations. That set of operations may map to one instruction or to a series of instructions executed sequentially within that state. Performance requirements dictate whether all the operations must be packed into one complex instruction or if a looser implementation of the state is acceptable. For each operation, write down the combination of basic C statements and intrinsic function invocations that represent those operations. Each unconditional sequence of states is represented as a block of C statements. Represent each conditional transition with a C if-then-else construct. A simple example, a subset of the state machine in Fig 6-20, with the C translation, is shown in Fig 6-21. For clarity's sake, the operations for each state are implemented with a new instruction (and the corre-

sponding intrinsic function), whose name is the same as the state. Even non-nested control flow finite state machines can be coded with C language "goto" statements, if necessary.

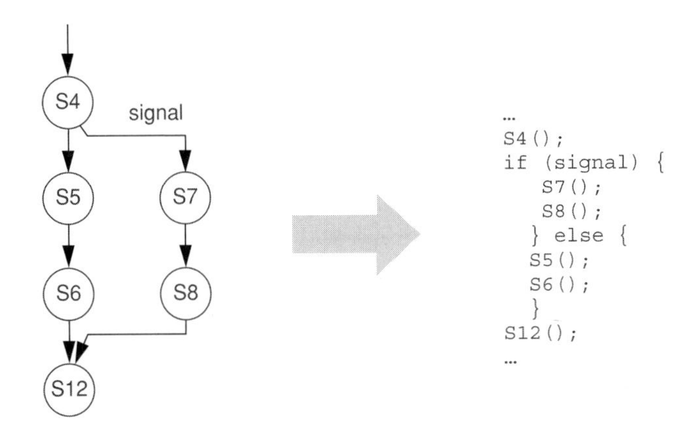

Figure 6-21 Translation of a six-state sequence to C.

When a developer uses a long-instruction-word processor, as described in Chapter 4, Section 4.5, a group of independent operations—typically a combination of general-purpose, base-architecture and application-specific operations—can be packed together into one instruction. For example, the state S7 might implement an application-specific operation, such as the `lookup.mul.mul` operation of Fig 6-19. That operation might be combined with the shifting of one general register and the load of a value into a wide, application-specific register file able to hold variables of type x. The operations corresponding to S7 might consist of this snippet of C:

```
int a;
my_wide_data x; *p;
a = a<<2;
x = *p;
lookup.mul.mul();
```

A long-instruction-word processor could combine these operations into the instruction and the cycle corresponding to S7.

5. Analyze the resulting performance to determine if processor translation performs acceptably. Pipelined processors sometimes run at a higher clock frequency than the non-pipelined logic they replace, potentially increasing performance. On the other hand, pipeline effects may increase the cycle count on some paths. It is often important to optimize the code and data path to reduce branch effects. Pipelined processors often have branch-delay

bubbles that waste one or more cycles for every taken branch. The typical RISC branch delay can often be avoided by using three techniques:

a. Reorder code to avoid taken branches on the most common or time-critical paths.
b. Use zero-delay loops or branches.
c. Replace conditional branches with conditional data-path execution.

These techniques are discussed in greater depth in Chapter 7.

6.4.2 Designing Application-Specific Processors for Flexibility

The methods for designing application-specific processors can map a sequence of hardware operations to a wide variety of implementation styles. The designer may choose among these different implementation styles to achieve the best fit between the implementation to the original hardware definition and to trade off between hardware-implementation efficiency (area, clock frequency, performance, and power) and flexibility after the SOC is built (the ability to fix bugs or to add support for new applications by changing the SOC's firmware).

Generally, the designer should fully populate the application-specific processor's available instruction-encoding space. Because instructions cannot be added after an SOC is built, the processor instruction set should be made as flexible and rich as possible within the constraints of silicon cost and power. Extra registers and operation units may consume valuable area, but extra instruction decoding is generally quite cheap, so implementing instructions with rich and varied combinations of existing operations and register operands makes sense. A few guidelines help the hardware designer balance the two universal objectives: minimum cost and maximum flexibility:

1. **Make constants programmable:** Move constant values from the data path into registers (more expensive and more flexible) or into encoded immediate fields in the instruction (less expensive and less flexible).
2. **Use all available register ports to provide additional operand flexibility:** For example, if a register in the base register file provides a memory address for a particular instruction, and one of that base register file's read ports is unused during that instruction, consider forming the memory-address operand from the computed sum of two base-register values (thus adding instruction flexibility at the possible cost of an adder). Similarly, consider providing versions of load-and-store instructions in which the calculated address (perhaps the sum of a register-file value and an offset) is written back to the general register file. This write-back is essentially free if a register file's write port is available during part of that instruction. The saved value of the calculated address may be useful later in the program, or it may come in handy for the next product upgrade.
3. **Encode many operational combinations:** If several independent operations are defined in the data path, create instructions that perform all combinations of the independent oper-

ations, even if the current algorithm only uses a subset of those combinations. The other combinations might turn out to be useful as the algorithm evolves. For example, if the application-specific instruction set defines a left-shift operation on wide registers `w1` and `w2`, and needs right-shift operations on wide registers `w3` and `w4`, consider defining left- and right-shift operations for `w1`, `w2`, `w3`, and `w4`. These instructions may add precisely the flexibility needed for future algorithm code. Or, if they don't need to be separate states, implement `w1`, `w2`, `w3`, and `w4` as a register file and define left- and right-shift operations for all register-file operands.

4. **Convert state registers to register files:** If encoding space permits, replace state registers with small register files. This guideline has two benefits. First, register files are automatically managed by the compiler, which moves values between the register file and memory as needed. Second, a register file allows the compiler to intermix operation sequences for multiple data streams, with each stream using different register-file entries.
5. **Encode memory operations together with computation operations:** Sometimes, an algorithm may appear to be completely executable just using the register state, with few loads and stores. As algorithms evolve however, their data requirements more often grow than shrink. By providing a rich set of load and store operations, including the combination of load and store operations with the most common computational operations, the instruction set may become much more useful as algorithms evolve. For example, suppose a developer defines four new vector computation operations, `add_v`, `sub_v`, `and_v`, and `or_v`, each with two vector-register sources and one vector-register destination. If instruction space permits, the developer could consider adding one vector load and one vector store merged with each of the four computation instructions, producing eight additional instructions: `add_v.load_v`, `add_v.store_v`, `sub_v.load_v`, `sub_v.store_v`, `and_v.load_v`, `and_v.store_v`, `or_v.load_v`, and `or_v.store_v`. The Xtensa LX processor's Flexible Length Instruction eXtension (FLIX) technology, described in Chapter 5, enables even easier encoding of multiple independent operations
6. **Evaluate useful conditions into Boolean values:** Most algorithms recognize that certain operand values have special significance. These values may not be directly used in one version of the computation, but conditional operations and branches based on these special values may become important as the algorithm evolves. Designers often find it convenient to record these special values or conditions as Boolean values. The Xtensa architecture, for example, incorporates an array of 16 1-bit Boolean registers and has bit-manipulation instructions, conditional branches, and conditional operations based on those Boolean values. Boolean values can also be encoded in results written back to the processor's base register file. The designer may also find it useful to make some operations of the extended data path conditional on these Boolean values. This design approach often reduces the number of computation cycles and eliminates branch-delay bubbles for some finite-state-machine transition paths. For example, the developer could define a set of SIMD compare

and select operations on vector registers, with one sample comparison and one sample conditional select operation shown in Fig 6-22.

```
1: regfile vec4x16 64 8 v
2: operation cmpltv {in vec4x16 x, in vec4x16 y, out AR cond4}{} {
3: assign cond4 = {28íh0,x[63:48]<y[63:48],x[47:32]<y[47:32],
4:    x[31:16]<y[31:16],x[15:0]<y[15:0]};}
5: operation seltv {in vec4x16 x, in vec4x16 y, in AR cond, out vec4x16 z}{}
6:    {
7:  assign z = {cond[3]?x[63:48]:y[63:48], cond[2]?x[47:32]:y[47:32],
8:    cond[1]?x[31:16]:y[31:16], cond[0]?x[15:0]:y[15:0]};}
```

Figure 6-22 Vector comparison and condition move TIE example.

The first line of the TIE description declares an eight-entry register file. Each entry is 64 bits wide (used as four 16-bit values). Lines 2, 3, and 4 define a vector-comparison operation, in which pairs of four 16-bit values are compared and the operation's condition results are written to the general register file (AR) as a 4-bit field. (The top 28 bits of the 32-bit register are always zero.) Lines 5 through 8 define a conditional-move operation, in which a 16-bit field is selected from one of the two source vectors, based on each of the four Boolean values in the source general register, creating a result vector of four selections. The combination of the `cmpltv` (compare less-than, vector) and `seltv` (select operand on true, vector) instructions produces a field-by-field minimum vector operation.

Hardware designers using application-specific processors to replace traditional data-path-plus-state-machine designs also benefit if design changes are likely. For example, popular standards for encoding and exchanging data often incorporate a wide range of modes or profiles. Even if the initial design specification targets only a subset of these modes, the designer can profitably examine the other modes and generalize the design to support additional modes and features. This is fairly economical to do and may rescue the design later in the project. In fact, adding on-chip resources in anticipation of future changes, especially when they don't increase hardware cost or push out the development schedule, can make the designer a hero when the inevitable design changes must be made.

6.4.3 Moving from Microcoded Engines to Processors

Microcoded engines are the big brothers of finite-state machines. These engines have been used in electronic systems for decades. Maurice Wilkes first proposed them for implementing computers in the 1950s, and IBM used microcoded engines widely in the System 360 family starting in 1964. Originally, microengines simplified the implementation of the early baroque mainframe processor architectures. These ornate and complex processor architectures have faded with time, and most popular processor architectures are now implemented directly with pipelined hardware.

Today's microengine design techniques now play a different role. The basic microcoding techniques have percolated down into embedded systems and represent a compromise between general-purpose processors and hardwired logic, often offering a more operational parallelism than general-purpose processors and a more flexible development process than hardwired logic. In fact, most microcoded engines are designed for specific applications and are therefore developed using many of the same goals as the application-specific processors that form the foundation for complex SOC design.

The typical structure of a microcoded engine appears in Fig 6-23. The figure sketches a simple microcoded engine and microcode storage system, with four function units and three state registers. The low-level control signals—function-unit controls, register-results selection, and register-write enables—are all directly expressed as fields of the microcode word. Every microword may also include a branch or conditional branch. The branch-target address is directly encoded in the microinstruction word, along with the selected branch condition.

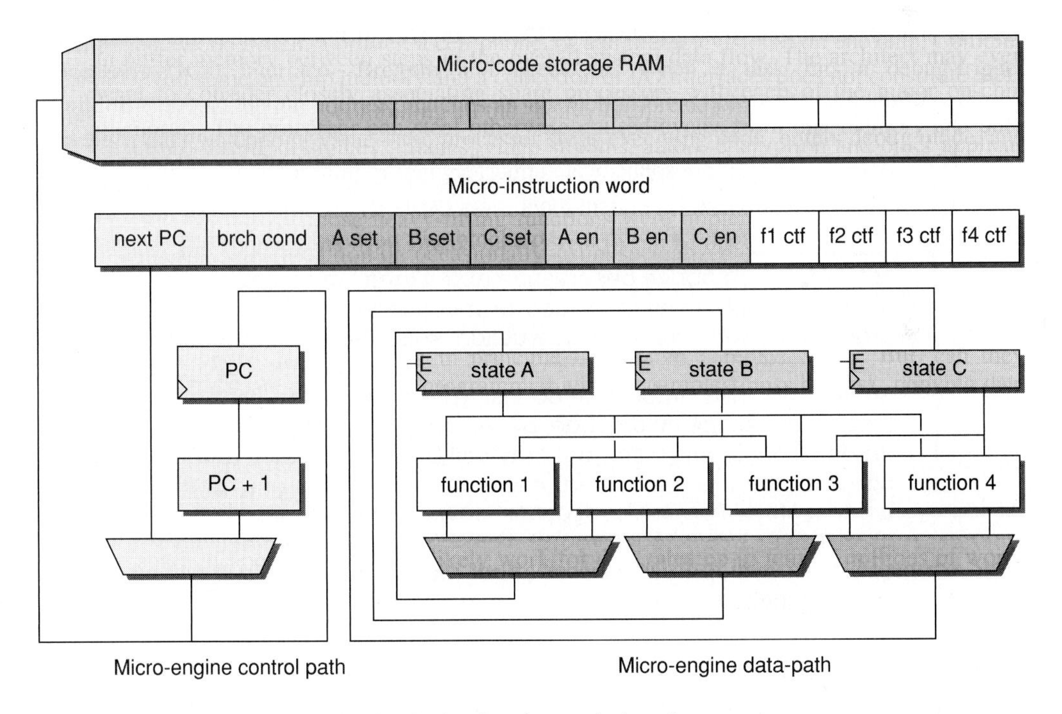

Figure 6-23 A simple microcoded engine structure.

Typical characteristics of microcoded engines include the following:

- **Wide, loosely encoded instruction words:** A microcode word contains numerous fields that directly correspond to the control signals for data-path registers, operation specifiers for function units, immediate values, operand specifiers for source and destination registers, microcode branch conditions, microcode target addresses, and other execution information. For simplicity and generality, these fields are largely independent. Only a few bits of the microcode word must be decoded to control a particular set of functions within the microcode engine. Many fields may go unused for any given microcode instruction. For example, a microcode word might include a shift-amount field, which exists even for instructions that do not specify a shift. Microcode word fields are not reused to save bits to reduce decoding complexity. The typical microcode word is wide, ranging from tens to hundreds of bits depending on the complexity of the data path being controlled and the sophistication of the instruction encoding.

- **Small program size:** Microprograms are typically modest in size, ranging from hundreds to a few thousand microinstructions. Two factors dictate this modest program size. First, microprograms are very difficult to develop and debug because the wide microprogram instructions are very complex, so large applications require much more engineering effort than do projects that use microprocessors with simpler instruction sets and efficient high-level language compilers. Second, compared to microprocessors, the typical loose encoding of microengines and the resulting wide instruction words make microprogram storage quite large relative to the application's complexity. Denser microinstruction encoding is certainly feasible, but more sophisticated microengines require more extensive verification and better software-development tools.

NOTE The above observation about small microprogram size does not apply to another use of microcode: implementation of processors with complex instruction sets. Historically, some of the most complex general-purpose instruction sets have been partially or fully implemented using microcode techniques. In these cases, the microcode development may represent hundreds of engineer-years, but may still be more flexible and manageable than a hardwired implementation of these complex processors. High-end-processor development projects usually merit much bigger funding than the typical SOC project, so the effort to develop huge bodies of microcode for these processors can be justified.

- **Unusual branch architectures:** A branch may be executable every cycle, in parallel with computational and data-movement operations. In some cases, the branch architecture can be complex, with multiway branches, microcode subroutine calls, and code-dispatch operations (using a data value from the execution units as an offset into a table of instruction addresses).
- **Parallel execution units:** Microcoded-engine architectures look nothing like RISC architectures. Microcoded engines implement multiple parallel function blocks, and many of these blocks can be active in each cycle. Each function block may have its own state regis-

ters, register files, execution units, and even data memories. Each function block typically has restricted communications with the state of other function blocks. Register files are typically shallow. All these differences make a general-purpose RISC architecture that executes one operation per instruction and has only one large centralized register file a poor substitute for a microcoded engine. Microcoded engines with distributed, parallel execution units can only be replaced by configurable processors with a similar ability to support large numbers of distributed parallel function units.

- **Shallow or exposed execution pipeline:** Microprograms for embedded engines are often small, so the selection and fetching of microinstruction words has low circuit latency. Low latency or low clock frequency obviates some of the need for pipelining. Decreasing the amount of pipelining reduces the latency of taken branches (discussed in Chapter 7), sometimes to zero. In other cases, the wide microcode word or complex microinstruction decoding makes single-cycle branches impossible. As a result, some RISC designers choose to implement *delayed branches*, branches whose effect is deferred for one or more cycles so that instructions that follow the branching instruction are executed whether or not the branch is taken. If the compiler can fit useful instructions into the cycles following the execution of the delayed-branch instruction and before the delayed branch takes effect, those cycles can perform useful work. However, delayed branches can create distinct programming and verification challenges for microprogram developers in the absence of tools to automatically schedule microcode operations.
- **Direct interfaces to surrounding logic:** Microcoded engines communicate directly with surrounding logic, so wires from other logic become direct data and control inputs to the microcoded engine. Data results from the microcoded engine and microinstruction fields used to directly control hardware become output wires going to surrounding logic. Consequently, microcoded engines may not need to communicate data or control information through shared memory.
- **Programmed with simple assembly tools:** The combination of small typical program size and complex instructions makes assembly coding possible or even mandatory with microcoded engines. A microcode assembler may only map from symbolic function and register names to microinstruction bit encoding and translate symbolic code labels to RAM addresses. For microengines with more sophisticated features, such as deep pipelining and delayed branches, the assembler might provide some support for code scheduling to hide the complexity.
- **Minimal support for source debuggers, compilation, and operating systems:** Many of the software tools used for large application development are difficult to create for architectures with wide, loosely encoded microinstructions, especially where the primitives of the data path bear little relationship to the primitives of high-level programming languages. Real-time operating systems may be unsupportable because of

  - limited system peripherals for interrupts and exception management,

- the inability to save and restore all machine state, and
- limited memory facilities.

Microengines with many small, isolated memory spaces present special challenges for system-software development, especially when the microengine is used in conjunction with another processor running more complex software. Microengines often lack the communications interfaces and runtime monitors to enable live connections to remote source-level debugging environments. The lack of a unified, global system-address map complicates efficient sharing of data and code between microengines and other processors. Debugging such systems is also more complicated. These limitations of microcode development typically restrict the use of homegrown microengines to deeply embedded and relatively isolated functions within SOC architectures.

- **Awkward integration into system-level interconnect and verification:** Historically, most microengine development has been ad hoc—designed to fit localized needs rather than as part of system-level communication and verification architectures. Microengines typically lack standardized bus interfaces; facilities for sharing memory; and fast, abstract simulation models suitable for incorporation into a larger system model. As a result, microengine integration into the verification process may only occur at the register transfer level rather than at the system-model level.

The benefits of the microcoded engine design style are flexibility and good performance; the liabilities are limited software (or firmware) tool support and poor suitability to large, complex applications. Using the infrastructure for automatically generating application-specific processors to create microengines retains the performance benefits of the homegrown microengine while avoiding the design and verification liabilities.

6.4.4 Microcode Data Paths

Most features of microengine data paths transfer directly into the data path and instruction set of an application-specific processor. Discrete registers become processor state registers. Register banks of arbitrary width and depth become processor register files of arbitrary width and depth. Function units described in RTL can often be transferred to an application-specific processor unmodified. The TIE language, for example, supports full combinational Verilog for description of data-path functions.

The control signals of the microengine data path correspond to fields in the microinstruction word. Application-specific processors can accommodate almost any arbitrary number and form of control fields, though the configurable processor may impose some restrictions on microinstruction length. Microcoded engine instruction widths are uniform, but may be an arbitrary number of bits, depending on the application.

Long-instruction-word processors, as outlined in Chapter 5, Section 5.7, typically support instructions that are a multiple of 8, 16, or 32 bits in width. For example, the Xtensa architecture's base instructions are either 16 or 24 bits wide. With Tensilica's FLIX (Flexible

Length Instruction eXtensions), instruction width can be extended to either 32 or 64 bits wide. These wide instructions permit a greater range of flexible instruction combinations, allowing extended instruction sets to emulate microcoded engine encoding.

In converting an instruction word from a microengine to an application-specific processor, additional unused fields may be added to pad out the instruction word to a supported width. Alternatively, microcode fields may be combined or encoded to compress the microinstruction so that it fits into the instruction widths supported by the application-specific processor. In practice, this field compression rarely imposes any limitations on the generality or performance of the microengine architecture. For example, a version of the simple microengine in Fig 6-23 could be implemented by the TIE code in Fig 6-24.

The first line of the TIE description of the processor define a new 64-bit format with five slots: one for the branch; one for condition-code calculation; and one each for computation of the A, B, and C state registers. Lines 2 to 4 define the A, B, and C registers. Lines 5 to 10 declare the operations that fit into each slot. The branch slot allows the Xtensa processor's built-in branch-on-condition-code instructions to operate on the condition codes. The branch slot also encodes instructions that generate logical combinations of different Boolean condition codes. The condition slot allows one instruction that computes a vector of eight different conditions. The A, B, and C operation slots each allow two instructions. Lines 11 to 14 declare the function specifiers. Lines 15 to 29 define the A, B and C functions and the operations used for generating condition code. The definitions of `func1`, `func2`, `func3`, and `func4` are not shown here, but these would be defined as shared functions.

```
1: format format1 64 {br_slot, cond_slot, A_slot, B_slot, C_slot}
2: state StA 32
3: state StB 32
4: state StC 32
5: slot_opcodes br_slot {BT, BF, ORB, ORBC, XORB, ANDB, ANDBC, ALL4, ANY4,
6:    ALL8, ANY8}
7: slot_opcodes cond_slot {COND}
8: slot_opcodes A_slot {A_f1, A_f2 }
9: slot_opcodes B_slot {B_f2, B_f3 }
10: slot_opcodes C_slot {C_f3, C_f4 }
11: immediate_range f1ctl 0 3 1
12: immediate_range f2ctl 0 7 1
13: immediate_range f3ctl 0 3 1
14: immediate_range f4ctl 0 3 1
15: operation COND {out BR8 c8} {in StA, in StB, in StC} {
16: assign c8 = {StA==32íh0, StB==32íh0, StC==32íh0, StA==B, StA==StC,
17:     StB==StC, StA[31], StB[31]};}
```

```
18: operation A_f1{in f1ctl f1} {inout StA, in StB} {assign StA =
19:     func1(StA,StB,f1);}
20: operation A_f2{in f2ctl f2} {inout StA, in StB} {assign StA =
21:     func2(StA,StB,func2);}
22: operation B_f2{in f2ctl f2} {in StA, inout StB} {assign StB =
23:     func2(StA,StB,f2);}
24: operation B_f3{in f3ctl f3} {inout StB, in StC} {assign StB =25:
25:     func3(StB,StC,f3);}
26: operation C_f3{in f3ctl f3} {in StB, inout StC} {assign StC =
27:     func3(StB,StC,f3);}
28: operation C_f4{in f4ctl f4} {in StA, inout StC} {assign StC =
29:     func4(StA,StC,f4);}
```

Figure 6-24 Simple microengine TIE example.

6.4.5 Encoding Operations

In theory, a microengine with instruction storage of 1024 words (for example) requires no more than 10 bits to uniquely identify 1024 different instructions or combinations of operations. Additional instruction bits serve only to simplify instruction encoding and to generalize the instruction set to support additional operation combinations for other code sequences. The developer can often identify combinations of operands and operations that must always be used together or one-hot-encoded operand or operation specifiers that are always mutually exclusive. These fields or combinations may be encoded together, reducing the number of bits required to encode the instruction while somewhat increasing the number of gates used in the processor's decoder.

Data memory and interfaces present some special issues when adapting a microengine or RTL design to an application-specific processor. Processor pipelines, even in configurable processors, typically constrain the stage, width, number, and latency of data accesses per instruction. Most fixed-ISA embedded processors allow only one data reference per cycle, but application-specific processors can allow two or more. Processor memories are usually a power-of-two in bit width: 32, 64, or sometimes 128 bits. This limitation simplifies support of multiple memory data types (32-bit processors usually operate on 8-, 16-, and 32-bit data).

Some thought is required for application data types that are not a power-of-two in width. Two approaches to non-power-of-two data width can be used:

- **Packed data:** Application data words are packed as tightly as possible into power-of-two memory words, routinely crossing power-of-two memory-word boundaries. Fig 6-25 shows the packing of eight 24-bit data samples into six 32-bit memory words (shown with "little-endian" byte order—least significant byte in lowest memory address).

  Special load instructions are defined to extract partial data words and to reassemble them into processor registers of the correct data width. Special store instructions decompose data into subwords that fit into the memory-word width and insert them into the cor-

	Byte 3	Byte 2	Byte 1	Byte 0
Address: 0x00	Data-sample 1 [low]	Data-sample 0		
Address: 0x04	Data-sample 2 [low]		Data-sample 1 [high]	
Address: 0x08	Data-sample 3			Data-sample 2 [high]
Address: 0x0a	Data-sample 5 [low]	Data-sample 4		
Address: 0x10	Data-sample 6 [low]		Data-sample 5 [high]	
Address: 0x14	Data-sample 7			Data-sample 6 [high]

Figure 6-25 Sample packing of 24-bit data into 32-bit memory.

responding memory location. With the right data-alignment and address-manipulation instruction set, application-specific load and store instructions can handle arbitrary bit-level addressing. A load or store alignment buffer can be used to sustain the full data rate on sequential access to unaligned data, even data with non-power-of-two width.

A load alignment buffer is shown in Fig 6-26. A load that brings in one value may also bring in the low-order part of the next value. This partial value is placed in an alignment buffer. When a second load brings in the word that contains the high-order part of that next value, the low- and high-order parts are merged into a complete 24-bit value and placed into the destination register file. A similar method is used in Chapter 7 for aligning the loading of unaligned SIMD vectors.

- **Unpacked data:** Where only moderate amounts of application data must be stored in memory, unpacked data storage may be appropriate. The application data is aligned to the next larger power-of-two bit width, so 12-bit data is aligned on 16-bit boundaries and 24-bit data is aligned on 32-bit boundaries. The application uses standard power-of-two load and store data widths, but application-specific registers and data paths store and manipulate the narrower data. Fig 6-27 shows the unpacked mapping of eight 24-bit data samples into eight 32-bit memory words

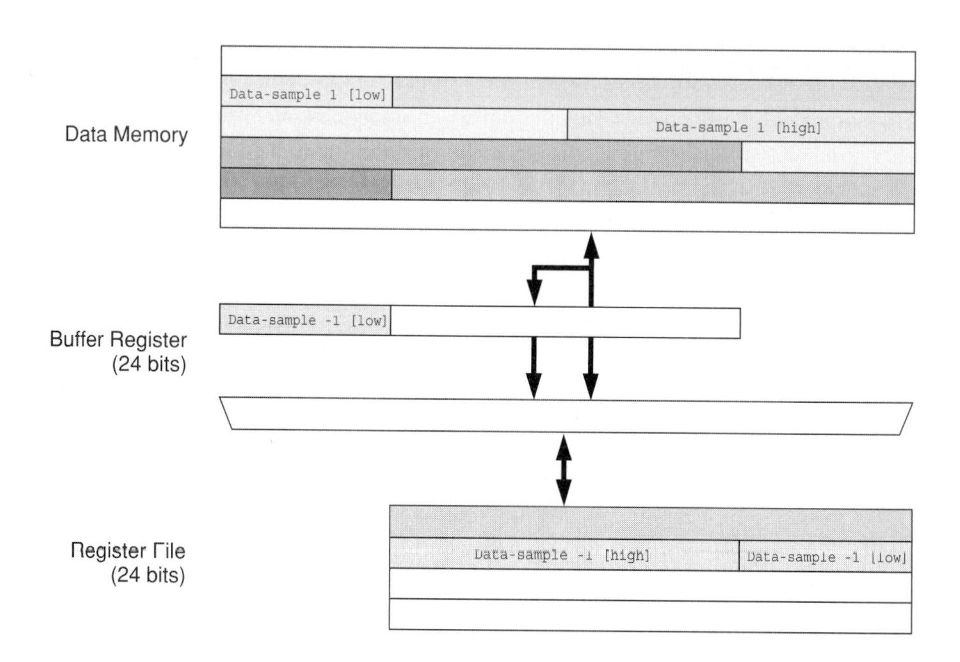

Figure 6-26 Use of alignment buffer to load packed 24-bit values from 32-bit data memory.

	Byte 3	Byte 2	Byte 1	Byte 0
Address: 0x00		Data-sample 0		
Address: 0x04		Data-sample 1		
Address: 0x08		Data-sample 2		
Address: 0x0a		Data-sample 3		
Address: 0x10		Data-sample 4		
Address: 0x04		Data-sample 5		
Address: 0x18		Data-sample 6		
Address: 0x1c		Data-sample 7		

Figure 6-27 Unpacked 24-bit data in 32-bit memory.

Fig 6-28 shows a simple instruction set written in TIE that performs 24-bit signed-integer multiplication-with-accumulation (with 24-bit saturated results) on a 24-bit, 64-entry register file with loads and stores from a 32-bit data memory.

```
1: regfile XR 24 64 x
2: operation l24i {in AR base, in AR offset, out XR dst} {out VAddr, in
3:     MemDataIn32} {
4: assign VAddr = base + offset;
5: assign dst = MemDataIn32[23:0];}
6: operation s24i {in AR base, in AR offset, in XR src} {out VAddr, out
7:     MemDataOut32} {
8: assign VAddr = base + offset;
9: assign MemDataOut32 = {8{src[23]},src};}
10: operation muladd24 {in XR m1, in XR m2, inout XR macc} {} {
11: wire [48:0] mad = (m1 * m2) + macc;
12: assign macc = (mad[48:24] != 25{mad[23]})?24{mad[23]}:mad[23:0];}
```

Figure 6-28 24-bit load, store, multiply-accumulate TIE example.

External interface design plays an important role in microcoded engines, because overall throughput is often governed by the sustained throughput of the wires at the boundary of the microcoded engine block. Depending on the data format and bandwidth, passing I/O data through shared random-access memory may be appropriate. In other cases, where data rates are high and random access in not important, other interface schemes such as hardware queues and simple registers at the block boundary may be a better design choice. In general, configurable processors can closely reproduce the full range of input and output interfaces used with microcoded engines including dual-ported memories, single-port shared memories with arbitration, memory-mapped queues, direct connection to input and output data queues, and direct connection to external function units and registers. Interface design is discussed in greater depth in Section 6.5.

6.4.6 Microprograms

The close similarity between microengines and application-specific processors means that it's usually a simple task to convert microprograms into processor programs. Developers can achieve something close to a one-to-one correspondence or *transliteration* for most microinstructions, depending on the nature of the original microengine pipeline and the set of function units.

The developer may choose to convert microinstructions into either assembly code or C code. Assembly code may have the more familiar syntax and provides the most precise control over the number of instructions (and number of cycles) along any path through the code. C code is more portable and provides automation of register allocation, code scheduling, and operation packing. C code may also enable optimizations such as software pipelining, loop unrolling, and interprocedural analysis that are awkward and time-consuming to perform with assembly-language programs. These optimizations can yield additional performance compared to simple assembly code.

Differences between pipelined processors and microengines in branch and memory-access architecture deserve special attention. These topics are discussed in Chapter 7.

6.5 Designing the Processor Interface

Application-specific processors offer the high performance and efficiency traditionally associated with hardware blocks, but the benefits of a small, fast, programmable computational core are lost without the right interfaces. Successful use of processors in traditional hardware tasks may require novel processor interfaces for two reasons: performance and familiarity.

First, consider performance: The typical 32-bit processor bus interface has a peak performance of no more than four bytes per cycle. The overhead of initiating bus transfers and the latency between access request and data response may cut this useful bandwidth by a large factor, particularly if the application randomly accesses memory for application data. At typical clock frequencies associated with 0.13-micron SOCs, the sustained useful bandwidth of a 32-bit processor external bus may be well below 300 Mbytes/sec. Consider a processor that performs a series of 32-bit loads from addresses that are not associated with the processor's local data memory, but designate another fast on-chip memory, as shown in Fig 6-29.

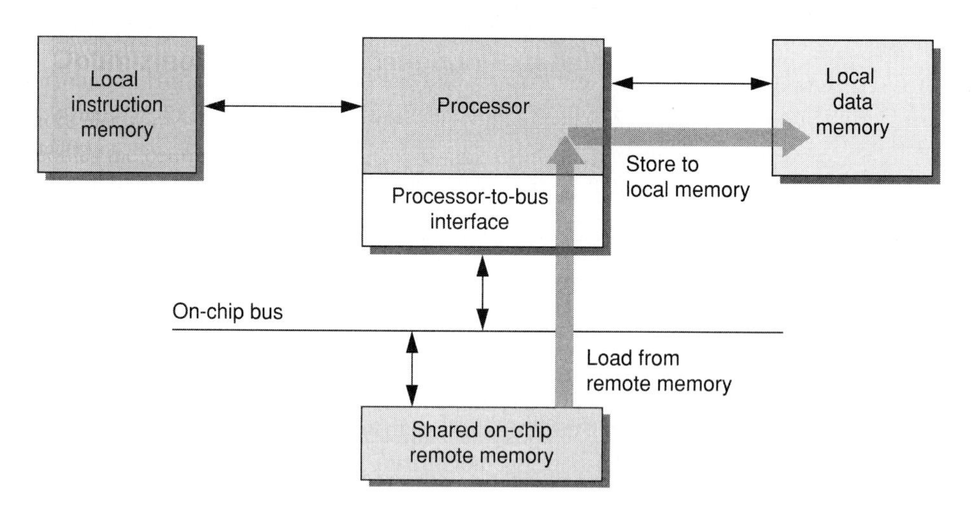

Figure 6-29 System structure for remote-to-local memory move.

For each load, the processor must acquire control of the external bus interface, send the read address, wait for the data on the bus, then move the data from the bus to the destination register. The pipe stages for an optimal bus read might look like that shown in Fig 6-30.

Cycle 1	Cycle 2	Cycle 3	Cycle 4	Cycle 5
Memory Stage (address from pipeline to bus)	Read Address on External Bus	Memory Data on External Bus	Restart (data from bus to pipeline)	Data available for register write

Figure 6-30 Read overhead for on-chip bus operation.

This external load transaction requires three extra cycles compared to an internal load. Moreover, a general-purpose RISC architecture combines no other operations with its load and store instructions, so even the simplest use of the retrieved data adds cycles and reduces the achievable I/O bandwidth. The most trivial data operation would be to mask some bits of each data word as it is moved from remote memory to local memory. The breakdown of cycles for this move-with-mask operation is shown in Fig 6-31, yielding a sustained bandwidth of just 0.67 bytes per cycle.

Operation	Cycles Consumed	Bytes Processed
32-bit load instruction	1	4
Overhead for remote memory access	3	
32-bit AND instruction (masking)	1	4
32-bit store instruction (local memory)	1	4
Total cost of move-with-mask	**6**	**4**

Figure 6-31 Bandwidth calculation for move-with-mask operation.

Input and output limitations are key reasons why many hardware designers dismiss RISC processors as a possible design alternative to hardware for high-bandwidth tasks.

By contrast, an application-specific processor with an instruction set that includes wide load and store instructions and an external interface that resembles a single-cycle local-memory interface can sustain data rates of more than four bytes/cycle (if loads and stores move data 64 bits at a time) and more than eight bytes/cycle (if loads and stores operate 128 bits at a time). These higher sustained data rates result from the reduced memory-reference latency, the increased transfer width, and the ability to merge masking operations into the load and store instructions. If the application-specific processor incorporates dual memory ports, the move-with-mask data rates double to eight and 16 bytes/cycle, respectively. Using wide data interfaces, direct connections from external logic to processor execution units can push sustained input and output bandwidth to tens of bytes/cycle.

Second, consider the ease in mapping from a hardware interface to a processor interface: Some hardware tasks operate on large data blocks—hundreds or thousands of operands per data block—accessed in complex and changing sequences. A typical example of such a task is image compression, where the application treats each image frame as a set of image arrays (one image array for each color component). Each image array is a 2D arrangement of pixel component values, which may be accessed in row, column, or zigzag (diagonals starting from one corner of the block) order. For such hardware tasks, loading input data from an external memory is necessary (due to the large size of the arrays) and generating the correct sequence of data addresses is a significant task within the overall image-compression subsystem.

Other tasks, however, operate on data presented sequentially at a hardware interface—on a set of wires. This interface may take the form of a queue, with a simple "pop" operation to access successive values, or the interface may require a more complex handshake in which the data consumer must execute a series of steps to cause the data producer to present the next data value.

In theory, the design could latch the value of these wires into a register and then connect that register to the processor over an on-chip bus, but this approach may be neither natural nor efficient. Mapping the data register onto the bus is unnatural and inefficient for three reasons:

1. The width of the data rarely matches the on-chip bus width (which is often 32 bits). If the data is wider than the on-chip bus width, several cycles are required to transfer and reconstruct the data in the processors. If the data is narrower than the on-chip bus width, the extra, unused bits of the bus transfer are wasted (lost opportunity, wasted power).
2. The transfer requires execution of a load instruction, which consumes instruction, memory, and bus bandwidth. This sort of data transfer also requires the computation and transfer of an explicit memory address pointing at the target register, even though that address is the same for all transfers to or from that data register.
3. The data transfer may require additional instruction, memory, and bus cycles to perform the data-transfer request/acknowledge handshake.

   - **Request:** The processor may need to poll for "data ready" by reading a status register mapped onto the bus. Polling loops waste code space and time.
   - **Acknowledge:** Once the processor reads the data, it may need to write to a control register mapped onto the bus to indicate to the data source that the current data has been read and new data can be prepared. In some cases, the acknowledgment of acceptance can be combined with the reading of the data.

Due to these three complications, the processor's bus overhead and memory-mapped I/O model obstruct the hardware designer. Application-specific processors provide a much richer palette of interface options than do conventional processors, allowing the hardware designer to mold the processor's interface to closely fit the interface needs, often even the exact signaling protocol of the application and the surrounding blocks. In fact, a properly configured processor should be able to provide a nearly perfect "plug-replacement" for a well-designed hardware block. Ideally, the hardware designer would simply specify the original hardware block's interfaces as features of the processor's interface.

An application-specific processor offers three primary forms of interface: memory-mapped RAM, memory-mapped queues and registers, and direct connection of wires to new processor ports and queues.

6.5.1 Memory-Mapped RAM

Shared memory is the natural interface for many applications. One task writes data into RAM as it is created; another task reads the RAM and operates on it. The two tasks can take a variety of forms: each could be a hardwired logic block, each could take the form of software running on different processors, or both could be software tasks running on one time-sliced processor. Shared RAM is the preferred interface under any of the following conditions:

1. **Data ordering:** Input data is used in different order than the creation order, or output data is produced in a different order than the consumption order.
2. **Bursty data:** The data-creation rate for input (or data-consumption rate for output) is not closely matched to the processing rate of the block in question, so large amounts of buffering are required to hold data waiting to be processed. When two software tasks on one processor must communicate, shared memory is particularly appropriate because the mismatch between data creation and data consumption grows with the time-slice duration of each task.
3. **Data source and destination uncertainty:** The source and destination of data are uncertain, so flexibility is mandatory. Buffering input and output through a memory that is accessible to a large number of processing blocks maximizes later choices in data flow among blocks. This data-flow flexibility goes hand-in-hand with the programmability offered by application-specific processors.

Three types of processor hardware interface underpin the flexibility of shared memory. Chapter 7 discusses the range of options available for establishing communications between software tasks through shared memory.

6.5.1.1 Bus-Based RAM

Shared memory accessed across an on-chip bus (as shown in Fig 6-32) is often the most appropriate design alternative when the input or output bandwidth is moderate—tens or hundreds of processor cycles per byte of transferred I/O data—and many different agents might need access to the data. The processor accesses this memory through its general-purpose bus interface. Other requestors gain access through their own bus interfaces. Arbitration for the bus includes arbitration for the block of shared memory because the entity that controls the bus controls the shared memory as well. A single-ported memory can therefore be shared by any number of function blocks or processors.

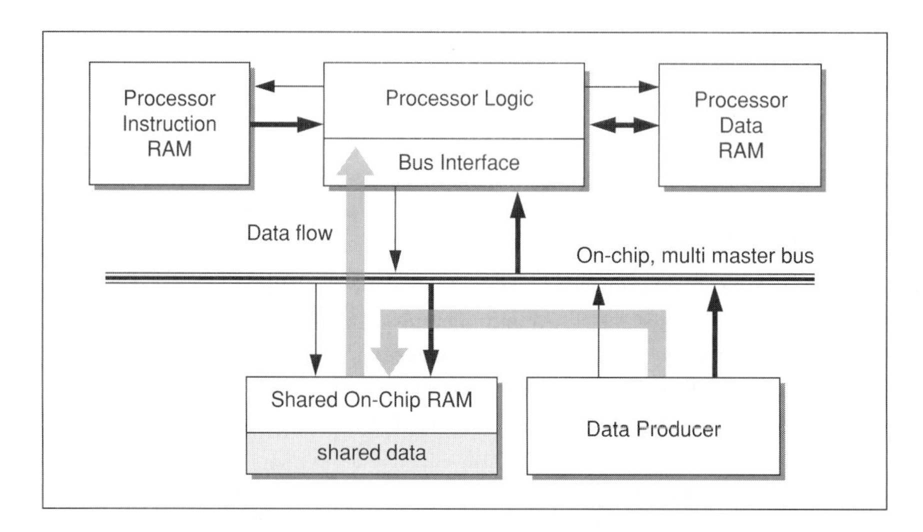

Figure 6-32 Shared on-chip RAM on bus.

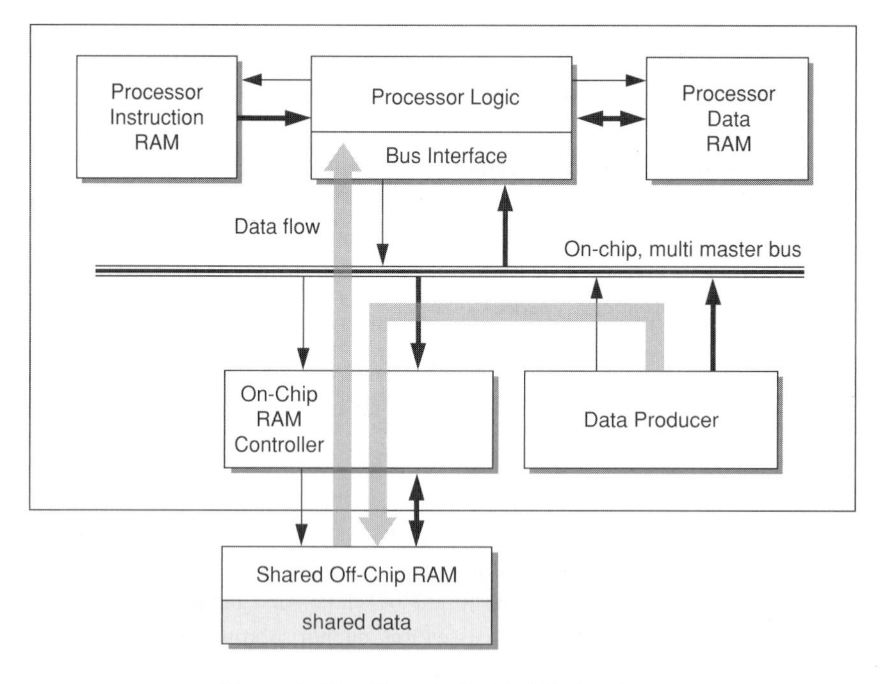

Figure 6-33 Shared off-chip RAM on bus.

The shared RAM itself may be entirely on-chip, or the RAM-control interface may be on-chip with large static or dynamic RAM devices off-chip, as shown in Fig 6-33.

When the size and number of I/O buffers is uncertain, mapping all buffers into one RAM may improve utilization and promise more capacity headroom for ongoing software changes. Uncertainty in buffer size is particularly common when the specification of software-based tasks is incomplete and when both the data-block sizes within each task and the relative data-processing rates among tasks are yet to be finalized.

On the other hand, centralizing all input-output data buffering in one RAM also increases the likelihood that the buffer RAM (and the bus used to access it) become performance bottlenecks as I/O demands increase.

Note that Fig 6-32 and Fig 6-33 show shared, multimaster, on-chip buses, but these figures do not imply that physically shared, multidriver, three-state wires are actually used on the chip. Chip-level buses routinely use only multiplexers and single-driver wires to improve device testability and timing predictability at the cost of somewhat greater on-chip wiring congestion. For SOCs, single-driver wires and multiplexers have displaced the three-state buses that were commonly used in board-level system designs. However, these on-chip wires and multiplexers still operate collectively like a bus, with a single master in control of the bus during any one cycle.

6.5.1.2 Shared RAM on the Extended Local-Memory Interface

Application-specific processors always have some fast local memory—data RAM or data cache—to hold intermediate results. Local-data memories typically range from a few kilobytes to a few tens of kilobytes, though memories up to a megabyte may fit the timing requirements of the processor pipeline. The address/data interface used to connect this modest size block of dedicated RAM can be extended to allow access to shared RAM blocks. The extension includes adding signals for arbitration (e.g., an output signal that indicates that the processor wants access, an input signal that stalls the processor when the shared RAM is not available to it), so that several requestors can share one RAM access port.

As long as the arbitration is simple, sharing RAM on the local-memory interface does not reduce processor clock frequency because the RAM control circuits add little logic to the critical data, address, or control paths. Some designers put true dual-ported RAM on the local-memory interface to ensure that both the processor and the other requestor have guaranteed single-cycle access. However, dual-ported RAM blocks may be twice as big as single-ported RAM blocks on a per-bit basis. Input to the processor using the shared-RAM approach is sketched in Fig 6-34.

6.5.1.3 External Access to Processor Internal RAM

If the processor has some local memory, efficient sharing also may be possible by allowing external access to the processor's "private" data memory through the processor's main bus. If the processor's bus interface is appropriately designed, external requestors on the bus can treat the processor as a slave, which gives other bus masters read and write access to the processor's internal memories. Local-memory access arbitration between the application running on the local processor and an external agent can be built directly into the processor pipeline, so sharing

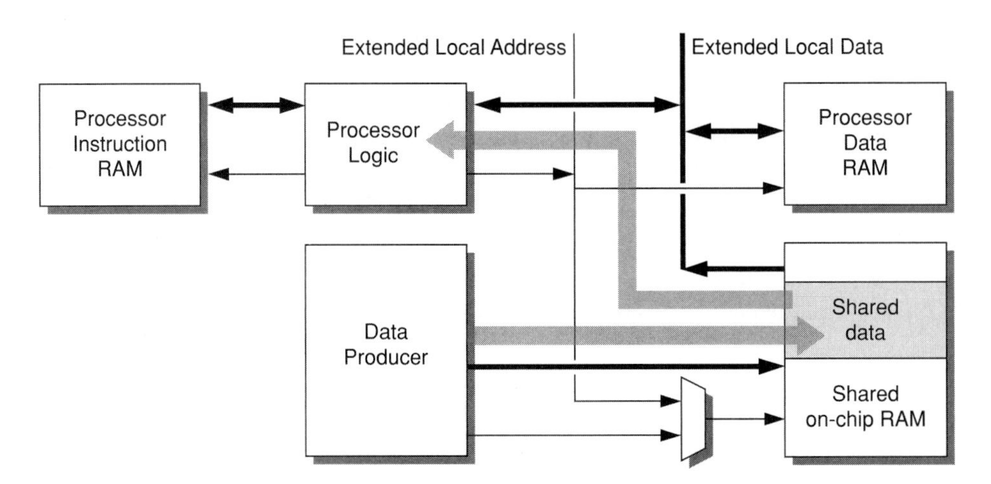

Figure 6-34 Shared RAM on extended local-memory interface.

the single-ported local-data RAM in this way is quite efficient from the processor's perspective. The burden of latency tolerance is on the external bus master, which must arbitrate for the processor's general-purpose bus interface and then wait for the processor's internal arbiter to grant access to the local-data memories.

External read and write requests can often safely wait a few cycles for an unused load/store slot, maximizing the local-RAM port's bandwidth utilization and minimizing the disruption of processor execution. The processor will see noticeable performance degradation only from very heavy external I/O traffic or when the application running on the processor is fully consuming the local-memory bandwidth. With some memory arbiters, the external agent that is generating external read and write requests can specify the priority of requests and obtain maximum bandwidth to the processor's local memory. In this case, the processor ends up stealing cycles from the external agent when cycles are available and stalling during periods when the external agent is taking all available memory cycles.

A system using this approach to input data is sketched in Fig 6-35. This method also allows the processor to operate solely as a bus slave, with an external agent (another processor or DMA engine) responsible for loading the instruction RAM with the application-task software as well as supplying input data to and taking output data from the processor.

6.5.2 Memory-Mapped Queues and Registers

A queue-based interface may be the most appropriate interface when the required input or output data sequence is simple or stable. The data producer pushes data into a queue and the processor pops data from the queue as needed by reading from an address that maps to the queue head. The queue head is mapped onto the extended local memory interface, as shown in Fig 6-36. This arrangement eliminates the request/acknowledge handshake—the read from the

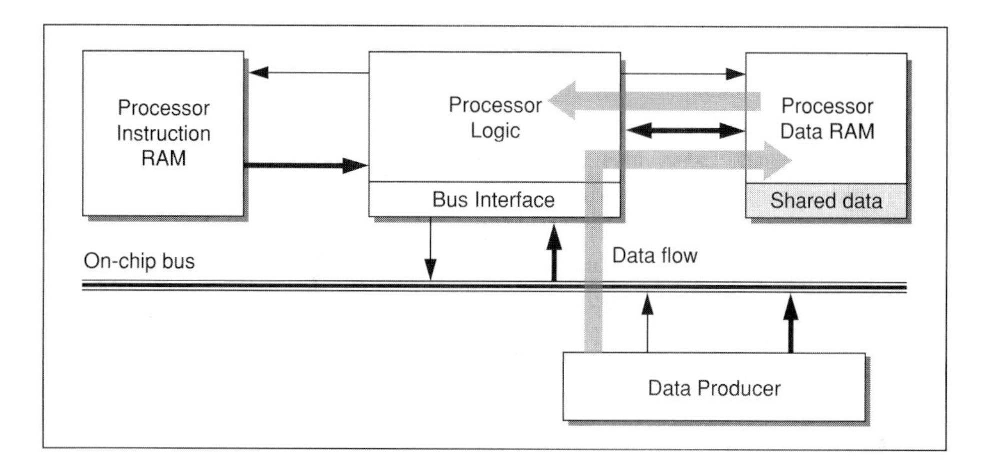

Figure 6-35 Slave access to processor local RAM.

queue has the side-effect of automatically advancing the queue so the next data word is immediately available. This method is particularly attractive when the required data rate is one byte of data every few cycles, or faster.

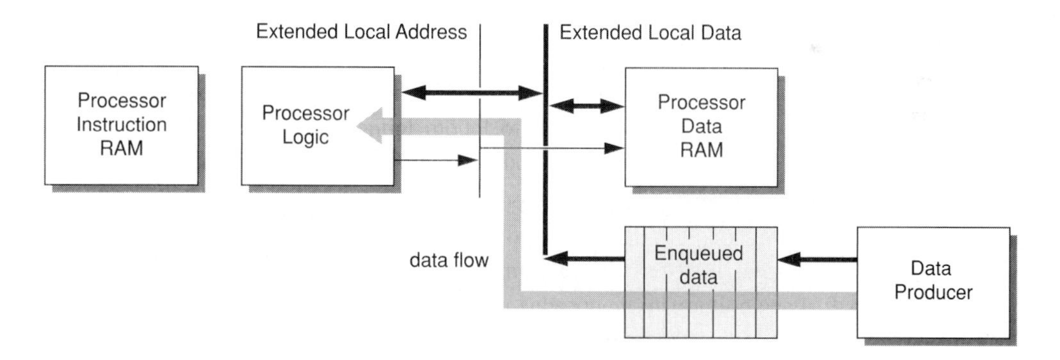

Figure 6-36 Input queue mapping onto extended local-memory interface.

Some I/O interfaces do not require the high bandwidth of a queue-based interface. A simple memory-mapped data register may be sufficient. A data register is really just a degenerate, one-word-deep queue, though the request/acknowledge handshake may not be implicit in reading the register. For input operations, the processor reads the data word, then signals to the data producer via a write to a control register that the current data word has been consumed. The producer may notify the processor when the next data word is available through a status bit in a register or via an interrupt. Output operations are similar: the processor writes to a memory-mapped data register and notifies the data consumer that data is available. The consumer reads the data and signals back that the next data word can be sent.

In general-purpose RISC architectures, the load and store operations for memory-mapped I/O consume useful computation cycles. Configurable processors can eliminate this I/O overhead by combining loads and stores with other operations, especially with application-specific operations on the data stream. For an application-specific processor where data words may be quite wide (up to 128 bits or more), loads and stores can be combined with data-manipulation operations to achieve sustained I/O rates of several bytes per cycle.

Consider the simple hardware block example of Section 6.2 and the input and output data flow through individual state registers shown in Fig 6-37 (similar to Fig 6-8). In this example, adding the load operations for input and store operations for output may require no additional instruction-encoding bits. Input and output become implicit parts of the operation.

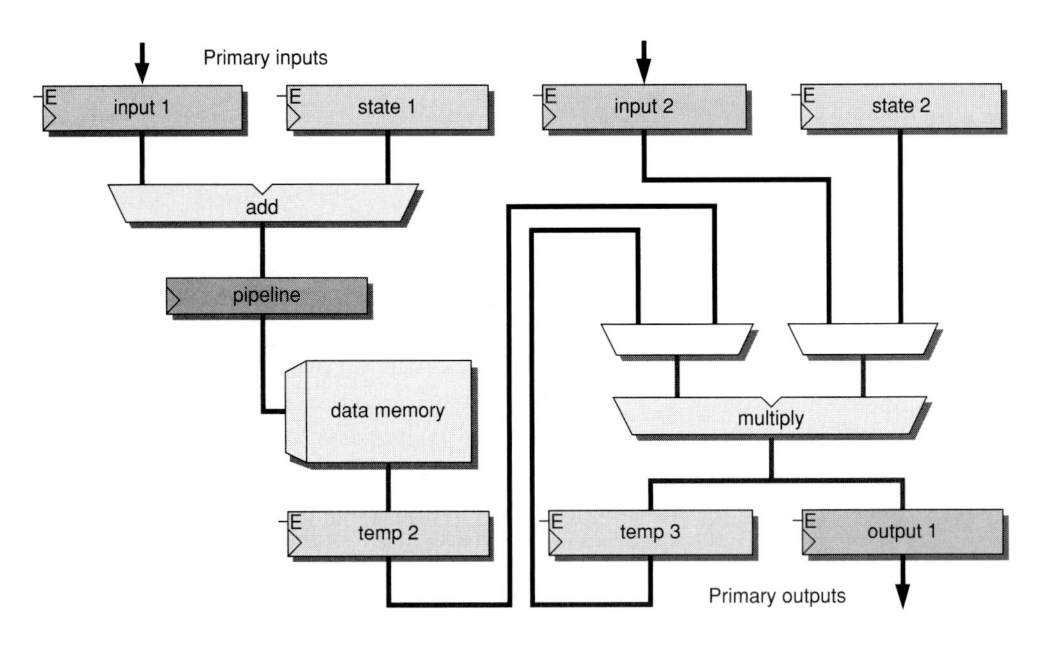

Figure 6-37 Optimized data path with input and output registers.

The combination of input loads and output stores with other operations is shown in Fig 6-38 (similar to Fig 6-12). All loads have a latency of two cycles. The store for the previous value of `output1` starts in cycle 2 and is placed in an output queue mapped into memory at `output1_address`. The load for the next value of `input1` starts in cycle 3, and the data comes from an input queue mapped into memory at `input1_address`. The load for the next value of `input2` starts in cycle 4, and the data comes from a queue that is mapped into memory at `input2_address`.

Cycle	Function
1	`lookup(input1,state1) [first cycle]` `input2 = read(input2_address) [second cycle]`
2	`temp2 = lookup(input1,state1) [second cycle]` `store(output1_address, output1)`
3	`temp3 = multiply(temp2,input2)` `read(input1_address) [first cycle]`
4	`output1 = multiply(temp3,state2)` `read(input2_address) [first cycle]`
5	`input1 = read(input1_address) [second cycle]`

Figure 6-38 Cycles for execution of pipelined function, with input loads and output stores.

The designer may also allocate the input and output registers in a central register file, as shown in Fig 6-39 (similar to Fig 6-13).

Additional register-file ports may be required to create the necessary register bandwidth because register reads (for address generation and memory-store operations) and register writes (for memory-load operations) must be able to occur in parallel with the register reads and writes generated by the computational data path.

In this example, two new register ports—a second write port and a third read port—are required to satisfy worst-case read and write combinations. The worst-case combination of reads occurs during cycle 3 with the reads of `temp2`, `input2`, and `input1_address`. The worst-case combination of writes occurs during both cycles 1 and 4. Use of a central register file may also increase instruction size because the instruction word must include operand fields that specify the input destination register, the output source register, and the register that holds the memory-mapped queue address.

To relieve some of the overhead of the large numbers of required read and write ports, values involved in address calculations (`input1_address`, `input2_address`, `output1_address`, `input1`, `state1`) might be mapped to a general-purpose register file, while other data values (`temp2`, `temp3`, `state2`, `output1`) might be mapped to a second, application-specific register file. In this case, the general-purpose register file would need two read ports and one write port, and the second data register file would also need two read ports and one write port. This design would be particularly attractive if the data values (`temp2`, `temp3`, `state2`, `output1`) are either narrower or wider than 32-bits, because the existing general-purpose register file is sized appropriately to handle the 32-bit address values (`input1_address`, `input2_address`, `output1_address`, `input1`, `state1`) while

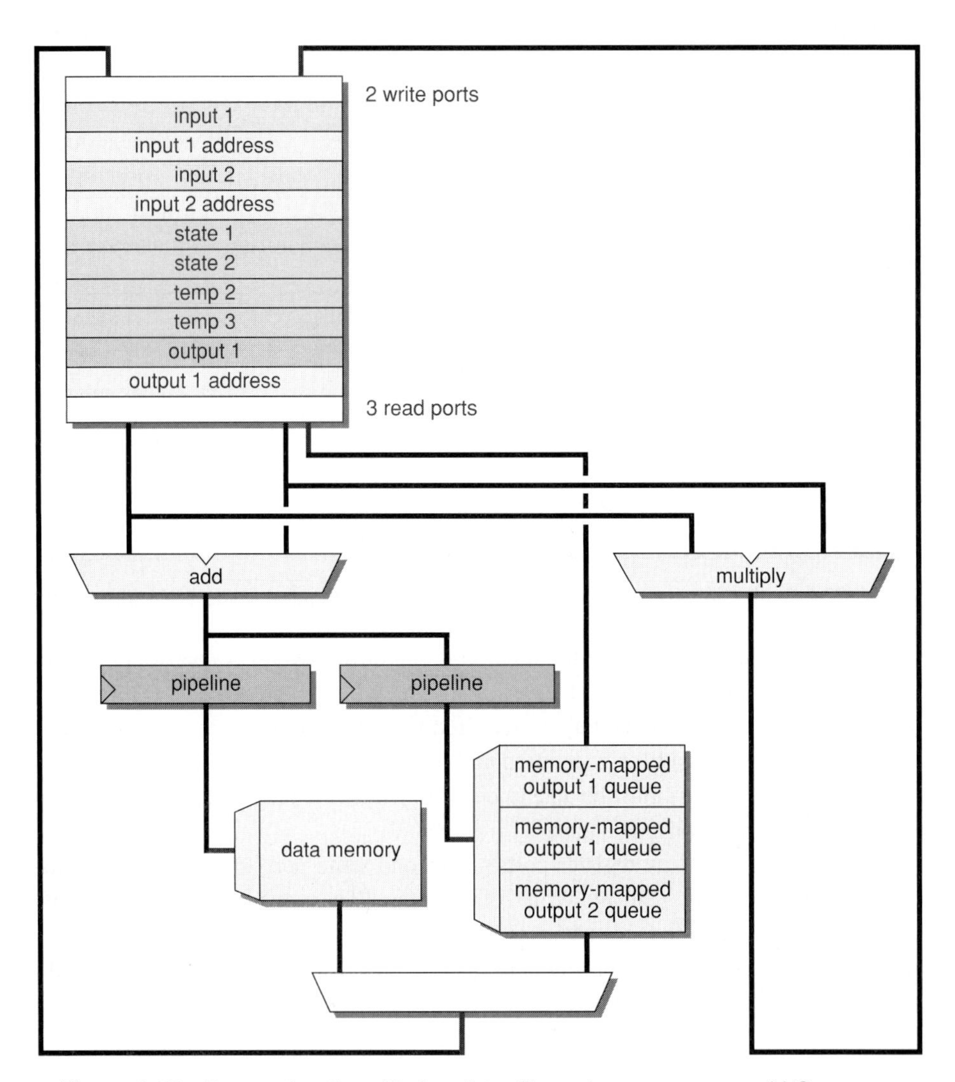

Figure 6-39 Data path with unified register file and memory-mapped I/O queues.

the oddly-sized data values would fit better in a custom-built register file of the appropriate size. Use of two separate register files is also attractive because this approach reduces the number of read and write ports needed for each register file. Large numbers of read and write ports connected to one register file cause long timing paths and reduce the processor's maximum operating clock frequency.

Memory-mapped input and output still require available memory ports and may still be constrained to execute in the pipe stage allocated to memory operations. As a result, an input operation cannot be combined with another operation that uses memory (unless multiple memory ports are specified). Instead, the input read should be combined with an earlier instruction, and the output operation should be combined with a later instruction.

Multiple memory ports create the opportunity for even higher bandwidth, but introduce additional pipeline complexity. Chapter 7 discusses this topic in more detail.

6.5.3 Wire-Based Input and Output

Mapping all input and output interfaces to memory addresses is neither necessary nor efficient for some applications. Sometimes the mapping of I/O interfaces to memory addresses permits the programmer or compiler to dynamically choose among several sources and destinations for computations. If this dynamic addressability is not important, direct connections from external signals to processor execution units can further accelerate performance and reduce complexity. Wire-based interfaces are also more familiar to many RTL designers and often allow processors to substitute for hardware blocks without even changing the block-interface ("pin") definitions of existing RTL blocks.

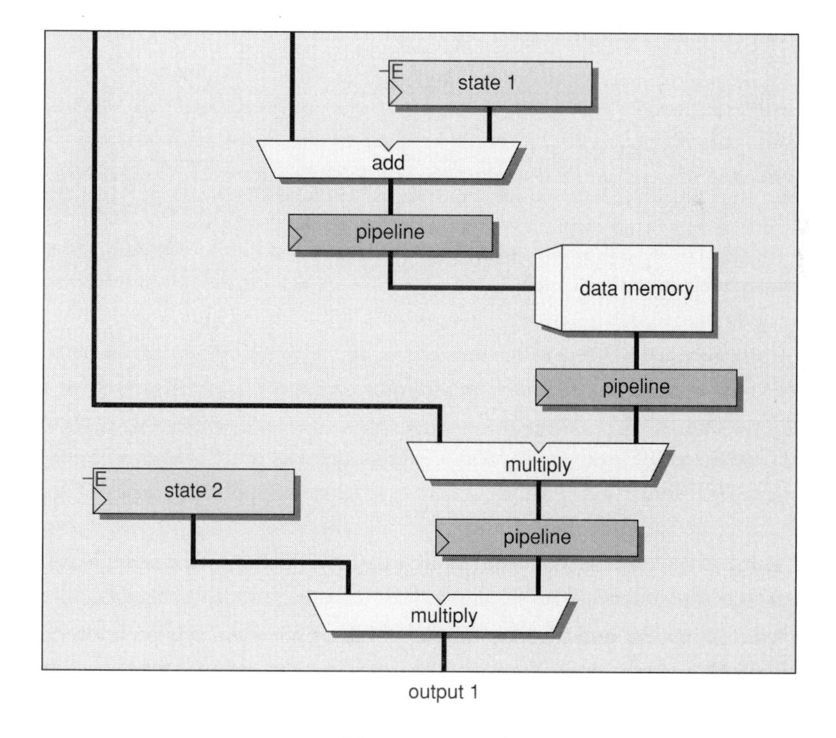

Figure 6-40 Fully pipelined instruction implementation with direct I/O from block.

Two basic styles of interface handshake serve the different input and output environments for direct connection of processors to external signals:

- Input and output queues
- Import of values and export of states through ports

Consider the simple hardware function of Section 6.2. The primary inputs and outputs may be simplified to wires at the boundary of the block, as shown in Fig 6-40 (similar to the pipelined implementation with registered inputs and outputs shown in Fig 6-18).

Input and output ports may serve as source and destination operands for configurable processor operations, enabling fast and flexible interfaces.

Current Xtensa implementation limits earliest use of input queue values to the memory stage (no **use** before schedule stage 2), and limits the latest definition of output queue values to the write-back stage (no **def** after schedule stage 3).

Fig 6-41 shows a sample implementation of the function block written in TIE.

```
1: state state1 24 add_read_write
2: state state2 24 add_read_write
3: state lastinput1 24
4: state nextoutput1 24
5: queue input1 24 in
6: queue input2 24 in
7: queue output1 24 out
8: operation lookup.mul.mul {} {in input1, in input2, in state1, in state2,
9:    inout lastinput1, out output1, inout nextoutput1, out VAddr, in
10:    MemDataIn32} {
11: assign VAddr = {8'h0, lastinput1 + state1};
12: assign lastinput1 = input1;
13: wire [23:0] mulout = MemDataIn32[23:0] * input2;
14: assign output1 = nextoutput1;
15: assign nextoutput1 = mulout * state2;}
16: schedule inst_sched {lookup.mul.mul} {use state2 4; use nextoutput1 3; use
17: input1 2; use input2 2; def lastinput1 3; def nextoutput1 4; def mulout
18:    3; def output1 3; }
```

Figure 6-41 Data path with input and output queues TIE example.

In this listing, the **use** and **def** arguments in the **schedule** statement on lines 16 to 18 specify the pipeline stage where the Xtensa processor's input queue interface has data (**use**) and where the processor's output queue interface accepts the output data from the pipeline (**def**). The input queue interface has data available in the pipeline's memory stage (stage 2), and the output queue interface accepts data during the pipeline's write-back stage (stage 3). The states `lastinput1` and `nextoutput1` allow the late input queue value to be used in the following instruction and the previous instruction's late computational result to be sent to the output queue.

Queue inputs and outputs use direct connection of wire structures. Accesses to the corresponding queue structures automatically pop data from the input queues and push data into the output queue. The queue-control mechanism is aware of instruction cancellation, as discussed in Chapter 7, and ensures that no excess data is popped or pushed even if processor encounters unexpected error conditions.

Queues are one form of instruction-mapped connection. They are particularly appropriate for streaming operand data through an application-specific processor because the request/acknowledge handshaking is already part of the queue interface, as shown in Fig 6-42.

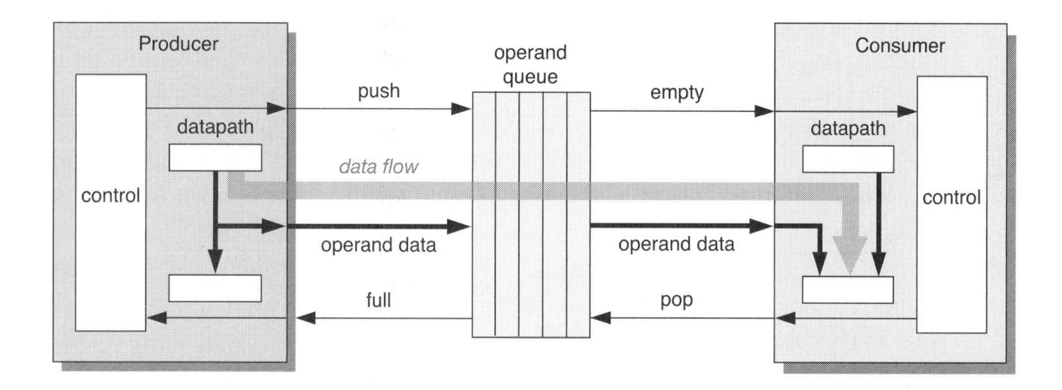

Figure 6-42 Basic handshake for direct processor connections.

Queue inputs represent a stream of data values to be consumed by the application running on a processor. Sequential executions of the consuming instruction should see sequential values. Similarly, queue outputs represent a sequence of values being produced by the application processor. The consumption and production of values can be managed in hardware so that all of the effects of possible instruction execution cancellation are hidden and no explicit request acknowledge handshake is needed. The processor consuming these operands stalls if not enough data has been produced. The processor producing the operands stalls if the consuming processor falls behind and allows the input buffer to fill. These queues form a highly efficient data-streaming connection between processors, especially where several processors comprise a large-scale computational pipeline.

The second form of the direct interface is based on ports: import of values and export of states on a set of wires. These ports are especially useful for tasks that test external status or condition information or control other logic functions.

Application-specific instructions use imported values on an input port just like other input operands. The wires do not need to be explicitly declared, so they do not consume instruction-encoding bits as would register-address specifiers. When the corresponding instruction executes, the instruction simply senses the value on the associated wire. The processor provides no external indication that the corresponding instruction is executing, and there is no acknowledgment that value on the wire has been used. If the application must signal that an input value has been used, it does so explicitly via a store to an external address or by executing an instruction that writes to an output queue or to an exported state (on another wire).

Exporting states creates output ports that deliver information from instructions to external logic or to other processors. The output signals do not change until the processor executes an instruction that explicitly modifies those signals. The normal implementation of this type of signal hides the speculative nature of modern processor pipelines, as discussed in Chapter 7. For pipelined processors, error conditions, conditional branches, cache misses, and other unexpected conditions may cause the premature termination of some instructions that the processor has started to execute speculatively. Speculative execution is a performance-enhancing processor-design technique, but it requires that the system be able to tolerate early instruction termination. Processor hardware that maintains the simple programmer's model of atomic, in-order instruction execution and avoids any unintended output glitches that might otherwise be caused by the processor's microarchitectural operations prevents externally visible state changes from occurring for instructions that do not complete.

The example shown in Fig 6-41 can also be implemented with imported values and exported states, as shown in Fig 6-43. This implementation uses an explicit output signal, `next_data`, to indicate that new values for `input1` and `input2` are required and that a new output value is available on the wire `output1`. It is the developer's responsibility to ensure that external logic has sufficient time to respond to the exported `next_data` signal before the next use of the `input1` and `input2` ports. This guarantee is easily achieved for moderate-performance applications with tens of cycles of latency between one input set and the next. However, queues are generally faster and simpler for very high input rates. Note that the instruction set and the program must explicitly assert the `next_data` wire to request new input data and to signal the availability of new output data. Also note that current Xtensa implementation makes the value of imported wires can be used as early as the ALU stage (`use` as early as schedule state 1), though exported states must be defined by the write-back stage (no **`def`** after schedule stage 3).

```
1: state state1 24 add_read_write
2: state state2 24 add_read_write
3: state nextoutput1 24
4: state output1 24 24íb0 add_read_write export
5: state next_data 1 1íb0 add_read_write export
6: import_wire input1 24
7: import_wire input2 24
8: operation lookup.mul.mul {} {in input1, in input2, in state1, in state2,
9: out output1, inout nextoutput1 out next_data, out VAddr, in MemDataIn32} {
10: assign VAddr = {8'h0, input1 + state1};
11: wire [23:0] mulout = MemDataIn32[23:0] * input2;
```

```
12: assign output1 = nextoutput1;
13: assign nextoutput1 = mulout * state2;
14: assign next_data = 1íh0;}
15: operation assert_next_data {} {out next_data} {assign next_data = 1íh1;}
16: schedule inst_sched {lookup.mul.mul} {use state2 4; def output1 3; def
17:     nextoutput1 4;}
```

Figure 6-43 Data path with import wire/export state TIE example.

This example also compensates for the fact that the computational result is available in a pipeline stage later than that required for state export. The operation therefore exports the result of the previous operation and saves the new computation result in nextoutput1 for export during the next cycle.

6.6 A Short Example: ATM Packet Segmentation and Reassembly

Internet packet processing lies at the heart of data communications. Packet processing encompasses a wide range of tasks, including routing, error checking, conversion among network-packet formats, and management of traffic with different priorities and service levels. These are hard problems with moderate complexity and very high performance demands. This kind of data-intensive problem can be well served with an application-specific processor. Many analogous high-data-rate problems in signal-processing, image-processing, cryptographic, and protocol-processing applications are addressable with similar design methods. The data types and computational details in signal, image, and encryption applications are quite different from those of packet processing, but the need for high data throughput, application-specific computation and interface, and easy reprogrammability are common across all these applications.

Many forms of data traffic are transported over ATM (asynchronous transfer mode) networks. Conversion between the long and variable lengths of different networking data types and fixed-length ATM packets is a common protocol-processing problem, known as ATM SAR (segmentation and reassembly). The hardware required for conversion between these variable length PDUs (Protocol Data Units) and 52-byte ATM cells (48 bytes of payload plus four header bytes) is not complex, so hardware data paths controlled by finite-state machines are often used to implement this function. A particular variant of ATM SAR, known as AAL5 (ATM Adaptation Layer, type 5), supports connection-oriented, variable-bit-rate data traffic and signaling messages, including quality-of-service functions. AAL5 services are suitable for internetworking with many data networking protocols such as Frame Relay, SMDS (Switched Multimegabit Data Service), Ethernet, and IP. Fig 6-44 shows the basic process of segmenting large data blocks into ATM cells and then reassembling the block from cells after the packets traverse the ATM network.

As data networks evolve, the variety of layered protocols, the importance of service guarantees, and the rapid evolution of networking standards all invite a more programmable

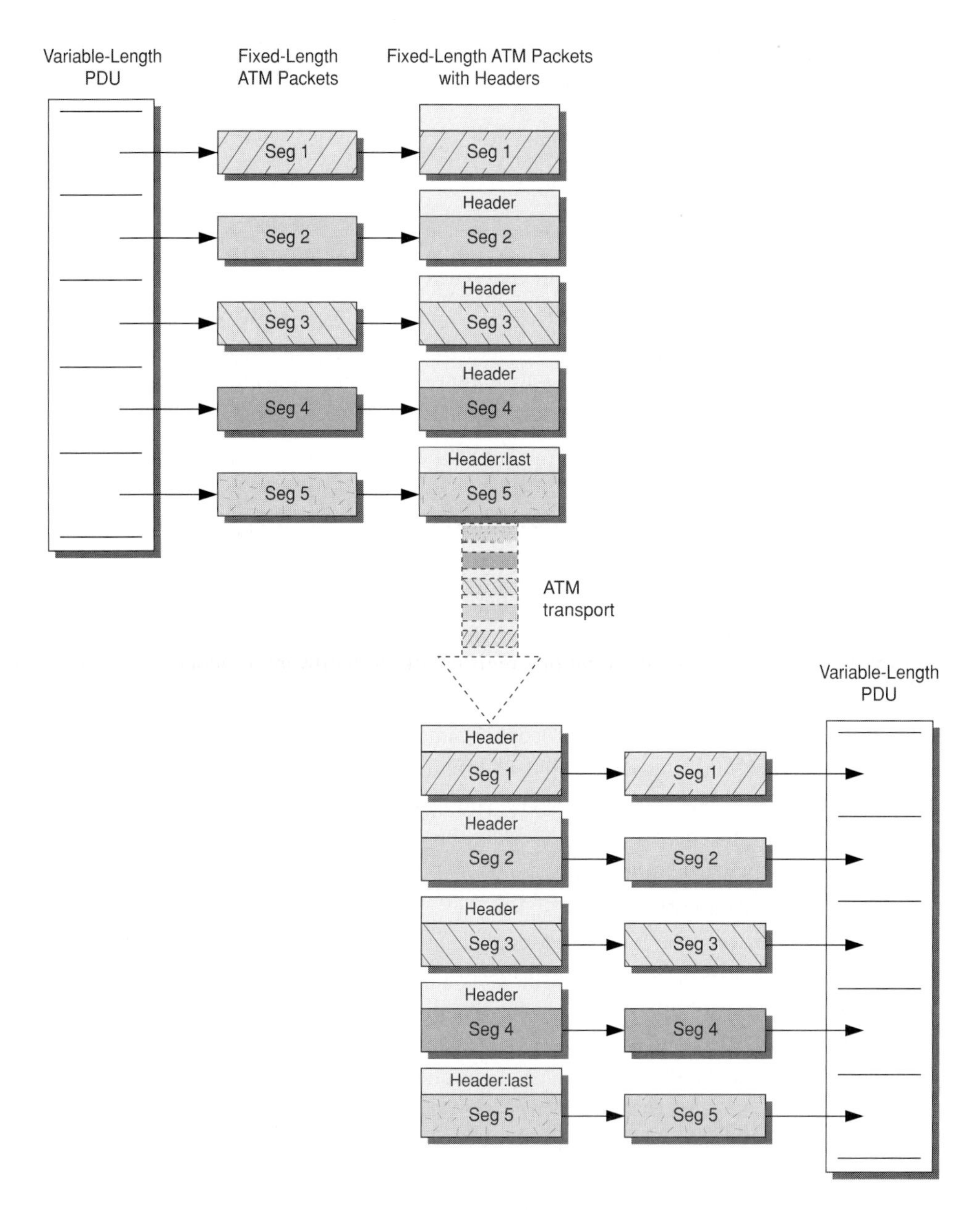

Figure 6-44 ATM segmentation and reassembly flow.

approach to ATM SAR implementation. If a processor can fully implement AAL SAR with significant processing headroom, additional value-added services for traffic management and monitoring, routing, and error checking might also be added using software. A programmable implementation enables differentiated networking products with very low incremental hardware and development costs because software provides the added differentiation.

A tuned implementation of the SAR function written in C for a RISC processor is almost two orders of magnitude too slow to meet the demands of current gigabit-per-second networks, so a general-purpose processor would appear to be out of the question. However, an application-specific processor can perform these ATM SAR tasks at the OC48c data rate (2.405 GB per second). Worst-case OC48c ATM streams (continuous 52-byte ATM cells) require the generation or processing of one ATM cell every 176 ns. For an SOC implemented in 130nm technology and running at 250MHz, 176 ns translates into a budget of 44 processor cycles per cell.

Hardware-design methods can define a state machine and data path capable of meeting these performance requirements. Fig 6-45 describes the basic algorithm for ATM segmentation. This algorithm assumes that both the incoming PDU packets and the outgoing stream of ATM cells are stored in a local-memory buffer. The hardware must also process every bit of the packet stream to perform a CRC (cyclic redundancy check) and other validity checks. A 128-bit memory interface ensures sufficient bandwidth into and out of the processor's local memory to handle OC48c rates.

This basic hardware architecture can be easily mapped onto an application-specific processor by implementing most of the unique registers of the hardware as added state in the processor and defining one new instruction for each of the unique cycles in the inner loop of the computation. Added state registers include the following:

- Three 16-byte registers to hold the data mask for ATM padding
- Three 16-byte registers to hold the current cell payload
- One 4-byte register each for the ATM cell header, the AAL5 trailer, and a counter for the number of AAL5 cells to generate and assemble
- One 2-byte register to help compute the PDU length and the conditional store mask
- Flag registers to report PDU-ATM header mismatch, packet-length errors, and identification of the last cell

Eight new processor instructions are needed: one to load the header template, three to bring the next 48 bytes of the PDU in from memory, one to apply the ATM header to the outgoing cell, and three to write the correctly structured ATM cell payload to the output buffer. Each instruction combines a load or store operation with a set of computations on the packet data and internal cells. Those computations include applying masks, checking CRC, checking packet lengths, and comparing headers. The instructions operate symmetrically. The same instructions and registers

- break up the PDU packets and insert headers to generate ATM cells, and
- check headers and combine ATM cells to regenerate the PDU packets.

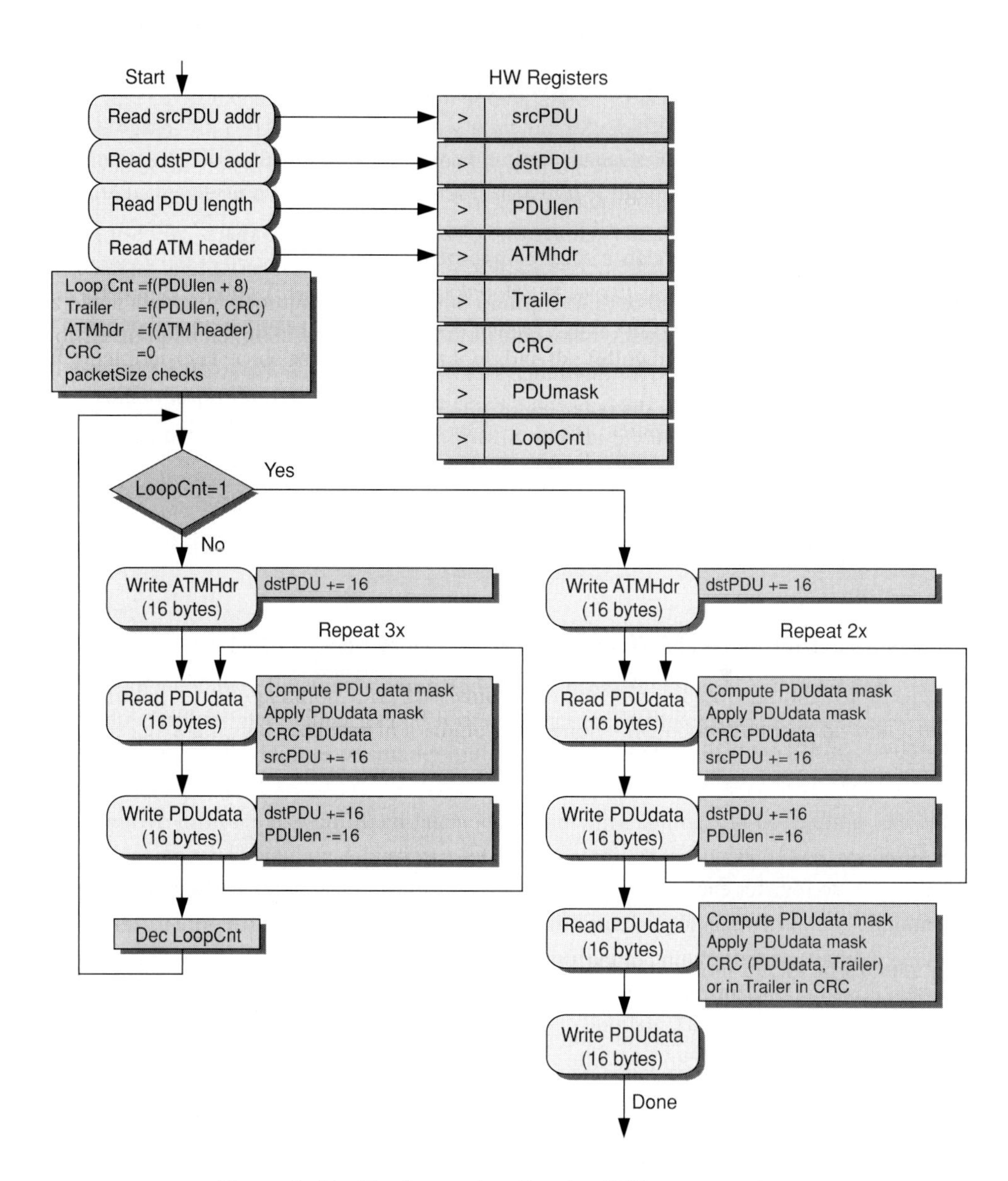

Figure 6-45 Hardware algorithm for ATM segmentation.

The registers and instructions added to an Xtensa processor to perform these tasks require about 34,000 gates, for a total processor gate count of 67,000 gates. This gate count requires much less than 1mm^2 of logic using 130nm standard-cell fabrication technology. A fully hard-wired design would be roughly comparable in gate-count to the added gates.

Once the instructions are defined, the application code is easy to write. The core segmentation routine consists of just 24 source lines including invocation of the intrinsic functions corresponding to the eight ATM cell-manipulation instructions. The compiler automatically uses the Xtensa processor's zero-overhead loop primitive, so no branch-delay bubbles occur. Fig 6-46 shows the pipelined implementation of the processor-based state machine. The core calculation, an inner loop that segments or reassembles each ATM cell, requires fewer than 10 cycles per ATM cell. One 128-bit data transfer occurs every cycle.

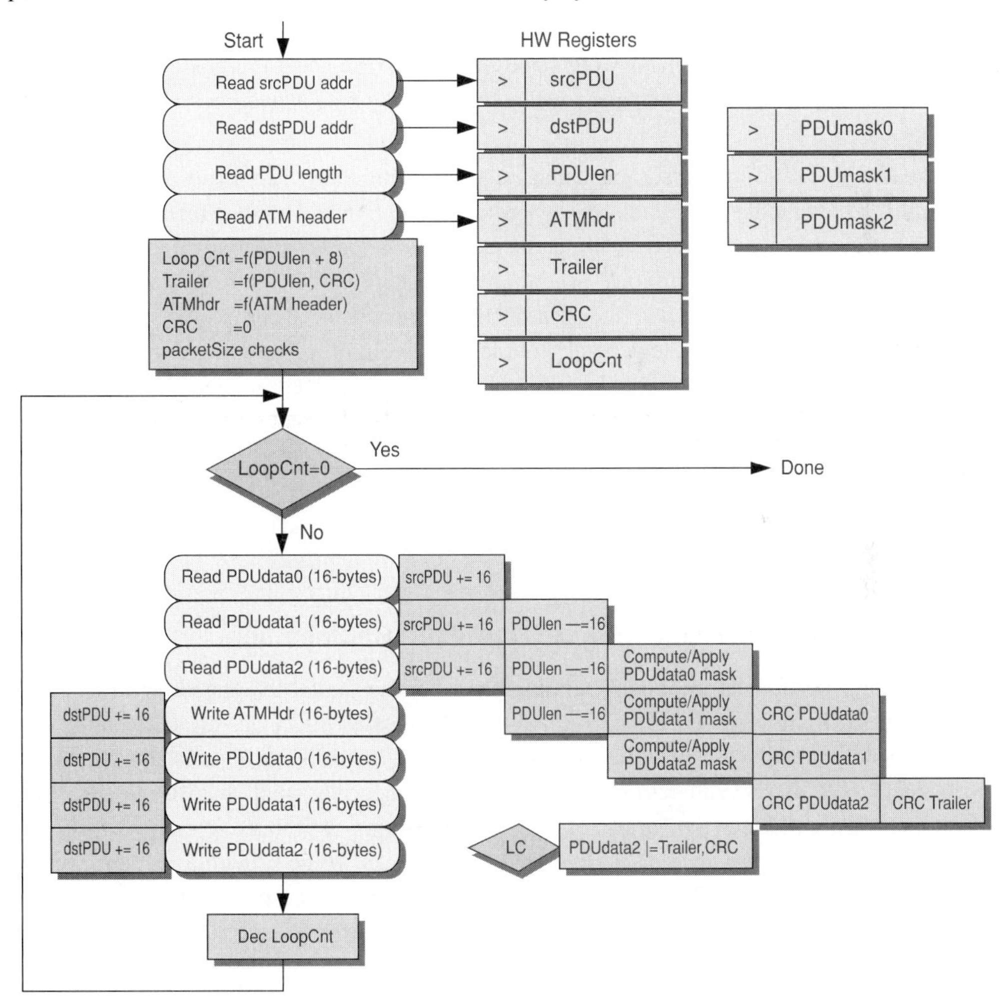

Figure 6-46 Pipelined processor Implementation of AAL5 SAR algorithm.

The processor-based design operates with the same high throughput as a hardware implementation. This efficient processor-based implementation leaves headroom of more than 30 cycles per cell—68% of the available processor cycles—for other operations or to support even higher network bit rates. Because both headers and payload are fully visible to software running on the AAL5 SAR processor, a wide range of additional functions could be added, using the processor's baseline instruction set and additional designer-defined instructions that perform more parallel or complex operations on fields of the current ATM cell. Added functions might include traffic management, protocol checking, compression, and encryption.

Moreover, using multiple memory ports, the inner loop could be shaved to an average of roughly five cycles per cell—reading 128 bits of the PDU and writing 128 bits of outgoing ATM cell every cycle. This enhancement would boost the performance to the point where OC-192 bandwidth (10Gbps) could be sustained, including some cycles for overhead and added functions on each cell.

Creating this set of processor extensions requires significantly less effort than does developing equivalent logic the traditional way using Verilog or VHDL. The state machine is entirely replaced by a short C algorithm. The data-path extensions are described at a very high level—in the form of TIE operation descriptions—and these extensions are automatically merged into the base processor's data path, which includes automatic connection to the processor's data memory and load/store unit. The automatic generation of memory-control logic further reduces the hardware-design effort typically associated with implementing embedded memories. The speed-up relative to a good C implementation running on a 32-bit RISC instruction set is about 80 times. The implementation of the ATM SAR as a processor does carry a cost—less than half a square millimeter—but gives the designer something quite valuable in return: complete reprogrammability. For many designs, this design approach reduces the initial design time and risk and therefore increases the SOC's value.

6.7 Novel Roles for Processors in Hardware Replacement

Processors have become an important building block for big digital integrated circuits in only the past decade. In most system designs, the processor remains the center of attention, driving design decisions about system buses, hardware/software partitioning, application tradeoffs, development tools, and overall system data flow. The emergence of small, highly adaptable processor cores vastly expands the variety of roles for processors, especially as their performance converges with the performance of complex hardware blocks, while providing easier development and programmability. Three added perspectives underscore the potential proliferation of processor uses in SOCs: the deeply buried task engine, the spare processor, and the system-monitor processor.

6.7.1 The Deeply Buried Task Engine

Application-specific processors often fill prominent places in system architecture and product structure, but processors are also percolating into lower profile, bare-metal hardware tasks because of the added flexibility they offer to SOC designers. Increasingly, the designer can put processors into deeply buried roles, characterized by four key shifts from the traditional uses of processors in SOC designs:

1. The deeply buried processor does not consume main-bus bandwidth or the memories in the SOC. A deeply buried processor might appear in the block diagram of an SOC as an application-specific "black box."
2. A single task, developed by chip designers, is loaded at boot-time (or runs from ROM) and executes without stopping as long as the chip is in operation.
3. The task is structured around data flow from input ports to output ports or to and from local memories. The task may be as simple as a series of loops and often does not include interrupt service.
4. A hardware team largely or entirely manages incorporation of the processor—selection, configuration, integration, and programming—into the SOC. From the software team's perspective, the deeply buried processor is indistinguishable from a traditional data-path-plus-state-machine hardware block. The block may even be controlled by device-driver software running under an operating system on another processor in the system.

Deeply buried task engines often focus on one complex data-transformation. A few of the many hardware-like tasks executed by deeply buried processors include

- Transformation of signals between the time and frequency domains.
- Color-space conversion for image and video representations.
- Packet validation, filtering, encapsulation, or reformatting, such as the ATM SAR (segmentation and reassembly engine) task explored in Section 6.6.
- High-data-rate DES or AES encryption.

The block diagram of a high-end set-top box in Fig 6-47 highlights likely candidates for implementing deeply buried application-specific processors. The obvious processor, the central CPU, is noted in bold outline. Sixteen other function blocks, each potentially a candidate for a task engine, are noted in dashed outline.

Sixteen additional processors do not necessarily constitute the optimal design solution for this application. In some cases, the designer might still find that an RTL-block is appropriate, especially where added flexibility doesn't add sufficient development and deployment leverage. In other cases, several functions can be combined. For example, a more flexible line interface—boot-time adaptable to QPSK, QAM, OFDM, and DSL signal-processing and forward-error-correction requirements—might actually cut hardware costs relative to four separate hardware interfaces. Once the hardware designer understands the versatility of the new class of application-specific task engines, novel roles for deeply buried processors will become apparent.

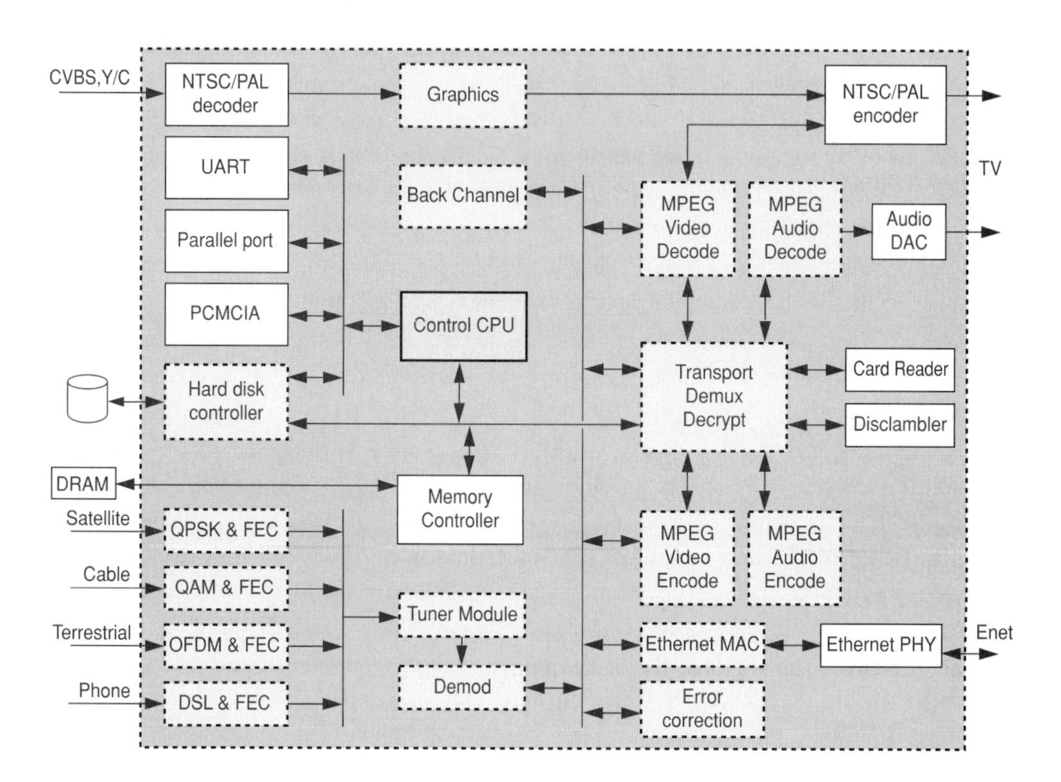

Figure 6-47 Opportunities for deeply buried processors in high-end set-top box design.

6.7.2 Designing with Spare Processors

The small size of application-specific processors, combined with the rising urgency for "first-time-perfect" silicon, has led some SOC architects to use a new tactic for reducing risk—incorporation of spare processors. On the surface, it might appear crazy to add undedicated processors whose use is undecided. However, designers have been adding "spare gates" to chips for more than two decades to allow bug fixes to be made with cheaper metal-only mask changes. Adding spare processors is the next logical step and may enable sophisticated bug fixes and feature additions with no mask changes at all.

Reprogramming of the existing processors is, of course, the first line of defense against function bugs, but sometimes there is insufficient excess computational power to handle desired enhancements with the existing processors.

Three considerations go into the decision to add spare processors to a design:

- *Does the silicon budget allow for the extra silicon area of the unallocated processor?*

  Many chips are pad-limited (the die area is set by the minimum chip perimeter, not by the area of logic gates and memory), so added logic is free so long as it requires no new pins. Note that "free" doesn't mean "without any cost." Even a disabled processor will contribute to leakage current. Also, the spare processor may add to the chip's test time, but it can also be used as a BIST (built-in self-test) processor when not used to boost application performance.

- *Can a processor configuration be defined that spans the likely range of "bug-fix" tasks?*

  Sometimes the necessary additions to computation are relatively simple, but processor bandwidth is critical. A fast, general-purpose integer- or signal-processing architecture with high-bandwidth interfaces may be a good compromise between flexibility and efficiency.

- *Is there sufficient bandwidth on the key buses and memories, such that the spare processor's traffic could be added without causing new bottlenecks to occur in the system design?*

  With extra bus and memory bandwidth, a spare processor can transform the data waiting in buffer memory between steps in the overall system data flow. The architect may even want to consider closely associating spare processors with each of the major on-chip memories. The processor can steal otherwise unused cycles from that memory and does not need to consume global-bus bandwidth.

Spare processors can even be used in-line to process data flowing through the SOC. Suppose a key stream of data between two blocks passes down a set of ports. A processor can be added where the wires from the data source serve as direct-connect inputs to the processor, perhaps through a queue. An equivalent set of output ports connects the new processor to the data destination.

Initially, the spare processor is programmed with the simplest possible task: copying data from the input ports to the output ports. If any additional transformations are needed, the copying task is upgraded to perform the transformations as well. Because the task upgrade cannot be clearly foreseen, it is unlikely that a single-cycle computation will serve. Several instructions, perhaps even tens of general-purpose instructions, per word may be required for the transformation. The spare in-line processor will likely work for data rates up to tens of millions of words per second, but not hundreds of millions of words per second.

However, if the spare processor is given instruction extensions that allow it to perform the work of either the upstream or downstream blocks, then the spare processor can share the processing load with either the data-producing or data-consuming blocks in the event that processing requirements increase beyond the design capacity of the other two blocks.

Fig 6-48 shows a system structure with two spare processors, one connected to the system bus, the other connected in-line on the primary data flow between a high-bandwidth data producer and high-bandwidth data consumer. The in-line processor and its input and output queues add some latency between producer and consumer, even when doing no useful computation. However, it consumes no global resources such as system-bus bandwidth, even when active.

The processor connected to the system bus can operate on any data in shared memory, either in the block of on-chip RAM or in the local memories of any of the other processors. When inactive, it consumes no bandwidth and adds no latency to other operations. However, when it is programmed to perform some added computation on data in some memory in the system, the spare processor connected to the on-chip system bus will consume bus bandwidth, potentially affecting the latency of other system-bus operations. On the other hand, this arrangement is more flexible because it augments the computational horsepower of any other processor on the chip as needed.

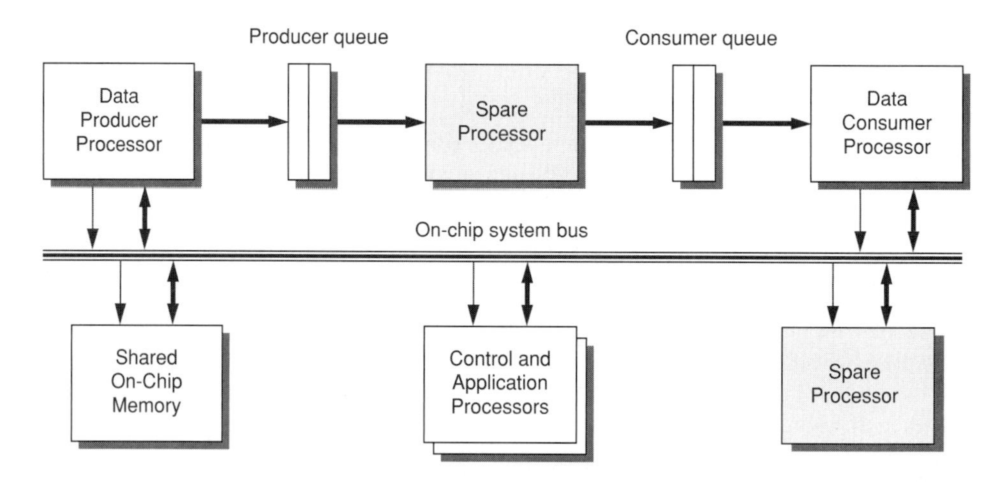

Figure 6-48 System organization with two spare processors.

6.7.3 The System-Monitor Processor

A complex SOC executing a number of application tasks in parallel is a complex dynamic system. Even the most diligent architectural design and modeling cannot anticipate all interactions among the subsystems. Initial system bring-up and ongoing system enhancement may severely strain software development, debug, and product validation. Dedicating a processor to on-chip debugging and monitoring accelerates SOC development and eases field updates. Sophisticated embedded debugging solutions exist, but most rely, in part, on using the same processors running the applications to provide trace and debug capabilities. This dual role makes debugging intrinsically intrusive, so the very act of monitoring application execution may influence that execution. Use of a separate debug processor helps minimize this "Heisenberg Uncertainty Principle" of debugging.

A system-monitor processor can perform many tasks, including the following:

- The system monitor can continuously check an application's progress and results to measure system throughput and correctness against requirements. It can accumulate the occurrences and save the results of anomalous application behavior for later analysis.
- The system monitor can analyze and filter all processors' debug states, especially in conjunction with real-time-trace hardware. Simple hardware can dump selected information of the cycle-by-cycle execution of each application processor into a buffer memory. A system-monitor processor can identify selected events, check properties, and send relevant application-performance data to external debug interfaces.
- On system boot, or at periodic intervals, the system monitor can load test routines into each application processor, monitor BIST results, and report failures to external agents. If the SOC implements redundant elements such as memories or processors, the system monitor can decommission nonoperative regions of memory or select operational processors to replace nonfunctioning ones.The potential uses of system-monitor processors are diverse. System and software architects are just awakening to their potential for providing rapid insight into complex system behavior.

6.8 Processors, Hardware Implementation, and Verification Flow

One central promise of application-specific processors is quicker, safer implementation. With the right tools, defining a processor is significantly easier than designing equivalent RTL hardware because most chip architects find that creating processor specifications (including instruction-set extensions) are more concise, more comprehensible, and less error-prone than traditional Verilog and VHDL design. Moreover, deploying an SOC that assigns processors to hardware roles is less risky because more of the chip function remains fully programmable. Nevertheless, it is important for hardware designers to understand the issues of verification and implementation of processor-based blocks. This section explains the basic implementation flow for synthesizable processors and highlights key opportunities to streamline overall SOC design.

The implementation and verification of each processor block is an important task within the overall chip-design flow. Fig 6-49 shows the three interacting tracks of hardware implementation and verification: the verification of the configuration, the physical implementation of the processor core, and the verification of the hardware implementation. All three tracks lead to final integration steps that complete the SOC design.

6.8.1 Hardware Flow

The application-specific processor generator is tuned for standard-cell implementation and delivers portable, synthesizable RTL code as the starting point for mapping a processor core into the target silicon technology and integrating that core into the full chip design. The processor design is intended to fit seamlessly into the normal synthesis and physical-design flows to ease the integration of the processor block into the normal hardware-design flow. The processor is represented using a synthesizable subset of Verilog or VHDL and is accompanied by tuned logic-synthesis scripts. The implementation support includes scripts for placement, clock-tree

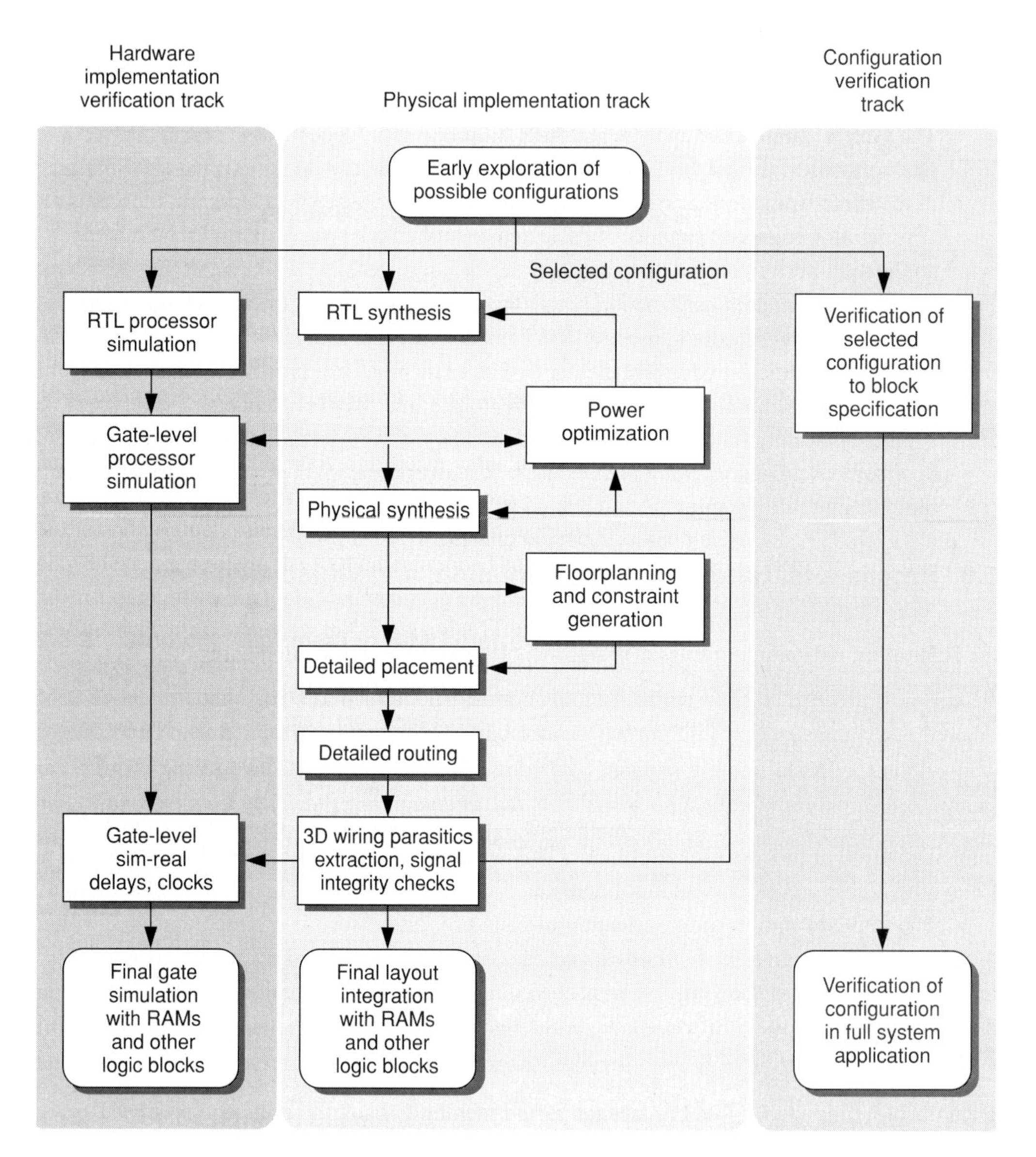

Figure 6-49 Hardware development and verification flow.

generation, routing, power optimization, detailed post-route timing analysis, and signal-integrity checking. A complete physical-implementation flow might follow a pattern similar to this:

1. **Initial synthesis**—synthesis of the configured processor using supplied synthesis scripts and a selected cell library for initial speed and area estimates. Scan flip-flops may be incorporated optionally into the net list. Scan-test vectors may be generated when the physical-synthesis results are stable.
2. **Initial power analysis**—characterization and analysis of power based on selected application code. Power-aware synthesis reduces switched capacitance (power) on heavily toggled nodes, especially nodes that are off the critical timing paths.
3. **Physical synthesis**—placement-and-routing-aware synthesis using realistic interconnect conditions to drive gate-level net-list generation. Physical synthesis may automatically balance processor pipe stages for optimal clock frequency by moving logic across flip-flop stages.
4. **Floor planning and layout constraint generation**—preliminary determination of block aspect ratio and pin location, based on surrounding blocks. This is a good point to generate RAM blocks due to tight timing and physical design interaction between processors and RAM blocks.
5. **Cell placement**—automated x,y placement of each cell in the net list
6. **Clock-tree synthesis**—generation of the clock circuit, including load-balancing based on final flip-flop placement.
7. **Routing**—detailed routing of all signals, clocks, and power/ground buses
8. **In-place optimization for timing improvement**—rough analysis of final layout and incremental improvement by cell re-placement or cell sizing to improve critical-path timing.
9. **Custom wire-load model generation and 3D parasitic extraction**—detailed 3D modeling of all wire resistance and capacitance to reverify timing, especially for clocks and other critical nets. Reverification of timing by static analysis.
10. **Gate-level power analysis and optimization**—more accurate power estimation with final optimization of gate sizes based on power.
11. **Signal-integrity checks**—checking for correct function and worst-case timing of signals where capacitive cross-coupling between wires introduces possible noise. Checking of voltage drops and noise in the power-supply grid may also be performed.
12. **Chip assembly**—integration of final processor layout with RAMs and surrounding blocks, including global buses, clocks, and power hook-ups.
13. **Final design-rule check (DRC) and connectivity check (LVS)**—final checking of complete chip compliance to target fabrication processor geometry rules and match reverification between desired net list and extracted transistor circuit.

Responsibilities for these stages vary. Sometimes, an ASIC vendor or design-service partner manages the back-end flow (post-synthesis). Sometimes the chip-design team carries the design through the entire flow. In some hardware organizations, the design team for a block is responsible for logical and physical design up to final chip assembly. In other cases, a separate

VLSI team takes over once the block's basic function, area, and timing specification have been confirmed by RTL simulation and initial synthesis. The design flow is largely the same, independent of the team organization.

6.8.2 Verification Flow

Block-level processor verification is generally easier than block-level verification of other RTL blocks with equal complexity for three reasons. First, fast and accurate instruction-set simulators provide a robust reference model for checking the correctness of the processor's RTL implementation. If the processor can be exercised, it can be checked. Second, processors more easily support complete test-bench environments that allow all processor features, including external interfaces, to be thoroughly exercised. Processors can naturally run self-checking code, so large portions of the test suite can be built independently of the surrounding logic. Third, the familiar processor abstractions—standardized buses, self-contained verification code, and a structured relationship between interface and internal function—all make chip integration easier because the interactions between the processor and other logic and memories are well-defined.

Hardware teams must recognize two dimensions of processor verification: verification that the processor configuration meets the system-performance goals and verification that the implementation of the processor hardware conforms to the configuration specification.

6.8.2.1 Configuration Verification

Configuration verification answers one central question: Does the configuration of the processor—instruction set, memory system, and interfaces—meet the requirements of the application and the expectations of the chip-architecture team? This high-level verification centers on the target application, as embodied in the chip-level specification. Sometimes this work is performed primarily by a software team. The hardware team may only be responsible for providing a useful reference platform.

Some application-verification teams prefer a simulation model as the reference platform and some prefer a hardware prototype, typically implemented with an FPGA (field-programmable gate array). The simulation approach is inexpensive to replicate, easy to update, and offers visibility to all internal states for debugging and performance analysis. High-level system-modeling tools built around fast processor emulators make large-scale multiprocessor system simulators much simpler to generate and analyze than hardware prototypes. The hardware-prototype approach, in contrast, offers higher performance and more realistic interfacing to external devices and networks. However, hardware prototypes involve higher expense, a longer engineering process, lower reliability, and poorer visibility for system analysis.

The configurable processor generator supports an additional configuration verification aid: native C implementation of extensions. Many software teams develop and test applications using native application compilation on their PCs. This approach offers convenience, high throughput, and a good debug environment, but it provides little insight into application performance because no code for the target architecture is generated and no cycle-accurate measurements can be made. This approach works so long as the target processor uses programming-language semantics that are a subset of PC programming-language semantics (i.e., no new datatypes or functions). Configurable processors derive much of their power from native support of extended datatypes and functions, so native compilation on the PC would appear impossible. The processor generator automatically creates extended programming libraries for each instruction-set extension, effectively adding the new datatypes and operators to the PC compiler. With these libraries, the developer can compile and run any extended application in a bit-exact form using standard C or C++ programming environments.

Often, configuration verification works best as part of a continuum of application analysis and system refinement. Configuration verification is preceded by architectural exploration—trial of a wide range of chip architecture topologies, algorithms, and configuration types—and is followed by detailed application development and post-silicon hardware/software integration.

The hardware or software team that specifies the processor configuration usually takes responsibility for configuration verification too. Sometimes, hardware teams use the traditional methods for verifying hardware blocks: test data streams are fed into the processor, and simulated output streams are checked against reference output streams. In other cases, the hardware team creates an executable reference specification in C and compares the results of this reference specification against the results of the configured processor executing the target firmware. This approach allows the hardware team to take fuller advantage of fast instruction-set simulation and realistic test benches built using C or SystemC.

6.8.2.2 Implementation Verification

Implementation verification answers another central question: does the sequence of hardware transformations—from configuration description down to final mask geometry—correctly maintain the initially specified architecture (as verified by the application and chip-architecture team)? The mixture of complex (and sometimes immature) EDA tools in the physical-implementation flow with occasional human manipulation of intermediate databases creates opportunities for many transformation errors. The cost of failure is so high that healthy paranoia demands good end-to-end checking of the configured processor definition against the final silicon design.

The implementation/verification flow typically uses both formal verification and simulation-based methods. Formal or static verification methods are used at several stages:

- Comparison of instruction-set representations (in TIE, this test consists of formal verification of abstract-but-bit-exact "reference semantics" against optional RTL-like "implementation semantics").
- Property and assertion checking of standard protocol implementations such as bus interfaces.
- Static-timing analysis, including timing analysis with extracted interconnect parasitics.
- Static-power analysis.
- RTL-to-gate checking.

Even with static and formal checking, complex hardware blocks still demand heavy reliance on simulation to meet the requirements of implementation verification. The processor generator provides four coupled components:

1. Source RTL for the configured processor, in Verilog or VHDL.
2. A reference test-bench for the processor configuration, including a test infrastructure for all external interfaces: buses, interrupts, memory interfaces, and other boundary signals. Monitors, checkers, and programmable external-event generators may be built into this test-bench.
3. A cosimulation interface to run the RTL simulation and the fast instruction simulator in lock step and to compare all relevant processor state, cycle-by-cycle.
4. Diagnostics to thoroughly exercise all processor features. Diagnostics may include both human-designed tests and random instruction sequences intended to fully cover all unique sequences and states. Diagnostics should also cover all configured features, including memory and interface configurations, plus, to the greatest extent possible, user-defined instruction-set extensions. Diagnostics for the extensions typically rely on a combination of automatic test-sequence generation and test vectors included in the instruction-set description.

Implementation verification often starts with RTL simulation of the processor alone. Because the processor RTL has not been transformed or modified at this stage, this simulation should not uncover any major issues. It does provide, however, a good initial sanity check of the entire verification infrastructure. As the processor RTL is mapped into a particular cell library, standalone processor simulation becomes a crucial check of the full tool flow and the libraries. Gate-level simulation is much slower than RTL simulation, but it may uncover important issues in reset signal, clock, and memory interfaces.

Implementation verification extends to hardware-level verification of the interaction between the processor and other logic. Many developers use mixed-level simulation, where the processor's fast ISS (instruction set simulator) is encapsulated in an RTL-like wrapper to create a bus functional model (BFM). Developers use the BFM in conjunction with other blocks that are traditionally designed in RTL, such as bus logic and peripherals, to accurately simulate the processor-logic combination. Healthy paranoia sometimes mandates covering a dense matrix of simulation combinations, such as that shown in Fig 6-50.

Processor Implementation		Non-processor Implementation	Purpose
ISS	with	**RTL**	Verification of non-processor logic under stimulus from processor-based software
RTL	with	**RTL**	Verification of interface logic design
Gate	with	**Gate**	Final verification of reset, clocks, scan-chains and memory hook-up (this simulation may be very slow)
RTL	with	**Test-bench only**	Verification of basic processor package
Gate	with	**Test-bench only**	Final verification clock and memory hookup

Figure 6-50 Typical combinations of processor and non-processor logic simulation.

6.9 Progress in Hardware Abstraction

On the surface, moving from RTL block design to processor-centric design might seem like a big change. In fact, most of the principles valued by hardware architects—precise control of cycle-by-cycle behavior, exploitation of latent parallelism with hardware, and performance through optimized data paths—are fully supported by application-specific processors. In fact, this transition increases the hardware developer's ability to tackle and predictably solve tougher problems by allowing small teams to design, verify, integrate, and maintain complex application tasks using many familiar design and development techniques. Use of configurable processors in traditional hardware roles extends the reach of the hardware team's skill set.

The transition from RTL-centric to processor-centric design is just one more step in a long evolutionary sequence for chip designers. Once upon a time, chips were designed polygon by polygon and transistor by transistor, with manual layout methods. Gradually, gate-level design with electronic schematics and automated placement tools caught on. In the late 1980s, RTL languages become the preferred medium of expression. Logic simulation and synthesis proliferated. With each evolutionary step, the hardware designer started to work at an increasingly higher level of abstraction to gain productivity and to leverage new characteristics of each new building block.

From a historical perspective, the time is ripe for another evolutionary transition. Automatic generation of processors from high-level instruction-set and application descriptions fits the evolutionary design model. This approach works with a new basic building block and a new form to capture design intent. The approach raises the abstraction level for the hardware-development team and delivers the flexibility benefits of post-silicon programmability. Automatic generation of processors constitutes a new way for chip design teams to intelligently balance speed, cost, power, design flexibility, development effort, respin risk, and the size of addressable markets.

6.10 Further Reading

Many of the important techniques and hardware design issues of SOC development derive from related methods and issues in VLSI design. A number of books and articles on ASIC and VLSI design may serve as good background:

- Wayne Wolf, *Modern VLSI Design: System-on-Chip Design* (3rd ed.), Prentice Hall, 2002.
- Jan L. Rabaey, Anantha Chandrakasan, and Borivoje Nikolic, *Digital Integrated Circuits* (2nd ed.), Prentice Hall, 2002.
- A good, basic introduction to VLSI design issues, including circuit-design issues, is found in Michael John Sebastian Smith, *Application-Specific Integrated Circuits*. Addison-Wesley, 1997.
- A discussion of current SOC hardware design methodology, with particular focus on design reuse, verification and test can be found in Rochit Rajsuman, *Top of Form System-on-a-Chip: Design and Test Bottom of Form*. Artech House, 2000.
- Most hardware designers are familiar with logic synthesis. A reasonable discussion of hardware design flows using the most common synthesis flow from Synopsys can be found in Himanshu Bhatnagar, *Advanced ASIC Chip Synthesis: Using Synopsys Design Compiler, Physical Compiler, and Primetime* (2nd ed.), Kluwer Academic Publishers, 2001.
- Processor performance has many dimensions. This book emphasizes the tailoring of processor architecture to each specific application as a basic technique to improve performance, power efficiency, and cost. These techniques are largely complementary to design techniques that emphasize circuit-level optimizations to achieve performance. Chinnery and Keutzer have written one of the best recent books on optimizing digital circuits within the context of automated design methodology: David Chinnery and Kurt William Keutzer, *Closing the Gap Between ASIC & Custom: Tools and Techniques for High-Performance Asic Design*. Kluwer Academic Publishers, 2002.

Power is an increasingly central concern for processor design. Discussion of the key issues can be found in the following papers:

- C.-L. Su, C-Y Tsui, A. M. Despain, "Low Power Architecture Design and Compilation Techniques for High-Performance Processor." In *Proceedings of the IEEE COMPCON*, February 1994.

- J.-M. Masgonty et al., "Low-Power Design of an Embedded Microprocessor Core." In *Proceedings of the European Solid-State Circuits Conference 96*, pp. 17–19, September, 1996.
- A. Chatterjee, R. K. Roy, "Synthesis of Low-Power DSP Circuits Based on Activity Metrics." In *Proceedings of the. International Conference on VLSI Design*, India, pp. 265–271, January 1994.

CHAPTER 7

Advanced Topics in SOC Design

Chapters 4, 5, and 6 present the basic methods of complex SOC design with configurable processors. All electronic systems are complex, and large SOCs are especially complex. This chapter deals with some important subtleties of processor-centric SOC design, including more details on microprocessor pipelines, advanced techniques for replacing hardware with processors, and special issues in memory-system design and synchronization. The last section of the chapter also gives a more detailed introduction to the Tensilica Instruction Extension (TIE) language. This tutorial gives designers greater insight into the TIE examples used throughout the book and suggests additional techniques for boosting application performance and efficiency.

7.1 Pipelining for Processor Performance

In a simpler world, all processor operations would complete in a single cycle. The number of instructions executed would always equal the number of execution cycles. Unfortunately, this constraint precludes many important but complex operations from appearing as processor instructions, forces the processor to run at much lower clock frequency and throughput because of the latency of complex execution units or requires duplication of execution units, even if they are not used heavily.

Chapter 6 introduced a simple RISC pipeline. One RISC instruction starts every cycle and passes from pipe stage to pipe stage, with specific steps in the instruction's execution taking place in each stage. One instruction completes every cycle, at least nominally.

Distributing instruction execution across pipe stages improves overall clock rate by reducing the amount of logic in each stage. Pipelining improves processor performance with a small increase in cost and power compared to performing all the work of the instruction during one clock cycle. New register files and execution units added to the pipelined hardware implementation extend the processor architecture. These additions to a processor's hardware are simple, but they affect the processor pipeline, modifying the behavior of existing processor blocks and pipe stages and possibly adding new pipe stages with new blocks and functions. The basic RISC pipeline with additional application-specific execution units, register files, and pipe stages is shown in Fig 7-1.

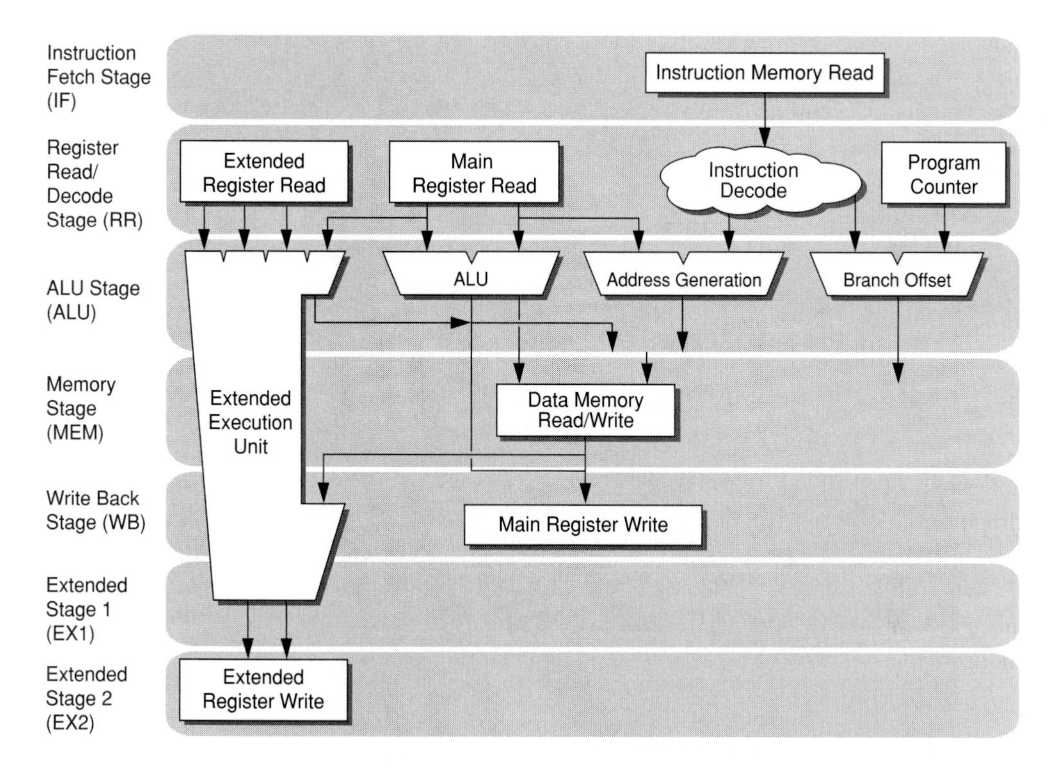

Figure 7-1 Basic and extended RISC pipelines.

Sooner or later, every developer using an embedded processor encounters a design challenge that requires a better understanding of the processor's execution mechanisms. This section outlines some of those mechanisms and some pipeline-optimization techniques.

7.1.1.1 Instruction Schedules

By default, when new function units are added to the processor, they are added to the ALU execution stage. This positioning ensures that extended instructions can use the results of previous ALU instructions, and that extended-instruction results are available for immediate use by all succeeding instructions, including succeeding invocations of the new instruction. Packing the entire function of this new instruction into the existing ALU cycle can have two undesired effects. First, a complex logic function forced into one pipeline cycle may require increasing the cycle time of the processor as a whole. Extending the new instruction's execution across several pipe stages alleviates pressure to increase the processor clock cycle, but delays succeeding instructions waiting for the results. Second, if the new instruction uses results from earlier instructions as source operands and those results are available only after the ALU execution cycle, the extended instruction may need to wait for those results. Deferring source usage to a later pipeline stage avoids the delay.

These problems can be avoided by improving the new instruction's operand-use schedule. The schedule determines the cycle of use of each source operand and the availability cycle for each destination operand. By definition, all of the logic of the instruction must fit between the arrival of sources and delivery of results. The processor generator distributes the logic for the new instruction's execution unit across the available cycles to minimize the new instruction's the impact on cycle time. The TIE `schedule` construct gives the designer explicit control over the `use` pipeline stage and `def` (definition or result) stage for each source and destination operand.

The designer naturally starts with the default pipelining of each new instruction. Direct generation of extended-instruction RTL code by the processor generator allows the designer to quickly synthesize and evaluate the cycle time for extended instructions. If the instruction introduces a long critical path between a source and destination, the processor's estimated cycle time may increase unacceptably. The designer should try alternative pipeline descriptions to shorten these critical timing paths.

Adding more cycles between source and result spreads the logic across more clock cycles and reduces the likelihood of a critical circuit path that increases processor cycle time. On the other hand, inserting too many cycles between source and result may force other instructions to wait, increasing the total number of execution cycles along the sequence of dependent operations. This latency may degrade application performance. Target application simulation with explicitly scheduled instructions informs processor extension.

7.1.1.2 Reorganizing Long-Latency Computation

When the extended instruction involves long-latency instructions or instruction sequences, the designer can use a number of optimization techniques to reduce the latency and improve performance. In particular, overlapping the execution of different sequences often boosts performance significantly. The designer should identify opportunities to pipeline computation, especially where the long-latency operations (for example, loads or multiplications) of one loop iteration can overlap with operations of another iteration. When data recurrence occurs (one iteration uses data from a preceding iteration), the designer can reorganize the computation to reduce that operand's use-to-definition delay. For example, consider the following fragment of C code, representing the multiplication of an array element by a scalar to form a result stored in a second array:

```
int a[N],b,c[N],i;
for (i=0;i<N;i++) {c[i] = a[i] * b;}
```

The critical path through this code involves the loading of the a[] element, a 32-bit multiplication, and the storing of the result. Loads and multiplication are both multicycle operations, so a series of loop iterations might be represented in time as shown in Fig 7-2. This picture assumes that other operations (computation of array indices, loop-termination condition checking, and branch back) are hidden, so only three operations per iteration remain critical. Fig 7-2 shows two cycles for the load of `a[i]`, three cycles for a 32-bit multiply, and one cycle for the store to `c[i]`. This non-pipelined version requires six cycles per loop iteration.

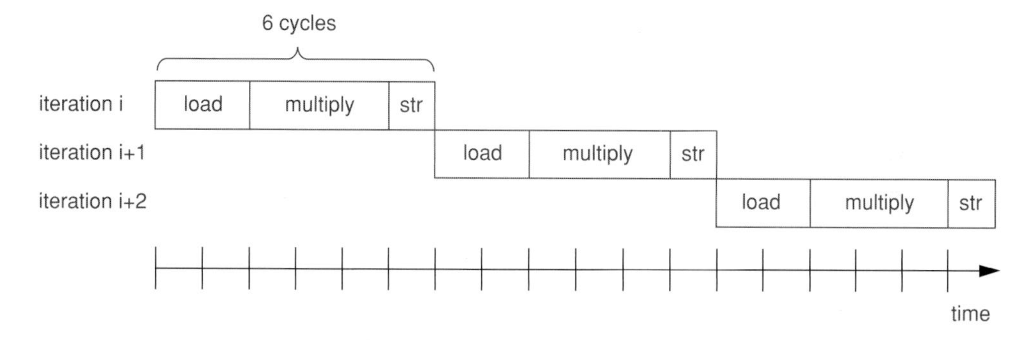

Figure 7-2 Non-pipelined load-multiply-store sequence.

Iterations can overlap to reduce the effective latency of the sequence by rearranging and combining operations, as shown in Fig 7-3. This rearrangement cuts the average number of cycles needed per iteration in half. Note that the store operations in one iteration overlap the second cycle of the load operation of another iteration. This overlap doesn't cause a resource conflict because the processor's load/store unit is pipelined.

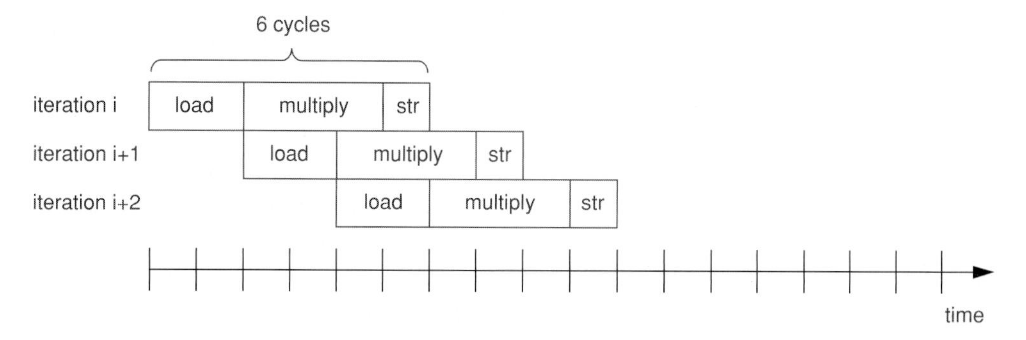

Figure 7-3 Pipelined load-multiply-store sequence (one load/store per cycle).

If two independent memory operations can be executed each cycle (for example one load and one store), the iterations can be fully pipelined, as shown in Fig 7-4, which achieves an effective rate of one cycle per loop iteration and a sixfold increase in the loop's execution speed (from the original six cycles/loop to one).

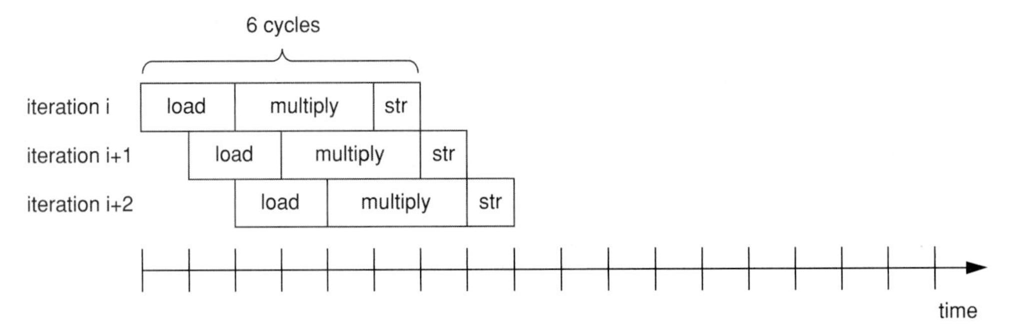

Figure 7-4 Pipelined load-multiply-store sequence (one load and one store per cycle).

These three implementations of one simple loop highlight the benefit of overlapping operation execution by overlapping the execution of individual operations (hardware pipelining) and overlapping separate iterations of the algorithm (software pipelining).

SIMD operations magnify the importance—and the benefit—of instruction pipelining. Several iterations of one loop can be performed in parallel by defining instructions that perform the same operation on *N* data operands in parallel. SIMD instructions may require widening the computational data-paths, registers, and memory interfaces. For short datatypes such as 8-bit video pixels, SIMD operations may be effective even for 32-bit data words (which store four 8-bit operands per word). By combining SIMD operations with instruction overlap, a simple loop might collapse from half a dozen cycles per iteration to multiple iterations per cycle.

7.2 Inside Processor Pipeline Stalls

Permitting the processor to stall or delay execution of operations under some circumstances makes the processor smaller, faster, and more power-efficient overall. This section explains the common sources of processor stalls and their impact on performance and on application-specific processor configuration. Understanding the causes of stalls can often help the designer further improve processor configurations, achieving higher performance or efficiency.

Understanding the sources of stalls is particularly important when the designer is attempting to squeeze the last bit of performance from a design. Within the main data flow of a critical application, one extra cycle of latency or one extra processing cycle per block of data may make the difference between meeting and missing a major design goal.

Pipeline stalls reflect sophisticated interactions between the processor's instruction set and the application code, but these interactions are entirely manageable. This section deals with three common types of pipeline stalls: data-dependency stalls, resource-conflict stalls, and branch delays. It explains how to mitigate the negative effects of pipeline stalls on application performance.

7.2.1.1 Data-Dependency Stalls

Often, the results of one instruction are not immediately used by others, so pipelining of execution and hiding of the operation latency will speed overall computation. When a succeeding instruction does need the result before it is ready, then that instruction (and possibly others) must wait for the data. The instruction extension techniques that drive application performance—more register files, more operands per instruction, and deeper pipelining for complex operations—all exacerbate the need for efficient bypassing and interlock structures. When new instructions are added, the interlock and bypass structure of the processor must accommodate the added pipeline interactions. Without automatic generation of result bypassing and register-interlock logic, the useful performance of the processor suffers and programmers might mistakenly get stale values from the register files.

In a simple pipeline, the result of an ALU operation is available by the end of the ALU pipe stage and read data is available by the end of the memory pipe stage. If the values are written into the register file as soon as available, two related problems arise. First, if results are written as soon as available, the important principle of sequential execution may be difficult to maintain. The result of a later instruction with a short latency might be written before the result of earlier instructions that have longer latencies.

Out-of-order instruction completion produces unexpected program behavior. This problem is particularly dramatic if a later instruction has written a result, but an earlier instruction fails to complete because it triggers an error that makes completion of the error-causing instruction impossible. This scenario creates an inconsistent processor state.

The pipeline avoids this problem by deferring all register writes until the write-back stage. In that pipe stage, results are written in the order that the corresponding instructions are executed to maintain the expected sequential model of computation. All conditions that might trigger errors on an instruction must therefore occur before the write-back stage is reached.

Second, reading from and writing to the register file takes time, much of a clock cycle. Consequently, the instruction result might not be available as a source value for subsequent instructions for at least an additional cycle. The pipeline tracks the intended destination register of each result value waiting to be written. If a subsequent instruction needs that register's value as a source, that instruction may be stalled until the value is ready (*the pipeline is interlocked*). When the value becomes available, it is directly transferred (*the value is bypassed*) to the execution unit and the value, perhaps stale, supplied by the register file is not used.

Bypassing allows the result value from one ALU instruction to be used as a source value for an immediately succeeding instruction. For some instructions such as loads with more than one cycle of latency, bypassing alone cannot make result values available to the succeeding instructions. When an interlock is triggered due to dependencies between one instruction computing a result and another that uses that result as an input, a `data-dependency stall` occurs.

Fig 7-5 shows a simple interlock and bypass example. A load, whose result goes into register a2, is immediately followed by an addition, one of whose sources is `a2`. In the register-read stage, as the addition is getting ready to start, the source register number, a2, is checked against register numbers whose results are pending. The load value is not yet ready, because the load takes two cycles to complete (the load starts in the ALU stage and completes by the end of the MEM stage). The pipeline marks the load result, a2, as busy in its ALU stage, so the pipeline's interlock check dictates a stall cycle. The subsequent add instruction must wait while the load proceeds. In the next cycle (at the end of the addition's stall cycle), the interlock is checked again. This check shows that the value destined for a2 is ready, so no further interlock is required. However, the value is not yet available in the register file, so the pipeline moves (*bypasses*) the newly loaded value directly from the load data path to the adder's input. This scheme allows the addition operation to continue with correct data and with no further delay.

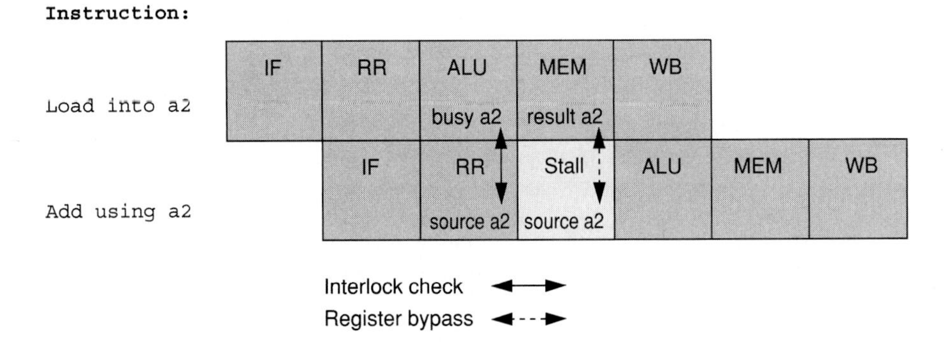

Figure 7-5 Simple bypass and interlock example.

The bypass and interlock structures for a basic RISC pipeline are simple because

1. One register file holds all operands.
2. No instruction uses more than two source operands or creates more than one result value.
3. Operation latency is short: one or two cycles.

Common instructions with long latency include loads (especially loads that also experience cache misses) and complex arithmetic such as multiplication. Application-specific instructions that combine several primitive operations often have more than one cycle of latency, so data-dependency stalls are an important issue, especially as the sophistication of processor extensions grows.

Interlocking on destination-source conflicts (especially for longer-latency operations) and bypassing of results between execution units become complex as application-specific register files are added. The processor generator must incorporate new interlock and bypass structures to maintain the programmer's simple sequential execution model and to allow designers to correctly add new, deeply pipelined instructions and multiple register files without comprehending the details of the pipeline's implementation.

Software Interlocks

Some processor architectures, especially long-instruction-word architectures, use *software interlocks* instead of *hardware interlocks*. The software tools and sometimes even the programmer must be aware of the implementation's bypassing rules and strictly conform to these rules. To do otherwise creates programs with unpredictable behavior. The assembler checks that no dependent instruction follows the instruction that generates the data too closely and warns the programmer when a suspected error is made. Data dependencies are resolved by reordering instructions. Independent instructions are executed between the instruction that computes a result and the instructions that need the result. If no independent instructions are available, NOP instructions must be added, which may increase overall code size. Errors in programming these software interlocks result in use of the wrong data. Subtle problems caused by missing software interlocks may be hard to detect.

7.2.1.2 Resource Stalls

Another related category of processor stalls is worth noting. Often a processor shares some resource among several instructions or pipeline execution stages. In theory, the resource could be duplicated to avoid the need to delay one use while another use is underway. In practice, too much hardware may be required to provide duplicate resources for all possible use conflicts. For example, many small processors implement multiplication without a fully pipelined multiplier. In such a case, if a second multiply instruction is fetched before an earlier multiplication has finished, the second operation must wait.

When resource conflicts are rare, providing resource arbitration is more efficient than duplicating resources. One common example of a resource conflict is the contention for the bus interface between the instruction cache and data cache. Normally, data and instruction cache misses do not occur at the same time, so one interface is sufficient. On rare occasions when cache misses occur on both caches, one cache miss must wait while the other is serviced by the bus interface. In this case, which is infrequent, the resource arbitration causes extra stall cycles for the associated instructions.

7.2.1.3 Branch Delays

Processor pipelining also affects control-flow changes. By default, a processor executes instructions stored sequentially in memory. When a branch instruction executes, more than one cycle is required to fetch the instruction from memory, decode it, compute the target address for the branch, and decide whether the branch is taken. Meanwhile, the processor is fetching and decoding instructions sequentially following the branch. If the branch is taken, subsequent sequential instructions are in pipeline limbo. If the branch is taken, most processor architectures choose to abandon the partially executed instructions following the branch. The effects of the instructions are canceled by disabling the write of any state by these instructions. This scheme creates an execution bubble where no useful work occurs for two cycles following a branch. Two instructions (an add and a subtract) are speculatively fetched and decoded, but then cancelled, as shown in Fig 7-6. (A few architectures use a "delayed-branch" model and choose to always execute these instructions in limbo, whether or not the branch is taken.).

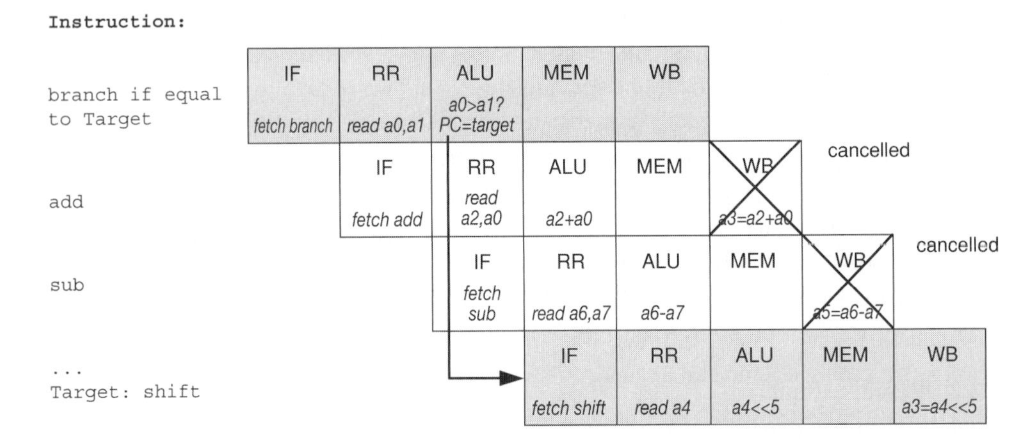

Figure 7-6 Branch delay bubble example.

Some processors avoid discarding these instructions by detecting an impending branch that will likely be taken. They start the fetch of instructions at the target address earlier to avoid any branch-delay bubbles. Both zero-overhead-loop instructions and branch-prediction mechanisms fit this class.

Managing branch delays may be important to performance, especially when migrating a function unit from an RTL-based, data-path-plus-state-machine design style to the configurable-processor design style. In this case, the developer may want to get as close as possible to a cycle-by-cycle match between the sequence of states in the original state machine and the equivalent operation sequence in the processor running a C program.

7.2.2 Pipelines and Exceptions

The principles of atomic, sequential instruction execution must be maintained even in the presence of unexpected events. Hardware and software developers should understand the basics of processor exceptions for two reasons. First, the processor's ability to handle exceptions offers big benefits to the developer. It enables precise application debugging.

It permits rapid response to external events such as interrupts. It provides an elegant recovery model for rare events such as hardware faults during processing. Second, exceptions are a standard feature of most embedded processors, so understanding the effect of unanticipated interruptions on the flow of instruction execution may help the developer avoid, or at least cope with, the negative ramifications of exceptions.

When an unexpected event occurs, one or more of the instructions in the pipeline will be terminated without completing. No results are written for these cancelled instructions. A number of exception conditions might trigger cancellation of instructions:

1. **Illegal instruction**—the fetched instruction is not recognized. Identifying and aborting undefined instructions aids code debugging and prevents unpredictable processor behavior.
2. **Memory address or data error**—the address is ill-formed or refers to a protected region of the address space. Data-integrity checking (parity or error-correcting code) shows that the data is corrupted in some form. Memory-management logic such as a TLB (translation look-aside buffer) might signal an exception when an address-mapping entry is missing.
3. **External interrupt**—external logic signals that the processor must stop what it is doing to perform some task on behalf of that logic, such as reading an input buffer.
4. **Computation error**—some rare, but important, computational condition such as arithmetic overflow or underflow has been detected.
5. **System call**—the application has reached a point where special processing by the operating system is required. From this point, the operating system can execute code on behalf of the application, often with broader and more trusted access to system resources such as I/O devices and memory.
6. **Debug breakpoint**—the application has reached an instruction designated as a stopping point at which the debugger can examine the state of the application execution.

Regardless of the exception type, all instructions at or beyond the exception point terminate without writing any results. Code execution then switches to an address specified by an *execution vector*, a code address corresponding to that exception type or interrupt signal. Most exceptions are recoverable—the original application will resume where it left off after the exception handling code has done its work—so the *exception handler* code saves current register values as appropriate so that its own code can use some registers for its own task. The handler restores the application code's values just before it returns control to the application and resumes execution at the address of the first instruction that was aborted when the exception occurred.

All exception conditions must be detected and action taken before the first stage in the pipeline where results (registers or data memory) are irrevocably written. Once an instruction reaches that point of no return, all of the changes to processor state will occur. All exceptions must be reported before the pipeline's commit stage. No write can occur before that stage, but writes may occur after that stage. This rule is irrelevant for RISC pipelines where instructions write only one word of state to the register file. In extended processors, which can write any number of states, writes may occur at any stage from the commit stage onward.

The possibility of generating exceptions introduces an issue to consider in connecting the processor to other logic. Any externally visible signal associated with an instruction that has not yet reached the pipeline's commit stage is considered tentative. External logic should not irrevocably change its state based on that speculative signal alone. The processor's cancellation and restarting of instructions mandates careful treatment of instructions that may affect system state outside the processor (instruction with external side effects). For example, reading data from an I/O queue outside the processor could be disastrous if the possibility of exceptions were not taken into account. Suppose the processor performs the read, popping a value from the queue, then the read instruction is cancelled, discarding the value just taken from the queue. When execution resumes, the read would be repeated, popping a new value from the queue and forever losing the first value read. Three standard design methods avoid this fault:

1. External interfaces that could connect to I/O structures such as queues are active only after the commit stage. An I/O read or write is never cancelled. This method is simple, but may limit performance, because I/O reads and writes may take extra cycles compared to the fastest local-memory reads and writes.
2. External interfaces that connect to I/O structures are permitted to perform speculative operations (for example reads and writes before the commit stage) are also equipped with signals that report cancellation once the commit point has passed. This signal may lag behind the read or write, but the side effect of the read or write (the popping of the queue) can typically be deferred by the external until the possibility of late cancellation is resolved.
3. The external interface can directly implement special functions, such as queue behavior. Even if an instruction that pops data from the queue is cancelled, the value popped is held internally and the reexecuted operation gets the same value from the queue as the first, cancelled execution.

The general characteristics of the local memory interface—wide bus width, high clock rate, and low latency—are very attractive for connections among processors or between processors and logic. With these mechanisms to handle the possible exception-processing side effects, flexible processing with fast interconnect and I/O is easily achieved.

7.2.3 Alternative Pipelining for Complex Instructions

As introduced in Chapter 6, multistate logic can be implemented as a series of single-cycle operations or as intrinsically pipelined operations. As designers of high-performance hardware understand, managing the pipeline is important for overall throughput. In theory, pipelining increases the overall operation rate—with respect to a simple machine that performs only one operation at a time—by a factor equal to the pipeline depth. Moreover, a pipeline can be in implemented two forms:

- Direct form, with state-register reads occurring only at the beginning of the execution pipe, one pipe stage directly feeding the next through simple pipeline registers, and state-register writes occurring only at the end of the execution pipe. This is sometimes called a "data-stationary pipeline."
- Exposed form, with the inputs and output operands of each pipe stage coming and going from explicit state registers. The operations of one pipelined operation may be encoded in several different instructions. This is sometimes called a "time-stationary pipeline."

Fig 7-7 and Fig 7-8 show a comparison of a three-stage execution pipeline implemented using these two forms. The operations of a directly pipelined execution (six-stage pipe) are highlighted in Fig 7-7 and the operations for the equivalent exposed pipeline (three independent operations per instruction, four-stage pipe) are highlighted in Fig 7-8. The op1 operation in instruction i writes results used by op2 in instruction i+1, which writes results used by op3 in instruction i+2.

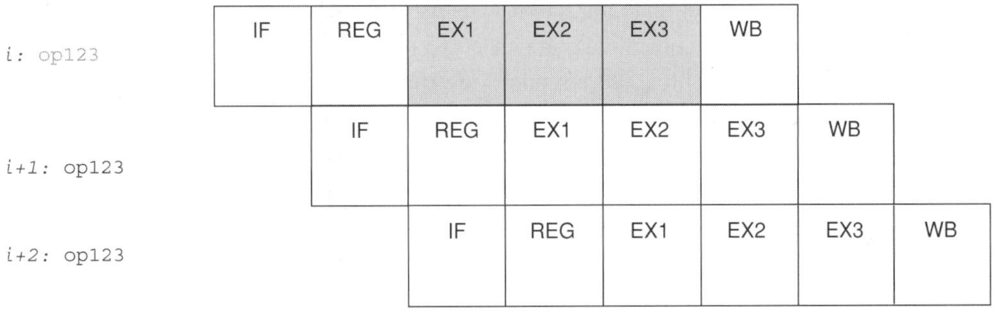

i: op123	IF	REG	EX1	EX2	EX3	WB		
i+1: op123		IF	REG	EX1	EX2	EX3	WB	
i+2: op123			IF	REG	EX1	EX2	EX3	WB

Figure 7-7 Direct pipelining model.

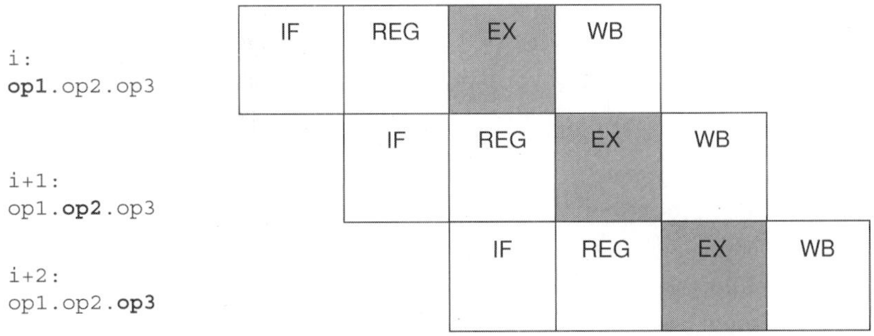

i: **op1**.op2.op3	IF	REG	EX	WB		
i+1: op1.**op2**.op3		IF	REG	EX	WB	
i+2: op1.op2.**op3**			IF	REG	EX	WB

Figure 7-8 Exposed pipelining model.

In the direct pipelining example, the three operations are executed in three successive cycles associated with a single instruction. In the explicit pipelining example, the first operation is associated with first instruction, the second operation with the second instruction, and the third operation with the third.

The principal advantage of the direct pipelined form is the efficiency of expression and implementation. Only the pipeline's source and destination registers need to be expressed directly and included in the instruction-set specification. The pipeline registers between stages are inferred and inserted at the optimal point in the logic to balance the delay among the stages. A new operation can be started every cycle without confusion over potentially conflicting uses of results in one stage and definition of results in another stage.

The principal advantage of the exposed form of pipelining is flexibility, especially when exposed pipelining is combined with the independent encoding of operations into compound instructions or with long-instruction-word technology (described in Chapter 4, Section 4.5). If the instruction encoding supports all possible combinations of the operations for each pipe stage, then software has full control over the pipeline. The pipe stages can be rearranged, other operations can be inserted in the middle of the execution pipeline, and operations can be omitted—all under programmer control and long after the SOC has been fabricated.

Some designers choose a combination of the two methods—building sections of the data flow as direct pipelines and putting these sections together, thus exposing the sources and destinations of the pipelined sections as explicit processor registers. Combinations of operations that are most stable make the best direct pipelined instructions. The boundaries between pipelined operations may correspond to points in the computation where uncertainties or options in the algorithm remain.

7.3 Optimizing Processors to Match Hardware

Chapter 6 introduces the basic techniques for migrating from hardwired logic and simple microengines to application-specific processors. Two differences between microengines and pipelined processors sometimes deserve special attention: branch architectures and memory access. This section suggests some basic methods for overcoming these differences.

7.3.1 Overcoming Differences in Branch Architecture

As discussed in Section 7.2, pipelined instruction execution aids processor performance but may introduce operational latency, particularly for operations that involve instruction- or data-memory references. In most processors, this latency includes one or more cycles for control-flow changes (a delay in taking a branch or jump). For most embedded architectures, this latency dictates one or more stall cycles, while a new instruction address—the branch target—is fetched from instruction memory.

In contrast, most microengines either do not pipeline instruction fetches or they implement delayed branches, as discussed in Chapter 6. Without special consideration for branches, the code migrated from a finite-state machine or microprogram may have execution "bubbles" where the processor stalls due to branches. These bubbles degrade performance on some code sequences relative to the original state machine or microengine.

Four techniques can mitigate these effects:

1. **Use zero-delay branches:** Many architectures offer some branches without a delay for a change in control flow. For example, the PowerPC architecture offers a special Count register to eliminate the branch delay for some branches. The Xtensa architecture offers zero-overhead loops to handle common loop-iteration cases without branch instructions or taken-branch bubbles in the inner loop code. Some complex processors employ dynamic branch prediction to reduce bubbles. Branch prediction may not be suitable for some microcoded applications because it introduces performance uncertainty and variability into application execution.
2. **Organize code to reduce taken branches:** The code developer can often organize the code so that common or performance-critical sequences require few taken branches. Pipeline bubbles are seen only by code executed off the main path. This approach is particularly appropriate where rare and exceptional cases must be tested and handled in special code. Exception-handling and error-recovery code often have looser performance demands than the primary flow of data processing because this code is executed infrequently. Simply inverting a branch condition and swapping the target and flow-through code blocks can eliminate performance issues. The flow charts in Fig 7-9 and Fig 7-10 contrast a code sequence with several branch-taken bubbles with one that uses zero-delay branches and branch reversal to eliminate branch delays. Simple branch optimization reduces the number of taken-branch bubbles from three to zero.
3. **Replace branches with conditional operations:** Branches are often used to choose between two short or similar operation sequences. In many cases, mainline code can perform both operation sequences, followed by conditional selection or storage of the results, based on the condition that would have been used for a branch. This approach typically eliminates two taken-branch bubbles (one on each path or two on one path).
   A very simple example is the C statement:

```
if (a){x = b} else {x = c}
```

This statement might be implemented by the following sequence of Xtensa instructions, where a, b and c are reside in the registers a0, a1, and a2 and the value x is computed and stored in register a3.

```
              beqz a0, Lfalse
              mov a3, a1
              j Ldone
Lfalse:       mov a3, a3
Ldone:
```

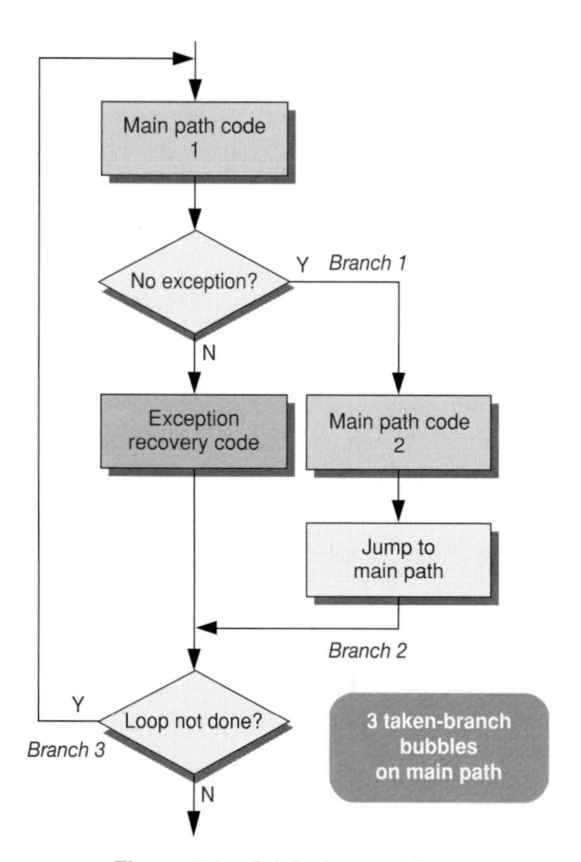

Figure 7-9 Original control flow.

This instruction sequence typically takes four or five cycles to execute on a processor with two bubbles for each taken branch (two instructions—`beqz, mov`—and two bubbles when `a` is false; three instructions—`beqz, mov, j`—and two bubbles when `a` is true). By contrast, using a conditional-move operation, this sequence is reduced to two instructions and two cycles.

```
mov             a3, a1
moveqz          a3, a2, a0
```

With a configurable processor, for example, the developer could define a new instruction with three source operands and compute, result in one cycle. Flexible support for condition codes and conditional moves in the base instruction set, as well as extended instructions with conditional results and instructions that use condition codes as inputs,

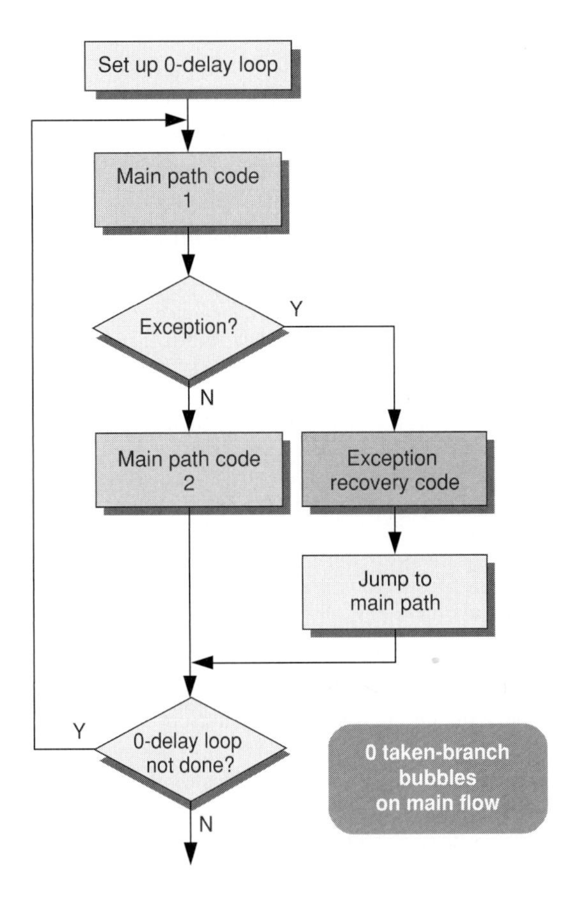

Figure 7-10 Optimized control flow.

make conditional operations part of the basic vocabulary of application-specific instruction sets.

Processor architectures that support arrays of condition codes, such as the Xtensa architecture, make assertion and use of condition codes for conditional operations quite simple. This method, commonly used in VLIW processors, is also known as *if-conversion.* The method is especially important for VLIW processors because multiple pipelines make computation instructions relatively cheap, while branches are expensive. The method works well with configurable processors because new computation instructions that perform many operations in parallel can be added very easily.

4. **Combine or pipeline operations to save cycles:** In some cases, a taken-branch bubble is unavoidable. The code, however, can compensate for the branch bubble by rearranging the operations. Operations that would have been executed following a microengine's zero-delay branch (operations at the branch target) or following a microengine's delayed branch

(operations that logically precede the branch) can be pulled earlier or pushed later in the code sequence. Operations that immediately depend on a preceding operation and can be executed in parallel with the branch can be added as pipeline stages following the first operation. Then, the first operation is effectively executed during the taken-branch bubble, and its results are available to the first instruction at the branch target.

Consider, for example, a microengine with a delayed-branch architecture, as discussed in Section 7.2.1.3, and a branch delay of two cycles. To execute a sequence of four dependent operations (before the branch: `op1`, `op2`, `op3`; after the branch: `opt`) the original four-cycle code sequence might look like the following:

```
        branch <cond> Label op1   //op1 always executed, 2 delay slots
        op2                       //op2 always executed, delay slot 1
        op3                       //op3 always executed, delay slot 2
        ...
Label:  opt                      //opt executed only on <cond>
```

If no appropriate zero-delay branch or delayed branch is available, this code cannot be transliterated to a processor with the performance of the original code. However, by defining a new operation, `op123` (which executes the three operations, `op1`, `op2`, and `op3`) as a pipeline sequence in a single instruction—the code sequence becomes the following:

```
        branch <cond> Label op123 //op123 always executed
        ...
Label:  opt                      //opt executed only on <cond>
```

Like the microcoded engine code, this code requires four cycles to execute, but it requires fewer instructions (and often few total microcode bits).

Other techniques can also significantly reduce branch overhead and improve performance. For example, a sequence of conditional branches can sometimes be replaced by a dispatch operation based on a jump to an address held in a processor register. Fig 7-11 shows the flow chart of an application that performs a series of tests on one value (x) where each test is implemented as a finite-state-machine state transition or a microcode branch. If the microengine can perform one conditional branch per cycle, the sequence of four tests takes from one to four cycles to complete, depending on the value of x.

To avoid the cost of branches, even not-taken branches without branch-taken bubbles, the sequence of branches can sometimes be converted to a dispatch to the appropriate block of code based on an instruction that jumps to a variable address. That address is typically the sum of a base address (indicating the first block of code) and an offset (the product of the variable and the

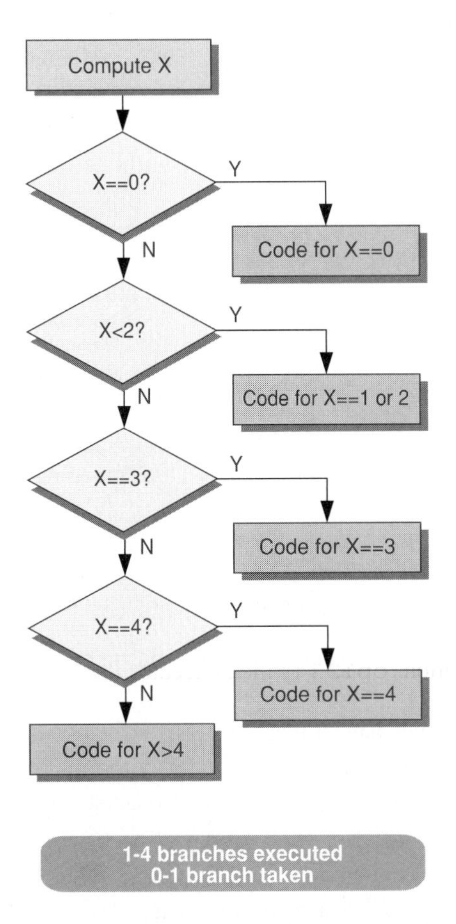

Figure 7-11 Sequence of tests of variable x.

size of the code block). A version of the dispatch code method is shown in Fig 7-12. This approach assumes that the block of code for each value of the variable is fixed in size (n bytes). Alternatively, the code could load an instruction address from a table, where each address points to the block of code for that value of the variable.

This technique can be used with general-purpose RISC instructions alone, but application-specific instructions that conditionally compute a target code address often prove very useful in reducing branch delays and accelerating performance.

7.3.2 Overcoming Limitations in Memory Access

Microengines often use data memories to hold operand arrays and other temporary values. When the microengine is converted to an application-specific processor, these arrays are generally assigned to some region of the processor's data memory and accessed via data loads and

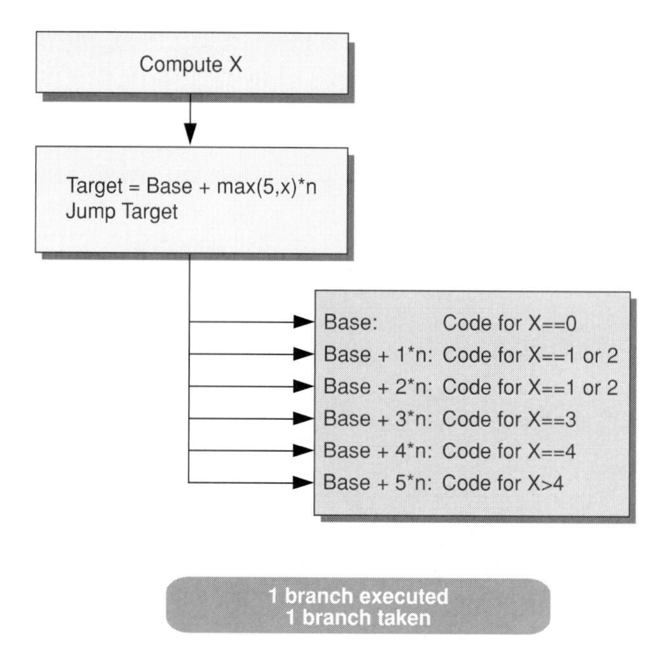

Figure 7-12 Multiway dispatch based on variable x.

stores. Some microcode engines have a number of small memories that can theoretically all be accessed in parallel during one cycle. If the number of simultaneously accessed memories exceeds the number of processor data-memory ports, some change is required. Three techniques may apply:

1. **Map small memories into register files:** Register files are a good alternative for modest memory depths—up to, perhaps, 128 words. Their arbitrary width and support for multiple read and write ports make register files quite attractive for small data memories, especially where operands are shared via simultaneous access by multiple parallel operations. Register files are often generated as arrays of latch cells. Latch arrays are not as dense as RAM for large arrays, but they are more flexible. This added flexibility comes both in the logical definition of the memory (arbitrary width, depth, and number of ports) and in the physical implementation (automatic placement and routing of the register-file cells allows the register-file to be seamlessly intermixed with the surrounding logic).
2. **Rearrange code to reduce peak number of simultaneous memory references:** Large memories are much better implemented as RAMs because the area and power per bit of RAMs are significantly better than that of latches or flip-flops. Multiported RAMs are relatively expensive to implement, especially as the number of read and write ports grows. The number of ports must match the peak number of simultaneous memory references that might occur in the microinstruction stream.

Minor reordering of the memory references allows port sharing across different uses and typically reduces the number of required ports. For example, if the original code makes three memory references in just a fraction of the inner-loop cycles, and zero or one memory reference in others, the memory-reference pattern can be smoothed out so that most cycles make two memory references. Smoothing may require that the application "fetch the data ahead" of the actual operand use. Increasing the number of load-result registers enables fetch-ahead, and may also make the instruction set more flexible, because the same registers can be used to hold important variables from one loop iteration to the next without a load.

3. **Use wider memories to increase total bandwidth:** In many applications, the total transfer bandwidth between processor and memory is more important than the number of independent references per cycle. For example, an application that makes three 16-bit memory references per cycle (perhaps two loads and one store) can often be improved by substituting a memory system capable of two 32-bit memory references per cycle (any combination of loads and stores). So long as a significant fraction of the 16-bit operand references are sequential, the 32-bit wide memory provides more useful bandwidth to the application.

7.4 Multiple Processor Debug and Trace

Much of the performance and sophistication of advanced SOCs comes from the opportunity to exploit high degrees of parallelism, especially concurrency among processors. Without good MP debugging tools, however, the development of correct, efficient software solutions might be impossible. A short discussion of MP debugging and program tracing highlights the opportunities and pitfalls.

7.4.1 MP Debug

Normally, tasks running on different processors in a system execute asynchronously of one another. The computations of different processors are independent except when the tasks communicate with one another via messages, interrupts, or shared-memory references, as discussed in Chapter 4. If tasks were always independent, the process of debugging the system would be fairly simple. As soon as each task is debugged in isolation, the system debugging is done. In reality, of course, every system has interactions among tasks—some intended, some unintended—so coordinated debugging of all of the tasks together may be critical to system correctness and performance. In addition, hardware support for MP debugging must be considered. Common debug-access hardware for all debugging operations, such as a JTAG port, reduces the hardware cost to a few pins on the chip and one external interface device.

Debugging the most complex tasks sometimes requires two forms of debugging information. Interactive examination of processor state from a source-level debugger gives the developer a complete picture of task state—all registers and memory—at one point in time. Examination of the complete state may take considerable time, so task execution must be stopped, and the developer will not see any history of computation that might explain how the task got to its current state. Stopping the processor at each program step, however, may so perturb the interaction of that processor's task with other processors and I/O interfaces that no useful debugging insight is possible. This situation creates a familiar dilemma in real-time systems debugging: the process of looking for a bug makes the bug go away! As an alternative, an execution trace showing the sequence of instruction execution and state values can give critical insight into the cause for unexpected task behavior. Modern embedded-processor cores often provide both a JTAG-based debug interface and trace outputs. These debugging interfaces can be daisy-chained in multiple-processor systems, as shown in Fig 7-13. This figure shows a schematic view of a complex SOC with four application-specific processors.

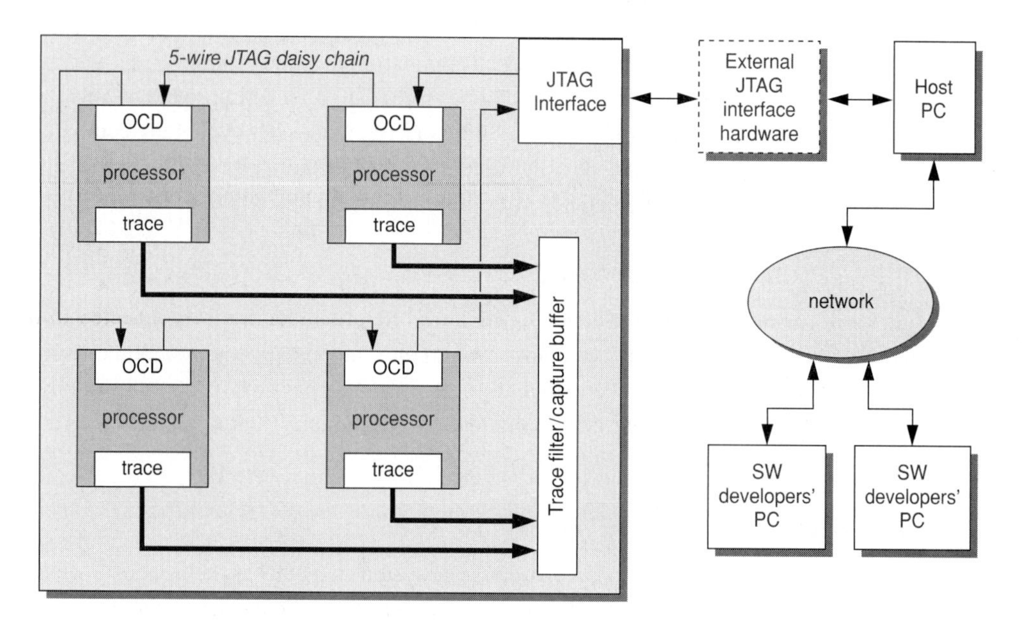

Figure 7-13 Four-processor SOC with JTAG-based on-chip debug and trace.

Each processor core in this example has a five-signal interface to its internal on-chip debug (OCD) module. These five signals—clock (`TCK`), control (`TMS`), data-in (`TDI`), data-out (`TDO`), and reset (`~TRST`)—follow IEEE Standard 1149.1 for the Test Access Port (TAP). The data-in and data-out signals are connected serially in a daisy-chain so one five-signal JTAG interface serves any number of processor cores. The host PC software controlling the JTAG interface can read and write any processor's state and is controlled from the multiple-processor source-level debugger running on the developer's desktop.

The interactive experience of JTAG-based debugging can be identical to the experience for pre-silicon debugging using a multiple-processor simulation model. Three differences between simulation-based and hardware-based MP debugging stand out:

1. Hardware generally runs faster than simulation, so larger test cases can be debugged.
2. Hardware fully implements interfaces to the system's external devices and network connections, so more realistic system behavior is possible. Simulation models of external interfaces are necessarily simplistic.
3. Even with good on-chip debugging support, hardware offers lower visibility into the details of processor state than simulators. Simulation models can give extremely detailed pictures of pipeline states and even individual modeled wires.

The serial connection among processors is only active when the developer is actively probing and setting processor state, and does not need to operate at processor speeds.

The ideal MP-debugging model should cope with varying levels of coordination among tasks. Sometimes, the developer wants to debug a single task while all others continue to run; sometimes, the developer wants to examine the state of some (or all) of the tasks at the same time.

The Xtensa processor supports this need by providing two additional signals at each processor's OCD interface—BreakIn and BreakOut. When a task hits a debug-trigger condition—the instruction address matches an instruction-breakpoint value or the data address matches a data-watchpoint value—the processor stops executing code, enters debug mode, and asserts the BreakOut signal. The BreakOut signal from one processor can be connected to the BreakIn input of other processors. The assertion of BreakIn immediately forces a processor into debug mode. The connections from BreakOut to BreakIn can be direct, if the debug relationship among processors is static, or through a masking table to provide software control over the sensitivity of each processor to the debug events of other processors.

7.4.2 MP Trace

Fig 7-13 also illustrates the connection of processor trace interfaces to an on-chip trace-capture mechanism. The processor's trace port emits a detailed cycle-by-cycle record of instruction execution and, optionally, data computation results. These ports operate at the full speed of the processor and sustain much higher information bandwidth.

In the case of the Xtensa processor, the trace port consists of three signal groups:

1. A status word of 8, 13, or 16 bits (depending on the processor configuration) that reports the status of the instruction in the completion stage of the pipeline, stalls, exceptions and other cycle-by-cycle execution status
2. A 32-bit address word that reports either the instruction (for most instructions, especially jumps and branches) or a virtual data address (for load and store instructions)
3. An optional 32-bit data word reports computation results being written to the general-purpose register file

This additional trace information, as much as 10 bytes per cycle, potentially gives crucial insight into code behavior in particularly complex code sequences or when a task is interacting in unexpected ways with other tasks and I/O devices. On the other hand, it may not prove technically or economically feasible to add enough signal pins to the SOC to bring these high-clock-rate signals off the chip for visibility. This limitation is particularly true for multiple-processor SOCs. A single 350MHz processor, for example, produces as much as 3.5 GB/sec of trace data, so a 20-processor SOC would nominally need an extra 70 GB/sec of output bandwidth.

Two choices are available for handling this huge volume of trace data. First, the trace can be restricted and compressed. Trace information is relatively compressible because instruction addresses are sequential except at branches and jumps. For most branches, the essential dynamic information is simply "taken" or "not taken" for each branch. However, exceptions and jumps to addresses held in registers require a full instruction addresses be emitted.

Trace-compression logic on the SOC can implement a simple state machine to reduce the number of bits required to transmit the most important execution-trace information to a few bits per cycle. A few signal pins on the chip can output these bits for each processor, to be captured by a logic analyzer or special trace-capture hardware. External hardware can easily capture thousands or, if necessary, millions of cycles of trace history. This approach may be acceptable for a handful of processors, but does not scale to larger processor counts, especially if the traces of many processors must be captured together.

Moreover, the compressed trace usually omits valuable information such as the data trace, which is much harder to compress.

Alternatively, the trace can be captured in an on-chip trace-capture buffer. On-chip RAM is expensive, so trace-buffer depth is limited, typically to hundreds or thousands of cycles. To maximize the effectiveness of the limited trace-buffer depth, the on-chip logic can filter the trace according to programmer-selected criteria, so only the trace data of certain cycle types or address ranges is put in the buffer.

Most importantly, on-chip trace buffering is effective for large processor counts. The trace buffer can be very wide to capture either the traces of many processors in parallel or more highly detailed trace information from a dynamically selected subset of the SOC's processors. If an on-chip trace buffer is implemented, that memory is connected either to the general-purpose bus, so software running on one of the on-chip processors can later transfer the data off-chip, or to the JTAG chain for access through the serial interface debugging mechanism.

Whether the traces are captured in an on-chip or off-chip buffer, they are later transferred to the host PC for analysis. A trace decompression and interpretation tool takes the trace data and the program code as inputs and creates a comprehensible history of program behavior for the trace's slice of time. If the trace contains full data results, the trace interpreter can even create a virtual source-level debugging session, allowing the developer to step through the program execution, examining the process state at each step, just as if the debugger were connected live to the running processor. Sophisticated embedded debug and trace capabilities are increasingly important to successful SOC development. Multiple-processor design makes trace and debug more emphatically critical. Note, though, that despite the challenges of multiple-processor debugging, a processor-based approach is inherently more debuggable than a more hardwired approach. No RTL design tools and methods give the level of state visibility and controllability routinely available with embedded processor debuggers and trace tools.

7.5 Issues in Memory Systems

Part of the power of configurable processors comes from the opportunity to tune the memory system to exactly match the requirements of the target applications. For data-intensive applications, performance is often limited by data bandwidth, either within the heart of the computation or in moving data in and out of the processor. Moreover, memory arrays and their control functions may comprise a major portion of the SOC's silicon area and operating power. This section discusses some advanced techniques for optimizing SOC memory systems.

7.5.1 Pipelining with Multiple Memory Ports

Many data-intensive applications rely on streaming data through the processor. Data is read, processed by a small sequence of operations, and written out. Little of the processed data is reused by subsequent computation. Application throughput is often governed by two basic factors: the rate data can move through the processor using load and store instructions and the degree of overlap between memory operations and computation. Configurable processors can have multiple, wide memory ports, so several different techniques are available to optimize the code around pipelined memory and computation.

For example, suppose input data is brought into a processor with a load operation (usually two cycles of latency), processed by a complex operation (in this example, with two cycles of latency), and sent out with a store operation (usually two cycles of latency). A single instruction, `load.op.store`, might encode all three independent operations. A long sequence of these load/operation/store computations could be pipelined as shown in Fig 7-14, where the computation for data item *i* is combined with the store for item *i-2* and the load for item *i+2*. The load, op and store operations for data item *i* are highlighted.

This aggressive pipelining achieves high bandwidth by configuring the processor to perform two full memory operations per cycle. With 128-bit input and output paths and a clock rate of 350 MHz, the sustained total data-memory bandwidth of this instruction sequence is 11.2 Gbytes/second:

$$\left(16\,\frac{\text{input bytes}}{\text{cycle}} + 16\,\frac{\text{output bytes}}{\text{cycle}}\right) * 350\text{M}\,\frac{\text{cycles}}{\text{second}} = 11.2 \text{ Gbps total I/O}$$

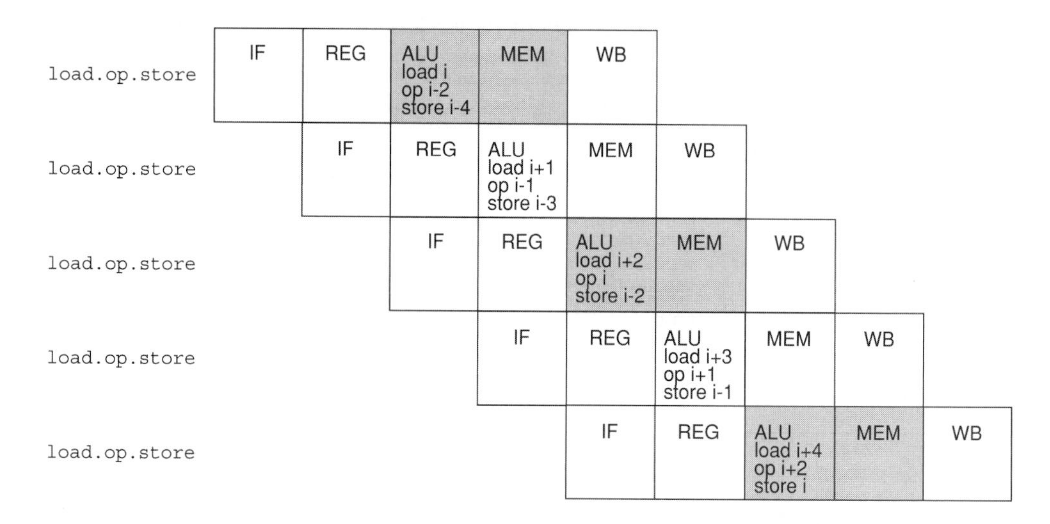

Figure 7-14 Pipelining of load-operation-store.

7.5.2 Memory Alignment in SIMD Instruction Sets

Memory operations also present additional complications when the vectors in memory cannot be forced to fall on natural address boundaries (the address is a multiple of the vector size). To keep the hardware cost down and performance up, embedded-memory systems usually can only operate, or operate most efficiently, on operands that are aligned on natural-word boundaries where the memory address of the vector is a multiple of the number of bytes in the vector. Otherwise, the vector crosses a memory-word boundary and requires more than one memory cycle to load or store.

This alignment problem is readily addressed, however, by defining load and store operations that use an alignment buffer to load the desired series of unaligned vectors into registers from a series of aligned memory references and, correspondingly, to store the desired series of unaligned vectors as a series of aligned memory references.

The basic operation of a load-alignment buffer appears in Fig 7-15. An initial load puts a word that contains part of the first vector into the buffer. A second load brings in the word that contains the rest of the first vector and part of the second vector. The merge multiplexer assembles the first complete vector puts it into the vector register file. The balance of the second word goes into the buffer register to be combined with the third word when it arrives from memory. In this manner, S unaligned words can be efficiently loaded and aligned with S+1 load operations.

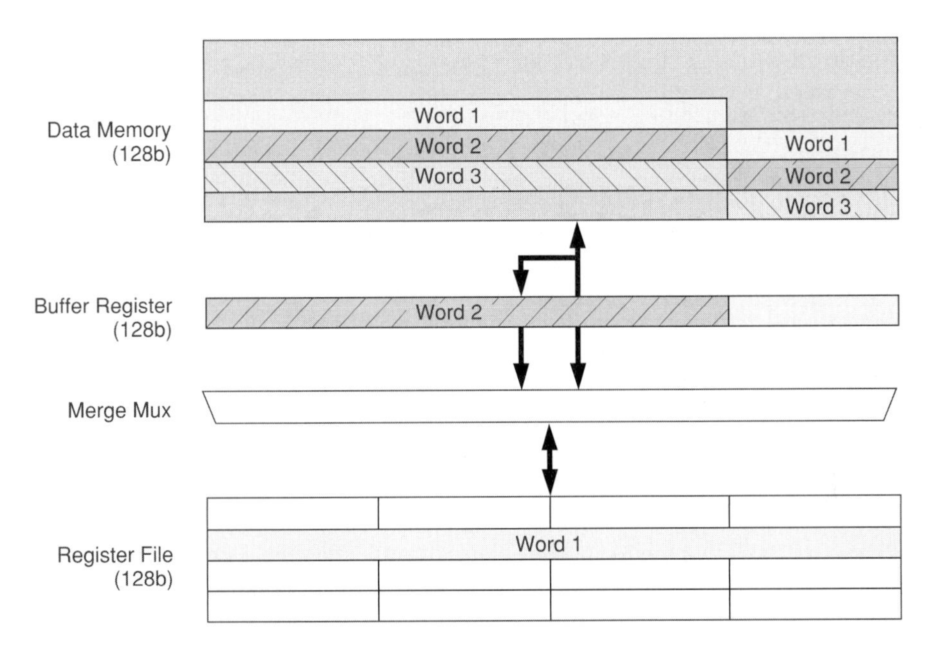

Figure 7-15 SIMD alignment buffer operations.

7.5.3 Synchronization Mechanisms for Shared Memory

As discussed in Chapter 4, when several processors communicate through shared memory, they must have some means to synchronize accesses to critical data. Otherwise, more than one processor may attempt to modify the data simultaneously. In the some simple cases, a binary flag or lock in memory can arbitrate access. In other cases, a more complex synchronization structure with special hardware support is required. Even in the simple cases, however, it is necessary to understand the basic concepts of memory ordering and atomic operations.

7.5.3.1 Memory Ordering and Locks

On the surface, memory ordering seems obvious. Instructions execute in program order and the effect of each load and store is seen in that same program order. This assumption is important to the behavior of locks and semaphores. Chapter 4 introduced the basic handshake between a data source and destination. Fig 7-16 shows the basic interaction between two tasks communicating through shared memory.

The correctness of this sequence relies on the ordering of operations performed on the ownership flag in memory. The programmer assumes that memory-load and memory-store operations occur in the same order that they appear the source code. The source/destination handshake will fail, however, if the stores in the source task somehow do not take place in order, especially with respect to loads at the destination task. If the store to the data location

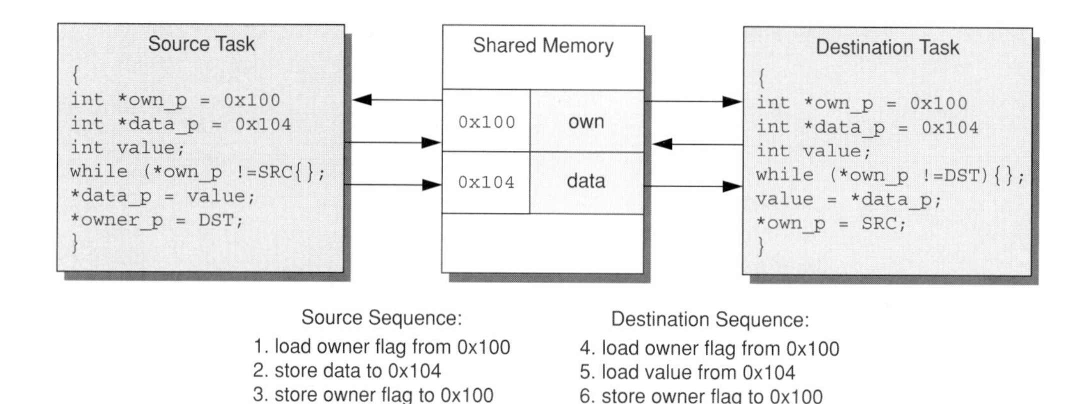

Figure 7-16 Shared-memory communications mode with ownership flag.

(`*data_p = value;`) is somehow delayed until after the store to the flag location (`*owner_p = DST;`), the destination task may grab the data before it is ready.

Similarly, the handshake will fail if the destination task's load ordering is changed. Specifically, if the load of the ownership flag in the while condition (`*own_p != DST`) is somehow delayed until after the load of the data (`value = *data_p;`), then the destination task might use the wrong version of the data. Moreover, if the data load is delayed until after the store that resets the ownership flag (`*own_p = SRC;`) then the source task could see the changed flag and modify the data value before the destination task can load the old value. This simple example demonstrates that the ordering of loads and stores between processors sharing data through a common memory often matters.

There are a number of useful formal definitions of memory-ordering requirements. One popular definition, and the one adopted by the Xtensa architecture, is *release ordering*. Release ordering employs *acquire loads* and *release stores* with the following properties:

- Before an ordinary load or store access is **performed** (as seen by any other processor), all previous *acquire loads* must be **performed**.
- Before a release store is **performed**, all previous load, store, acquire loads and release store accesses must be performed.
- Before an acquire load is **performed**, all previous acquire loads must be **performed**.

For a release-consistency programming model, no stricter memory ordering can be assumed by the programmer, though the system hardware may implement stricter ordering. Stricter memory ordering generally results in cheaper, simpler hardware. Looser memory ordering generally results in memory operations with higher performance.

The Xtensa architecture defines instructions to implement release ordering. These loads and stores should properly be used for lock operations such as loads and stores on the ownership flag in the simple example of Fig 7-16, (though ordinary loads and stores may work fine, so long as the memory system and the associated interconnection logic maintain strict ordering without any instruction-level support.

The Xtensa processor's load and store instructions that implement release ordering are as follows:

`L32AI`	32-bit load acquire	This load is typically used to test the synchronization variable that protects mutual-exclusion memory regions (for example, to acquire a lock).
`S32AI`	32-bit store release	This store is typically used to write a synchronization variable to indicate that this processor is no longer using a mutual-exclusion memory region (for example, to release a lock).

This store is typically used to write a synchronization variable to indicate that this processor is no longer using a mutual-exclusion memory region (for example, to release a lock).

7.5.3.2 Exclusive Access Instructions for Multiple Producers or Consumers

Load-acquire and store-release instructions permit interprocessor communication, as in the single-source/single-destination example above, but they are not efficient for guaranteeing exclusive access to data when several sources or several destinations are involved. For these cases, systems implement some mechanism for atomic updates of memory-based synchronization variables. One of the common synchronization mechanisms is a semaphore, which counts the number of active uses of a particular shared resource.

It is essential that every task accessing the shared resource know exactly how many others are using it and that each task updates its own usage indicators without corruption by updates from other tasks. For example, a system with *N* producers and *M* consumers might share a common data-passing queue among producers and consumers. Addresses of the current head and current tail of the queue must be updated atomically for correct operations. If each of these pointers is a semaphore, no producers and no consumers will become confused and accidentally update the wrong queue entry in memory.

In some cases, atomic update mechanisms are directly supported in the processor instruction set. In the case of the Xtensa architecture, this support takes the form of the Store Compare Conditional (S32C1I) instruction and a corresponding register (SCOMPARE1):

`S32C1I`	32-bit store	Stores to a location if and only if the memory location contains the same value as the SCOMPARE1 register. The comparison between the value in the designated memory location and the compare register value is atomic. The instruction also returns the old value of the memory location to a processor register.

This atomic operation may be used to implement a number of more complex mutual-exclusion primitives. For example, an atomic increment, where a processor must load, increment, and store a memory semaphore, might be implemented by the following snippet of Xtensa assembly code, where register a2 points to the synchronization variable in memory:

```
        l32ai    a3, a2, 0          // load current value of sync variable
loop:   wsr      a3, scompare1      // put current value in SCOMPARE1 register
        addi     a4, a3, 1          // increment synchronization variable
        s32c1i   a4, a2, 0          // attempt to store
        bne      a4, a3, loop       // if value has not changed, try again
```

The S32C1I instruction is defined for operation on both cached and uncached memory locations, making code using this mechanism portable across different system types. The implementation is different for cached and uncached memory. When operating on cached addresses, the instruction will maintain exclusive access to the word in cache while it does the compare and the possible store. For uncached operations, the processor sends the old and new data to the memory, where the compare and possible store are performed atomically.

The implementation of the S32C1I instruction is a good general access-protection mechanism because other widely used synchronization primitives such as test-and-set or fetch-and-add can be synthesized from it. This instruction requires special support from the memory controller, but that support is similar to the support required for bus locking.

Moving the atomic operation from the bus (where bus locking is a common implementation method) to the memory increases parallelism (because different processors can be performing synchronization operations in different memories attached to the same bus) improves worst-case bus latencies (because synchronizations are not necessarily waiting for exclusive, sequential control of the bus), and reduces the risk of system deadlocks (because the absence of global bus locks removes a possible source of catastrophic error).

7.5.3.3 Interrupt Handlers

Interrupt-driven synchronization can also improve processor-to-processor communication. One possible liability of synchronization based on memory-resident locks and semaphores is excessive memory traffic. When a processor is waiting for the lock, it may "spin," polling the lock address location continuously in hopes of claiming the lock as soon as it is available. Those accesses are fruitless for the waiting processor, and they consume memory bandwidth, a valuable global system resource. Note that some cache-coherency schemes allow the processor to spin-wait with accesses only to its local cache in anticipation of a write-update or line invalidation triggered by the releasing write of the processor holding the lock. For systems with a small number of locks, cache coherency is often expensive in terms of hardware overhead, relative to the performance or programming benefit.

Instead, interrupt-driven synchronization can provide efficient communication between processors and incurs less memory-bandwidth overhead. Shared data still passes through globally accessible memory locations, but the locking mechanism is handled via interrupts. The producer processor creates the data, writes it to shared memory, and asserts an output signal (perhaps via a configured wire) that is connected to an interrupt input on the consumer processor. The consumer processor responds to the interrupt and reads the data from shared memory. The consumer's interrupt handler then asserts its own output signal connected to an interrupt on the producer processor, which signals that the data has been accepted. The basic structure of this mechanism appears in Fig 7-17.

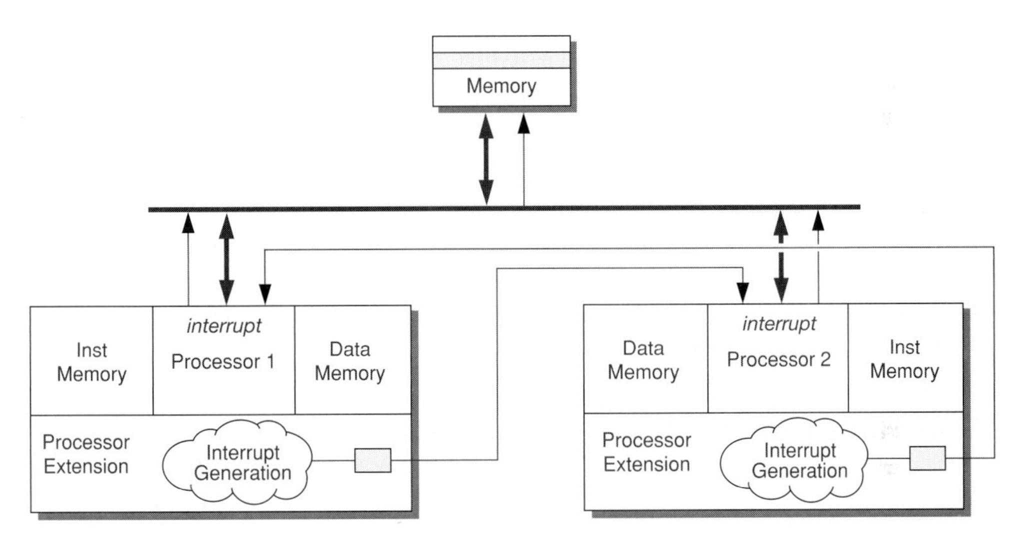

Figure 7-17 Interrupt-driven synchronization of shared-memory access.

Software for producer and consumer data transfer can reside entirely within the two interrupt handlers. Some care must still be taken in managing the relative delays between memory writes and interrupts. If the write of producer data is delayed more than the interrupt signaling write completion, the consumer's interrupt handler could read old data, not the newly written data.

7.5.3.4 Hardware Synchronization

Synchronization between data producers and consumers can also be performed in hardware, so that the processors automatically wait for data availability. Chapter 4 introduced processor-to-processor queues. These queues are commonly mapped into memory so that the producer writes its data to the address of the queue's tail and the consumer reads its data from the address of the queue's head.

Configurable processors may also map queue pushes and pops directly into the definition of selected instructions. Direct queue operations eliminate synchronization overhead and permit higher producer-to-consumer transfer rates. This is particularly true for wide data operations and instruction-mapped queue operations. Nevertheless, queues may be more expensive per data word held, so they may not be appropriate for all producer consumer communications, especially when very large data blocks are being shared.

7.5.4 Instruction ROM

Some embedded applications require use of low-cost instruction ROM to save a few cents or a dollar of manufacturing cost. Often, companies move software that was held in a flash-memory device into on-chip ROM after the product definition stabilizes. This approach can actually be critically important for very competitive markets where selling price is the most important product feature.

Moving code into on-chip ROM would appear to sacrifice the reprogrammability of that device, but the flexibility of the processor-centric SOC design approach is enormously valuable until the code is moved into ROM. Further, the option to squeeze that next dollar out of the product's manufacturing cost out by converting to hard ROM or even to a hard embedded ROM on the SOC is actually an important condition of all of the embedded market space covered by SOC designs.

7.6 Optimizing Power Dissipation in Extensible Processors

As more electronics becomes portable, power is no longer just one of many design issues—often it becomes the central issue in an SOC's design. Configurable processors have a deep influence on low-power design in two ways.

First, compared to hardwired logic, software-based design allows for more sophisticated algorithms and control of operating modes. In many applications, the software can be much smarter than hardwired logic about when to run and how fast. These characteristics can save far more power than the best lean logic design.

Second, application-specific processors can reduce the energy requirements for data-intensive computation by an order of magnitude or more compared to general-purpose RISC processor cores. Application-specific instruction-set extensions pack the same work into far few cycles, allowing the SOC to run at a lower clock frequency (and a lower operating voltage). Extending the processor does increase the power dissipation per clock cycle, but the corresponding power increase is usually far smaller than the increase in application performance. The power savings will improve, in line with the cycle-count reduction.

This section gives some insight into power optimization in the processor core, energy improvement through processor configuration, and power optimization in memory systems.

7.6.1 Core Power

Energy use due to active switching in CMOS logic circuits for computation of a complex function can be approximated using the following equation:

$$E = \alpha C V^2 n$$

where
α is the average fraction of circuit nodes switching between one and zero each cycle (weighted by capacitance)
C is the total capacitance of all the switched nodes in the circuit
V is the voltage
n is the number of cycles required to execute the function

The impact of a good processor configuration is to sharply reduce n, while increasing C only slightly relative to the baseline processor due to the increased number of logic gates and the increased area needed to compute the function. In fact, for some functions, it may be possible to reduce n and to reduce the switched capacitance at the same time. In addition, configurable processors can be quite smart about activating execution units only when necessary. The processor generator can determine the combinations of logic blocks that must be active at each stage of the pipeline and create logic for fine-granularity clock gating thereby reducing α. A processor with complex instruction-set extensions may have hundreds of clock-gated domains to reduce activity and power.

In the case of the Xtensa processor, automatic power optimization reduces energy requirements significantly. Larger processors do dissipate more power, but as a guideline, the power required to run a compute-intensive application grows at less than 1.4nW/MHz per gate for typical 130nm CMOS logic powered by 1.0 volts. This rule of thumb is a good first-order tool for overall power estimation.

A minimal processor core built from 12,000 gates dissipates about 30μW/MHz. A typical processor core built from 30,000 gates therefore dissipates about 60μW/MHz, and a full-featured quad-MAC DSP dissipates about 350 μW/MHz. These figures assume that the processor is running applications that use the extended execution units heavily. Clocks to extended execution units are typically disabled while executing code that does not use these units, so the power impact of unused extensions is quite modest; they still have a second-order impact on power by increasing the total processor size and increasing average wire length and capacitive loading in all processor circuits.

7.6.2 Impact of Extensibility on Performance

Clearly, processor configuration holds the potential to reduce the number of cycles to execute an application, but what does extending the processor do to energy efficiency? Does the reduction in cycles more than compensate for the increase in power dissipation associated with adding application-specific hardware?

The energy impact varies widely, depending on the nature of the application and the appropriateness of the extensions to the application. The most dramatic energy improvements come in particularly data-intensive arithmetic algorithms in signal and security processing, where general-purpose integer RISC architectures lack the optimized computation units. Fig 7-18 shows the results for optimization of four algorithms: the dot-product of two 2048-element vectors, Advanced Encryption Standard (AES) security coding, Viterbi trellis decoding for wireless communication, and a 256-point complex fast Fourier transform (FFT).

Application		Reference Processor Configuration	Optimized Processor Configuration	Energy Improvement
Dot-Product	Area (mm^2)	0.9	1.3	
	Cycles (K)	12	5.9	
	Power (mW/MHz)	0.3	0.3	
	Energy (mJ)	**3.3**	**1.6**	**2**
AES	Area (mm^2)	0.4	0.8	
	Cycles (K)	283	2.8	
	Power (mW/MHz)	0.2	0.3	
	Energy (mJ)	**3.3**	**1.6**	**82**
Viterbi	Area (mm^2)	0.5	0.6	
	Cycles (K)	280	7.6	
	Power (mW/MHz)	0.2	0.3	
	Energy (mJ)	**65.7**	**2**	**33**
FFT	Area (mm^2)	0.4	0.6	
	Cycles (K)	326	13.8	
	Power (mW/MHz)	0.2	0.2	
	Energy (mJ)	**56.6**	**2.5**	**22**

Figure 7-18 Impact of processor optimization on energy efficiency.

The optimizations are not intended to show the maximum possible improvements, but rather the range of tradeoffs in effort and efficiency. The reference configurations are individual, general-purpose RISC-processor configurations targeting each application domain. The dot-product reference configuration, for example, includes a fully pipelined multiplier, but no other

DSP support features. The degree of configuration aggressiveness for each the optimized processors listed in Fig 7-18 also varies. The dot-product configuration, for example, adds three more 16-bit multipliers with 32-bit accumulation, but no other DSP-like features. The optimized AES configuration, in contrast, adds multiple wide execution units and can perform parallel loads and stores to dramatically improve application performance (more than 100x).

In many ways, extensible processors provide an ideal environment for power optimization. Like hardwired logic, they can precisely incorporate the deeply pipelined execution units and memory-access paths required for efficient data computation without extraneous data shuffling among registers and memory. Good pipelining has a positive influence on power efficiency, because it reduces the amount of sequential logic in each pipe state and permits high computational throughput at lower voltage. Just like conventional processors, configurable processors can implement higher level power management policies. Moreover, automatic processor generators like Tensilica's generator analyze pipeline interactions among instructions and generate optimized clock gating circuits. This combination—global power management, deep pipelining, detailed clock gating, and optimized configuration—provides an effective power-management platform at all levels.

7.6.3 Memory Power

Processor logic consumes only part of the total processor-related power. Embedded processors always include some amount of local memory for speed and power efficiency. Memory contributes to processor power dissipation in some unexpected ways. Memory configurations vary widely, with memories ranging from a few thousand bytes to hundreds of thousands of bytes in capacity, and from 32 bits to hundreds of bits in width. Memory width, depth, organization, and application-reference patterns all have significant influence on a processor's power dissipation.

Configurable processors are smart enough to cycle instruction and data memories only when access is likely. Instruction memory is accessed in the majority of cycles, because a new instruction is issued almost every cycle. Data memory access is highly application dependent. A general-purpose RISC processor running a typical application instruction stream will access memory for about 40% of instructions in the instruction stream. Loads are more common than stores. Application-specific processors access data memory much more often because they can combine with memory accesses and because processor optimization typically reduces the number of computation operations more than it reduces the number of memory operations. In fact, processors configured with multiple load/store ports to memory may sustain more than one data-memory reference per cycle.

The following list of observations is based on a power analysis of 50 combinations of processor configuration and application, using the EEMBC benchmark results for the Xtensa processor shown in Chapter 3.

Observation 1: Core power, except in extreme memory systems, dominates total power. Core power dominates the total power dissipation (about 70% of total power) and instruction-memory power dominates the total memory power dissipation (about 70% of memory power).

Observation 2: RAM power dissipation depends more on memory access width than on memory depth. Doubling the access width while keeping the capacity constant increases power by 30 to 60%. Doubling the capacity while keeping the width constant increases power by only 10 to 30%.

Observation 3: Using set-associative caches or combining multiple memory types (e.g. data cache combined with data RAM) increases the effective memory-access width. The combination of a 128-bit cache-line refill and four-way set associativity creates a particularly power-hungry cache. For example, a 64-Kbyte, 128-bit, four-way set-associative cache consumes 178µW per million reads/sec, but a 64-Kbyte, 64-bit, direct-mapped cache consumes only 49 µW per million reads/second.

Observation 4: Instruction memories are typically used more heavily than data memories, so instruction-memory sizing is important. Keep instruction memory width as narrow as possible, consistent with performance requirements. For example, using a 32-bit cache-line-refill width instead of 128-bit refill width on the instruction cache saves about 35% in memory power.

Observation 5: Data reads consume approximately 70% of data-memory power. At the level of the RAM array, writes cost about 10% more than reads. When configured as a set-associative cache, writes are less expensive than reads because only one cache way is written.

Observation 6: Parallel access between data cache and data RAM has roughly the same effect on power as adding an extra way to the cache. For small RAMs, consider adding a way and using cache-line locking.

Observation 7: Leakage is becoming a significant problem for deep-submicron and nanometer SOC designs. According to the available estimates for memory and core-logic leakage, this problem is still manageable even with 130nm, low-threshold-voltage processes. Both core logic and memory have leakage current. Core-logic leakage scales with core-logic dynamic power (and clock gating has no effect). Leakage can be treated as "extra MHz" in the power calculation. For integrated circuits fabricated with 0.13µ lithography, the rule-of-thumb is that core-logic leakage is 4.5 times the per-MHz dynamic power. For example, leakage adds about 5MHz of power dissipation (i.e., the total power at 300MHz is approximately equal to dynamic power at 305MHz) for the EEMBC benchmarks.

Observation 8: Memory leakage power is quite variable, but it generally tracks total RAM capacity. RAM power leakage and core-logic power leakage are comparable in scale (RAM is roughly 40% of the total).

7.6.4 Cache Power Dissipation Guide

Looking at the most current memory-power data on a design-by-design basis is essential in any power-sensitive design. The following charts, shown in Fig 7-19 and Fig 7-20, may be useful for quickly estimating power dissipation for instruction and data caches in 130nm, 1V CMOS fabrication technology. In all cases, the figures give active power in µW per million references/second. Data caches use a blend of read and write power based on average benchmark behavior. Instruction caches only service reads, using the assumption that cache refills are uncommon. The data-cache chart (Fig 7-20) assumes a normal mix of reads and writes. Note that associativity is somewhat less power-hungry for data caches than for instruction caches. The figures assume 64-bit cache lines throughout. Shorter line sizes increase power slightly (less than 2% for halving the line size) due to larger tag arrays.

Power is an important consideration in processor-centric SOC design. The first step in minimizing SOC power dissipation is to start with an extremely efficient base processor. The Xtensa processor for example—with base instruction set, minimal interface, and small instruction and data memories—dissipates less than 50µW/MHz even when running at hundreds of MHz. Even without application-specific extensions, these processors can exceed 15,000 general-purpose MIPS per watt, including memory.

Application-specific configurations can increase this power efficiency more than tenfold by performing more of the work in parallel and cutting the operating frequency accordingly. The combination of lean base processor configuration, automatic power management in the processor logic, application-specific instruction-set extension, and appropriate memory configurations make configurable processors a very suitable building block for energy efficient SOCs.

7.7 Essentials of TIE

The TIE language provides the designer with a concise way of extending the Xtensa processor's instruction set through instruction-set-extension descriptions. This section gives an introduction to TIE. More exhaustive documentation is available from Tensilica.

Instruction-set-extension descriptions written in TIE serve as inputs to the TIE Compiler, which generates Verilog and VHDL descriptions of the processor extensions; tailored software tools (compiler, assembler, simulator, and debugger) for that processor; and other support such as test benches, diagnostics, and documentation. The TIE Compiler is a major component of Tensilica's Xtensa processor generator and can be used separately during development of application-specific instruction-set extensions.

A typical TIE description specifies the assembly instruction mnemonic, operands, and semantics of the instruction extensions. A TIE description consists of basic description blocks to delineate the attributes of new instructions. By default, the processor generator assigns instruc-

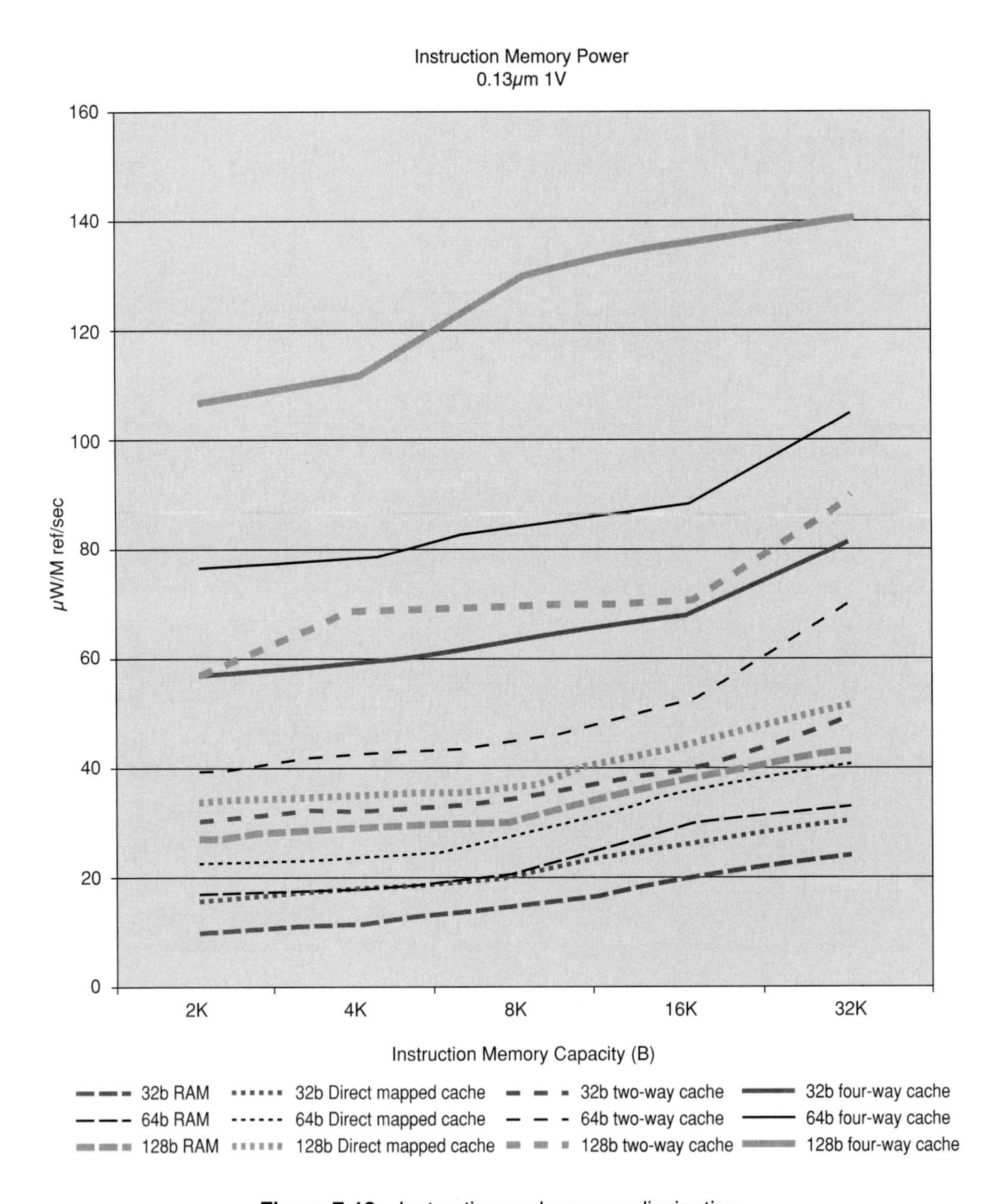

Figure 7-19 Instruction cache power dissipation.

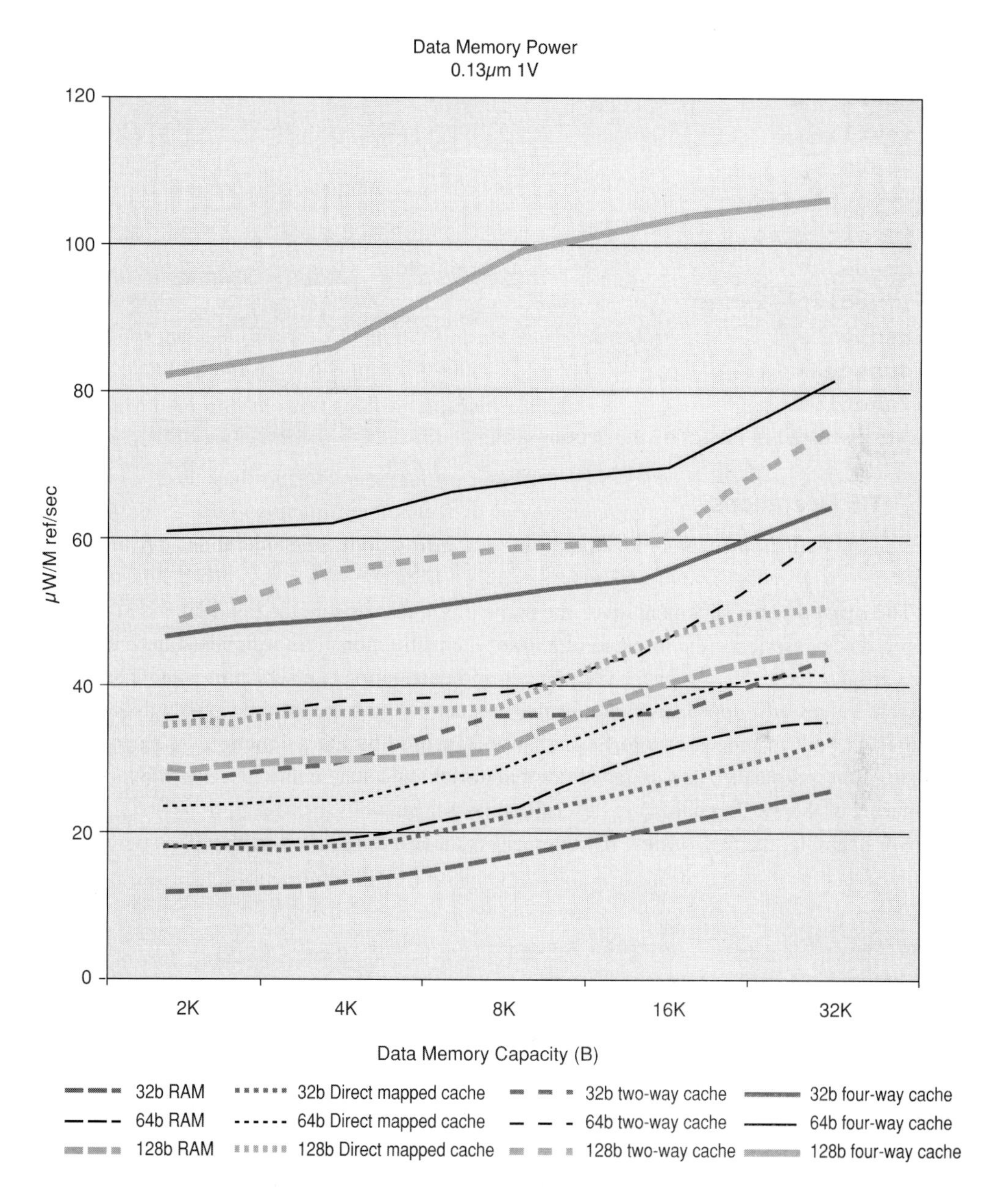

Figure 7-20 Data cache power dissipation.

tion encoding and, where appropriate, creates supporting C functions, datatypes, and data-movement instructions. TIE has the following description blocks,

```
operation
regfile
state
export state
import_wire
queue
immediate_range
table
schedule
function
```

which are discussed in the following sections.

7.7.1 TIE Operations

operation defines a mnemonic, operations for instructions, and operations on instruction inputs:

The **operation** statement gives the name of the instruction, the instruction's arguments, and a reference describing the implementation of the instruction. The arguments are broken into the two groups: the explicit register-file sources and destination (for which register specifiers or immediate values will appear as fields in the instruction) and the implicit operands—the state registers and built-in processor interfaces that will be used by the instruction. The operation of the instruction is described as a series of combinatorial logic statements expressed in a combinatorial subset of the Verilog hardware description language.

For example, the following TIE statement is all that is required to specify a new instruction:

```
operation ADDX2I { out AR a, in AR b, in simm8 c } {} {
 assign a = (b << 1) + c; }
```

This **operation** statement defines an instruction called ADDX2I that has three arguments. Each argument is specified with a direction (**in**, **out**, or **inout**), an argument type, and a name used to refer to the argument. An arguments type may be a register-file name, a table name, an **immediate_range** name (see below), or an operand name.

The ADDX2I instruction has one output argument and two input arguments. The output argument and one input argument each designate a register from the Xtensa processor's general-purpose AR register file. The remaining input argument designates a value defined by the simm8 operand, an immediate value.

The body of the **operation** statement is a series of assignment statements. For each assignment, the left-hand side is an **operation** output operand, a built-in processor signal, or a wire declared within the **operation** and used as a temporary value. The right-hand side is an expression using Verilog combinational logic syntax. The expression may use the names of input and output operands, locally declared wires (similar to a temporary local variable, but in realized hardware), and built-in processor signals. The ADDX2I could use a local wire to simplify the expression:

```
operation ADDX2I { out AR a, in AR b, in simm8 c } {} {
  wire [32:0] bshift = b << 1;
  assign a = bshift + c; }
```

TIE operators are the same as Verilog logic operators, as shown in Fig 7-21. (Note: All arithmetic operations in TIE are defined as unsigned.)

Operator Type	Operator Symbol	Operation	Number of Operands
Arithmetic	+	Add	2
	-	Subtract	2
	*	Multiply	2
Logical	!	Logical negation	1
	&&	Logical and	2
	||	Logical or	2
Relational	>	Greater than	2
	<	Less than	2
	>=	Greater than or equal	2
	<=	Less than or equal	2
	==	Equal	2
	!=	Not Equal	2
Bitwise	~	Bitwise negation	1
	&	Bitwise and	2
	|	Bitwise or	2
	^	Bitwise ex-or	2
	^~ or ~^	Bitwise ex-nor	2

Operator Type	Operator Symbol	Operation	Number of Operands
Reduction	&	Reduction and	1
	~&	Reduction nand	1
	|	Reduction or	1
	~|	Reduction nor	1
	^	Reduction ex-or	1
	^~ or ~^	Reduction ex-nor	1
Shift	<<	Left shift	2
	>>	Right shift	2
Concatenation	{ }	Concatenation	any

Figure 7-21 TIE's operators are the same as Verilog.

7.7.2 TIE States and Register Files

state declares processor states realized as registers. These TIE state registers are in addition to the predefined states in the core Xtensa ISA and architectural options.

```
state ShiftRegister 5
```

States can be used in any operation description. When a **state** register is used, it should be included in the list of implicit operands in the operation statement. State registers do not appear in the argument list of the assembly code for the instruction and do not require a specification field in the instruction encoding. They are implicitly encoded in the instruction. If the modifier **add_read_write** is included after the **state**-register-width argument, the processor generator automatically generates instructions for transferring this state to and from general registers. It also adds this new state to the processor state that is visible to the debugger, and the new state is saved and restored by RTOS context-switching code, which is also created automatically by the Xtensa processor generator. The `ADD2XI` might be extended into a more general `ADDNXI` instruction by holding the shift amount in a state register.

```
state ShiftReg 5 add_read_write
operation ADDXNI {out AR a, in AR b, in simm8 c} {in ShiftReg} {
 assign a = (b << ShiftReg) + c; }
```

regfile defines a new register file with the specified depth and bit width.

An argument to an operation can also be a user-defined register file. The register file is declared using the TIE regfile construct and is referred to by name in an operation argument (just like AR register file in the above example). For example,

```
regfile XR 16 16 x
```

The **regfile** state includes the name of the new register file (the long name), the bit width (any width), the depth (a power of two), and a short name (usually a single letter) to be used as the prefix for naming the individual registers in assembly code.

```
regfile XR 16 16 x
operation ADDX {out XR a, in AR b, in simm8 c } {} {
 assign a = b + c; }
```

The Xtensa processor generator automatically creates a C data type for each register. The name of the created type is that same as the register file's long name. Also, compiler-required load, store, and move instruction prototypes are created automatically for each C type. In the above examples, the processor generator creates a C type called XR to represent all the bits in a register of the XR regfile.

7.7.3 External TIE Ports and Queues

TIE includes a flexible mechanism for defining and using new external interfaces. These interfaces allow more intimate coupling between the processors and surrounding functions and attain higher bandwidth communications among processors. This mechanism includes a means to import values on input ports, export state register values on output ports, and operate directly on input and output data queues.

export state outputs the committed value of a TIE state register on wires specifically generated for this task by the Xtensa processor generator. Exported states are specified with the keyword **export** in the state declaration. The value of this output changes whenever the architectural copy is updated—after the corresponding state register has been written or after the commit stage of the pipeline, whichever is later. Exporting a state adds an output port with the same width as the state to the Xtensa processor.

```
state mul16 32 32'b0 add_read_write export
state mul32 64 64'b0 add_read_write export
operation twomultiplies {in AR a, in AR b} {out mul16, out mul32} {
assign mul16  = a[15:0] * b[15:0];
assign mul32  = a * b;}
schedule mul {twomultiplies} {def mul16 2; def mul32 3;}
```

Exported states must also have a reset value—the initial value assumed by the external interface when the processor is reset. For example, the following TIE defines two state registers whose values are exported.

The result of the 16 x 16-bit multiplication is internally available one cycle after the ALU pipe stage, but it doesn't appear externally until after the write-back stage—the pipeline's commit stage. The result of the 32 x 32-bit multiplication is available internally and externally two stages after the ALU stage.

import_wire defines a port whose value is taken directly from a set of automatically generated wires at the interface of the processor. Using an input wire as an operand in a computation has no sideeffect. External logic sourcing the value cannot directly sense that the value has been read. The input is available to TIE instruction logic in the ALU stage. The input is always registered before use in any processor computation. The following example imports a normalization value (a shift amount) used in a multiply-accumulate operation.

```
import_wire shift_amt 5
regfile XR 64 2 x
operation mul_shift_add {in AR x, in AR y, inout XR acc} {in shift_amt} {
assign acc = acc + ((x * y)>>shift_amt); }
```

queue creates an input or output queue interface to the processor. This construct allows an input queue to serve as a source operand and an output queue to serve as a destination operand. Queues have a name, a width, and a direction (**in** or **out**). The width is unconstrained, so operations using queues can sustain up to tens of bytes per cycle of useful I/O bandwidth. An input **queue** name can appear on the right-hand side of an assignment, and an output **queue** name can appear on the left-hand side of an assignment in an operation/reference/semantic statement. A simple send and receive might be implemented as follows:

```
queue qo 34 out
queue qi 32 in
operation send (in AR mes) {out qo} {assign qo = mes;}
operation send (in AR mes) {out qo} {assign qo = mes;}
operation recv (out AR mes) {in qi} {assign mes = qi;}
```

High-bandwidth data streaming can be implemented using multiple queues. The following TIE code implements a rounding operation on an input **queue** of vectors. It writes the result vectors to an output queue, where each queue operand is treated as eight 32-bit values.

```
queue vin 256 in
queue vout 256 out
operation scalev {in AR a} {in vin, out vout} {assign vout =
 {(vin[255:224]+a)>>1, (vin[223:192]+a)>>1, (vin[191:160]+a)>>1,
 (vin[159:128]+a)>>1, (vin[127:96]+a)>>1, (vin[95:64]+a)>>1,
 (vin[63:32]+a)>>1, (vin[31:0]+a)>>1};}
```

At 350MHz, this method sustains an aggregate data bandwidth of 22GB/sec when the `scalev` operation is executed every cycle.

The declaration of an input queue named `iqueue` with a width of 1024 bits, for example, creates three signals in the processor-interface hardware implementation, shown in Verilog as well as in the instruction-set simulator:

```
input [1023 : 0] TIE_iqueue
input TIE_iqueue_Empty /* Empty causes stall on attempted use */
output TIE_iqueue_Pop  /* Advances external queue on data use*/
```

If the instruction reading from an input queue gets killed, by an exception, for example, the data read from the queue is saved in an internal buffer within the processor, and the next instruction that reads the queue uses the saved data from this buffer instead of reading new data from the queue. This buffering hides the effects of exceptions and guarantees that **`queue`** operations can take place at the rate of one per cycle.

The declaration of an output named **`oqueue`** with a width of 500 bits, for example, creates the following processor ports in Verilog:

```
output [499 : 0 ] TIE_oqueue
input TIE_oqueue_Full/* Causes a stall on attempted use if full */
output TIE_oqueue_Push  /* Shows new data is available */
```

Queue outputs are always written when the instruction reaches the commit stage in the pipeline. Output data is registered and pushed onto the queue.

7.7.4 TIE Constants

`immediate_range` defines the range of values that an instruction's immediate operand can take. The TIE compiler automatically determines the number and the choice of bits required for the instruction opcode to specify the valid immediate values. The **`immediate_range`** statement gives the name of the immediate range, the size in bits of each value in the immediate range, the low value in the range, the high value in the range, and the step. For example, the following **immediate_range** specifies the immediate values -16, -14, -12, ..., 10, 12, and 14:

```
immediate_range my_ir -16 14 2
```

An immediate range can be used as the type of an operation argument, just like a register file. For example,

```
operation ADDII { out AR a, in my_ir b, in my_ir c } {} { assign a = b + c; }
```

table declares a table of constants indexed by an immediate field in the instruction. The individual values are listed, and the successive values of the instruction field indicate successive values in the table.

```
table mytbl 32 4 { -1, 5, 7, 12 }
```

When the instruction field has the value 0, this operation uses –1 as input operand value; when the field has the value 1, the operation uses 5 as input operand value, and so forth. Tables are especially useful when the table values are fixed in a stable industry standard or when they represent mathematical constants (low-precision trigonometric function values, for example). Similarly, a table name can be used as the type for an operation argument:

```
operation ADDII { out AR a, in my_ir b, in my_ir c } {} { assign a = b + c; }
```

Notice that the same table name can be used for two arguments. There is no need to declare two identical tables.

The TIE language's sophisticated encoding of immediate values and tables can play an important role in reducing code density. Most processors do not encode immediate fields—the immediate value in the instruction binary is the value available to the application. Consequently, general-purpose instruction sets need large immediate fields to encode relevant immediate values. An application domain's range of relevant immediate values is often well known before the application-specific SOC is built, so capturing that foreknowledge in the processor's instruction set allows a small field in the instruction word to express all or most of the interesting constant values in the application. This ability reduces the size and number of instructions and can sometimes eliminate data tables stored in memory. Reducing instruction and data storage reduces chip size.

7.7.5 TIE Function Scheduling (use and def)

schedule specifies the timing of operand uses and definitions for multicycle instructions. By default, all source operands are used in pipe stage 1 (the ALU execute pipe stage), and all result operands are available at the end of pipe stage 1. A **schedule** can be written for a list of operations or instructions with a common set of operands. Any source operand may have a **use** cycle; any destination may have a **def** cycle. For example, the following **schedule** indicates that the ADDX2I (from the example above) instruction's output has a latency of 2 cycles, where msched is simply the name of this **schedule** statement:

```
schedule msched {ADDX2I} { def a 2; }
```

The **schedule** for a more complex instruction—say, a multiply/accumulate function—might involve both using and defining the accumulator value in a later pipe stage, where x and y are the multiplier and multiplicand operands, and acc is the accumulator operand.

```
schedule macsch {MAC} { use x 1; use y 1; use acc 3; def acc 3; }
```

This multiply/accumulate operation is automatically pipelined across three pipe stages.

7.7.6 Using Built-in Registers, Interfaces, and Functions with TIE

The Xtensa processor core makes a number of signals and state resources available to user-defined instruction sets. These resources permit design of high-bandwidth instructions that include multiple loads and stores, conditional branches, and other operations with multiple source and destination operands.

The Xtensa architecture provides two built-in register files accessible in all instructions. The general-purpose AR register file consists of sixteen 32-bit registers. The optional BR register file makes sixteen 1-bit Boolean registers available. The AR register file is present in all Xtensa processor configurations, and the base Xtensa instruction set uses it heavily. The optional BR register file is present only if configured. The Xtensa processor generator automatically adds a set of Boolean manipulation, conditional-move, and conditional-branch instructions to the generated processor's instruction set when the BR register-file option is selected.

Built-in TIE signals allow efficient connections between TIE instructions and the data-memory system, as shown in Fig 7-22.

Name	Width	Direction	Purpose
`Vaddr`	32	out	Load/store virtual address to memory port
`MemDataIn8, MemDataIn16, MemDataIn32, MemDataIn64, MemDataIn128`	8, 16, 32, 64, 128	in	Load data from memory port in various widths
`MemDataOut8, MemDataOut16, MemDataOut32, MemDataOut64, MemDataOut128`	8, 16, 32, 64, 128	out	Store data to memory port in various widths
`LoadByteDisable`	16	out	The 16-bit byte-disable signal for conditional load or store from memory port
`StoreByteDisable`	16	out	The 16-bit byte-disable signal for conditional stores to memory port

Figure 7-22 TIE built-in memory interface signals.

Tensilica FLIX configurations optionally support multiple memory ports. If multiple memory ports are configured, all of the memory-address, data-input, data-output, and byte-lane-enable signals for each memory port are duplicated.

The TIE language also supports a range of optimized combinatorial functions, shown in Fig 7-23. These functions are especially useful for defining large arithmetic functions.

Built-in Function Syntax	Function Definition
`sum = TIEadd(a, b, cin)`	Addition with carry-in.
`sum = TIEaddn(A0, A1, ..., An-1)`	Efficient n-number addition, there must be at least three arguments.

Built-in Function Syntax	Function Definition
`{carry, sum} = TIEcsa(a, b, c)`	Carry-save adder (csa).
`{lt, le, eq, ge, gt} = TIEcmp(a, b, signed)`	Signed and unsigned comparison, producing a vector of four Boolean results.
`o = TIEmac(a, b, c, signed, negate)`	Multiply-accumulate; the multiplication is signed if signed is true and unsigned otherwise. The multiplication result is subtracted from the accumulator c if negate is true and added to the accumulator c otherwise.
`prod = TIEmul(a, b, signed)`	Multiply operation. The multiplication is signed if signed is true and unsigned otherwise.
`{p0, p1} = TIEmulpp(a, b, signed, negate)`	Partial-product multiply operation. This module returns two partial products of the multiplication. The multiplication is signed if signed is true and unsigned otherwise. The sum of the two partial products equals the product. If negate is true, the sum equals the negative of the product. The definition does not give specific meaning to the individual partial product.
`o = TIEmux(s, D0, D1, ..., Dn-1)`	*N*-way multiplexer. This module returns one of the n data inputs depending on the value of the select signal. The number of data, n, must be a power of 2.
`o = TIEpsel(S0, D0, S1, D1, ..., Sn-1, Dn-1)`	*N*-way priority selector. This module selects one of n data inputs according the values and priorities of the select signals. The first select signal has the highest priority and the last the lowest. If none of the selection signals is active, the result is 0.
`o = TIEsel(S0, D0, S1, D1, ..., Sn-1, Dn-1)`	*N*-way, one-hot selector. This module selects one of n data input values according to the values of the select signals. The select signals are expected to be one-hot. If none of the selection signals are active, the result is 0. If more than one select signal is active, the result is undefined.

Figure 7-23 TIE built-in functions.

7.7.7 Shared and Iterative TIE Functions

function creates a block of continuous assignment statements with explicitly declared input arguments and return size. This statement can be used as syntactic shorthand, where each invocation of the function instantiates a new copy of the corresponding logic, or it can be used to share access to a single copy of the block. For example, a complex function such as a multiplier is often used by several different processor instructions. The most efficient way to implement this structure is as a shared function. The processor generator will identify all instructions using this block and build appropriate routing and multiplexing, so one copy of the block can operate on different operands at different times.

```
function [31:0] M16 ([15:0] A, [15:0] B, S) shared {
 assign M16 = {16{S & A[15]}, A} * {16{S & B[15]}, B};
}
```

The following statement might be used inside an operation to invoke the shared multiplier:

```
assign x = M16(a, b, s);
```

Often, a complex function is computed by repeated execution of a simpler function. This sequence is implemented with a special form of function. Iterative multiplication is a common example—a series of additions or smaller multiplications are combined to perform a larger multiplication. TIE supports this via iterative functions.

Iterative instructions with fixed latencies are just like multicycle instructions except that they share some computational hardware at different pipeline stages. As a result, iterative instructions cost less to implement but may not be fully pipelined. The TIE Compiler infers an iterative instruction from the way its semantic logic is described. If an instruction semantic logic uses multiple instances of a shared function, it is identified as an iterative instruction.

In addition to sharing the computational logic, an iterative instruction typically needs some flip-flops to pass partial results from one stage to another. These flip-flops are no different from the pipeline flip-flops the TIE Compiler inserts into the processor's data path for multicycle instructions. However, because iterative instructions are not issued every cycle due to logic sharing, some of these flip-flops are not used every cycle during the execution of the iterative instructions. The TIE compiler will try to identify these independent flip-flops and fold them onto one physical flip-flop in the implementation.

In this example, an iterative function serves as the heart of a 64 x 64-bit multiplier using 64-bit values from an extended register file:

```
regfile XR 64 8 x
function [127:0] mul64x8 ([63:0] s0, [63:0] s1, [63:0] a, [7:0] b) shared {
 wire [63:0] t0 = (a << 0) & b[0]; wire [63:0] t1 = (a << 1) & b[1];
 wire [63:0] t2 = (a << 2) & b[2]; wire [63:0] t3 = (a << 3) & b[3];
 wire [63:0] t4 = (a << 4) & b[4]; wire [63:0] t5 = (a << 5) & b[5];
 wire [63:0] t6 = (a << 6) & b[6]; wire [63:0] t7 = (a << 7) & b[7];
 assign mul64x8 = TIEaddn(s0, s1, t0, t1, t2, t3, t4, t5, t6, t7);
}
operation MUL64 {out XR r, in XR s, in XR t} {} {
 wire [63:0] s00, s10, s20, s30, s40, s50, s60, s70;
 wire [63:0] s01, s11, s21, s31, s41, s51, s61, s71;
 assign {s00, s01} = mul64x8(64'b0, 64'b0, s, t[7:0]);
 assign {s10, s11} = mul64x8(s00, s01, s, t[15:8]);
 assign {s20, s21} = mul64x8(s10, s11, s, t[23:16]);
 assign {s30, s31} = mul64x8(s20, s21, s, t[31:24]);
 assign {s40, s41} = mul64x8(s30, s31, s, t[39:32]);
 assign {s50, s51} = mul64x8(s40, s41, s, t[47:40]);
 assign {s60, s61} = mul64x8(s50, s51, s, t[55:48]);
 assign {s70, s71} = mul64x8(s60, s61, s, t[63:56]);
 assign r = s70 + s71;
}
schedule mul64_schedule {MUL64} {use s 1; use t 1; def r 8;}
```

Given the MUL64 schedule, the processor generator allocates each instance of the function `mul64x8` to a unique stage and inserts flip-flops between each stage. The same shared function is used in eight cycles, so the flip-flops for latching the shared function's outputs can all be folded into a single physical flip-flop bank (128 bits wide).

Furthermore, the same two source values (r and s) are used every cycle, so the generator uses a bank of flip-flops to hold the values for stages 2 to 7.

7.7.8 Multi-Slot Instructions

format specifies the instruction length and slots for a new instruction format. New formats may specify 24-, 32- or 64-bit lengths (though not both 32- and 64-bit in the same processor configuration). The **format** statement includes a list of slot names.

```
format longinst 64 {baseslot, slot1, slot2, slot3}
```

The number of slots is bounded only by the number of bits available for the automatic encoding of all the operations within the slot and all the slots within the instruction. The operations available within a slot are defined with a **slot_opcodes** statement.

```
slot_opcodes slot2 {op0, op1, op2, op3}
```

where the operations in the list can be any TIE-defined operation or almost any base Xtensa instruction.

7.8 Further Reading

Good introductions to the design of pipelined microprocessors, including superscalar implementations, can be found in the following:

- Mike Johnson. *Superscalar Microprocessor Design.* Prentice Hall, 1991.
- Harvey G. Cragon. Memory Systems and Pipelined Processors. Jones and Bartlett, 1996.
- Michael J. Flynn. *Computer Architecture Pipelined and Parallel Processor Design.* Jones and Bartlett, 1995.
- Harold S. Stone. *High-Performance Computer Architecture*, Addison-Wesley, 1993.
- One of the earliest well-known pipelined RISC microprocessors was the MIPS Machine from Stanford University. A basic overview can be found in the following: J. Hennessy, N. Jouppi, J. Gill, F. Baskett, A. Strong, T. Gross, C. Rowen and J. Leonard. "The MIPS Machine." In *Proceedings of IEEE Compcon, Spring 1982*, February 1982.
- The DEC Alpha microprocessor family paid particularly close attention to the issues of pipelining and clock frequency: D. W. Dobberpuhl, et al. "A 200-Mhz 64-bit Dual-Issue CMOS Microprocessor." *IEEE Journal of Solid-State Circuits*, SC-27(11): pp. 1555–1565, November 1992.
- A good discussion of pipeline interlock and bypass tradeoffs can be found in this comparison of two basic pipelines: Michael Golden and Trevor Mudge. "A Comparison of Two Pipeline Organizations." In *Proceedings of the 26th Annual Symposium on Microarchitecture (MICRO-27)*, 1994.
- Some of the most aggressive methods used to optimize code for performance in deeply embedded engines parallel the methods used in trace scheduling, which is a compilation technique used with long-instruction-word processors: See J. A. Fisher. "Trace Scheduling: A Technique for Global Microcode Compaction." *IEEE Transactions on Computers*, 30(7): 478–490, 1981.
- One of the first efforts to deal with branch delay in a pipelined microprocessor is found in the following: D. R. Ditzel and H. R. McLellan. "Branch Folding in the CRISP Microprocessor: Reducing Branch Delay to Zero." In Proceedings of the 14th Annual Symposium on Computer Architecture, 1987.

- See the following basic discussion of techniques for on-chip debug: Gert Jan van Rootselaar, Frank Bouwman, Erik Jan Marinissen, and Math Verstraelen. "Debugging of Systems-on-a-Chip." In *Proceedings of the ProRISC Workshop on Circuits, Systems and Signal Processing*, 1997.
- Commercial products for on-chip debug and trace of embedded processor cores are presented at *http://www.arm.com/products/DevTools* and *http://www.tensilica.com.*
- Handling complex data organization and alignment for SIMD architectures has stimulated a set of basic techniques. See Gwan-Hwan Hwang, Jenq Kuen Lee and Dz-Ching Ju. "Integrating Automatic Data Alignment and Array Operation Synthesis to Optimize Data Parallel Programs." In *10th International Workshop on Languages and Compilers for Parallel Computing*, 1997.
- For good discussions of memory-ordering definitions and tradeoffs, see Gharachorio et al. "Memory consistency and event ordering in scalable shared-memory multiprocessors." ISCAS 1990; and Sarita Adve and Mark Hill. "Weak Ordering: A New Definition." In *Proceedings of the 17th Annual Int'l Symp. on Computer Architecture*, 1990.
- The foundations and philosophy of the Tensilica Instruction Extension language are discussed in the following: Albert Wang, Earl Killian, Dror E. Maydan, and Chris Rowen. "Hardware/Software Instruction Set Configurability for System-on-Chip Processors." In *Proceedings of the 38th Design Automation Conference*, 2001.

CHAPTER 8

The Future of SOC Design: The Sea of Processors

In 1965, Dr. Gordon Moore famously observed and codified an exponential growth in the number of transistors per integrated circuit. Engineers now can put entire systems on one integrated circuit as a direct result of Moore's law scaling. In a generic 130nm standard-cell foundry process, silicon density routinely exceeds 100K usable gates per mm^2. Consequently, even a low-cost chip (50mm^2 of core area) can carry 5M gates of logic today, and inexpensive chips will likely carry 20M usable gates by 2007. This density increase has been consistently matched by improvements in power dissipation, circuit speed, and system form factor. Silicon scaling creates an enormous range of design opportunities, but it does not teach us how to exploit this potential.

At the same time, end products are evolving with daunting speed. Virtually every category of electronic product is getting more Internet-connected, more portable, and more complex. These products also face more cost pressures. This pressure drives every design toward higher integration levels—toward the use of SOCs—to get longer battery life, higher bandwidth, smaller physical size, and lower manufacturing cost. The functions to be integrated differ by product segment, but the pressure to "integrate or die" is universal. The theoretical benefits of SOC design are now well understood. SOC integration reduces product costs, boosts performance, and prolongs battery life compared to designs based on lower integration circuits. But the opportunities of silicon scaling and demands of the electronics marketplace alone don't tell us how to design SOCs quickly and cheaply.

Application-focused SOC design has become the dominant design style for high-volume electronics. It reflects the convergence of two phenomena:

- The market-driven need to pack richer functions into smaller cost, power, and physical-size budgets.

- The emergence of advanced semiconductor technology, especially at 130nm and 90nm geometries.

This book has shown a practical connection between these two forces, between the potential of the underlying silicon and the timely availability of end products. In this chapter, we review the essentials of the advanced SOC design methodology and the new role for processors, and then look to the future. Starting from the identification of a new foundation for SOC design, we can start to forecast something of the future of electronics. We look out over the next 15 years, projecting both the future role of processors in SOC design and the attendant changes in the structure of the semiconductor industry and the electronic design process.

8.1 What's Happening to SOC Design?

High-integration SOC design is clearly part of the answer to the big integration question. Across a wide range of product types, designers strive to combine all the major digital functions of the system—a network switch, a printer, a cell phone, or a digital television—on one chip. Combining a CPU with some on-chip memory and a collection of I/O interfaces does not constitute SOC design. SOC design involves combining all the "heavy lifting"—performance-critical, high-efficiency data processing of signals, images, video streams, audio, encryption and other application-acceleration—together on one chip.

Previously, most of those heavy-lifting functions were implemented using custom logic dedicated to each single function or, occasionally, using specialized digital signal processors. In theory, custom logic is ideal for those subsystems, but designing everything as custom blocks raises big questions. The fabs can build hundred-million-transistor chips, and the community of chip designers knows how—at least in principle—to design complex SOCs. But, can they develop these complex SOCs at reasonable cost? Can they design flawlessly and quickly enough to catch each market wave? Can they get sufficient volume and profit? In short, can they get acceptable return on the investment needed to create these complex designs?

Today, SOC design teams are struggling with the complexity of these multimillion-gate designs. Many teams have reported that they now spend as much as 70% of their hardware development effort on block-level verification. The research firm International Business Strategies projects that total SOC development cost is approaching $20M for 130nm geometry (130nm) designs and will perhaps triple for 90nm designs, as shown in Fig 8-1. Moreover, IBS projects that software's share of the development effort will grow rapidly. The IBS data also suggests that the block-level design effort represents a declining share of the total, though its absolute size continues to increase.

Despite significant EDA tool improvements, current SOC-design methodologies have failed to close the long-recognized and growing gap between logic complexity and designer productivity. The Semiconductor Research Corporation's famous 1997 warning about the growing design/productivity gap appears in Fig 8-2.

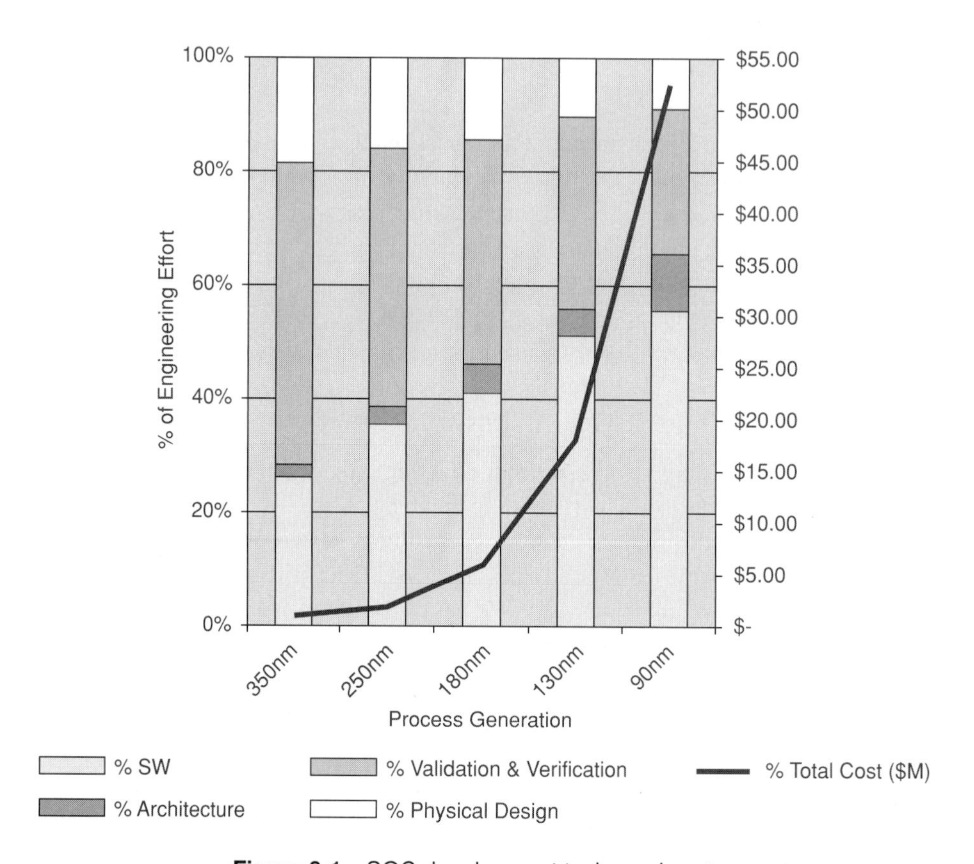

Figure 8-1 SOC development tasks and costs.

In addition to the problem of the growing hardware-design gap, software-development costs creep upward as developers try to build complex systems using multiple tool environments, one for each type of on-chip processor. Clearly, design costs will spiral out of reach without significant shifts in design methodology. This book's advanced design methodology for complex-SOC development uses configurable processors as basic building blocks to circumvent IBS's forecasted cost explosion.

8.1.1 SOC and ROI

The economic challenge of SOC design is not limited to rising development costs; chip volume, manufacturing cost, and design longevity also play a central role on the return on investment (ROI) of chip design. The increasing specialization of each SOC design makes the dilemma worse. As developers integrate more of the electronics of a system—say a digital camera—onto one chip, the SOC naturally becomes more narrowly optimized and dedicated to the features and price targets of that particular camera. The developers get the best possible technical solution in

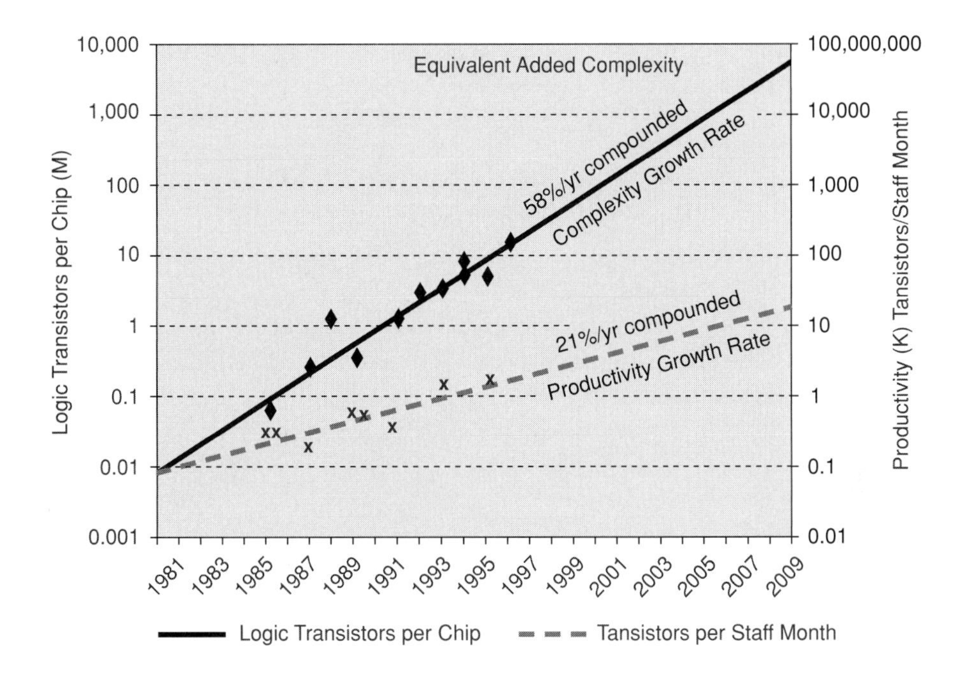

Figure 8-2 Design complexity and designer productivity.

terms of size, battery-life, and product size. In the limit, if the entire system function is committed to that one SOC, then every system needs its own new SOC design.

When SOC designs cost tens of millions of dollars, that's untenable. Chip developers cannot recover the design cost on any but the highest volume, highest margin systems. Furthermore, SOC design is slow and risky; two years from product concept to manufacturing ramp is considered brisk development today. That's plenty of time for the end-market standards to change, for the marketing group to change strategy, or for the engineering team to make mistakes. Clearly, building a new SOC for every single new electronic product may optimize the technical solution but it makes for poor economics.

A simple equation for calculating chip ROI appears in Fig 8-3. This equation shows ROI from a semiconductor vendor's viewpoint, but system companies must also consider ROI for their internal SOC-design decisions. For the system company, the chip sales price is not directly measurable. Instead, decision makers must consider the influence of the SOC's impact on the competitiveness and ultimate sales price of the equipment that contains the SOC. Nevertheless, the decision factors are roughly the same for semiconductor and systems companies.

$$\text{Return On Investment} = \frac{\text{Volume} * (\text{Sales Price} - \text{Manufacturing Cost})}{\text{Development Cost}}$$

Figure 8-3 Chip return on investment calculation.

Volume is the total unit shipments of the SOC across all the systems that use it. Volume is driven by success in high-volume systems and flexibility to serve in multiple systems with one chip design. The difference between unit sales price and unit manufacturing cost is the gross profit margin per chip. Manufacturing cost is principally governed by silicon area, packaging cost, and process-technology choice. Sales price is governed by the dynamics of the marketplace—if the chip adds significant value to the end product and is comparatively unique, it commands a higher price. If it is a commodity or adds little to the competitive differentiation of the end product, sales price and gross profit margin will be modest. SOC-development costs play a leveraged role in the ROI calculation. Those costs include the engineering time, the capital cost of engineering tools, nonrecurring engineering (NRE) fees paid for design and prototyping services, mask charges, and intellectual-property license fees. A major design project may also carry some indirect costs, including risks associated with late market entry and opportunity costs (which include the loss of other projects made impossible because key management or engineering resources are tied up on this SOC design project).

While this ROI calculation is rarely performed explicitly, the tradeoffs are constantly present in the minds of design and product managers. The potential benefits of SOC design are enormous—compelling pricing and category-leading system features—but the pitfalls are also substantial, especially the high development costs and the risks of missing important function and schedule goals. Processor-based SOC design promises to improve the tradeoff between technical optimality and economic feasibility. Pervasive implementation of major blocks as processors brings critical flexibility to individual function blocks and to the communications among blocks. This boost in flexibility increases the SOC's volume and profit margin while cutting development costs, development time, and most importantly, development risk.

8.1.2 The Designer's Dilemma

SOC designers want both efficiency (maximum performance, minimum silicon area, minimum power) and flexibility (easy mechanisms to alter the function), and designers want these at the same time. For a data-intensive subsystem needing maximum performance or minimum cost, SOC designers naturally gravitate toward custom logic (circuits built specifically for that single function). If the function is simple enough, it is routine to design something 10 or 100 times faster than achievable in a generic programmable block such as a RISC processor or field-programmable gate array (FPGA). But when the designer must make even a small change to the custom logic, it may require 6 to 12 months and millions of dollars to deliver that design tweak. Even a small design change forces a new chip design.

On the other hand, for functions where performance or efficiency are not as important as flexibility—in a digital camera's user interface, for example—a general-purpose processor may be just fine. The traditional one-size-fits-all processor can be reprogrammed in minutes and can theoretically handle almost any conceivable digital-camera interface function. The price for this unbounded flexibility is a block that is relatively big and slow.

FPGA technology plays a similar role. FPGA cost and clock rate are 5 to 20 times worse than a custom-logic implementation using the same fabrication technology, but the turnaround time for a design change is just hours or days. This flexibility means shorter turnaround time for changes, lower engineering costs, and wider applicability of one chip. The benefits are the reasons why FPGA usage has climbed steadily over the past several years. The higher unit cost and lower clock rates are the reasons FPGAs have not completely replaced ASICs and SOCs.

There's the dilemma: in the mission-critical tasks for SOC applications, you can get efficiency and performance or you can get flexibility and reprogrammability, but you usually can't get both.

This dilemma translates directly into inadequate ROI for SOC designs. Optimized nonprogrammable functions—often necessary for system performance and unit cost—lack the flexibility to support multiple system designs and the latest "hot" system features. This lack of flexibility cuts the chip volume and often reduces the profit margin on the chip. The longer design cycle for custom logic and the higher risk of redesign also raise the overall development cost and degrade the overall chip ROI.

This dilemma defines the opportunity for configurable or application-specific processors. These processors are tuned to the target application class, just like custom logic, so they are much smaller and faster than conventional processors. However, they are fully reprogrammable using the most widely adopted software tools and languages, so they are flexible just like traditional, general-purpose processors.

8.1.3 The Limitations of General-Purpose Processors

The ongoing performance improvement of high-performance, general-purpose processors is remarkable. The combination of improved microarchitecture, more aggressive circuit design, and improved transistor speed have combined to advance the performance of leading-edge microprocessors by 40 to 50% per year over the past decade. Much of the absolute performance increase has come at the expense of increased transistor count, relative die size, and power. Fig 8-4 shows the historical trend for the performance of Intel's microprocessors, normalized for transistor speed. It also shows the relative die size and power of Intel processors, based on Intel's estimates and assuming a common silicon process. The figure also shows the increase in raw transistor speed over the same period, as measured by gate delay.

The most important architectural innovations—multiple parallel pipelines, out-of-order instruction execution, deep speculative pipelines, sophisticated multilevel caches—all increase

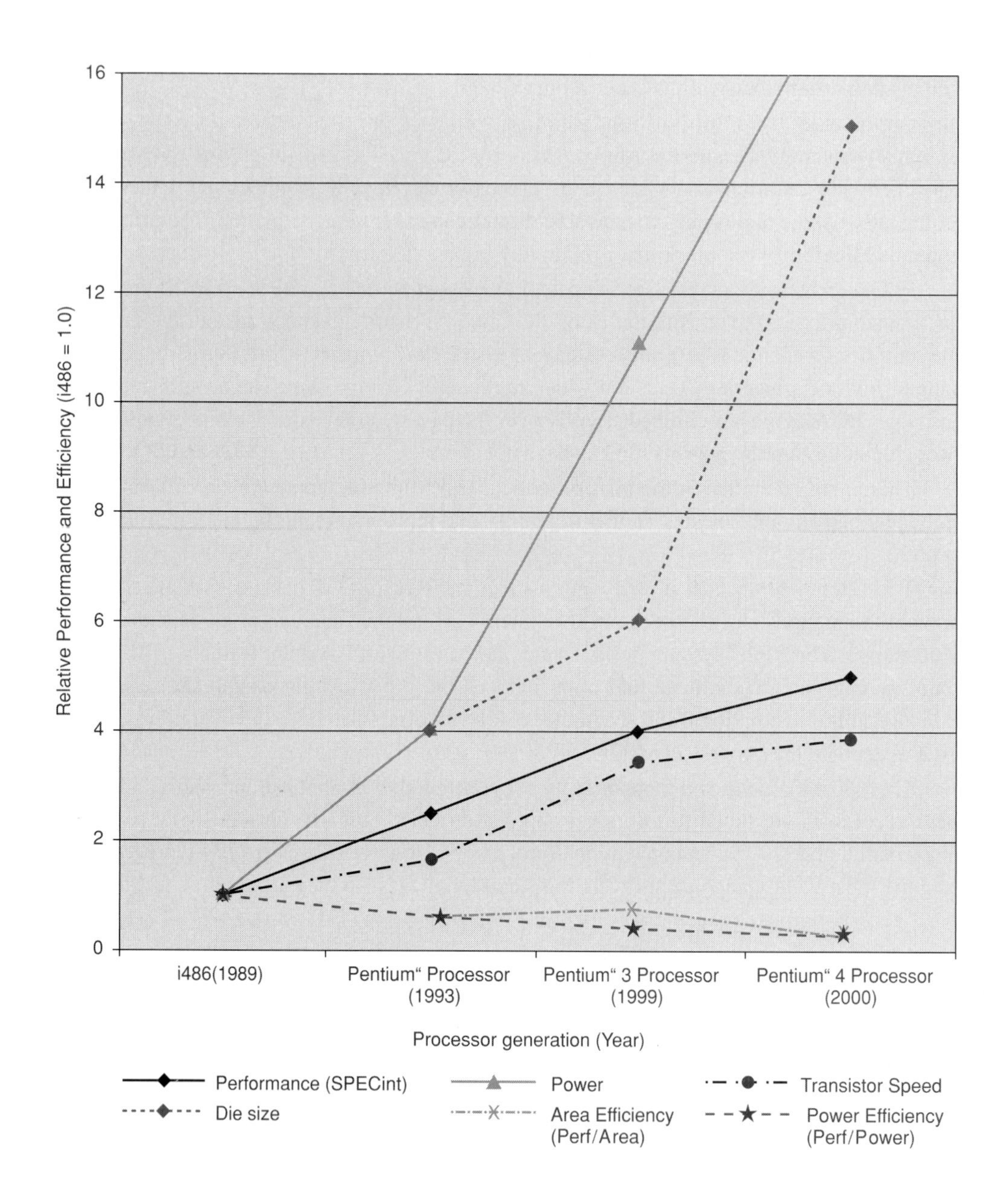

Figure 8-4 Intel processor efficiency trend independent of process improvement.

processor performance, but they incur very high transistor counts per processor. Despite these innovations in general-purpose microarchitecture and ongoing transistor-level improvements, general-purpose processors are becoming significantly less efficient. Over time, more transistors, more relative silicon area, and more power must be allocated to the processor to continue boosting performance at historical rates. If this trend continues, general-purpose processor performance gains will become limited by device- and system-level cooling and by processor silicon area and cost, relative to other electronic functions.

8.1.4 The New Processor

This book has described a new approach to SOC development: systematic design using the application-specific processor as the basic building block. The effectiveness of the methodology relies directly on two essential characteristics of application-specific processors:

- **Universality:** A family of application-specific processors can span the full range of datatypes, computation requirements, communication patterns, and software styles found in embedded systems. Processor universality simplifies SOC design by maximizing the range of problems that can be efficiently solved with a closely related family of configured processors. The individually generated processors themselves do not strive to be universal, of course, although these processors all incorporate the general-purpose functionality of the base processor architecture, so they can all run a common subset of tasks. However, a large part of their attractiveness comes from their specialization: a configured processor contains everything needed for optimal efficiency in one specific target application domain.
- **Automation:** Everything required by hardware and software teams is fully and reliably generated for every possible processor configuration. Automation simplifies SOC design by minimizing design risk and the skill set required of the design teams.

Chip designers have recognized the potential benefit of application-specific processors for years, but manual design of new processors is slow and expensive. Architects write specifications for new architecture extensions; hardware developers design and verify the logic; circuit designers lay out the transistors to get high speed and low power; the chip builder contracts with third-party developers to get software tools like compilers and real-time operating systems adapted to the new processor. This process typically takes years and millions of dollars. As a result, few chip designers bother to develop application-specific processors, and those who do build their own processors commonly struggle to get sufficient software support for their proprietary architectures.

By contrast, an automated processor generator takes a high-level electronic specification and produces both the optimized hardware design and the software tools in about 60 minutes. The hardware design incorporating all the instruction-set and interface extensions is delivered in standard HDL form to ease integration into the final chip layout. The software tools also fully support the application-specific architecture and enable rapid prototyping of complex systems. Fast turnaround for hardware and software lets the chip architect explore a wide range of possible architectures. A design team can typically develop the optimal set of processors for a complex application in a few weeks.

Interestingly, rapid processor generation also changes design-team demographics. A much larger population of software and hardware engineers can now do the job because the required skill set changes. Smaller teams with more diverse design skills can safely use tuned processors to develop the various blocks of an SOC design.

A quick review of the automated process, outlined in Fig 8-5, highlights the key attributes of these configurable, application-specific processors.

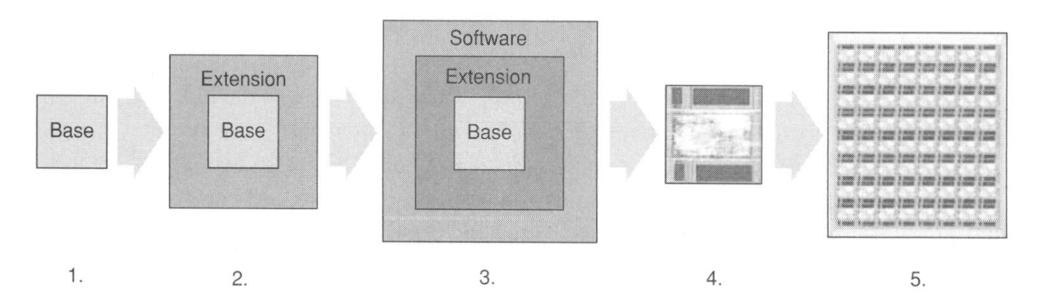

Figure 8-5 Basic processor generation flow.

1. Generation starts from a small base processor. The base architecture is quite fast and quite general purpose. The processor logic occupies less than $0.2mm^2$ using 130µm geometries. The manufacturing cost of the base processor is pennies.
2. The designer configures the processor by selecting and describing the instructions, the memories, and the interfaces to significantly improve the processor's throughput on the target applications. Sometimes, the configuration process is as simple as running a sample set of applications, so the processor generator can profile the code and select the optimal processor architecture for those applications. The generator then creates the processor hardware and the software-development tools directly from the extended architecture definition.
3. The software-development environment includes compilers, debuggers, simulators, integrated development user interfaces, and operating systems—all of the software tools normally associated with popular general-purpose architectures. The generated tools, however, support all the application-specific extensions, making the extensions immediately usable.
4. The generator produces efficient integrated circuit designs for the processor core using standard semiconductor libraries and compatible with the most widely used CAD tools so that the design is easily portable across fabs.
5. The resulting processor is small, flexible, and tuned to the task at hand. In fact, tuned processors are so efficient that they rival the performance of custom logic blocks and are so

small that they can be used in large numbers on a single chip. Both the processor interfaces and the software environment are tailored to make it easy for designers to put processors in roles in which processors have never before been fast enough or small enough.

8.1.5 What Makes These Processors Different?

The simple comparison in Fig 8-6 dramatizes the difference between a high-performance, general-purpose processor and an automatically generated processor that delivers the same high performance for a specific task.

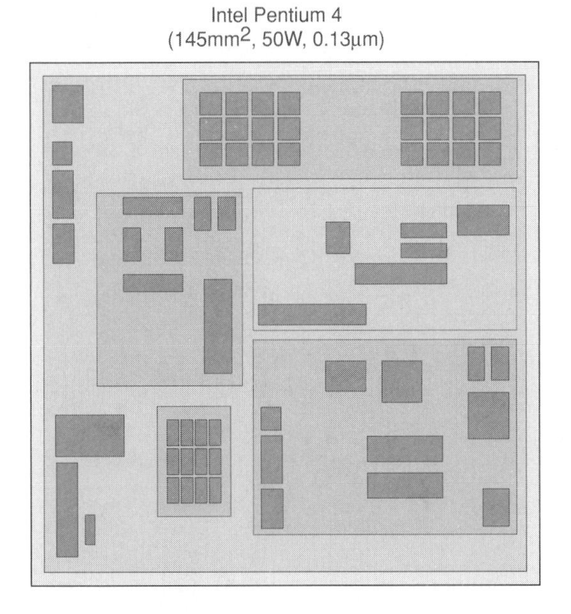

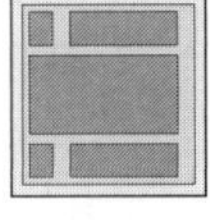

Figure 8-6 Comparison of Pentium 4 and configurable processor die.

On the left is a representation of the Intel Pentium 4 processor. It took hundreds of engineering years to develop, requires almost $150mm^2$ of silicon area, and consumes more than 50W of power. The Pentium 4 processor's large area and high power consumption reflect its "one-size-fits-all" mission. It achieves high throughput across a wide range of applications through its extremely high clock rate (several gigahertz) and a rich set of architectural and microarchitectural features (integer, floating-point, and multimedia instruction sets plus sophisticated out-of-order superscalar execution methods). The Pentium 4 processor succeeds in being pretty good at everything (Web browsing, word processing, spreadsheets, database applications, image processing, 3D graphics) but not exceptionally fast or efficient at any one task. It's probably the right answer for a general-purpose PC, which is built without a specific application in mind.

On the right of Fig 8-6 is a typical Tensilica Xtensa processor, complete with its extensions and memories. In the same fabrication geometry as the Pentium 4, it's about 100 times smaller and may consume 500 times less power. It doesn't need the high clock rate because it is tuned for the application. The tailored Xtensa processor can run one of those heavy-lifting applications, say image processing, faster than the Pentium processor. This tiny size, low cost, and narrow application focus are not appropriate for legacy desktop applications—but efficiency and application optimality are highly relevant for digital cameras, networking equipment, digital television, smart cell phones, and most other embedded electronics.

Detailed examples of the performance impact of processor extensibility are shown throughout this book, especially in chapters 3 and 6. A quick summary highlights the important issues. The aggregate efficiency of configured processors on three benchmark suites—EEMBC Networking, EEMBC Telecom, and EEMBC Consumer—is shown in Fig 8-7.

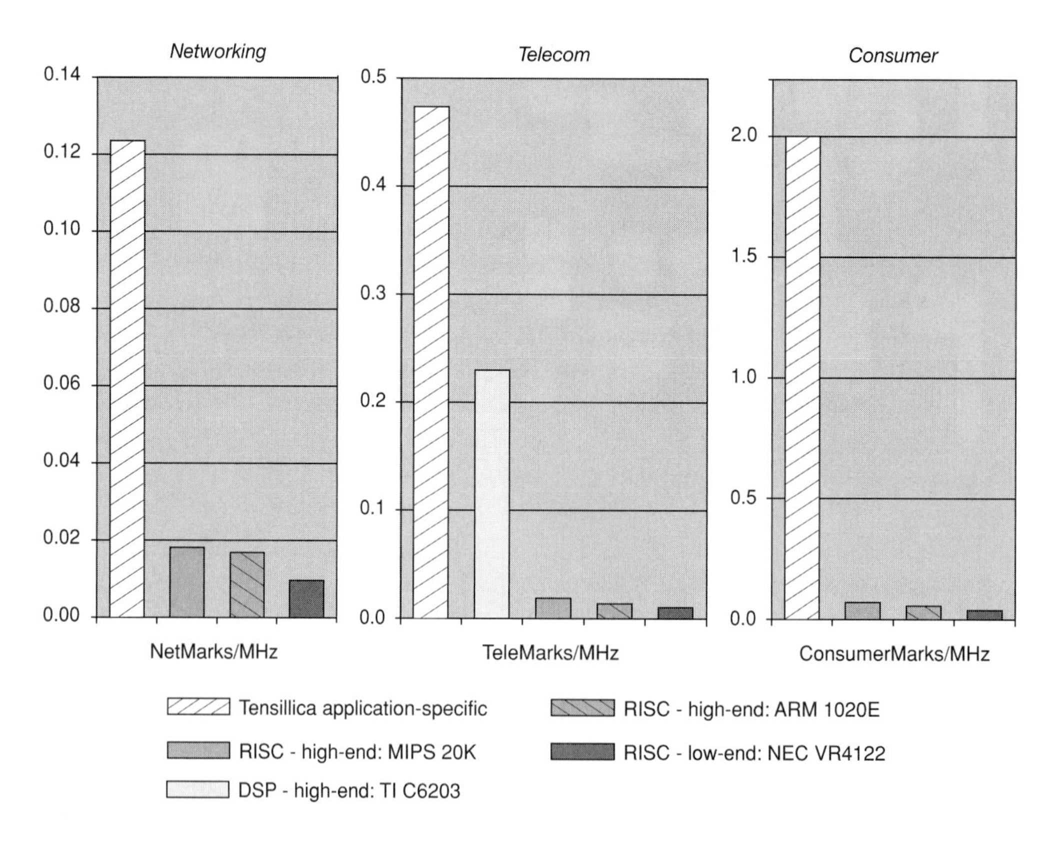

Figure 8-7 EEMBC summary performance: configurable processors versus RISC and DSP.

The short bars in each of the three graphs show the performance of general-purpose 32- and 64-bit RISC processors (ARM and MIPS cores). The second-tallest bar on the Telecom-suite graph represents the performance of a high-end DSP optimized for telecom. The tallest bars in each of the three graphs show the performance of three tuned Tensilica Xtensa processors—one for consumer, one for telecom, and one for networking. For each of the three EEMBC benchmarking suites, the tuned application-specific processor is 10 to 30 times faster than the general-purpose RISC processor cores. Improvement in power efficiency generally tracks the improvement in performance. The application-specific processor cores consume 90% less energy than the general-purpose processors while performing the same work.

Application-specific processors reach performance and power-efficiency levels previously attained only by specialized custom logic blocks but offer the added benefit of full programmability to adapt to ongoing changes in requirements and standards.

8.1.6 The SOC Design Transition

The emergence of the configurable processor and the processor-centric SOC design style is driving an important transition in large digital chip designs. A hypothetical complex SOC, sketched in Fig 8-8, includes not only a generic CPU, memory, and I/O subsystems, but also a range of application-focused subsystems for signal, media, protocol, security, and other application processing. Historically, these tasks have used hardwired logic methods (or simple, specialized DSPs) to reach throughput and cost goals.

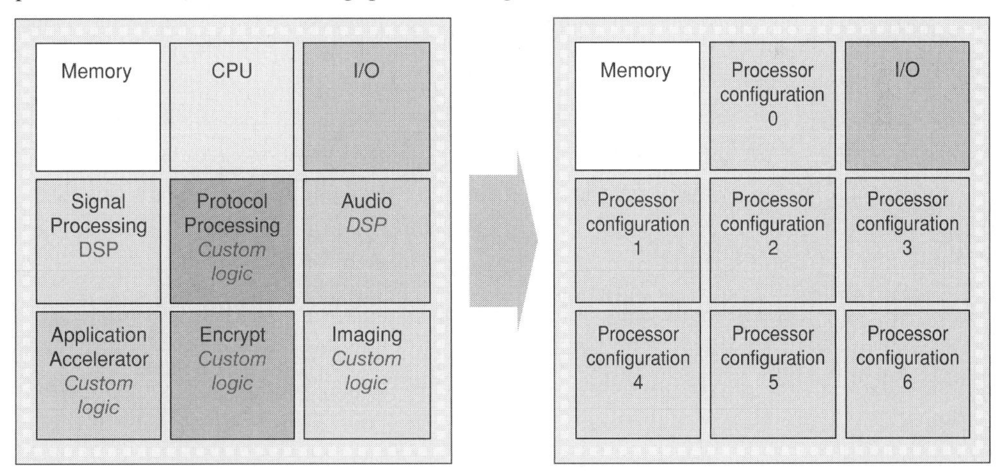

Figure 8-8 Transition of SOC to processor-centric design.

Configurable processors are well-suited to these heavy-lifting tasks—the signal, protocol and image processing (in fact, the lower two-thirds of the chip shown in Fig 8-8). Specifying a processor is much simpler than designing custom logic, and configurable processors deliver comparable performance. Functional blocks in SOCs are increasingly complex and performance intensive, so tuning processors becomes more compelling over time. Each of a wide range of different blocks can be implemented with a tuned processor from one processor family, so the various subsystems can all be integrated together with a common set of compatible compilers, simulation models, operating systems, and debuggers.

This situation is a big improvement over "Tower-of-Babel" integration that occurs today when designers lash together unrelated DSPs, image processors, audio engines, custom accelerators, and other blocks. This unified development environment for hardware and software design also sustains configurable processors successfully in traditional control-processor roles. In these roles, the performance demands alone might not be enough to justify switching from a legacy CPU, but the opportunity to simplify the overall SOC design, verification, and programming process often is enough to drive this switch.

The result is this: application-specific processors are already being used in a wide range of roles and used in large numbers. The SOC that uses automatically-generated processors already employs more than half a dozen processor cores per chip, with some using more than 100 processors per chip. Over time, the trend toward using processors for more roles on an SOC merges with the trend of putting more functions on a chip.

We already see Moore's law scaling in transistor density playing out in processor density. Fig 8-9 shows the general evolution of processor-intensive SOCs over one process-technology generation, where more functions, and therefore more processors, are integrated on each device. The interface-pin count will probably increase in absolute numbers, global memory will increase in absolute capacity, and the general-purpose CPU will get faster and consume more transistors. However, these increases will not keep up with silicon capacity. The dominant change will be the increase in the number of processors, including their embedded memories, and the increase in the fraction of total silicon area dedicated to processors. This trend underscores the need both for smaller processors and for processors with the communications flexibility and tool sophistication required for large-scale multiple-processor programming, interconnect, debug, and field support.

Having small, fast, highly adaptable processors available to be used as SOC building blocks is not sufficient. The design team must also know how to combine the processors into systems quickly and correctly. The adoption of application-specific processors as the default building block is eased by adoption of a multiple-processor SOC design style. Conversely, SOC design is significantly simplified by using these high-efficiency processors as a "universal logic element." Designers develop both communication and computation models within a common programmable processor-based framework. Functional prototyping, performance modeling, software development, and block-level verification all use common simulation, debugging, and analysis tools. Abstract tasks evolve by gradual refinement from high-level functional programs into sets of programs mapped onto optimized processors and memories. A few tasks evolve all the way down to hardwired logic blocks, but only when application-specific processors are not appropriate. The essential processor-centric SOC design flow is shown in Fig 8-10.

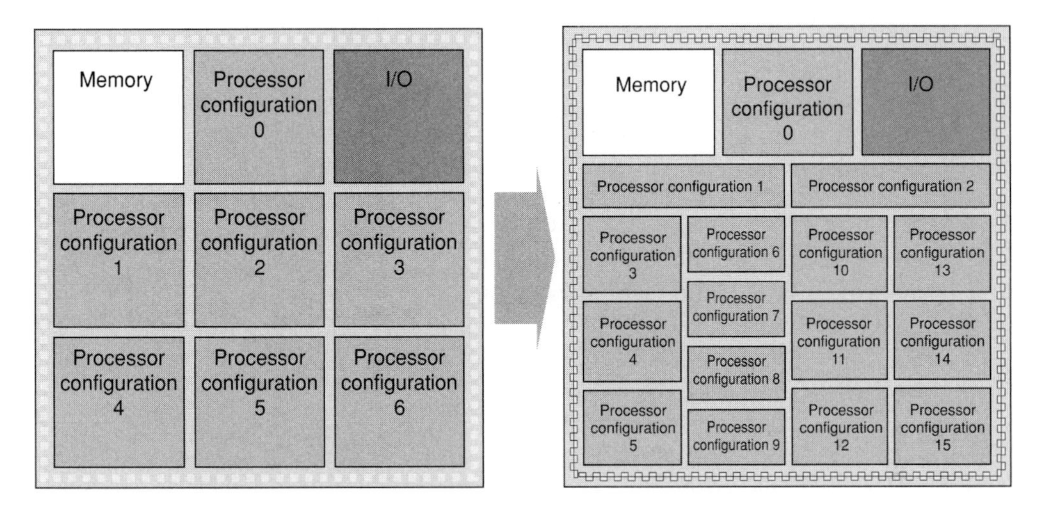

Figure 8-9 Influence of silicon scaling on complex SOC structure.

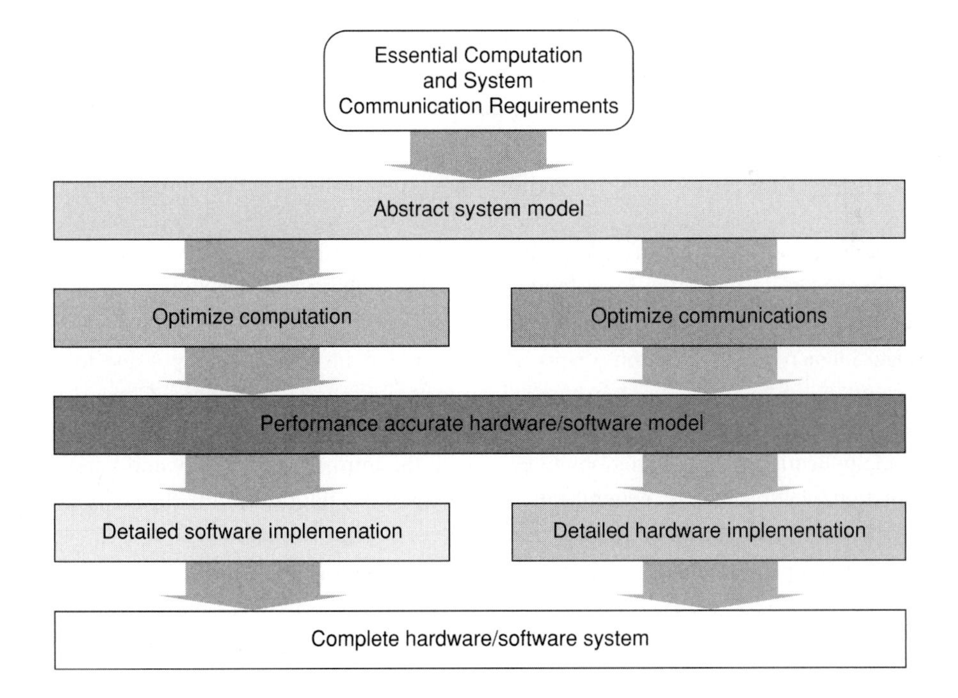

Figure 8-10 Advanced SOC design process.

Since Dr. Moore made his first projection back in 1965, the industry has done a remarkable job of scaling both integrated circuit density and performance at a blistering pace. As a result, we now blithely assume breakthroughs in battery life and performance with each new product generation. Unfortunately, design productivity has not kept pace with chip complexity, so the design costs of big chips threaten to explode. Some fundamental shift in design method is necessary, or the resulting deterioration in ROI will prevent designers from realizing the benefits of Moore's law silicon scaling. If the industry continues to try to design big SOCs using the design approach of the past, the brutal discipline of the marketplace will curb the number of successful projects. The shift to a processor-centric SOC design style changes the ROI equation and holds the promise for faster and less risky design, higher volume per chip, and greater profitability in SOC investments.

8.2 Why Is Software Programmability So Central?

One central premise of this book is this: designers must make programmability, especially software programmability, more central to their SOC design jobs. This premise is built on two facets of the SOC design challenge. First, the functional requirements of emerging embedded systems are so complex that only software-based methods can scale to manage the necessary complexity. The user interfaces, network protocols, encoding algorithms, and visual-display methods are all growing beyond the expressive power of hardware designers using RTL. Second, the variety and instability of system requirements is so high that post-silicon programmability, the opportunity to change system functions in software, becomes mandatory. These two facets of the challenge mandate substantial changes in how we think about software as a basic ingredient of SOC design.

As a result, SOC complexity drives the embedded processor core into an increasingly pervasive role in SOC design. High-level-language programmability enables both more rapid functional development and more agile adaptability to changing system requirements.

Data-intensive SOC functions, especially functions with high-throughput and low-power requirements, have historically been handled by inflexible, hardwired logic. General-purpose embedded RISC cores have long handled lower-performance user-interface, system-management, and application-control functions to cope with the intrinsic complexity and variability of these functions. Unfortunately, general-purpose processors cannot reach the throughput or efficiency achieved by hardwired logic—or by application-specific processors.

General-purpose embedded processors face two limitations:

1. These processors evolve slowly because the processor hardware and the software-development tools for new variants of these general-purpose processors are developed by manually, reinforcing the need to adhere to the one-size-fits-all nature of these architectural designs. There are not enough processor designers and there is not enough time to manually design the many processor variants needed to match the wide range of embedded applications. Processor architects omit features that are key to specific embedded applica-

tions, and yet they also burden every processor implementation with extraneous features not needed by all.

For example, just seven architectures and variants of the ARM architecture have appeared over the past decade or so: v4T, v5TE, v5TEJ, v6, v6Z, v6T2, and Intel's Xscale. Moreover, conventional architectural evolution is almost always cumulative—each new processor variant is a superset of the previous one so that legacy code is not abandoned.

As a result, features accumulate and are carried as baggage over time, even when the market trend for which the features were added has run its course. The high cost and effort required to manually develop new processor implementations and software environments inhibit more precise tuning of these general-purpose processor architectures to fit the target applications.

2. When the architectures are unable to evolve, advances in general-purpose processor performance increasingly depend on Moore's law silicon scaling. General-purpose processors do benefit from improvements in transistor speed (raising clock frequency directly) and improvements in transistor density (raising processor throughput through more concurrency in the processor implementation). However, hardwired logic also receives the same scaling benefit from Moore's law, so generic embedded processors are hard-pressed to close the throughput and efficiency gap.

Existing general-purpose, one-size-fits-all processor architectures fall short as universal SOC building blocks for four fundamental reasons:

1. Existing processors lack the efficiency and throughput to handle the most data-intensive and efficiency-sensitive SOC tasks.
2. Existing processors evolve slowly because architecture, hardware design, and software environments are decoupled from each other, so development support for new architectural features may lag when new embedded-computation features appear in the processor hardware.
3. Existing embedded processors evolved in an environment of board-level, single-processor designs, and they typically lack essential low-latency, high-bandwidth, and processor-to-processor communications features.
4. When new features are added to a manually-developed architecture, they cannot easily be omitted from other variants causing implementation size to grow and implementation efficiency to fall.

This book concludes not only that processors must play a bigger role in SOC design, but also that many or all of those processors will be configured to the task at hand. Legacy code and legacy processor architectures remain important for some SOC designs, but their liabilities—excess silicon area and power and insufficient performance—may eventually obsolete the currently popular processor architectures in favor of configurable architectures.

8.3 Looking into the Future of SOC

Given the basic message of this book, that configurable processors can and will become the basic building block within a new systematic SOC design style, then the big question becomes this: How far can the processor-centric SOC trend go?

Chapter 3 presented the fundamental efficiency advantage of concurrency in semiconductor-based systems: whenever a task can be split into multiple tasks running simultaneously, the combination of the power dissipation, task latency, and task throughput can usually be improved through concurrency. For many tasks, the concurrent implementation is also no bigger than a more performance-optimized, single-threaded implementation. Therefore, designers will seek good parallel implementations for any performance-critical or efficiency-critical system task. Designers will gravitate to the use of more concurrent SOC designs if they can find efficient solutions. Designers will gravitate to concurrent processors for SOC design if that efficiency can be achieved more quickly and flexibly with less design risk.

If we could predict our success in this quest for efficient parallel designs, we could paint a more detailed picture of future semiconductor designs. We would know how silicon real estate was allocated, what levels of computation performance were reached, which application types were particularly favored, and what new design issues were emerging.

These issues lead to four basic questions:

1. How will processor-centric SOC designs scale with underlying semiconductor technology?
2. How will system-application characteristics be reflected in the SOC designs used to implement those applications?
3. How will the design process change in response to the opportunities and problems of very large-scale SOC-design projects in the future?
4. How will the electronics sector, especially the semiconductor industry, evolve in response to these fundamental technology changes?

The first of these questions drives to the heart of processor use, which may expand over time and is the focus of the following section. Succeeding sections address the other three questions—the evolution of applications, the evolution of methodology, and the evolution of the industry.

8.4 Processor Scaling Model

The economic motivation to use processors, if they meet efficiency and performance requirements, is clear. All things being equal, a programmable block eases correct design, simplifies simulation and integration into the rest of the hardware/software structure, and lends greater flexibility to the final design after the chip is built. But what are the technical limits?

The International Technology Roadmap for Semiconductors (ITRS) represents the semiconductor industry's consensus view for silicon device density and performance over the next 15 years. Its use as a target for technology planning has historically driven progress in meeting Moore's predictions. The continued scaling of the underlying semiconductor technology stands in sharp contrast to projected diminishing returns in improving high-performance, general-purpose processor architecture. The legacy of computer system technology and performance scaling over the past 15 years cannot simply be applied to the next 15 years of embedded SOC design. The ITRS roadmap serves as the foundation for a further set of predictions about the role of processors in SOC design.

This section focuses on the implication of this extended model of processor scaling. The model provides crucial reassurance that density growth at the transistor level as predicted by Moore's law can be effectively exploited to provide continued improvement in the performance, efficiency, and flexibility of electronic products.

Our model uses the 2003 edition of the roadmap and accounts for the following expected transitions:

Gate-level scaling: Shrinking device size and increasing device density provide the technical ability and economic motivation to rapidly integrate electronic system features into SOCs. As typical production-volume SOC device complexity grows from roughly 10 to 500 million usable ASIC gates over the next 15 years, technical and economic factors will force structural changes in SOC designs. Fig 8-11 predicts typical gate density and worst-case clock speed for standard-cell logic using industry-standard logic-synthesis and gate-level place-and-route tools. Our projected clock frequency of typical synthesized designs lags behind the ITRS logic "local clock" frequency predictions for hand-optimized microprocessor circuits. The circuit design style of high-end microprocessors is unlikely to proliferate into typical embedded SOCs both because it is too labor-intensive to justify the investment and because high-end microprocessors employ clock frequencies and power densities that are inappropriate for high-volume consumer electronics and communications products.

Mandatory programmability: Rising SOC design cost will drive the incorporation of programmability into more SOC functions to both limit development expense and increase potential manufacturing volume. Increased programmability allows one integrated circuit to serve the needs of many more products and also allows design errors to be fixed quickly and inexpensively with software. End-product demands for increased functional complexity favor software-based processor programmability over other configurability and programmability mechanisms.

Processor-based execution of software is not the only form of programmability destined to play a role in SOC design, however. Blocks of FPGA logic, derived from standard programmable-logic devices, will also appear, though limitations in the FPGA programming model (via hardware-description language rather than high-level programming language) and circuit efficiency will probably restrict FPGA use to programmable I/O-signal interfaces and special-purpose computing structures. According to IBM researchers, FPGA-based logic suffers roughly a 25 times clock-speed disadvantage, a 50 times density disadvantage and a 100 times power disadvantage relative to standard-cell logic in the same integrated circuit process technology. How-

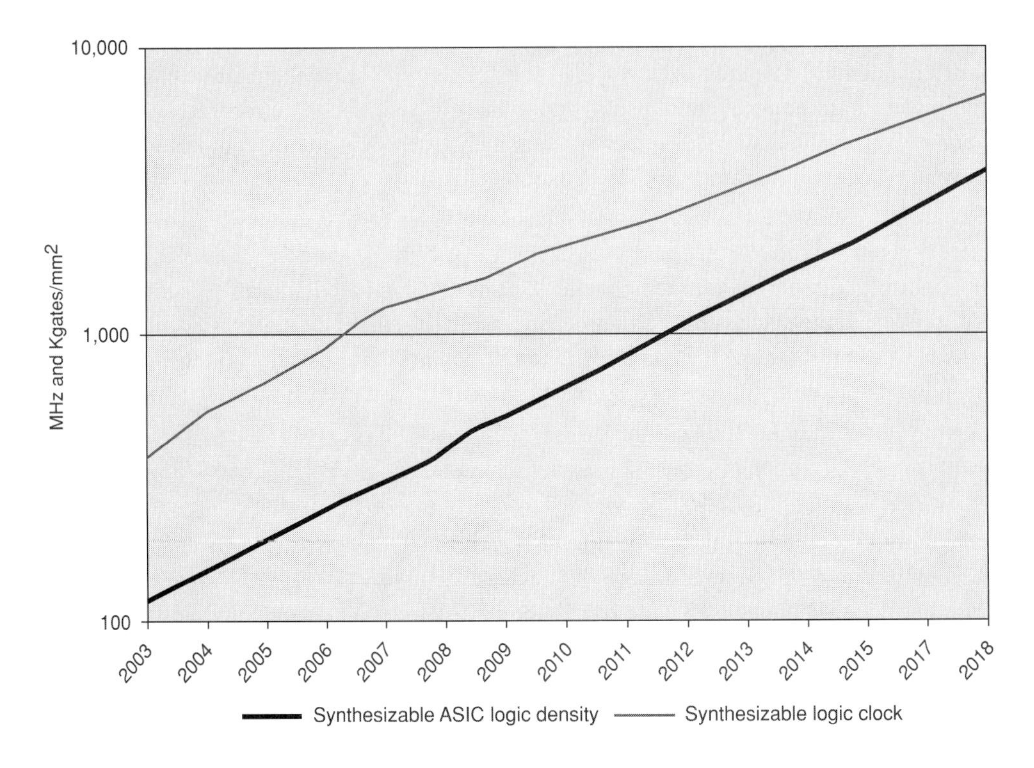

Figure 8-11 Standard cell density and speed trends.

ever, continued scaling of line widths from just under 130nm today to roughly 18nm in 2018, as projected by the ITRS roadmap, does not close this gap due to many factors, especially the high sensitivity of FPGA architectures to interconnect delay.

Configurable processors: Convenient generation of tailored, application-specific processors will allow processors to efficiently replace hardwired logic blocks across a broad range of tasks. In some cases, the tailored processors will be as fast and small as the hardwired blocks they replace. Replacement of logic blocks by application-specific processors is possible because the processors will implement data-path logic that closely resembles the data-path logic of the hardwired blocks and because the control function of the processors (instruction storage, decode and issue) is small compared to the data-path functions.

In fact, processor flexibility will often allow a rich variety of related functions to be time-multiplexed on one processor, where a hardwired logic approach would require a number of different blocks, each underutilized. In other cases, tailored processors will not quite match the area efficiency of hardwired logic, but the savings in development time and the reduction in project risk through programmability will compensate for this added area. After all, with continued Moore's law scaling over the next decade, transistors will often be "free," while design time and design costs will dominate development concerns. Moreover, fully automated generation of new processors from the application source code alone holds significant promise for both reducing the effort and skill level and for increasing the optimality of generated processor architectures. We project that tailored processors will soon be the dominant form of programmable element in SOCs.

Concurrency in applications: Growing intrinsic concurrency in system applications and improved methods for exploiting that parallelism will make massive use of small, application-specific processor cores the natural fabric of advanced SOC design. The level of available concurrency in future SOC applications is one of the most controversial issues in forecasting the future of chip design. Much of the historical analysis has assumed a single task written in a convention sequential language such as C and attempts to discover the degree of latent concurrency extractable from the task in that form. This approach is likely to understate the concurrency that could be realized with substantial re-architecting of the computation.

Nevertheless, studies from researchers such as David Wall and Linda Wills have shown intrinsic concurrency ranges into the hundreds of operations per cycle for individual, data-intensive tasks. With substantial restructuring of the computation, such high concurrency might be achieved across a wider range of applications. That algorithmic and architectural restructuring is only likely to be attempted once appropriate processors, small processors with very high-bandwidth and low-latency communications, are available as the basic building block. Moreover, as data resolution increases (especially for audio, image, and video media types) and communication block sizes increase (especially for high-bandwidth wireless communications), the intrinsic concurrency in the algorithms also increases. While finding more concurrency per task will not be easy, application trends give reason for optimism.

At the same time that individual tasks grow in concurrency, SOCs are integrating more diverse functions. For example, systems that today implement simple audio interfaces are adding still images, 3D graphics, and streaming video as well. Systems that have historically used simple user interfaces based on keyboard input and text output are incorporating voice recognition, face and gesture recognition, and context-based graphical user interfaces. This aggregation of complex functions into one system, increasingly into a single chip, also increases SOC concurrency. The combination of the two forms of concurrency—single-application parallelization and multiple-task integration—gives considerable hope for long-term scaling of SOC designs. Performance for many applications will be limited only by the ability to integrate many processors onto one device with appropriate high-bandwidth, low-latency, interprocessor communications.

The obvious challenge to this projection is the scarcity of examples of such high concurrency. However, companies such as Picochip and PACT are already integrating hundreds of processor elements into their SOC designs. While the commercial success of these platforms is uncertain, the technical feasibility is clear. The limitation of the success of these devices is more likely to stem from the one-size-fits-all generic communications structures (the lack of a tight fit to a target application domain) than from some fundamental lack of concurrency in embedded systems.

Moreover, if one argues that embedded electronics lack the latent concurrency to exploit hundreds or thousands of parallel processors, then how can embedded electronics possibly exploit hundreds of millions of parallel logic gates? The success of high-complexity logic in embedded systems strongly suggests that much concurrency already exists in applications and that designers are endlessly creative in discovering and exploiting that concurrency.

Our model of processor scaling tracks two processor categories: very small processors with modest degrees of internal concurrency (used in parallel in large numbers) and fatter, higher performance processors, implementing much richer instruction sets that employ SIMD and long-instruction word techniques for higher internal concurrency (used in parallel in relatively modest numbers). These two processor categories leverage both forms of concurrency outlined in Chapter 4, Section 4.1. Small processor configurations best exploit coarse-grained concurrency; high-performance processor configurations best exploit fine-grained concurrency.

The processor-scaling model predicts that small, extended processors will be used in massive numbers, where a leading-edge design may incorporate hundreds or thousands of communicating cores. Alternatively, dozens of high-performance, data-parallel SIMD long-instruction word cores may be used per chip.

Fig 8-12 shows the model for the maximum practical number of processors per chip for both large numbers of small processors (for applications with the most coarse-grained concurrency) and for smaller numbers of larger processors (for applications with the most fine-grained concurrency over the 15-year span of the ITRS 2003 technology projection). This figure assumes 140mm^2 die area, a common reference die size in the ITRS roadmap.

SOC architects will exploit both instruction-level and task-level parallelism. Application-specific processor architectures exploit instruction-level parallelism through vector (SIMD) and long-instruction-word techniques to increase throughput and efficiency within a single algorithm. One processor may execute 10, 20, or more basic operations per cycle, especially for data-intensive tasks in signal and media processing. By contrast, a simple processor optimized for coarse-grained parallelism may execute only one or two operations per cycle. Parallelism with this processor type comes from higher level concurrency among different instruction streams operating on different data streams (and on different processors, sometimes with different configurations).

Integrated development tools and processor generators will allow architects to explore both the spectrum of individual processor extensions from simple, enhanced RISC processors to fat, long-instruction-word vector architectures. They will also be able to explore a wide range of processor quantities, interprocessor communication patterns, and system topologies.

Aggregate performance of multiple processors: Tasks with high degrees of coarse-grained concurrency will use large numbers of relatively low-throughput processors. Tasks with high degrees of fine-grained concurrency will use more modest numbers of high-throughput processors. Our performance models for large numbers of simpler processors and for smaller numbers of more complex processors yield similar overall throughput predictions. Aggregate performance throughput is shown in Fig 8-13 (which assumes 140mm^2 die area). In the cases of

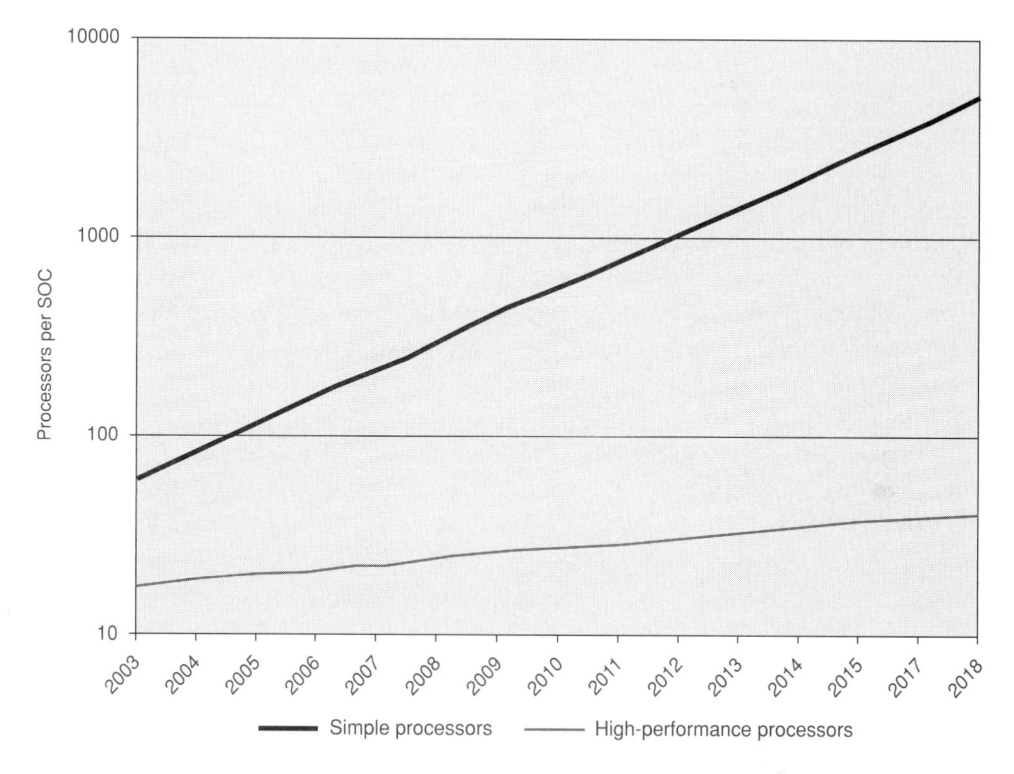

Figure 8-12 Processors per chip for 140mm^2 die.

both smaller, minimally extended processors, and fatter, data-parallel processors, the growth rate (65% per year) and absolute performance (10^{13} operations per second by 2015) are similar.

The processor count and corresponding aggregate potential performance results in Fig 8-12 and Fig 8-13 lead to two key predictions about future SOC designs, which might be called fundamental trends of SOC processor scaling:

Trend 1: *Software-centric multiprocessor SOC design will become the dominant chip design methodology, with maximum on-chip processor count rising by 30% per year and ranging into the thousands of processors by 2015.*

Trend 2: *The aggregate maximum computing power of a typical processor-based SOC will grow at roughly 65% per year, reaching 1 trillion operations per second by 2009 and 10 trillion operations per second before 2015.*

The allocation of chip real estate: One subtle but important trend underlies these results: the reallocation of silicon real estate to favor processors and their memories. In this model, processors and their memories are assumed to occupy one third of the die area at the current state of the art. This figure reflects the fact that processors are today used primarily for complex control tasks with modest contributions to signal, media, and network processing. The balance of the die

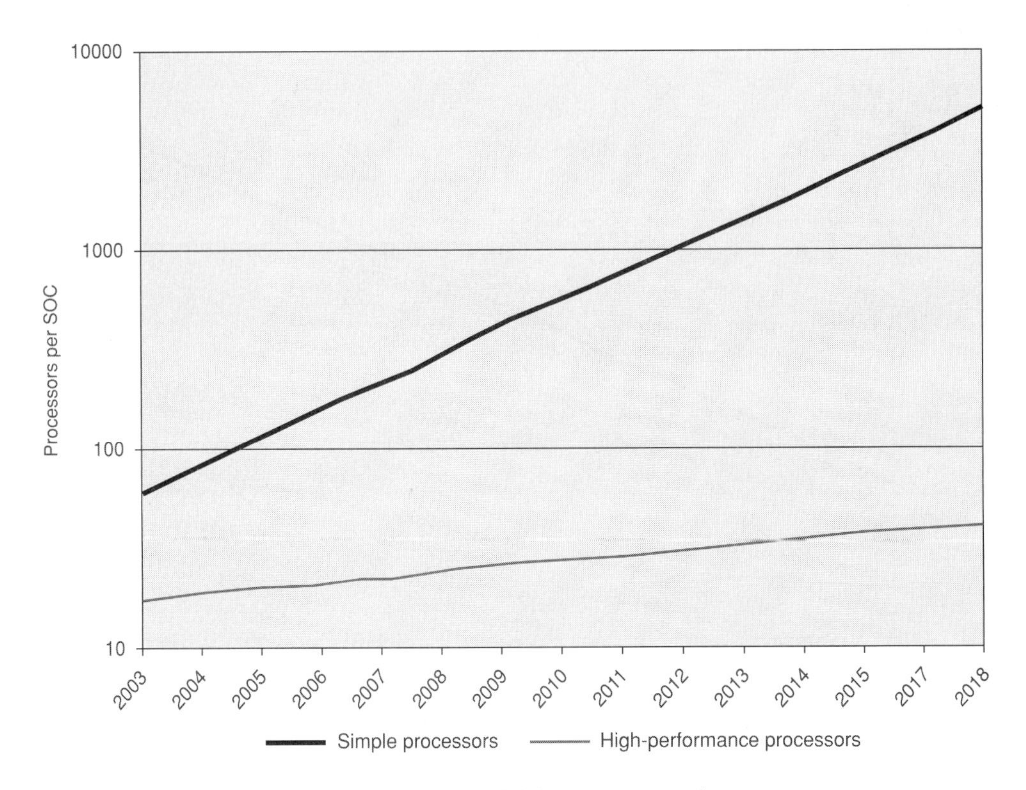

Figure 8-13 Aggregate SOC processor performance.

area is used for hardwired logic, I/O circuits, and bonding pads. The fraction of the die area for I/O interfaces and pads is not likely to change dramatically, though the trend to flip-chip packaging may reduce constraints on the number and location of bonding pads in the design. The big shift toward greater programmability, enabled by configurable processor generation, drives a reallocation of chip area away from hardwired logic area in favor of processor area. The model predicts that the typical fraction of SOC area dedicated to programmable functions (programmable processors and associated memory) will also grow to more than 75% by 2018.

This trend moves functionality and area from random logic blocks with their distributed memories toward processors and their attached instruction and data memories. Instruction memories can be considered an alternative to the logic gates implementing finite-state machines in hardwired logic and to RAM-based lookup tables used in FPGA-like fabrics. Processor data memories are directly equivalent to the embedded memories found in traditional ASICs today, but they provide a more structured approach to logic hook-up, testing, allocation, and sharing.

The processor-scaling model predicts that the fraction of a chip dedicated to processor memory grows to 55 to 60% by 2018, depending on the type of processors. This result closely matches the prediction of Virage Logic Corporation, that memory will occupy more than 50% of future SOC area. Our key observation is that most on-chip memories will be closely coupled to processors. High-performance processors typically require more memory and represent the high end of our predicted allocation of silicon area to memory.

Also note that increasing the amount of on-chip memory is not likely to reduce the need for additional off-chip memory. External flash memory and DRAM are likely to remain important technologies under this scenario. In many cases, the on-chip RAMs will serve as the first level in a memory hierarchy that extends off-chip and even to storage resources accessed across a network. Interface circuits including both analog functions and configurable interfaces based on FPGA logic will remain important consumers of the remaining SOC real estate. Non-processor core logic and memories will become much less significant.

One caveat on memory technology trends is worth noting. The drive for higher density and yield, lower power, and flexible non-volatility make memory technology an area of rich innovation. On-chip RAM's evolution from today's common six-transistor static RAM cell to one-transistor pseudostatic RAM, to magnetic memory, or even to organic storage technologies is hard to predict accurately.

The processor-scaling model also projects that on-chip instruction storage will grow dramatically over this period as well. Total on-chip instruction storage (the aggregate capacity of all instruction caches and other local processor-instruction memories) is probably the best measure of the complexity of the set of tasks running together on a chip at any point in time. Off-chip instruction storage, in the form of flash memory, local disk, or remote storage accessed over the network, may be useful for booting the system, for holding infrequently executed code, for storing error-handling routines, or for storing application code for operating modes or features not currently in use. Off-chip code requires tens (and sometimes hundreds) of cycles to fetch, so processors cannot sustain high performance while executing from off-chip instruction memory. The designers must make instruction caches and local memories sufficiently large to hold all or most of the expected performance-critical code. Therefore, the capacity of on-chip memory tracks the design team's expectation about the code "working set," the amount of code that must reside locally to run important functions. Simple scaling suggests that this code working set will grow from a few hundred thousand instructions today to more than ten million instructions by 2018. However, the working set may represent just a small part of total code ported to the system.

8.4.1 Summary of Model Assumptions

The SOC processor-scaling model presented above is built on top of the ITRS model. It makes a modest number of important assumptions about the degree of processor concurrency; about performance scaling of automated design flows; and about the relationship between processor operations, data-path gate count, and processor memory requirements. The assumptions highlight important characteristics of extensible processors for SOC designs.

Fig 8-14 summarizes key assumptions used in this model.

Parameter	Value	Rationale
Synthesizable standard-cell logic area utilization	75%	Typical for automatic synthesis/place/route flow.
De-rating of on-chip clock frequency (typical custom logic) to worst-case synthesizable logic clock frequency	8x	Synthesized logic typically has more gates delays per clock, and synthesis/place/route flow sacrifices circuit and layout efficiency for automation. Also typical ASICs lag at 1—2 years behind custom microprocessors in exploitation of leading-edge transistors.
Base processor gate count	20,000	This number is larger than the smallest Tensilica Xtensa processor, but typical for a speed-optimized, general-purpose integer processor configuration.
Rich processor gate count	150,000 + 20%/year	This number is typical of a Tensilica Xtensa processor with long-instruction word and complex SIMD data paths today. Growth assumes evolving architecture support and increasing automation and concurrency of multiple pipelines.
Rich processor gates per operation	20,000	Currently, a typical rich processor executes about 10 RISC operations/cycle.
Local memory per base processor	12Kbytes	Typical total instruction and data RAM or cache for small configurations.
Local memory per rich processor	32Kbytes + 30%/year	Richer processor configurations typically demand more data RAM per processor.
Local RAM Density	24Kbytes/mm^2 + 26%/year	Small memory blocks from RAM generators are less dense than ITRS assumption (38KB/mm^2) and improve slightly less quickly (28%/year).
Fraction of chip for processor logic and memory	33% + 2%/year	Gradual adoption of multiple-processor SOC design methodology, but processor logic occupies only 22% of area by 2018.

Figure 8-14 Processor scaling model assumptions.

This model of processor scaling drives a concrete vision of advanced SOC architecture, where the typical design is constructed from a large number of processors used in a huge diversity of roles. A "sea-of-processors" SOC design approach makes the fully programmable, application-tuned processor the essential building block for integrated systems, just as the silicon transistor became the basic building block of integrated circuits in the 1960s and the ASIC logic gate became the basic block in the 1980s.

8.5 Future Applications of Complex SOCs

The second major question for the SOC future considers the impact of application trends on design trends. The seemingly universal hunger for greater electronic systems sophistication, lower cost, and greater portability provides the underlying drive toward SOC integration. Over time, a wider range of systems will need to combine the memory, logic, processor, and I/O subsystems on one chip. Wireless communication, in particular, is an especially important application driver for processor-centric SOC design. Despite the vast variety of integrated applications, a few trends stand out.

First, all communications products and most consumer products are becoming connected to the Internet via broadband connections. These connections not only enhance the role of products as explicit voice and data portals, but enable new productivity and entertainment applications. High-bandwidth data channels enable real-time audio and video transfers; high-speed access to large, continuously updated databases; and continuous validation of high-volume transactions. This trend toward ubiquitous broadband connectivity implies that future SOCs must support high data rates and complex protocol stacks, including significant security functions. Each of these requirements puts additional demands on the internal communication and computation capacity of the SOC.

Second, the data rates and resolutions for existing data streams are continuously increasing. In just the past two years, for example, we have seen a fourfold increase in typical digital-camera image resolution, from 1.3 to 5 Mpixels. Similarly, high-definition television broadcasts are becoming mandatory by governmental fiat, and digital television broadcasts are now starting to target handheld receivers. Wide-area network bandwidths are gradually stepping up from 2.5Gbps (OC-48) to 10Gbps (OC-192) and will eventually move to 40Gbps (OC-768). Internet protocols are moving from IPv4 to IPv6 to expand addressability and support advanced services on very-large-scale networks. In almost every application domain, the aggregate data rates are growing just about as rapidly as Moore's law transistor scaling.

Third, many more products are relying on wireless communication. The combination of wireless communication and data-bandwidth growth creates extraordinary pressure for efficient data encoding. The electromagnetic spectrum is effectively finite, at least for any given generation of transistor technology. Within this finite spectrum, an ever-increasing number of users are attempting to pump ever-faster data streams between handheld devices and communications-access points. This trend suggests that two basic changes must occur. First, wireless-cell standards must continuously evolve to make cells

- Smaller—so users can be distributed across a larger number of cells.
- Provide higher capacity—so a cell can handle more users.
- Operate at higher data rates—so each connection sustains support fatter data streams.

Second, data encoding must become denser so more useful information can be pushed through the available electromagnetic spectrum.

Improved encoding comes in two forms: encoding for error tolerance and content-specific encoding. The goal of error-tolerant coding is to approach the information-theoretic limit for communication in the presence of noise—often called the Shannon limit. Innovation in encoding systems continues to bring communication systems ever-closer to this limit in the presence of various forms of noise.

Content-specific coding exploits knowledge of the type of information being transmitted to encode more useful information in fewer bits. For example, video encoding takes advantage of the typically high similarity between sequential image frames to transmit just the frame-to-frame differences. Audio coding takes advantage of our growing understanding of human audio perception to minimize unimportant components of the sound stream and maximize accuracy in those audio components for which human sensitivity is highest.

The rising computational requirements for wireless communication present a particularly potent challenge for SOC designers. Fig 8-15 shows the trend in the computational demands of wireless communication, both for the evolution of major cell-phone generations and as an overall trend, as estimated by Ravi Subramanian. The figure also shows the historical improvement in transistor speed and the consensus forecast for future speed improvement made by the semiconductor industry (see the International Technology Roadmap for Semiconductors, 2001).

By this analysis, the rapidly increasing computational complexity of advanced wireless communications far outstrips the ability of simple transistor-level innovation to keep up. Moreover, improvement in general-purpose processor performance, which takes advantage of both faster transistors and greater microarchitectural concurrency, still falls far short of the projected 69% annual increase in computing demands. The industry must look to new methods to sustain this pace. Application-specific processors start out roughly 10 times faster and more efficient for channel coding than general-purpose processors, and groups of application-specific processors can be expected to deliver aggregate performance increases of 65% per year. Application-specific processors can deliver the levels of performance and programmability demanded by communications over available electromagnetic spectrum.

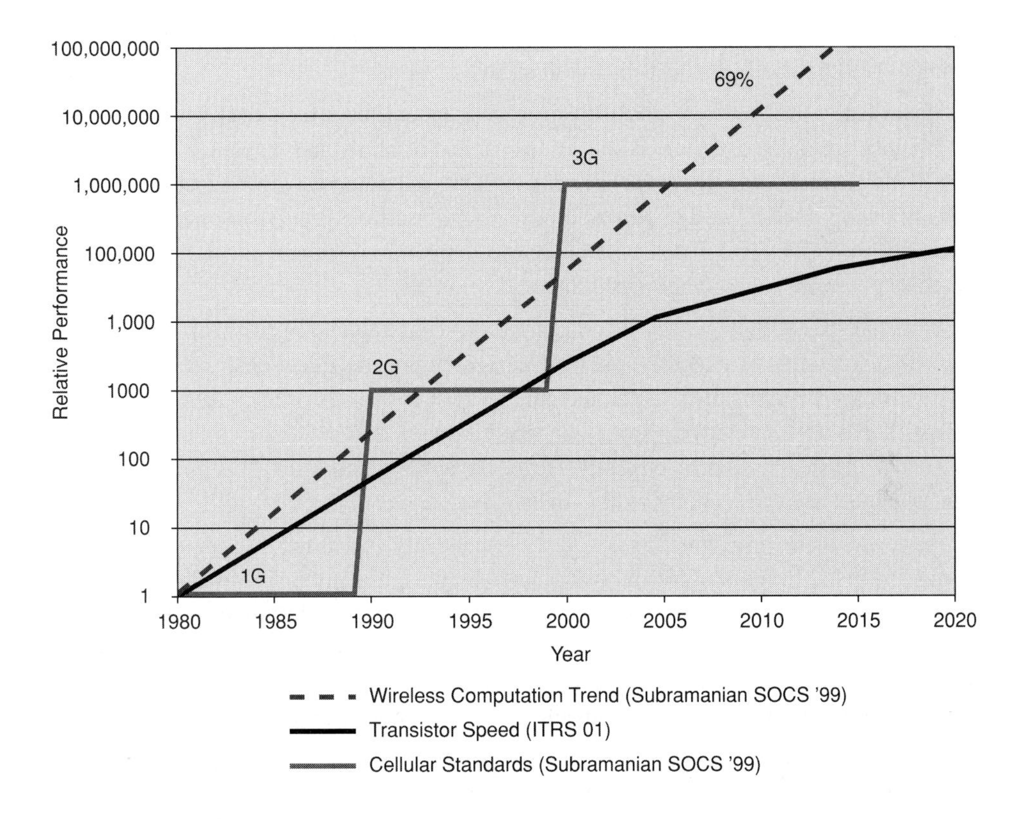

Figure 8-15 Wireless computation complexity outstrips transistor performance scaling.

8.6 The Future of the Complex SOC Design Process

This book has outlined a new way to speed development of multifunction, multimillion-gate SOCs. First, extensible processors replace both conventional embedded processor cores and hardwired logic functions designed at the Register Transfer Level (RTL) to accelerate design time and bring full programmability. Software-development tools, simulation models, and optimized hardware design are generated from a single-source representation to speed development and guarantee completeness and correctness. Second, these application-specific processors are customized to run the functions almost as efficiently as the rigid RTL logic blocks they replace while still retaining full programmability.

High performance and easy adaptability in data-intensive applications allow these tailored processors to play a broader role as a basic fabric for SOC design. This multifaceted role for processors brings more complete and universal programmability to both control and data functions. A single programmable SOC can be used for multiple applications and customers. As standards change, new functions can be added to the software, saving development time and money, and lowering overall development costs by avoiding SOC respins.

Historically, the hardware and software-development tasks have each relied on different sets of highly specialized design skills. The diversity of skills and challenge of coordination among hardware and software teams have made SOC design increasingly expensive, risky, and slow. Unification of SOC design methodology around multiple processor cores speeds system partitioning, simplifies subsystem design, eases hardware/software integration efforts, and yields SOC platforms with higher reusability and better ROI.

Processor-based SOC design leverages the growing abundance of cheap transistors relative to the availability of expensive, skilled design engineers. Functional specification at the software-task level is typically simpler and quicker than the design of equivalent hardware logic functions. Some hardware developers argue that hardware design is more robust than software development. This assertion is typically based on adhering to strict "keep-it-simple-stupid" design principles and on a commitment to exhaustive verification.

However, these principles do not apply exclusively to hardware design. Any function, implemented in hardware or software, that is both simple and vigorously verified, is likely to be quite robust. Software's lower bug penalty (in terms of the cost and time required to fix design bugs) allows software to be used to implement complex functions that make a hardware-based design approach either prohibitively expensive or prohibitively risky.

A software-centric design approach offers the potential for much higher designer productivity. Moreover, the use of application-specific processors retains the advantages of software-based design methodology but allows processors to perform many of the high-performance, data-intensive tasks previously realizable only in hardware. The rationale for software-centric design is further explored in Transition 3 below.

A series of transitions in the design process seem likely in the next decade:

Transition 1: SOC designs will use an increasingly large number of processors and the processors will play a greater range of roles. This transition is the dominant theme of this book and triggers a number of the other transitions forecasted below. This transition both exploits the underlying growth in system parallelism as more functions are integrated together and leverages new processor capabilities, especially the breakthrough in performance per mm^2 and performance per watt of application-specific processors. Moreover, processor hardware-interface enhancements, such as direct instruction-mapped wire and queue connections, allow the processor to directly and easily replace blocks of hardwired logic, which adds programmability to the task block while retaining the existing pin interfaces and high-bandwidth handshakes shared with other logic blocks, memories, and processors.

The ability to automatically generate optimized processors makes these small processors a universal building block, at least for functions that require the equivalent of tens of thousands of logic gates. (Smaller logic functions either remain as hardwired logic designed with conventional RTL tools or become tasks running on processors that also implement other functions.)

As the processor increasingly becomes a universal tool for developing logic functions, we see new designer behavior. Architects will drop in spare processors “just in case” new functions are required late in the design process or after SOC fabrication. Designers will consciously balance instruction-set generality against the silicon area benefits of highly specialized instruction sets. Hardware developers will wrap processors around otherwise standalone memories and logic blocks to maximize design flexibility and improve debuggability. Automated processor-configuration tools will let SOC designers focus on the big picture—the top-level hardware structure and software content of the SOC—instead of on implementation details: instruction-set encoding, block model development, RTL coding, software-tool development, and block debugging.

Transition 2: As processor generation simplifies and automates the creation of building blocks, the design team’s focus can and must shift toward the overall system structure. This structure spans the software view (how do tasks communicate?) and the hardware view (how does data move among blocks?). The advanced SOC design methodology puts development of the communication structure front and center. As SOCs scale in complexity and concurrency, the greatest opportunities for functional innovation and the greatest leverage on system performance, power efficiency, and cost derive from good communications-structure choices.

This increasing visibility into communications issues and the sheer degree of available concurrency will drive the development of new communications-network concepts. Increasingly, SOC architects will not think in terms of buses, but in terms of on-chip networks. Concepts borrowed from computer systems and global communications networks such as packet-based protocols, quality of service guarantees, robustness in the face of node failures, high cross-section bandwidth, and optimization for different traffic classes will become commonplace. Architects will choose between datagram (dynamic routing), virtual circuit (static routing), and direct-connect communications. Rightly or wrongly, the daunting richness of network topologies, layering models, and analysis tools will be applied to on-chip networks.

Some global network concepts are clearly inappropriate to some SOCs. For example, the high reliability of integrated circuits and limited system scalability of many end applications may limit the need for robust node-failure management and continuous on-chip network reconfiguration to SOCs used in products that must be highly fault-tolerant.

Transition 3: One central motivation for increased processor use is the attractiveness of software-based development over hardware-based development. This motivation is not driven by intrinsic superiority of software languages like C over hardware languages like Verilog. (Arguably, the Verilog language is actually better than C for expressing concurrency.) Instead, three factors underlie the motivation for a software-based approach:

a. Many more engineers and developers know software programming than know hardware design.

b. The available tools and methods for software development in standard programming languages scale to very complex systems more readily than the available tools and methods for hardware development in RTL languages.
c. Software methods provide the easiest and most obvious methodology continuity among initial algorithm development, detailed SOC design, and post-silicon product debug, enhancement, and evolution. The use of data abstraction and thread-level concurrency in language variants such as SystemC creates the opportunity to use the software description as an "executable specification," suitable for incremental refinement into complete implementations.

If the silicon efficiency of a software-language-based approach can come even close to the silicon efficiency of a hardware-language-based approach, the greater flexibility and system-development seamlessness of software-based system-development methods will win the day. Prior to the emergence of application-specific processors, closing the hardware-software efficiency gap was difficult. Automatic processor generation changes the picture.

Transition 4: One by-product of a more processor—and software-centric—approach deserves special mention. Historically, hardware and software developers have worked with quite different tools and viewing models, especially during the analysis, debug, and verification phases. Hardware developers use synthesis tools to assess gate count, circuit performance, and power dissipation; to run RTL simulators to verify functionality; and to employ signal-waveform viewers to understand unexpected behavior. In contrast, software developers use compilers, assemblers, and binary tools to generate and statically assess code, and they use prototype boards or single-processor simulators to run that code and measure cycle counts. Software developers employ C source debuggers to understand unexpected behavior.

A new class of hardware/software integrated development environment is lowering the barrier between these two developer use models. One single environment is used to create both programs and processor configurations, including new instruction-set descriptions. The environment also generates and runs multiple processor simulations, with cycle and pin-accurate behavior for each processor, plus as much detail in non-processor modeling (especially modeling of interconnect and I/O behavior and performance) as desired. The development environment estimates key hardware characteristics such as the gate count and clock frequency of processor extensions. System debugging, including detailed signal-level debugging of extended hardware, all takes place within the source-level debugger, with direct linkage between the simulated state of each processor, the program source code, and the extended instruction set. A common profiling environment aids analysis of hardware-centric performance characteristics (interconnect traffic, handshake stalls, pipeline stalls) and software-centric performance characteristics (overall application performance by line and function, cache performance, and communications stalls).

This unification of views welds the hardware and software perspectives, creating not just a common vocabulary and tool environment but shared simulation models, useful for both large-scale application development and detailed VLSI implementation and verification. The unification creates a "closed-loop" flow that guarantees that the software and hardware teams are using the same assumptions and validating the same design.

Transition 5: This convergence between hardware design and software design triggers a downstream transition in hardware engineering. No amount of automation and abstraction in the system-level hardware design eliminates the complexity and importance of VLSI design, especially management of the increasingly important deep-submicron effects. Circuit design, timing closure, signal integrity, low-level power optimization, and design for manufacturability remain critical. Emerging VLSI design-automation tools target these issues but will require significant ongoing engineering effort on a tool-by-tool basis and as part of a comprehensive back-end design flow.

We expect current chip hardware engineering to split into two increasingly distinct disciplines: *SOC design,* which will be closely tied to software, applications, and system architecture, and *VLSI-flow engineering*, which will be closely tied to large-scale data management, automated circuit optimization, and semiconductor fabrication. The SOC hardware designer may never see a net list or a gate, much less a transistor. RTL will be increasingly treated as an intermediate representation, much as assembly code is treated by software developers today. The designer must be able to read and understand it, and may even have to write it occasionally.

The VLSI flow engineer, by comparison, may be deeply involved in high-level floor planning and technology planning, but may play no role in the functional description of the chip. This distinction between SOC front end and back end may eventually become more important than the historical distinction between hardware and software development. Despite the increasing distinction between the SOC design and VLSI engineering jobs, each discipline needs to understand the other discipline well because of the strongly complementary roles of SOC architecture and VLSI implementation.

Transition 6: As the abstraction level of SOC design creeps upward, the nature of intellectual property IP will also undergo a transition. In the "virtual component" IP reuse model of the late 1990s, a wide variety of hardware blocks—for interfaces, buses, complex logic functions, and processors—were expected to reside in libraries to be selected and instantiated by the hardware designer as needed. This model has worked reasonably well for some general-purpose processor cores and for particularly stable standard interfaces such as PCI and USB. It is already being displaced in many cases by IP generators, which create optimized hardware blocks according to specific needs. The popularity of both memory generators and processor-configuration systems reflect this trend.

The original IP-reuse model failed to live up to inflated expectations because much IP either failed to conform to a high standard of design for reusability or the interfaces and documentation made it more difficult to instantiate a block than to redesign the block from scratch. Early IP-reuse efforts also failed because of the short "shelf-life" of many IP blocks. Hard IP—blocks mapped to a particular process technology—requires substantial redesign and reverification for each process technology port. As a result, hard IP reuse happens largely within a single family of derivative products using the same processor technology.

Soft IP cores—blocks designed and distributed as synthesizable RTL—are more tolerant of process technology changes, but give up some efficiency in density, clock frequency, and power efficiency relative to process-optimized hard IP cores. Even "standards-based" soft IP requires redesign and verification every time the standard undergoes even minor revision—a surprisingly frequent occurrence.

As more of the SOC design is created using processors and software, the reusable element changes. The configurable processor retains the abilities and use model of a general-purpose processor and that component is suitable for reuse independent of any particular software package. The natural synthesizability of configurable processors extends the IP shelf life, as do the well-documented, debugged tools and the thorough IP documentation. (Even vendors of general-purpose, nonconfigurable processors are more universally choosing synthesizability to improve their IP's shelf life.) More importantly, the infrastructure of configurability creates a new natural form of IP—the combination of a processor configuration (for example, TIE source code) and the set of applications and other software that exploit that configuration (for example, the application source or binary that uses those TIE-based instruction extensions).

Processor configurations may represent hardware, but the usage model for processor-configuration IP looks more like the usage model for applications software for two reasons. First, the configuration IP captures significant systems insight, especially about the computation and communications algorithms and performance bottlenecks. From the application developer's viewpoint, the configuration is the "distilled essence" of hardware differentiation. Second, configuration IP is easy to mix and match. It can be mapped onto a wide range of compatible processors without modification. Multiple packages of configuration IP can also be combined and mapped onto a single processor without modification, allowing the combination of corresponding applications to run on the superset processor.

System-configuration IP—the topology of processors, the configuration of interconnects, the memory organizations, address maps and I/O interfaces—is also valuable IP. System topologies are naturally packaged with multitask application suites because they jointly reflect the communication structure and behavior of a system. By partitioning the IP into system-configuration IP and processor-configuration IP, the communication and computation details can be optimized and reused with some independence. A wide range of cost and performance options can be covered with a reusable family of system organizations.

The emergence of this new form of IP—combining processor configuration and applications—is already creating new business models. Configuration IP can be licensed and exchanged with fewer technical impediments than lower-level RTL or gate-level IP. It is more concise and comprehensible than RTL IP (improving its supportability) and configuration can be combined with complementary IP (even configuration IP from other vendors) into a single processor. These abilities create the potential for a lively market in configuration IP as an important part of the future processor-centric SOC value chain.

These six transitions will not occur simultaneously or immediately, although the changes have already started. The initial uptake of configurable processors has already demonstrated all six trends. Design teams that understand and prepare for these design transitions will be early beneficiaries of improved SOC design productivity.

8.7 The Future of the Industry

This book's view of SOC design raises some provocative questions about the business of building complex chips. Semiconductor doomsayers may assess the demands of the system market, the increasing engineering cost of chip design, and the rising capital cost of fabrication facilities and jump to some extreme (but erroneous) conclusions:

- The technical imperative for single-chip systems becomes so pressing that the number and scale of unique chip designs must grow dramatically to match the number of unique system designs. However, there are not enough skilled engineers to design all these chips. Semiconductor growth becomes resource-limited by growth of trained manpower.
- The cost of a production semiconductor fabrication will spiral well past today's $2 to $3 billion price tag to the point that only a tiny number of fabs will be built to fabricate circuits at 45nm and below. Semiconductor growth becomes limited by fabrication capacity.
- The cost of SOC design will continue to grow exponentially so that the number of designs must trend to zero. Semiconductor growth becomes limited by skyrocketing design costs.
- When a single run of wafers through a fabrication can approach 100,000 good die, the minimum acceptable volume for an SOC product will grow beyond the reach of many applications. System designers must accept a lower level of product differentiation through consolidation of the semiconductor device market around a small number of economically viable but highly generic platform chips.
- The economic imperative to build more volume per design creates chips with such generic and low-level programmability that cost per function and power per function deteriorate despite continued process scaling.

These conclusions are built on simple extrapolations of current trends. If no business or technical innovation mitigates these trends, then the future of embedded electronics is indeed bleak. But similar past projections have failed to anticipate qualitative changes in the basic technology and business models on which semiconductor electronics are built. A few examples of these unanticipated changes include the emergence of automatic logic synthesis and chip layout, the appearance of chip foundries and large fabless semiconductor vendors, the growth of independent intellectual property suppliers, and the relegation of the RAM product segment—once viewed as strategically vital—to commodity status, making digital logic, especially microprocessors, the focal point of the market.

We see a set of nine important transformations that are likely to have a profound cumulative effect on the industry. These basic transformations—some subtle, some dramatic—drive this picture of a new SOC industry:

Prediction 1: Moore's law scaling continues, but improved raw transistor density and speed will increasingly be seen as a means to enable more abstract design, to save development time, and to attain higher SOC functionality rather than as a means to simply get more raw capacity. Increased silicon density becomes more visible as a lever for increased design abstraction and system functionality rather than just as a lever for system capacity. System designers will beg less often for simply bigger RAMs or processors with faster clock rates. Instead, they will beg for richer software environments, greater programmability, and easier scaling of designs across a cost or performance range. "Wasting" gates and even "wasting" processors to add functionality and boost productivity will become accepted mainstream tradeoffs.

Prediction 2: SOC platforms will become widespread, but rarely as true single-chip systems. The process-technology requirements for analog, memory, and logic are different and are likely to remain different over the long term. Instead of insisting that all these different types of circuits exist on one chip—an unnecessarily expensive and risky combination—the industry will evolve toward "systems on three chips": one for analog or RF circuitry (especially power amplifiers, line drivers, and power management), one for bulk memory (especially flash memory and dynamic RAM), and one containing all the digital electronics. Multichip packages for combining memory, analog and digital circuits will become common rather than exceptional. Chip-to-chip standards for interfacing among analog, memory, and digital devices may become more important than chip-to-chip bus standards for interfacing among multiple logic chips.

Prediction 3: The different processes and circuit technologies needed for analog, RF, and R. F. memory, and digital functions are likely to become a major structural distinction separating different classes of suppliers as well. While some analog and RF circuits have found their way onto digital chips, higher degrees of specialization in fabrication and design are likely. The business and process strategies of different vendors will become more highly differentiated over time, separating the industry into three segments—analog, memory, and logic—of which logic will be the largest segment, as measured by both revenue and diversity of products.

Prediction 4: The second major structural distinction in the semiconductor industry will be in fab operating strategy. Two types of fabs may emerge over the long term: one optimized for a small variety of high-volume products, including general-purpose processor ICs, FPGAs, and SOCs for consumer electronics, wireless, and PC applications, and one optimized for smaller lot sizes and faster turnaround times including ASICs and SOCs for communications infrastructure applications. The two fab types may use similar process geometries but will have different organization (and maybe even different equipment) to support different design processes and to build different product categories.

A new "lightweight fab" model with lower capital cost and capacity may tolerate somewhat higher manufacturing costs per wafer but enable lower costs per product at moderate volumes. The lightweight fab may permit proliferation of fabs into new regions of the world and will permit strong financial performance even with large numbers of products and small lot sizes. These fabs will give chip designers a greater range of manufacturing options.

The traditional fabs will be built for large lot sizes, long design life, and minimum manufacturing cost per wafer. Silicon foundries are likely to be found in both camps, and the largest foundries may offer both services: the fast-turnaround, small-lot-size service and the high-capacity, low-unit-cost service. There is little evidence from foundries and semiconductor equipment makers to support this prediction yet. However, the underlying pressure to improve overall manufacturing-flow efficiency for smaller lots makes this prediction plausible over the next decade.

Prediction 5: The third major distinction in the industry will lie between semiconductor companies differentiating on system expertise and companies differentiating on VLSI design expertise. The companies emphasizing systems expertise are likely to have deep vertical-market knowledge and close ties with leading system OEMs. They may either choose a fabless business model or use lightweight fabs. The VLSI-focused companies will emphasize their mastery of circuit design, including mixed-signal design. These companies are likely to gravitate toward close ties with traditional large-lot-size fabs. Many companies will attempt to be good at both systems-centric and circuit-centric design, and a few will succeed. Tensions, however, in the priorities and skills of the two specialties will cause many organizations to choose one path or the other.

Prediction 6: Differential global economic development, especially the continued rise of the Asian economies, will transform the electronics landscape. China and the developing economies of India, Indonesia, and Southeast Asia as well as South America and even parts of Africa will become large end markets. Many of these geographies, especially Asia, will become major manufacturing sites and important engineering centers. Products tuned to the needs of rapid industrialization and the emergence of a large growing, global middle class will become dominant trends. The speed of Asian economic development will also influence product definition. The relative paucity of existing communications infrastructure in these geographic regions will relieve many product categories of backward-compatibility requirements in communication protocols, user interfaces, and use models. These markets will permit more revolutionary products to become established, increasing the opportunity for new entrants at the expense of established leaders.

In addition, the engineering graduation rates already strongly favor Asia. This has two effects. Existing organizations will inevitably favor a gradual shift of engineering resources toward Asian development centers. More importantly, new indigenous engineering efforts will take on a greater share of all electronics development. This new cadre of Asian engineers may be less constrained by compatibility with the system architectures and user interfaces of historically successful products of North America and Europe. At the same time, design languages, design methods, and tools from Europe and North America are likely to remain central to engineering education and design practice in Asia.

Prediction 7: Two broad categories of programmable chip design will emerge over time.

a. The **general-purpose platform** type will use large arrays of generic programmable elements and new application-mapping tools to implement a complex set of tasks on a programmable, general-purpose chip. These devices are the cultural descendents of today's generic FPGAs and general-purpose processor chips. Each generic platform will satisfy the needs of a large number of modest-volume applications. We believe that processor-based platforms will dominate over FPGA-based platforms because of the greater scalability of programming languages over hardware-description languages for complex problems. Moreover, configurable processors will play a prominent role in processor arrays because of small size, good tools, and faster processor-to-processor interconnect. Nevertheless, the lack of application-specificity in either the architecture of the programmable elements or the architecture of the interconnections among elements will make these platforms significantly slower, more expensive, and more power-hungry than platforms tuned for an application domain.
b. The **application-domain platform** type will also use many programmable elements but heavily adapt the system topology and the nature of the element-programming model to fit the application. These devices are the cultural descendents of today's ASICs and application-specific standard products. Each domain platform will satisfy the needs of a modest number of high-volume applications. Application-specific processors will be used for many of these building blocks. The programming environment and tools will be uniform across the different processor blocks, just as with the more general-purpose platform above, even if the individual processors have very different configurations and roles in the platform architecture.

Prediction 8: The emergence of processor automation changes both the design process and the final SOC platforms themselves. Much more of the digital functions will be implemented as configured processors, both to accelerate the design effort and to embed programmability throughout the SOC. Automatic processor generation will take on a role analogous to the role today played by gate-level design automation (logic synthesis, standard-cell placement and routing, and physical verification). The chip designers will work at a significantly higher level of abstraction, able to directly understand and optimize the interactions between hardware and software and between computation within individual functions and communications among functions. New engineering job types and new organizational structures will evolve within major SOC-design companies to exploit the changes inherent in processor automation.

Prediction 9: Changes in the design process will also change the structure and nature of the design-infrastructure industry. The EDA industry as we know it—the companies and processes built around RTL-level, gate-level, and transistor-level design—will not go away. It will, in fact, continue to grow and thrive because the sheer number and sophistication of VLSI design projects will continue to grow. However, the traditional EDA tool providers will continue to focus on the back end of the SOC-design process (from RTL down to lithographic masks) and the accumulation of basic (application-neutral) IP blocks and IP capacity, while more of the focus of SOC architects shifts upward toward software, applications, and systems concerns. This split will create the opportunity for an entirely new set of players focused on the new design tools, building blocks, and design methodologies needed for system-centric design. These new companies will combine skills and perspectives derived from embedded software and intellectual property companies as much as from traditional EDA companies. The result will look like a new industry built on top of the existing EDA and embedded-software industries.

These nine predictions, like all predictions, involve going out on a limb. Nevertheless, each prediction is based on trends already underway and likely to accelerate as underlying technical and global economic trends evolve.

8.8 The Disruptive-Technology View

Professor Clayton Christensen of Harvard Business School has proposed a powerful model of how new technologies and companies displace existing technologies and incumbent market leaders. In studying the rise and fall of industry-leading companies, Christensen noted that leaders typically did not get displaced by companies with better technology, but by companies with technology that could be considered worse in some ways! Large incumbents actually tend to be very good at technology innovation so long as the new technology sustains their position with existing customers and product types. In the early days of a new technology, customers demand more performance, capacity and reliability than the technology can easily provide. Companies become leaders by successfully improving their technology to meet those high expectations.

Over the long run, however, the market leaders push sustaining innovation at a faster pace than customers can absorb it until product capabilities overshoot the real needs of the customer, at least on existing performance, capacity, and reliability criteria. This overshoot creates the opening for new entrants with new product alternatives. These new products often succeed because they offer much greater convenience or dramatically lower prices. By the previous measures, the new products may initially be considered as using inferior or even "lousy" technology; they are slow, unreliable, and limited in capacity compared to the technology they are displacing. But if the new products start to catch on in a narrower market that values cost and convenience over speed, reliability, and capacity, then those new products can move up the technology curve until they become good enough to displace products based on the old technology, even for the more demanding customers. At this point, the incumbent technology and the incumbent vendors are pushed out of the mainstream and forced into smaller and smaller high-end markets.

This process is sketched in Fig 8-16, where the heavy black line shows the gradual increase in customer demands for higher performance over time.

The customers' ability to fully use a product's performance increases gradually over time and is shown as a single line, but in reality different customers have some range of demands, as

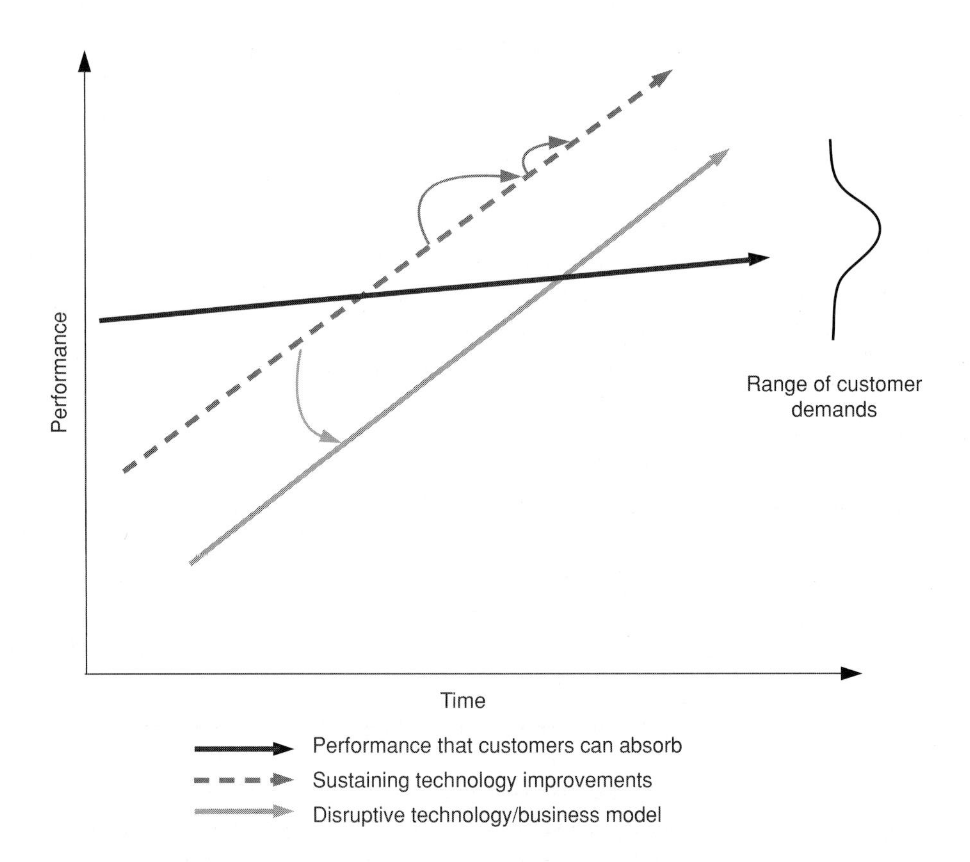

Figure 8-16 Christensen's technology disruption model.

shown by the bell curve on the right of the figure. An established supplier or group of suppliers typically follows the line labeled "Sustaining technology improvements," making their products better by a series of sustaining innovations. Sometimes, these innovations occur in small steps, and sometimes they appear as major technical breakthroughs.

The threat to these incumbents is shown by the line labeled "Disruptive technology/business model." Due to greater convenience and much lower costs, products built along this line are used in new ways. For example, these products may adopt a more modular structure to allow for more rapid customization or specialization to fit new market niches. They typically gain early market momentum by finding new classes of customers (competing with nonconsumption) rather than by directly displacing incumbents.

Along the way, disruptive technology changes the demographics of use. Users who could not use the incumbent product because it cost too much or required specialized skills now have direct access to a useful product. While the new market may initially be small, the ultimate population of users may dwarf the original market.

This pattern is easily recognized in the displacement of minicomputers by PCs, the replacement of large-scale photocopiers optimized for copy centers by personal desktop copiers, the appearance of disposable cameras, and the replacement of standalone microprocessors by embedded RISC CPU cores in ASICs.

Christensen has identified Tensilica's configurable processors as a prime example of disruption in the processor world, particularly as a smaller, cheaper, more modular alternative to traditional, general-purpose, one-size-fits-all microprocessor architectures. Looking at configurable processors through the "disruptive-technology" lens reveals more about the theory and the potential future for the technology.

The general-purpose microprocessor, as typified by Intel Pentium processors, emphasizes one set of competitive criteria. Performance and generality are crucial. We have seen the microprocessor industry evolve from products capable of delivering a few MIPS in the early 1980s to today's sustained performance of thousands of MIPS. General-purpose processors evolved by combining very aggressive circuit design and out-of-order execution of multiple instructions per cycle. At the same time, these processors have become capable of running very large and sophisticated software suites, including complex operating systems and many layers of libraries and protocol stacks.

To support this software richness, the processors have added memory management, extended addressing, and additional instruction-set features to support broad application areas such as multimedia. As a result, computers based on these processors have grown in performance and capability to support even extremely demanding customers, displacing earlier high-end systems such as mainframes, minicomputers, and engineering workstations for many tasks.

At the same time, these PC processors have also grown in cost, silicon area (even after compensating for dramatic decreases in transistor size) and power dissipation. High-end x86 processors, which were originally intended for cheap, low-power embedded applications, now consume nearly 100W of power, and their cost (including power supply and cooling requirements) approaches $500.

Measured by the criteria for these conventional, general-purpose PC processors, the configurable processor is pretty weak. In its basic configuration, it delivers only a few hundred MIPS of performance and runs only a limited range of well-known operating systems. For example, Tensilica's Xtensa processor runs open-source Linux and Wind River's VxWorks operating system, but not Microsoft Windows. Few off-the-shelf applications are ported and distributed in binary form for configurable processors. When configured for a specific application domain, the application-specific processor can be as fast as or faster than the Pentium, but with less generality—the sweet-spot for any one configuration of application-specific processor is much narrower than for the Pentium processor.

On the other hand, power dissipation for the configurable processor, including its small memories, is well below 100μ/MHz or less than 10mW in even performance-demanding applications. The core plus its memories can be built in less than 0.5mm^2 of 130nm silicon, equivalent to a few cents of cost. For an increasing range of applications, the performance of these application-specific processors is fast enough, and the cost, power, and focused application performance are compelling enough to enable embedding these processors in many places were no other processor could suffice. The penetration of these application-specific processors into roles previously reserved for hardwired logic fulfills the "disruptive technology" criteria of enabling new uses. These processors compete more with nonconsumption of processors than with consumption of the more established processor families.

Intel's pursuit of the most demanding standalone processor customers is echoed in the evolution of embedded RISC CPU cores. ARM Limited and MIPS Technologies are the established leaders in embedded RISC cores. ARM, in particular, originally focused on low-end, low power cores, cores that lacked the performance and features common in standalone processors of the day. They were "lousy" CPUs by general-purpose computing criteria. Over time, however, ARM's processor cores have grown in performance and features such that they can satisfy the needs of even very demanding embedded applications. The progression from ARM6 to ARM7 to ARM9 to ARM10 to ARM11, and soon to ARM12, demonstrates remarkable sustained improvements in absolute performance.

As a result of these improvements, however, the ARM11 has now accumulated a wide range of sophisticated features. It is a suitable choice for a broad range of control and user-interface applications. It also requires more than 10 times the silicon area of a configurable processor and dissipates almost 10 times the power per MHz.

A high-end embedded RISC core is clearly able to displace standalone processors in a wide range of general-purpose, personal-computing devices such as PDAs, and ARM processor cores are used in most of the high-end PDAs made today. On the other hand, the "improved" RISC core can no longer compete for lowest cost, highest efficiency computing functions within SOC designs. Many SOC designers must satisfy their data-intensive computing requirements with hardwired logic, not ARM processors.

Fig 8-17 shows how the disruptive technology model might be applied to two classes of customer needs—embedded-control tasks and personal computing—and how three types of processors (standalone microprocessor chips, RISC CPU cores, and configurable processors) compare. For this comparison, the performance axis is taken to be general-purpose performance and availability of a wide variety of applications.

The impact of the configurable processor is not just to lower silicon size and cost for embedded processing, but to change the basis of competition. Automatically generated processors serve in roles where embedded RISC cores cannot easily play. The adaptability to new applications, the high performance in data-intensive applications, and the more comprehensive support for multiple-processor SOC design all open new markets for processors where customers value criteria besides general-purpose performance and the availability of shrink-wrapped applications.

The configurable processor should also be seen as a disrupter to the tradition of home-grown processor design. The most sophisticated embedded SOC design teams have long recognized that only application-specific processors can combine the performance and flexibility

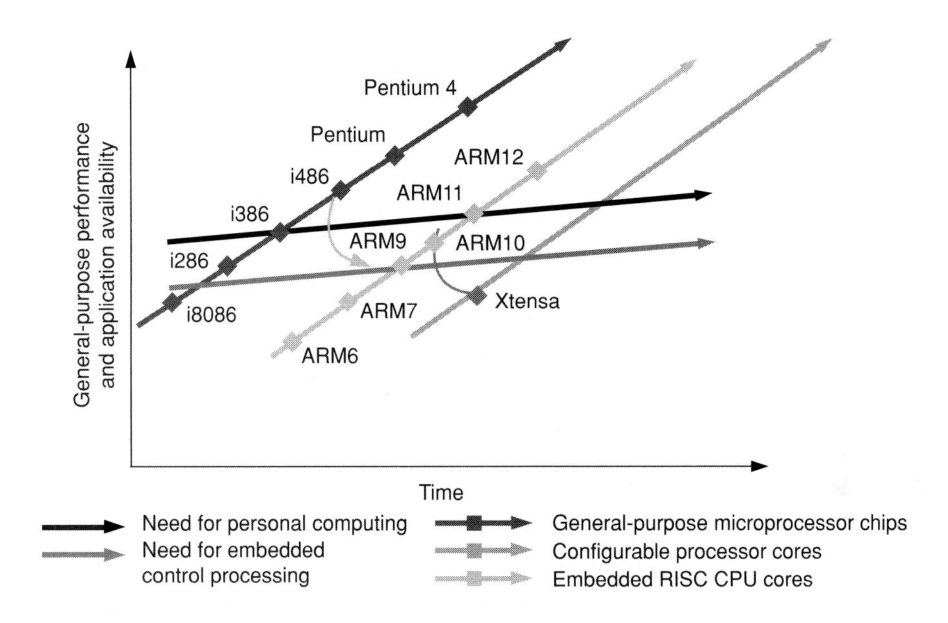

Figure 8-17 Applying the disruptive technology model to embedded processors.

required for high-performance embedded designs. However, the specialized skills in architecture development, processor hardware design, and software tool development limited the number and use of homegrown processor design. Automatic processor generation makes processor design much cheaper and faster for existing application-specific processor-design teams and makes application-specific processors a realistic option for less sophisticated teams.

The natural upward migration of processor architectures creates the opening for configurable processors. The fundamental new technology of automatic processor generation has significant disruptive potential for four reasons:

1. Configurable processors deliver cost, power, and application efficiency advantages of 10x over RISC cores, and 100x over standalone processors, in a wide variety of deeply embedded SOC environments.
2. Automatic processor generation effectively opens up embedded processor design to an entirely new population of users: software and hardware architects for whom processor design was too slow and risky and required too many specialized skills. The reduction in design time, design risk, and required design skills creates new opportunities for processors—new "sockets"—where existing processor types cannot be considered.
3. The intrinsic flexibility of processors—not just to add new features, but to strip the processor down to its bare-bones—limits the pressure on configurable processors to migrate upmarket in same way that traditional architectures do. This disrupter is less likely to become an eventual disruptee.

4. The small size and easy integration of these tiny new processors enables a new basis for innovation—the use of modular processors as basic building blocks for SOC development. An entirely new ecosystem of suppliers and customers may emerge out of the disruption.

Christensen's disruptive technology model is a powerful tool for analyzing market evolution. It suggests that major changes in the structure of SOC architectures will be mirrored by major changes in the processor marketplace.

8.9 The Long View

The electronics industry prides itself on a forward-looking view. When we look back over the decades, the cumulative change in technologies, products, applications, and players is startling. But the changes are rarely instantaneous or even unanticipated. Instead, fundamental new trends emerge from the fog over a period of years and quietly grow from vague familiarity into dominance. By the time a trend is widely accepted as important, it is often unstoppable. The world of SOC design is now undergoing that kind of transformation.

Automatic processor generation is now emerging from the fog and establishing itself as a new basic tool in the hands of the system architect, software developer, and hardware designer. It has been widely, but not yet universally, adopted both as a means to build better chips and systems—cheaper, faster, lower power, longer lived—and as a means to design systems more quickly and reliably. While the precise evolutionary path for configurable processors cannot be forecast, their potential to become ubiquitous justifies the investment in understanding.

As configurable processors become both smaller and more widely accepted as a basic building block, we can start to talk realistically about the sea of processors. This term reinforces the parallel to the emergence of the ASIC in the 1980s, when transistor density reached the point that a higher level of abstraction became not only possible but necessary to exploit the potential of tens of thousands of gates per design.

"Sea-of-gates" ASIC design triggered a number of major transitions in the industry. Chip design was opened up for the first time to a larger population of chip designers. Chip design was systematized and automated to hide the details and bring productivity to large-scale chip design. New hardware-description languages such as Verilog and VHDL and new tools such as logic synthesis and automated cell placement emerged as key standards. The EDA industry as we now know it emerged to serve this new population of IC-design engineers with simple, powerful tools for design creation, analysis, and enhancement. The idea of independently developed intellectual property appeared to exploit the new sea-of-gates architecture, languages, and tools.

The sea-of-processors SOC is poised to follow this pattern. Configurable-processor technology opens up processor optimization and multiple-processor system architecture to a large population of hardware and software developers. The details of processor design and processor-to-processor communication are automated and hidden. New description languages, such as the Tensilica Instruction Extension, language are proliferating into broader use, and new tools such as processor generators and automated instruction-set designers are significantly raising the abstraction level in SOC design.

These new ideas and tools may spur a new class of design-automation companies, drawing on some of the themes of the EDA industry but more centrally concerned with software and systems design and their embodiment in new programmable-hardware platforms. We have already seen the emergence of new forms of intellectual property—especially the combination of configuration IP and the corresponding application software—as future evidence of the basic shift toward advanced programmable SOC design and the sea of processors.

The frenetic pace of incremental improvement in electronics sometimes distracts the community from the pattern of long-term change. Taking a longer view—looking out 10 or 15 years—helps create a broader perspective on the opportunities inherent in innovation. The deep impact on the lives of end-product users and designers does not come from small increments in productivity within existing methods and applications, but from more basic shifts in thinking. Those basic shifts cannot be discontinuous. The user and the designer cannot be left behind by some massive shift in mandatory skills and knowledge. The shifts must gradually pervade many aspects of the user's or the designer's life.

This book advocates and explains the pervasive shifts driven by a new form of processor. These processors have already achieved substantial adoption in a range of important applications—from network routers to digital cameras—and will percolate into an increasing range of applications and into an expanding role in designers' toolboxes.

8.10 Further Reading

- Christensen's work on the impact of disruptive technology on companies and markets is truly significant. Applying Christensen's disruptive technology theory to microprocessors and processor-centric SOC design has markedly influenced our thinking about new processor markets and opportunities. His book, however, is recommended to technology managers or business strategists in any market. Clayton Christensen and Michael Raynor. *The Innovator's Solution: Creating and Sustaining Successful Growth.* Harvard Business School Press, 2003.
- The International Technology Roadmap for Semiconductors is the consensus review of trends and priorities for the semiconductor industry. It is revised annually and has become both prediction and target for major manufacturers. "International Technology Roadmap for Semiconductors—2003 Edition—Executive Summary," *http://public.itrs.net/*.
- R. Nair. "Effect of increasing chip density on the evolution of computer architectures." In *IBM Journal of Research and Development*, March/May 2002.
- R. Subramanian. "Shannon vs. Moore: Digital Signal Processing in the Broadband Age." In *Proceedings of the 1999 IEEE Communication Theory Workshop*, Aptos, CA, May 1999.

While the degree of available parallelism in applications is likely to remain a point of lively technical debate, the following articles suggest that some applications have a great deal of latent parallelism:

- David W. Wall. "Limits of Instruction-Level Parallelism." In *WRL Research Report 93/6*, Digital Equipment Corporation, June 1993.
- L. Wills, T. Taha, L. Baumstark, and S. Wills. "Estimating Potential Parallelism for Platform Retargeting." In *Proceedings of the 9th Working Conference on Reverse Engineering (WCRE)*, IEEE Computer Society Press, Richmond, VA, pp. 55–64, October 2002.

Information on commercial processor arrays can be found at *http://www.pactcorp.com* and *http://www.picochip.com.*

Scaling of different design approaches form an important area of research (and debate) as exemplified in the following:

- Intel's discussion of conventional processor performance and efficiency trends can be found in the following article: D. Marr, F. Binns, D. Hill, G. Hinton, D. Koufaty, J. Miller, M. Upton. "Hyper-Threading Technology Architecture and Microarchitecture: A Hypertext History." In *Intel Technology Journal*, 6 (1), February 2002. Available at *http://developer.intel.com/technology/itj/2002/volume06issue01/.*
- Historical transistor performance trends have been widely analyzed. See the discussion of gate delay in William J. Dally and John W. Poulton. *Digital Systems Engineering*. Cambridge University Press, 1998.
- For IBM's comments on FPGA vs. standard-cell ASIC efficiency, see Leon Stok, Ruchir Puri, John Cohn, David Kung, and Dennis Sylvester. "Pushing ASIC Performance in a Power Envelope." In P*roceedings of the 40th Design Automation Conference*, 2003. Also, see Paul Zuchowski, Christopher Reynolds, Richard Grupp, Shelly Davis, Brendan Crewen, and Bill Troxel. "A Hybrid ASIC and FPGA Architecture." In *International Conference on Computer-Aided Design 2002 (ICCAD 2002)*, pp. 187–194.

Detailed information on embedded operating systems can be found at vendor Web sites:

- For embedded Linux, see *http://www.montavista.com.*
- For VXWorks see *http://www.windriver.com.*

INDEX

A

D

E

F

G

J

L

M